Lecture Notes in Intelligent Transportation and Infrastructure

The series "Lecture Notes in Intelligent Transportation and Infrastructure" (LNITI) publishes new developments and advances in the various areas of intelligent transportation and infrastructure. Merging theoretical foundations, practical applications, and forward-looking insights, LNITI provides a comprehensive understanding of both the state-of-the-art and the future prospects within this dynamic field.

LNITI is designed to be an inclusive platform that covers an extensive array of topics including, but not limited to intelligent transportation systems, smart mobility, intelligent logistics, critical infrastructure, smart architecture, smart cities, intelligent governance, construction design, data security, operational analysis, optimal route planning, digitalization, autonomous vehicles, the evolution of transport systems as well as green and sustainable urban structures. The series contains monographs, conference proceedings, edited volumes, lecture notes and textbooks. Of particular value to both the contributors and the readership are the short publication timeframe and the world-wide distribution, which enable wide and rapid dissemination of high-quality research output.

Proceedings published in the series are indexed by INSPEC.

All books, including proceedings, published in the series are evaluated by Web of Science

Kardi Teknomo

Ideal Flow Network for Transportation Research

Unveiling the Hidden Secret of Nature to Solve Traffic Congestion

Springer

Kardi Teknomo
School of Urban and Regional Planning
University of the Philippines Diliman
Quezon City, Philippines

ISSN 2523-3440 ISSN 2523-3459 (electronic)
Lecture Notes in Intelligent Transportation and Infrastructure
ISBN 978-981-95-4726-5 ISBN 978-981-95-4727-2 (eBook)
https://doi.org/10.1007/978-981-95-4727-2

This Springer imprint is published by the registered company Springer Nature Singapore Pte Ltd.
The registered company address is: 152 Beach Road, #21-01/04 Gateway East, Singapore 189721, Singapore

To my beloved wife and children,
whose patience and love guided this journey.
This work is as much yours as it is mine.

Foreword

Reading Kardi Teknomo's *Ideal Flow Network for Transportation Research: Unveiling the Hidden Secret of Nature to Solve Traffic Congestion* is a rewarding experience. The book introduces the Ideal Traffic Assignment (ITA) framework and Ideal Flow Network (IFN) as theoretical tools for understanding and addressing traffic congestion, offering a fresh perspective grounded in mathematical clarity and systems thinking.

As a sociologist, I am drawn to how these models illuminate the entanglement between human behavior and built environments. In 2011, our research group collaborated with Kardi on a campus traffic study in Metro Manila. While IFN had not yet been formulated at that time, the project laid the groundwork for later innovations. The IFN framework itself was discovered in 2014, and our subsequent work on ITA—published between 2017 and 2019, demonstrated how stochastic matrices can guide efficient traffic schemes even in the absence of traditional demand data.

Kardi's work invites us to reconsider how traffic flows are shaped, not just by infrastructure or policy but by the intrinsic structure of networks themselves. His framework reveals that optimal flow is not merely a product of demand but of the network's geometry and connectivity. This insight challenges conventional paradigms and opens new avenues for modeling, simulation, and policy design.

The book's contribution lies in its ability to bridge disciplines: mathematics, engineering, and the social sciences. It encourages us to see traffic not just as movement, but as a patterned interaction between people, systems, and space. In doing so, it offers a compelling vision for how we might reimagine urban mobility.

I commend Kardi for this rigorous and visionary work. May it inspire scholars and practitioners alike to pursue more elegant, equitable, and efficient solutions to the persistent challenge of traffic congestion.

Dr. Rosita G Leong School of Social Sciences
Ateneo de Manila University
September 2025

Czarina Saloma

Preface

This book introduces the *Ideal Traffic Assignment (ITA) framework* for transportation research and analysis. The framework proposes a theoretical solution to traffic congestion by revealing the intrinsic relationship between a network's physical structure and its optimal flow distribution. By leveraging principles from graph theory, Markov chains, and information theory, this work demonstrates that a specific, optimal flow state is an inherent property of network topology. Its primary purpose is to help solve the persistent challenge of traffic congestion, offering a fresh perspective on how traffic flows through our cities.

Ideal Traffic Assignment is based on Ideal Flow Network (IFN). The IFN concept originated from a simple exploration of how flows behave on networks. My early simulations, tracking random walks on various network topologies, initially produced complex and seemingly chaotic flow data. At first, I didn't see any clear, stable pattern. The raw flow values simply kept increasing with simulation time.

A breakthrough occurred after careful observation and normalization. When I divide the link flows by the total flow (or by any first order statistics of the network), a unique and remarkably stable pattern emerged for each distinct network structure. This consistent pattern, which I call the *ideal flow*, crystallizes at the asymptotic limit of the simulation. It emerges reliably for every strongly connected network structure as the random walk simulations approach the asymptotic limit. This wasn't just a mathematical convenience. It offered a glimpse into something more profound. What emerged was a governing principle, a subtle law embedded within the network's design, manifesting as a harmonized structural equilibrium. Nature's concealed wisdom resides in this integrated balance equilibrium. Thus, IFN transcends conventional modeling. It embodies nature's quiet revelation, now brought to light. I've come to believe it is a divine gift bestowed upon humanity. Its mechanism aligns seamlessly with the network's intrinsic structure. This insight uncovered a latent order in the pathways of flow, guided solely by structure.

My original motivation for this deep dive was, and remains, to fundamentally understand and solve traffic congestion. Though IFN research has expanded across numerous domains, its origin lies in transportation science. This book returns to that primary goal, presenting a comprehensive theoretical solution rooted in the IFN framework. In many ways, it feels like a return home. By understanding the ideal flow, we have found a way to resolve traffic congestion. This solution does not always require more infrastructure. I call this solution *premagic equilibrium*.

Premagic equilibrium describes an ideal, highly efficient state for a traffic network. In this state, it's theoretically possible to maintain traffic congestion at any desired minimum level. The congestion is distributed uniformly across the network. More profoundly, as I will rigorously demonstrate, this ideal state uniquely aligns individual user optimization (Nash Equilibrium) and with system-wide efficiency (System Optimum), at least theoretically.

The ITA framework also provides a new metric for urban network health called *VarPi index*. This metric to measure the "distance" of a real-world network's performance from the balance ideal condition of premagic equilibrium, allowing planners to *objectively* evaluate the effectiveness of proposed traffic solutions and design interventions. Based on this metric, I derived a remarkably concise and analytically elegant expression, referred to as the *Network-Flow Identity*, which establishes a relationship between total network flow to its overall capacity, maximum allowable congestion, and efficiency metric as captured by the VarPi index. All of these demonstrate how the ITA framework can be applied to evaluate traditional urban designs such as the grid, and show how complex interventions can be formalized and analyzed.

What makes Ideal Traffic Assignment a truly unique and valuable contribution to transportation research? It offers a *parsimonious approach* that significantly reduces the reliance on costly, complex, and often uncertain Origin-Destination (O-D) surveys, a major bottleneck in traditional modeling. Instead, it leverages the inherent properties of network structure and can be effectively calibrated using more accessible data like link flow counts and microscopic turning movements. This framework provides unparalleled insights into how network structure intrinsically dictates flow distribution. The framework presents policy interventions as a "chemistry of policy," providing a formal grammar for synthesizing, simulating, and optimizing planning scenarios. This perspective enables a proactive, science-based methodology for urban planning, moving beyond reactive congestion mitigation. It offers a powerful lens for strategic planning, scenario analysis, and dynamic traffic estimation.

This work is intended for a diverse audience, including college students, transportation researchers, urban planners, and government officials. It aims to equip the reader with a new paradigm and powerful tool for transportation analysis, fostering a deeper understanding and inspiring innovative solutions for designing more resilient and equitable cities.

The book is structured to ensure an integrated and balance approach between theory and practice. It builds rigorously from foundational concepts, progresses through the detailed mathematical derivations of IFN and its idealized equilibria, and then transitions to practical applications, calibration techniques, and real-world considerations. This book provides the necessary theoretical foundation for Ideal Traffic Assignment. My goal is to equip the reader with a new and powerful tool for transportation analysis, fostering a deeper understanding and inspiring innovative solutions. This work is structured as a journey from core principles to practical implementation. It begins by introducing Ideal Flow Network (IFN) and traditional traffic assignment concepts within the four-step model (FSM). The paradigm shifts from demand-based models to structure-based flow inference from topology. Mathematical foundations are built using matrix algebra, graph theory, and information theory. These tools support the development of the core ITA framework for network analysis. Later chapters apply the ideal flow network theory to transport systems using data calibration. Final sections present premagic equilibrium and chemistry of policy modeling that demonstrate both theoretical and practical applications of the framework.

The organization leads readers from theory to application in a logical, layered progression. It starts with IFN as a complementary technique to conventional transport modeling approaches. Mathematical tools are introduced early to support later theoretical development. The IFN theory is then adapted for transport analysis and policy evaluation. Practical chapters cover calibration with observed data and real-world policy implications. Case studies illustrate how the framework performs under real urban conditions.

This work should be treated as a comprehensive, methodical guide to IFN theory and Ideal Traffic Assignment Framework. Readers are encouraged to progress sequentially for a full understanding of the framework. However, chapters are designed for standalone reading with minimal dependency on earlier sections. Each chapter is mostly self-contained, allowing focused reading without constant backtracking. This design ensures clarity and ease of use, even for selective topic exploration. Some repetition of definitions is intentional to support clarity and independent understanding. The structure favors *accessibility* and *clarity*, even if it sacrifices conciseness. When choosing between rigor and clarity, *clarity* is always prioritized for broader reach.

I am deeply indebted to many students and colleagues for their invaluable discussions, insights, and unwavering support throughout the years. I especially thank John Lee for proposing, back in 2014, the initial idea of using random walks to model traffic. That seminal idea was the seed for the entire IFN theory. I sincerely hope this work inspires new solutions to transportation challenges worldwide and contributes to the ongoing evolution of our understanding of complex urban systems.

Quezon City, Philippines
August 2025

Kardi Teknomo

Declarations

Competing Interests This book was not supported by any specific funding agency, grant, or institutional support related to its writing or publication. Previous research on Ideal Flow Networks has been funded over the years 2014-2018 by various institutions, including Philippine Higher Education Research Network (PHERNET), Institute of Philippines Culture, and The Philippine Council for Industry, Energy and Emerging Technology Research and Development Department of Science and Technology (PCIEERD-DOST). These earlier grants contributed to the foundational work but did not influence the development of this book.

The author declares no conflicts of interest relevant to the content of this book.

Ethics Approval This book does not involve primary studies with human or animal participants. As such, ethics approval was not required.

Contents

Acronyms

BSM	Behavioral Stochastic Matrix
IFN	Ideal Flow Network
IPLC	Identical Path Link Counts
ITA	Ideal Traffic Assignment
NE	Nash Equilibrium
NFI	Network Flow Identity
O-D	Origin-Destination
PoA	Price of Anarchy
PE	Premagic Equilibrium
SLC	Steady Link Characteristics
SO	System Optimum
UE	User Equilibrium

Chapter 1
The Ideal Flow Network: A Structural Paradigm for Urban Mobility

1.1 Introduction

This chapter offers a broad conceptual overview of modeling in general and introduction to the Ideal Flow Network (IFN) paradigm. It presents the rationale behind modeling traffic flows via Ideal Traffic Assignment framework. The exposition remains informal and illustrative, using examples and narratives rather than mathematical formalism, to build foundational intuition.

1.2 The Scientific Imperative to Model Complex Systems

Before we delve into the specifics of Ideal Flow Network (IFN) theory and its application for transportation research, it is essential to first understand the purpose, process and the reason of mathematical modeling.

1.2.1 Scientific Modeling Process

Science progresses through the creation of models. A model is any concept or narrative conceived in the human mind. When a model gains widespread acceptance as a shared understanding, it becomes part of reality and can influence the physical world. Foundational aspects of society, like law, education, religion, language, and even abstract ideas such as morality, money, and marriage are originated as models that evolved into collective agreements. Therefore, understanding models and the process of modeling is essential to grasp how human thought shapes civilization.

K. Teknomo, *Ideal Flow Network for Transportation Research*,
Lecture Notes in Intelligent Transportation and Infrastructure,
https://doi.org/10.1007/978-981-95-4727-2_1

People once used simple analogies to model life. Ancient elements were grouped as wood, fire, earth, metal, air, or water. Horoscope is a model based on uncontrollable external forces. Scientific modeling creates concepts and predictive tools. We build miniature versions of technical objects. Chemistry uses balls to model molecules. Models simplify systems to help us understand them. They improve our predictions and control of complex systems.

All human thought relies on *mental models* to navigate the world. We all use mental models to interpret our experiences, without noticing. They are inner images that help us think. Mental models use concepts we rearrange mentally. These concepts aren't the real things they represent. As children, toys support imagination and modeling. Mental models come from experience and education. Culture and context shape how we form them. Thinking depends entirely on mental models. However, these mental models are often ill-defined and prone to confirmation bias. Mental models are often unstable and oversimplification. We resist changing them when reality conflicts. This creates bias and rigid beliefs. In complex situations, we cling to dogma. Mental models lack clarity. Our assumptions are vague and hard to define and shift unnoticed. Communication becomes difficult and unclear. As the result, we struggle to adjust or use our models well.

A model is a purposeful abstraction of reality. We model to solve real problems better. A model is a simplified version of reality. It is a simple representation designed to help us understand, predict, and control the systems we are interested in. It helps us test ideas before acting. Trying every solution in real life is costly. Modeling saves time, money, and effort. It reveals how systems behave. It helps us predict and control outcomes. We use models to find optimal solutions. They guide us toward balance and efficiency. Models can be physical or abstract. Physical models are great for visual evaluation. But they're expensive and hard to optimize. Abstract models are flexible and easy to manipulate. They include images, descriptions, and rules. Mathematical models are the most powerful kind. They describe relationships and system behavior. Used with computers, they enable simulation and precision. Architects use scale models to test building strength. Physicists use equations to describe planetary motion. Models help us simplify complex systems. They turn hard problems into manageable ones. Modeling is central to understanding and solving complexity. Without them, we are left with intuition and guesswork, which often fail in the face of complexity.

Scientific modeling formalizes this process to overcome these limitations. It turns informal ideas into rigorous frameworks. Mathematical modeling helps us represent complex phenomena. We build logical structures from them using deduction. The result includes propositions, lemmas, and theorems.

People often think theory is just guesswork. In common language, theory is confused with unproven ideas. But in science, theory means proven results from theorems. We begin modeling with assumptions or hypotheses. All models begin with assumptions about the system. The assumptions are called axioms. These are accepted as true within the model. We define terms and relationships before stating axioms. Axioms are things we accept as true without proof.

A foundational aspect of science is its reliance on starting assumptions, or axioms. These axioms are fundamental truths accepted without proof, serving as the basis for a theory. Research often begins with hypotheses, which are informed propositions. A conjecture is an unproven idea based on limited evidence.

The validity of a scientific theory depends on its underlying assumptions. If a theory's axioms or proofs are shown to be incorrect, the theory itself may be disproven. This process of falsification, however, often leads to the development of improved theories.

While both science and religious traditions can similarly begin with foundational principles that accepted without proof, their methodologies differ significantly. Science employs skepticism as a core principle to rigorously challenge its assumptions. In contrast, religious traditions typically rely on faith and established doctrines. Science is defined by its objective search for verifiable evidence through systematic testing.

We build tools to observe and verify our hypotheses. We reject hypotheses when they conflict with observed facts. Even one counterexample can collapse an entire theory. Scientific education teaches honesty, objectivity, and skepticism. We accept facts, not opinion, and start over when needed.

Deductive modeling starts from general ideas and moves to specifics. It builds a closed system based on axioms. We use known models or invent new ones. Science turns guesses into proven theories through deduction. We define terms and state axioms we accept as true. From axioms, we create propositions, provable true or false statements. Next, we form lemmas, based on axioms and propositions. Corollaries are simple consequences of earlier results. Theorems are important results built on previous proofs. Proposition, corollary, lemma, and theorem differ by context. All must be proven within the logical system. Proofs develop new knowledge and verify truth. The final result of deduction is a mathematical model. Ideal Flow Network (IFN) is one such example. IFN uses structural and behavioral invariants to model systems.

Inductive modeling comes after deduction. It uses real data to generalize and refine the model. Calibration adjusts parameters to fit data. Validation tests the model using different data. Verification checks that model behavior looks reasonable. Inductive modeling ensures the model reflects reality accurately.

1.2.2 Purposes of Mathematical Modeling

A well-constructed model serves three primary purposes:

1. *Understanding:* By abstracting a complex system into a set of rules and relationships, we can gain insight into its underlying behavior. The model allows us to identify which factors are most influential. This level helps us isolate what's essential.

2. *Prediction:* A calibrated model allows us to conduct "what-if" analyses. This level lets us explore future possibilities. We can experiment with different scenarios and predict their outcomes without the cost and risk of real-world implementation.
3. *Control and Optimization:* Once we understand a system, we can use the model to find ways to control its behavior to achieve a desired goal. This involves identifying the optimal set of inputs to produce a target output. This level enables us to plan and steer systems.

Models help us simplify complexity by focusing only on relevant features and values. By applying the principle of parsimony, they filter out noise and highlight what truly influences a system. This gives us insight into internal mechanisms that aren't easily observable in the real world.

Models empower prediction and experimentation. They extract meaningful patterns from past data to anticipate future outcomes. By simulating alternative actions, models enable efficient comparison through performance metrics, offering valuable insights without costly real-world testing. This abstract model can be tested without real-world high cost. For example, instead of constructing the real road and highways, we can compute first in computer. We calibrate and validate using real flow data. We can find the optimizations from all options that we can think of, in simulated scenarios. Clearly, modeling reduces real-world high cost and risk significantly.

Models offer control by guiding interventions and system optimization. They reveal which variables affect outcomes, distinguish between fully and partially controllable factors, and expose a system's sensitivity and robustness under change.

A relation is a set of ordered pairs connecting two sets. A function is a specific kind of relation with unique outputs per input. Network models general relations and object transformations beyond simple functions. Models relate input data X to output data y. Input data is called independent variables; output data is dependent variable. A model can be thought of as a function $f(\mathsf{X}, \alpha) = \mathsf{y}$. The parameters α represent the underlying system behavior. Calibrating a model means finding the right parameters using input and output data. With a model, we can explain phenomena, predict outcomes, and optimize control. Prediction assumes the model and parameters remain unchanged. It uses known input and model to forecast output: $\mathsf{X}, \alpha, f \rightarrow \mathsf{y}$. Control aims to find input that produces a desired output: $\mathsf{y}, \alpha, f^{-1} \rightarrow \mathsf{X}$. These tasks rely on invariant assumptions in both model and parameters. Understanding a system means interpreting its parameters, values, signs, gradients, and significance.

1.2.3 The Principle of Invariant

Assumptions are the foundation and potential failure point of any model. Both explicit and implicit assumptions must be scrutinized, especially those involving invariants. A critical aspect of any predictive model is the *Principle of Invariance*: any quantitative model that can predict must have at least one invariant. To make a quantitative prediction, certain elements must remain fixed, and identifying these constants is key to properly applying the model. An invariant is something that stays constant in a model. This could be the model's structure, its parameters, or certain variables. Identifying these invariants is crucial for understanding a model's scope and limitations. Knowing these limits helps define the model's power and its usefulness. To understand phenomena, always look for what doesn't change in the model. For instance, the function itself may stay fixed between base and alternative scenarios. Parameters can be calibrated once, then held constant over time. Some variables are intentionally fixed to enable predictions. In hierarchical models, lower-level parameters may be constants. The model structure itself often remains unchanged across layers. The non-linear parameters might mask a deeper invariant structure. Constants may exist inside nested sub-models of a complex model. Effective prediction relies on invariants. A predictive model must rely on at least one invariant. Without invariants, predictions lose their scientific grounding. Changing models or parameters breaks the basis for prediction, and therefore we might not be able to compare them. Invariants are often hidden and assumed without being stated. Assumptions about invariants are key to scientific modeling. If assumptions shift, the model fundamentally changes.

1.2.4 Successful Modeling

A model is a simplified version of the real world. Simplification means the model can never be fully accurate. However, it can still capture essential behaviors of the system. Its value lies in clear communication and consistent structure. Usefulness is more important than absolute accuracy. As the father of System Dynamics, Jay Forrester [7] said, a model's validity is judged relative to mental or abstract alternatives. Clarity of structure and assumptions makes a model more effective. Time-based system behavior influences how well the model performs. Assessing the model means weighing its clarity, usefulness, and predictive consistency.

A model's utility is not predicated on a perfect isomorphism with reality. Rather, its value is assessed by its capacity to yield valid insights and predictions within its defined scope. Models aren't judged by perfection but by their ability to clarify. Their value lies in improving understanding and insight into systems. Diagrams and equations help make models visible and examinable. Stating assumptions enables better reasoning and clearer communication. Simulations reveal how models behave over time. If a model achieves these, it brings us closer to understanding reality.

Building upon these foundational concepts, one of the particularly insightful models presented in this book is the Ideal Flow Network (IFN) theory and its application to Transportation Research.

1.2.5 The Core Idea: Inferring Flow from Network Structure Alone

The foundational premise of the Ideal Traffic Assignment framework is that a network is not a passive container for traffic. Instead, it is an active system that channels, guides, and constrains flow in a predictable manner. In a conventional transportation modeling paradigm, the network is conceptualized as a static medium. This medium facilitates the movement of traffic based on a predefined *Origin-Destination* (O-D) demand matrix. This approach suggests that the demand for travel is an exogenous variable. The network's topology and capacities are viewed as fixed attributes that accommodate this demand. The focus is on how travel patterns, which are determined externally, are distributed across the network's links. The network itself is a passive stage for these predetermined travel patterns. The Ideal Traffic Assignment framework suggests a different view of traffic analysis. It suggests that the network's physical structure fundamentally directs flow. This structure includes its topology, capacity, and connectivity. This represents a paradigm shift. The focus moves from a demand-centric approach to a structure-centric, or supply-side, perspective. In this view, the network is not a passive stage. It is an active agent that shapes and directs traffic performance.

This shift in perspective changes the fundamental question we ask. The traditional model asks, *Given a known demand of 1,000 trips from Zone O to Zone D, how will these trips distribute across the available paths?* This question presupposes a complete and accurate knowledge of macroscopic demand, which is the traditional model's greatest practical weakness. The Ideal Traffic Assignment framework's microscopic, probabilistic approach has significant implications for calibration, enabling the use of modern data sources. The traditional method, which relies on a macro-level understanding of travel demand, often requires labor-intensive and costly surveys to establish an Origin-Destination (O-D) matrix. This OD matrix is the traditional model's greatest practical weakness due to its expense and potential for inaccuracy.

The IFN framework re-frames this problem by focusing on local, microscopic decisions. The core question becomes, "When a vehicle is at intersection i, what is the probability it will proceed to intersection j?". This approach makes the model's calibration process more compatible with modern Intelligent Transport Systems (ITS) [17].

Instead of relying on backward-looking questionnaires, the model can be calibrated using real-time data. This data is collected from sensors that capture local traffic behavior. For instance, GPS data can show the turning movements at intersections, providing the necessary probabilities for the Markov chain. Computer vision systems can also be used to count vehicles and track their paths through an intersection, offering direct empirical evidence for these local probabilities. This shift eliminates the need for a complete and accurate knowledge of macroscopic demand and leverages readily available, automatic dynamic data. By focusing on these local, probabilistic turning movements, the model conceptualizes demand not as a single, massive O-D matrix but as a distributed set of microscopic decisions that occur at every node in the network.

The global pattern of traffic flow is then understood as the emergent, steady-state consequence of aggregating these countless local decisions. This approach does not deny the existence or importance of travel demand. Rather, it proposes that the network's physical structure provides a powerful and robust proxy for the aggregate effect of that demand over time.

This idea is grounded in the principle of co-evolution. Urban transportation networks are not random constructs. They have evolved over decades in direct response to historical travel patterns and land-use development. A four-lane arterial road exists because, historically, there has been significant demand along that corridor. A small residential street remains small because it has primarily served local traffic. Therefore, the existing infrastructure, particularly the capacity of its links, serves as an embodiment of *past wisdom*. The network's current structure is a physical manifestation of long-term, aggregated demand patterns. By analyzing this structure, the IFN framework seeks to uncover the inherent, time-averaged flow distribution that the network was built to accommodate.

1.2.6 An Intuitive Analogy: Flow in Conduits and Circuits

The concept of a structure-driven flow distribution can be made more intuitive through simple analogies from the physical sciences. These analogies help to build a conceptual understanding of how a system's physical characteristics naturally and deterministically govern the flow within it, independent of the total volume being pushed through.

Consider a complex plumbing system in a large building, consisting of a network of pipes of varying diameters. If water is introduced into the system from a single source, it will not distribute itself evenly. The wider pipes, which offer less resistance, will naturally carry a greater proportion of the total flow than the narrower pipes. This distribution is a fixed property of the system's physical layout. If we double the total water pressure, the absolute flow in every pipe will increase, but the *proportion* of flow carried by each pipe will remain exactly the same. The distribution is determined by the conduit characteristics, not the total demand. A plumber, without knowing the exact water pressure, could analyze the blueprint of the pipe network and accurately predict which pipes would carry the most and least flow purely based on their relative diameters.

A similar principle governs the flow of electricity in a circuit. According to Ohm's Law, current distributes itself across parallel paths in inverse proportion to their resistance. A path with low resistance will draw a larger share of the total current than a path with high resistance. An electrical engineer can analyze a circuit diagram and, based solely on the resistance values of its components, predict the steady-state distribution of current throughout the network. This distribution is an intrinsic property of the circuit's design.

The Ideal Traffic Assignment framework applies this same logic to traffic networks. The capacity of a road link, a function of its number of lanes, width, and design speed, is analogous to the diameter of a water pipe or the inverse of electrical resistance. It represents the link's ability to carry flow with minimal impedance. Just as water favors wide pipes and electricity favors low-resistance paths, traffic will naturally favor high-capacity corridors. The IFN framework formalizes this intuition, proposing that a transportation network has an inherent, underlying equilibrium flow state that is determined by the relative capacities of its constituent links. This natural, structure-driven distribution is what we term the *Ideal Flow*.

1.2.7 The Discovery: Revealing the Ideal Flow Pattern

The Ideal Traffic Assignment (ITA) framework was developed from a rigorous theoretical foundation of Ideal Flow Network (IFN), not from empirical observations of chaotic traffic data. The breakthrough moment was the discovery of a stable and predictable pattern within an abstract mathematical space, not real-world traffic flows.

The initial research of Ideal Flow Network explored the behavior of *random walks* [1, 11, 18] on random networks. We examined various distributions of stochastic matrices. The key insight emerged from this theoretical analysis: the long-term, steady-state flow in these abstract networks was determined by the network's structure alone. This finding was not dependent on specific Origin-Destination (O-D) patterns or empirical data. It demonstrated that even with varying stochastic matrices, the asymptotic relative flows converged to a stable, structural equilibrium. This theoretical observation formed the basis for the IFN framework.

The critical step in the discovery was a shift from analyzing absolute flows (e.g., vehicles per hour) to analyzing *relative flows*. This involved a process of normalization, where the flow on each link was expressed as a proportion of the total flow across the entire network. This simple mathematical transformation changed the analytical focus from *how much* in absolute value to *how much, relative to the whole*.

The breakthrough occurred when these relative flow distributions were analyzed under a long-term, steady-state condition, known as the asymptotic limit. The researchers discovered that as the simulation time extended towards infinity, the relative flow distribution converged to a single, unique, and stable pattern. Remarkably, this convergent pattern was the same regardless of the specific O-D matrix used to generate the traffic in the first place. Whether the demand was concentrated between two specific zones or dispersed evenly across the entire region, the long-run *proportional* distribution of flow remained invariant. This discovery was profound. It demonstrated that underlying the transient, demand-driven fluctuations of daily traffic, there exists a fundamental, time-invariant flow signature that is a unique property of the network's physical structure.

This intrinsic pattern was named the *Ideal Flow* to denote its unique, invariant, and steady-state solution to the underlying stochastic process. It is 'ideal' not in a normative sense of being optimal, but as a representation of a fundamental, structure-driven equilibrium. However, its stability and universality also make it a reference point for evaluating interventions and pursuing equitable or efficient outcomes. The existence of this stable Ideal Flow provided the empirical and theoretical justification for building a new modeling paradigm, one that could predict this underlying distribution directly from the network's structure, thereby bypassing the need for a specific O-D matrix.

1.3 IFN in Brief

Ideal Flow Network (IFN) theory itself started initially in the field of transportation engineering. That's why terms like capacity, lanes, flow, capacity, and congestion remain central to the model. These terms are familiar because road traffic is part of everyday experience. As a result, modelers in other disciplines can easily grasp IFN concepts.

Ideal Flow Networks are powerful tools for modeling any flow-based systems. It works by defining fixed rules for structure and behavior. The elegance of Ideal Flow Network (IFN) theory lies in its generalizability. So long as the four axioms and core definitions are upheld, the same modeling principles apply across disciplines. Ideal Flow Network (IFN) is a versatile modeling approach that captures system dynamics across diverse domains, from traffic engineering to AI, data science, computer science and physics. Its applications span engineering, biology, economics, logistics, and even the social sciences. IFN serves as a tool for understanding equilibrium, flow, and structure within strongly connected networks. Depending on your background, IFN might resemble a simulation engine, a statistical model, or even a philosophical lens on systems in balance.

Even if your domain isn't traffic-related, studying IFN through transportation models offers practical insight: Road networks are intuitive, making abstract flow concepts easier to comprehend. Common terms like "flow" and "capacity" become vivid metaphors for other systems. It provides a low-barrier sandbox to test the theory's power before applying it elsewhere. Exploring IFN through transportation doesn't just build understanding. Ideal Flow Network (IFN) modeling fosters systemic thinking, policy-awareness, and empirical rigor. It serves as a foundational analytical lens capable of revealing structural insights across diverse domains. When a problem is representable as a flow equilibrium within a strongly connected graph, IFN may offer a suitable framework. However, several limitations must be acknowledged: IFN is not applicable to disconnected networks or systems lacking equilibrium. The framework fails when essential axioms are not satisfied. These include connectivity, equilibrium, and the premagic irreducible properties.

1.3.1 Fundamental Axioms of IFN

The behavior and application of IFN models are governed by four foundational principles, which we treat as axioms for model development. The theoretical foundation for the Ideal Flow Network (IFN) framework rests on this set of fundamental axioms. These axioms describe the long-term, asymptotic behavior of flows within a network. The subsequent model is rigorously developed from these core principles. The four axioms are irreducibility, flow conservation, steady state, and the Markov property.

Axiom 1.1 (Irreducibility [10]) *An IFN is a strongly connected directed graph. This ensures that the network behaves as a single, indivisible system.*

Axiom 1.2 (Flow Conservation) *Flow is conserved at every node within an IFN. The total inflow to a node must equal the total outflow from that node. This premagic [19] property ensures the system is in a stable, balanced state. This condition reflects a state of perfect balance, where no flow is created or destroyed.*

Axiom 1.3 (Steady-State Equilibrium [16]) *The* relative flows *in an IFN represent a long-term, steady-state distribution. They can be scaled by any positive constant without changing the underlying proportional distribution.*

Axiom 1.4 (Markov Property [14]) *Decisions within the network are local and memoryless. In each decision point, the agents as decision-makers take action based on their immediate local situation.*

In the context of mathematics or systems theory, irreducibility often refers to a state where a system cannot be broken down into simpler, independent subcomponents without losing essential properties. Due to the irreducibility axiom, the adjacency matrix A and the ideal flow matrix F of the network are therefore irreducible. This property ensures that a path exists between any two nodes in the network, allowing for system-wide flow. Transportation network is one of the example of such strongly connected network. When we go somewhere, we must be able to go back home.

Nature has special proportions that create balance and harmony. These include musical frequencies and the Fibonacci sequence in plants. In studying traffic congestion with math and simulations, we found striking patterns. Certain link capacity ratios in networks can balance congestion levels. These levels can theoretically be reduced to a minimum. We call this pattern *premagic* [19] due to its line sum symmetry [5]. That means the row sums of the capacity matrix equal the column sums when transposed. This pattern is based on the Ideal Flow Network. An IFN is a steady state of flow conservation in a strongly connected network.

Steady state refers to a long-term equilibrium condition. In the long run, aggregated flow values on links can grow extremely large. These values may approach infinity and become practically meaningless. In contrast, relative flow, called ideal flow, remains useful over time. Ideal flow converges to mostly unique and positive values. The flows within an Ideal Flow Network (IFN) represent a *steady-state equilibrium.* This *limiting long-term distribution* is the asymptotic state of the system. This state reflects a stable distribution of traffic over an extended period. It captures the system's equilibrium structure without scale distortion. This equilibrium is independent of the network's initial state. Because of this, we can rescale ideal flow globally without changing its meaning. The ideal flows numbers are relative flow. The model, therefore, predicts the *relative proportions* of flow, not the absolute magnitude. To obtain the absolute magnitude, we need to calibrate the model with real world flow data.

Ideal flow values are positive real numbers. In case when the ideal flow values are rational numbers, we can put them into fractions. Then, we can find the least common multiple (LCM) of the denominators of these fractions as our scaling factor. This also allows the use of integer values to simplify flow representation.

At each intersection (or node), agents can freely choose the next link. Their choice is influenced by a predefined probability distribution. This setup reflects the Markov property in decision-making. Drivers make local decisions at intersections without needing global path knowledge. Each random choice follows a Markovian stochastic pattern. These decisions depend only on current conditions, not the full network layout. Decisions within a system can adhere to the Markov property, meaning choices are based solely on the current state, not on the sequence of past states. This principle simplifies complex dynamics by focusing on local information rather than the complete history of an agent's journey. The Markov property applies to any decision context where choices are made sequentially and are memoryless. A decision-maker's next action is contingent only on their immediate situation. This simplifies the modeling of dynamic systems.

Here are some examples:

- In economics, a consumer's purchasing decision at a store might depend only on the products currently on the shelf, not on their past shopping habits.
- In social networks, a user's decision to "like" a post may depend only on the post's content and not on the previous posts they have liked.
- In robotics, a robot navigating a maze might decide to turn left or right based only on the sensory data from its current position, ignoring the turns it made to arrive there.
- In transportation, a driver's choice of turn depends only on their current intersection, not their origin or path. The routing decisions at each node within an IFN adhere to the Markov property. A traveler's path choice depends only on the current location and local information, not on the complete history of their journey.

The Markov axiom provides a powerful simplification. It reduces the computational complexity of predicting future states by eliminating the need to track an entire history. This focus on local, rather than global, information is a core assumption in many mathematical models, including the Ideal Flow Network, and is central to understanding a wide range of systems.

In Ideal Flow Theory, the *flow itself is undefined.* An Ideal Flow Network (IFN) represents a steady-state distribution of flow, but the nature of that flow is not intrinsically defined by the theory itself. The IFN model is a parsimonious mathematical framework that describes how *flow proportions* are distributed throughout a network. It does not specify what constitutes the flow.

The definition of "flow" is determined by the specific application of the model. For instance:

- *Traffic Engineering:* Flow can represent the number of vehicles per hour on a road segment.
- *Logistics:* It might signify the volume of goods transported through a supply chain link.
- *Epidemiology:* Flow could denote the number of individuals moving between locations, potentially carrying a disease.
- *Computer Science:* It can be the rate of data packets transmitted between network nodes.

In each case, the underlying mathematical principles of the IFN, such as flow conservation and steady-state distribution, remain constant. However, the units of the flow are specified by the particular problem being modeled. The theory provides the structure for the network, but the modeler must provide the specific interpretation of the flow.

1.3.2 Philosophical Synthesis

The philosophical implications of the Ideal Flow Network (IFN) model are profound. The model's success hinges on two fundamental properties: premagic and irreducibility. These properties have conceptual significance beyond their mathematical definitions.

Irreducible [10] means the system is unified, with all parts connected and functioning together. Irreducibility ensures the system is strongly connected. Every part is linked to every other part. This property prevents the system from being compartmentalized or isolated. It reflects a holistic, unified perspective, feedback loops. The system is not a collection of separate components but a single, integrated whole.

The premagic property signifies perfect balance. It asserts that flow is conserved at every node. In a broader sense, this represents a state of long-term stability and equilibrium. It implies a balance where the total inputs equal the total outputs, ensuring no accumulation or deficit in the long run. This concept embodies a natural justice, where every element receives what it contributes.

Together, these two properties suggest a harmonious system. The premagic condition of flow conservation reflects a form of equality and fairness. The irreducibility of the system highlights the importance of interconnectedness. Viewing a system through the lens of IFN encourages a deeper, integrated understanding. It promotes a perspective where harmonious relationships and long-term causal effects are recognized as essential for equilibrium. This approach moves beyond fragmented analysis to a more unified, profound view of complex systems.

1.4 The Persistent Challenge of Traffic Congestion

1.4.1 The Global Crisis of Urban Mobility

The modern city breathes; its lifeblood is the constant, pulsing flow of people and goods through a vast circulatory system of streets and pathways. When this flow is efficient, the city thrives, appearing productive and vibrant. But when the flow falters and its arteries clog, the city suffers. Gridlock, wasted time, and economic loss are the symptoms of a system in distress. For decades, planners and engineers have battled this urban ailment, yet the problem of congestion often seems intractable. This challenge is not local but global, manifesting in familiar vignettes in almost any city.

Consider these familiar scenes, these problem vignettes from the daily life of almost any modern city:

- *The Gridlocked Heart [4]:* Imagine a major downtown intersection, a marvel of civil engineering with multiple lanes in every direction. Yet, every evening at 5 PM, it transforms into a parking lot. Horns blare, engines idle, and the air thickens with exhaust. The very infrastructure designed to maximize throughput becomes a symbol of systemic failure. The conventional solution? Add another turning lane, re-time the signals. But often, this only moves the bottleneck 500 m down the road, a phenomenon known as "congestion transfer."
- *The Path of Most Resistance:* Two blocks away from that gridlocked heart runs a parallel street. It is clean, wide, and strangely empty. A few cars trickle through, their drivers enjoying an uncongested journey. Why don't more drivers use this route? Perhaps the entry points are awkward, or it's perceived as being "out of the way." The network has spare capacity, yet the system as a whole fails to utilize it. The supply is there, but the demand is pathologically concentrated.
- *The Neighborhood Speedway:* A quiet residential street, lined with homes and trees, has become a high-speed shortcut for commuters seeking to bypass a congested arterial road. Residents watch with frustration as cars race past, endangering children and destroying the neighborhood's tranquility. The street was designed for local access, but its position within the larger network has turned it into an unwilling and unsuitable artery.

These scenes reveal a crucial point: traffic is a classic example of a complex adaptive system. Our transportation networks are complex systems, and their problems are often systemic, not local. The overall pattern of flow is an emergent property that arises from the local, independent decisions of millions of individual agents. In such systems, cause and effect are often decoupled, and simple solutions can lead to counter-intuitive, negative outcomes.

Rigorous modeling is crucial in urban transportation. Traffic congestion is a complex adaptive system. Overall flow results from many local decisions. Simple, intuitive solutions can have negative effects. The infamous Braess's Paradox [2, 3] is a classic example of this phenomenon. Adding a new road can sometimes increase travel times for everyone. This paradox highlights the inadequacy of intuition alone. It is a stark reminder that our intuitive mental models are insufficient for managing these dynamics. We must rely on mathematical modeling to address congestion effectively.

The result is a crisis of mobility defined by inefficiency and, critically, by unreliability, which degrades the function of the entire urban ecosystem.

For too long, our primary tool for fixing them has been the hammer of "more capacity." We widen roads, add lanes, and build bigger intersections. Yet, this often induces more demand, leading to a frustrating cycle of construction and renewed congestion. What if this approach is fundamentally flawed? What if, instead of focusing on individual links, we focused on the health of the entire network? What if the solution isn't always *more*, but sometimes *better*, or even *less*?

1.4.2 Economic and Environmental Costs

Traffic congestion is a pervasive feature of urban life that imposes significant and compounding inefficiency costs on society. These costs are not merely inconveniences. They represent a direct drain on economic vitality and a tangible threat to environmental health. The economic losses are multifaceted. Wasted time spent in stalled or slow-moving traffic represents billions of dollars in lost productivity and forgone leisure. For commercial transport, congestion disrupts finely tuned supply chains, increasing fuel consumption, vehicle wear, and labor costs, which are ultimately passed on to consumers.

The environmental costs are equally severe. Idling engines in gridlocked traffic burn fuel inefficiently, thickening the air with exhaust and releasing excess greenhouse gases that contribute to climate change. This also creates localized hotspots of harmful pollutants, such as particulate matter and nitrogen oxides, which have documented negative impacts on public health, particularly respiratory and cardiovascular wellness. The constant noise and stress associated with congested environments further degrade the quality of urban life. These substantial economic and environmental burdens create a clear imperative to develop more intelligent and effective mitigation strategies.

1.4.3 The Core Need for Prediction in Planning and Operations

To address the systemic problems of congestion, we must move beyond intuition and embrace the power of mathematical modeling. Effective interventions in complex systems require a predictive understanding of how the system will respond to change. Rigorous models provide a purposeful abstraction of reality, allowing us to test hypotheses and evaluate the potential consequences of decisions in ways that are impossible or prohibitively expensive in the real world. This predictive capability is fundamental to the work of planners, engineers, and policymakers.

Urban planners use predictive models to assess the long-term impacts of structural changes to the network. When evaluating a proposal to add a new bridge or approve a major residential development, a model can forecast how flow patterns will shift and where new bottlenecks might emerge. This allows for the proactive design of more resilient and efficient urban forms.

Transportation engineers require predictive tools for operational management. To optimize traffic signal timing or plan for detours during road works, engineers can use models to simulate the effects of their interventions. This prevents changes that, while locally beneficial, might create counter-intuitive negative outcomes for the network as a whole.

Policymakers rely on prediction to formulate and justify strategies like congestion pricing or changes in land use patterns. Models provide a quantitative basis for setting policy parameters and communicating their expected benefits to the public, transforming political debate into evidence-based discourse. In all these domains, prediction is the essential tool that enables a shift from reactive problem-solving to proactive, intelligent system management.

1.5 Introducing the Ideal Traffic Assignment Framework: A Paradigm Shift

The limitations inherent in conventional traffic models are not superficial; they are fundamental. The critical dependence on costly, static, and often inaccurate Origin-Destination (O-D) matrices, coupled with significant computational and calibration burdens, reveals the need for more than just an incremental improvement. What is required is a paradigm shift: a fundamental change in the underlying assumptions and analytical perspective used to model traffic flow.

The Ideal Flow Network (IFN) model has broad applications. Its primary use, however, is analyzing urban traffic flow and congestion. This transportation application of IFN is called Ideal Traffic Assignment (ITA). The ITA framework provides a structured approach for studying congestion. It also allows for testing interventions and exploring systemic impacts.

Many people propose solutions to traffic problems. These ideas often lack a rigorous method for testing. The IFN provides a tool to test these theories. It empowers citizens to model the effects of policy changes. This democratizes the process of traffic policy evaluation.

The Ideal Traffic Assignment framework represents such a paradigm shift. It moves the analytical focus away from external, aggregated demand and onto the intrinsic, structural properties of the network itself. It proposes a radical yet intuitive idea: that the physical layout of a transportation network, its topology, connectivity, and capacities, is the primary determinant of how flow is distributed within it. In this next section, we only introduce the core concepts of this new paradigm, building an intuitive understanding of its foundations before the rigorous mathematical treatment in later chapters.

Beyond its practical use, ITA has inspired new ways of thinking about what an *ideal* network might look like. One such concept is called *Premagic Equilibrium (PE)*, a state where the two classic goals in traffic modeling, *user equilibrium* and *system optimal*, are perfectly aligned. In this ideal state, the inefficiency caused by selfish routing [15], known as the *price of anarchy*, drops to one, meaning no loss at all. To measure how close a real network is to this ideal, ITA introduces a metric called the *VarPi index*.

Using this index, we uncovered a surprisingly simple yet elegant formula that we call as the *Network-Flow Identity*: $\kappa = \tilde{c}\xi\varpi$. This equation links total flow κ in a network to its total capacity $\tilde{c}$, its maximum allowable congestion ξ, and its efficiency ϖ as measured by the VarPi index. It assumes a proportional flow distribution (Markov Split), and offers a powerful way to connect physical infrastructure with demand and performance, using only the network's structure.

Through this framework, policy interventions can be designed almost like chemical reactions, combining elements of the network like atoms to form new molecular configurations that improve flow and reduce congestion. It is a fresh and intuitive way to think about transportation planning, grounded in both theory and practical insight.

1.5.1 Ideal Traffic Assignment as a New Paradigm

The Ideal Traffic Assignment framework [20, 21], built upon the discovery of Ideal Flow [18], is not merely an incremental improvement on existing methods or a different algorithm for traffic assignment. It represents a distinct and innovative paradigm for traffic analysis. However, it is crucial to position this paradigm with clarity and intellectual honesty, acknowledging both its power and its proper domain of application.

If interpreted rigidly as a tool that can, from a static road map alone, predict the absolute number of vehicles on every street at any given time of day, then the Ideal Traffic Assignment framework ventures into wishful thinking. Real-world traffic is too complex and too heavily influenced by dynamic, short-term demand patterns for such a simple, uncalibrated model to be universally predictive in that manner. The claim is not that network structure is the *only* factor, but that it is a profoundly powerful and often overlooked *primary* factor. The true value and utility of the IFN paradigm are realized when it is understood and applied in its four distinct but related capacities.

First, Ideal Traffic Assignment is *a powerful theoretical paradigm*. It offers a unique and elegant mathematical framework, rooted in the theories of random walks and maximum entropy [9], to understand the intrinsic relationship between network structure and flow distribution. It provides a *physics* for traffic networks, allowing us to derive fundamental properties from first principles. This theoretical lens provides unparalleled insights into the conditions required for optimal network performance, such as the convergence of User Equilibrium and System Optimum, which will be rigorously proven in later chapters.

Second, it is *a parsimonious tool for relative analysis*. This is perhaps its most immediate practical application. An urban planner can use the IFN model to rapidly compare the structural efficiency of alternative network designs, for example, evaluating the impact of a new bridge versus a new ring road. By analyzing how each structural change alters the network's Ideal Flow distribution, the planner can gain crucial insights into which design promotes a more balanced and efficient use of the entire system. This can be done without requiring a new, expensive, and time-consuming O-D survey for each alternative, making it a remarkably agile tool for strategic, high-level planning.

Third, Ideal Traffic Assignment is *a calibratable and adaptable model*. The framework provides a robust bridge between its elegant theory and messy reality. While the core model produces a relative flow distribution, this distribution can be scaled to predict absolute flows through calibration against a small, readily available set of real-world link flow counts. This process, which will be detailed in later chapter, makes the model practically useful for quantitative assessments. Furthermore, this calibration can be performed sequentially over time. By integrating with real-time data streams from ITS sensors, the model can be dynamically recalibrated to *nowcast* current network conditions, making it an adaptive framework well-suited to the modern data environment.

Finally, Ideal Traffic Assignment serves as *a guide for design philosophy*. Its idealized concepts, such as the state of Premagic Equilibrium, provide clear, albeit aspirational, targets for designing highly efficient transportation networks. Even if a perfect *ideal* state is unattainable in the real world, it provides a theoretical benchmark, a North Star, that can guide the decisions of urban planners and property developers. It helps to answer the question: *What does an inherently efficient network look like?* This guidance is invaluable for long-range planning and the design of new urban areas.

We may need to distinguish the IFN from the Ideal Traffic Assignment. The capacity and congestion matrices are parts of Ideal Traffic Assignment framework but they are not components of IFN itself. However, they both serve important roles in modeling IFN behavior. The capacity matrix helps approximate inputs to the flow or stochastic matrix. The congestion matrix can approximate output performance of the system. This output depends on both the flow matrix and the capacity matrix.

One basic input for Ideal Traffic Assignment framework is the capacity matrix. It estimates the maximum flow that can pass through infrastructure. Link capacity refers to the maximum flow a single link can handle. This is often determined by lane count or road width in a network. The (0,1) pattern of the capacity matrix matches an irreducible adjacency matrix. The ideal flow matrix also shares this same (0,1) structure. Its input comes from a stochastic transformation of the capacity matrix. What matters most is each link's capacity relative to others, not its absolute value. The absolute capacity just serves as a global scaling factor. Therefore, we can simplify the matrix using lane counts per link. Each lane is assumed to have a fixed, unchanging capacity.

Hadamard division [8] refers to element-wise division of one matrix by another. In this operation, division by zero is defined to result in zero. Applying Hadamard division to the ideal flow and capacity matrices gives the congestion matrix. Each positive entry in the congestion matrix represents a flow-to-capacity ratio. This ratio quantifies the level of congestion in each link. Congestion reflects the demand relative to the available supply in the network. When link length and free-flow speed are known, flow/capacity ratio helps estimate link speed. It also allows computation of travel time and delay through nonlinear relationships. To improve network performance, analyzing congestion ratio or the flow/capacity distribution is more useful than speed or delay. Speed, travel time, and delay are merely functions derived from congestion level.

Analyzing patterns between the capacity matrix input and the congestion matrix output reveals unexpected insights. Adding a lane to the most congested link lowers congestion locally. However, this adjustment can raise the minimum congestion level across the entire network. The relationship between congestion and speed, travel time, and delay is nonlinear. This global effect on minimum congestion is often invisible in the short term. Over time, congestion may return to the same link due to changes made elsewhere.

In IFN, equilibrium represents a stable distribution of flow across the network that balances supply and demand. If we have policy to change the network structure, it may create *equilibrium shifts* in Ideal Traffic Assignment context. For example, when we add a lane to a congested link, we're modifying the input capacity matrix. This changes the stochastic structure of the system, resulting in a new set of relative flows. The redistribution of flow is nonlinear because some paths attract more agents due to perceived lower congestion. This doesn't just relieve the bottleneck. It perturbs the entire flow balance. The system finds a new equilibrium, but that shift may inadvertently raise congestion elsewhere, especially the minimum network congestion level.

For incorporating route choice behavior and network adaptation, we introduce feedback loop. The key mechanism is the indirect feedback loop: a local intervention creates global behavioral consequences. Agents respond to reduced local congestion by rerouting, increasing flow on previously underutilized links. Over time, this leads to congestion creeping back, sometimes into the same link. The system behaves like a self-organizing adaptive network, adjusting flows until a new stable (but not necessarily optimal) distribution emerges. These feedback loops are not explicitly coded but emerge from the probabilistic decision-making of agents and the structural constraints of the network.

Interesting tie-in or invariants in motion might happen in congestion rebalancing by changing capacity. Even when flow magnitudes shift, the underlying structure such as (0,1) link accessibility patterns would remain invariant. The congestion level acts as a dynamic indicator, revealing where the system resists equilibrium due to changing agent behavior. Ideal flow highlights this: although scaled, its relative distribution reveals deep patterns of systemic stress. Even though minimum network congestion may rise initially, repeatedly adding lanes to the most congested links will eventually balance congestion, as long as we do not overshoot.

Every irreducible (0,1) pattern matrix may have a certain premagic proportion. This suggests a distinct capacity structure tailored to each network topology. Such premagic proportions could reflect the organic growth of real cities. Historical and environmental factors might shape these equilibrium patterns.

The Ideal Traffic Assignment framework provides a method to predict network flow from the physical network structure alone. This distinguishes it from traditional models that rely on extensive data, such as Origin-Destination surveys. This enables the model to generate the most probable flow distribution under conditions of uncertainty.

The Ideal Traffic Assignment model offers a scientific basis for evaluating solutions that may seem counter-intuitive. For example, it can demonstrate how a reduction of capacity for one mode of transport, such as cars, can result in a more balanced and efficient overall system. The model can also show that a well-designed network of alternative transport paths, like bike lanes, is essential for a high-performing urban transportation system.

Ideal Traffic Assignment (ITA) offer foundational principles for addressing urban mobility system challenges. A new framework rethinks transport design by moving beyond reactive congestion mitigation strategies. Congestion is inevitable in cities but can be managed by approaching a systemic lower bound. This lower bound emerges when traffic is *balanced*, with uniform congestion across all active links. Such a state, called *Premagic Equilibrium*, unifies *User Equilibrium* and *System Optimum*. At this point, individual choices align with collective goals, eliminating the *Price of Anarchy*.

Traffic distribution is a core part of transportation planning. Two principles, defined by Wardrop [22], are fundamental to this field. The first principle is User Equilibrium (UE). UE is a state where no driver can find a faster route by changing their path. This is also known as a Nash Equilibrium [12, 13]. The second principle is System Optimum (SO) [22], where the network's average travel time is at a minimum. This is a type of Pareto Optimum. A solution is Pareto optimal [6] if no individual can be made better off without making someone else worse off. In most real-world networks, these two states of UE and SO don't align. This happens because individual driver choices often lead to a traffic pattern that isn't optimal for the whole system. This book will show a theory where these two states can coincide *under special conditions* for Premagic Equilibrium.

Traffic management benefits from aiming at a mathematically coherent ideal: a state in which individual choices and system goals naturally align. We develops this ideal, called *premagic equilibrium*. Starting from a simple structural condition on capacity (premagic), and two mild design axioms (steady per-link characteristics and identical path link counts), we show that Wardrop's User Equilibrium and the System Optimum coincide (UE = SO = PE). Equal congestion across all links yields maximum network efficiency. This equal distribution minimizes the overall congestion level absolutely. Such distribution aligns with the maximum entropy principle in flow systems [23]. Congestion becomes equal when link capacities follow a premagic proportion. A premagic capacity matrix has rows and columns summing to the same value. This structure ensures uniform flow/capacity ratios across the entire network. Maintaining premagic proportions helps achieve minimal congestion equilibrium. We call this balanced state the *premagic equilibrium* (PE).

To operationalize this target, we introduce the VarPi efficiency index to measure distance to the ideal Premagic Equilibrium (PE) state and derives the *Network Flow Identity* (NFI), a compact relationship linking total flow, capacity, and congestion.

$$\kappa = \tilde{c}\xi\varpi \tag{1.1}$$

This equation, derived later in the book Chap. 13 Sect. 13.5, links total flow in a network κ to its total capacity $\tilde{c}$, its maximum allowable congestion ξ, and its efficiency as measured by the VarPi index ϖ. This general formula applies to any strongly connected network under the assumption of linear proportional capacity (Markov Split). At first glance, it appears deceptively simple, linking physical input $\tilde{c}$ with total demand κ, maximum congestion ξ, and a performance indicator derived purely from the network's capacity matrix ϖ.

But its true power emerges when applied across network scenarios. For example, consider a baseline network that undergoes a molecular policy transformation, such as a road diet or signal reconfiguration, resulting in a modified network structure. If we hold certain invariants (e.g., total demand κ remains constant in the short term) and introduce inequalities (e.g., reduced capacity $\tilde{c}$ due to intervention), the Network Flow Identity allows us to explore the theoretical limits of such changes.

By comparing pre- and post-intervention VarPi indices, we can quantify how far the new network deviates from ideal performance. This opens the door to evaluating policies not just empirically, but analytically, revealing whether a proposed intervention can maintain equilibrium, reduce congestion, or even approach Premagic Equilibrium.

In this way, the identity becomes more than a formula, it becomes a lens for understanding the deep structure of network behavior, and a tool for designing interventions with precision, much like combining atoms into molecules in a controlled reaction. It bridges abstract mathematical formulations with actionable policy design by introducing a formal grammar that enables systematic, reproducible evaluation of transport interventions. This contribution elevates Ideal Flow Networks from a descriptive framework to a prescriptive methodology for urban transportation planning, forging a direct connection between abstract mathematical structures and actionable policy design. By introducing a formal grammar for transport policy, it enables planners to systematically compose, evaluate, and refine interventions with reproducibility and precision. Rather than limiting analysis to a handful of predefined scenarios, the framework opens a combinatorial space of design options possibilities, allowing exploration of both immediate impacts and cascading effects across the network. It supports reasoning about both direct impacts and cascading network-wide effects of interventions. It offers not just a tool for analysis, but a language for creativity and a disciplined approach to discovery, empowering planners to act as architects of future mobility, wielding mathematical precision to shape systems that reflect both technical insight and societal values.

Practical guidance is offered for applying these principles to real-world transportation planning. The Ideal Traffic Assignment (ITA) framework enables proactive, data-driven policy design and evaluation. Policy-making becomes a transformation process on structural matrices, like molecular chemistry in networks. Expansion strategies should focus on improving flow balance, not merely adding road capacity.

The VarPi index quantifies how close a network is to its balanced configuration. This index provides a clear, auditable benchmark for guiding infrastructure and policy decisions. Future research challenges include universal balance metrics and generative design tools for ideal networks.

From Network Flow Identity, we obtain actionable laws and design guidance. Networks with *Eulerian structure*, symmetry, and *premagic capacity* are optimal under ideal conditions. Uniform grids physically embody the theoretical ideal proposed by Ideal Traffic Assignment principles.

Although the framework presented in this volume diverges from conventional transportation modeling paradigms, such divergence is both intentional and methodologically grounded. Some experts may question the reliance on idealized assumptions within the Ideal Flow Network (IFN), arguing that such simplifications fail to capture the dynamic and heterogeneous nature of real-world traffic systems. This critique is acknowledged and addressed directly: the IFN is not proposed as a substitute for empirical models but as a normative benchmark. By establishing a theoretical optimum, the framework enables systematic quantification of deviation from ideal conditions. Metrics such as the VarPi index formalize this comparison, offering planners a structured, data-driven basis for evaluating network performance and guiding intervention strategies. The model's utility lies in its ability to illuminate inefficiencies and support policy design, not in replicating empirical complexity.

A second line of critique may concern the framework's treatment of demand, particularly its departure from the traditional Origin-Destination (O-D) matrix. While it may appear that the model circumvents demand altogether, it instead reframes it through a structural lens. The underlying premise is that the built form of a network encodes long-term travel patterns, allowing flow to be inferred from capacity and topology. This inversion of the conventional approach is especially pertinent in contexts of uncertain or evolving demand, such as anticipatory planning or scenario testing. Where empirical data are available, the framework supports calibration procedures that reconcile inferred flows with observed O-D distributions, thereby maintaining empirical validity without requiring exhaustive trip matrices.

The book also introduces the concept of a *chemistry of policy*, which may be perceived as metaphorical or overly academic. However, this analogy serves a precise methodological function: it formalizes policy formulation as a composable and auditable process. By representing interventions as atomic and molecular units, the framework enables systematic enumeration, simulation, and evaluation of policy combinations. This structure supports reproducibility, sensitivity analysis, and outcome forecasting, thereby enhancing transparency and accountability in planning workflows. Far from being abstract, the approach equips practitioners with a rigorous toolset for evidence-based scenario evaluation and decision support.

Finally, the proposition that a road diet may alleviate congestion could appear counterintuitive and politically sensitive. Critics may question the empirical basis for such claims. The framework addresses this through the Principle of Congestion Rebalancing, which demonstrates that strategic capacity reduction on overburdened links can redistribute traffic more equitably across the network. This reallocation has the potential to reduce peak congestion system-wide, even if localized delays increase. The model provides quantitative tools to assess these trade-offs, enabling planners to evaluate whether the net benefits justify the intervention. In doing so, it offers a defensible rationale for policies that challenge conventional intuition but yield improved systemic outcomes.

The theory provides a useful foundation for understanding network-level relationships between capacity, flow, and congestion. However, its application requires careful consideration of assumptions and integration with real-world data and models. Future research should focus on making the theory more dynamic, behavioral, and validated across diverse networks.

1.6 Summary

In conclusion, Ideal Traffic Assignment should be understood not as a single tool, but as a comprehensive framework that offers a new way to see, analyze, and design transportation systems. The parsimonious framework uses network topology to determine a steady-state flow distribution, contrasting with traditional models that rely on external demand. It offers advantages in data flexibility and scalability, providing a new approach to transportation modeling. It is a theoretical lens for understanding fundamental principles, a practical tool for rapid structural analysis, an adaptable model for the modern data era, and a philosophical guide for building the efficient cities of the future.

References

1. David Aldous and James Allen Fill. *Reversible Markov Chains and Random Walks on Graphs.* Available at: https://www.stat.berkeley.edu/~aldous/RWG/book.html.2002.
2. D. Braess, A. Nagurney, and T. Wakolbinger. "On a paradox of traffic planning". In: *Transportation Science* 39.4 (2005), pp. 446–450.
3. Dietrich Braess. "Über ein Paradoxon aus der Verkehrsplanung". In: *Un-ternehmensforschung* 12.1 (1968), pp. 258–268. https://doi.org/10.1007/BF01918335.
4. Carlos F. Daganzo. "Urban Gridlock: Macroscopic Modeling and Mitigation Approaches". In: *Transportation Research Part B: Methodological* 41.1 (2007), pp. 49–62. https://doi.org/10.1016/j.trb.2006.03.001.

5. B. C. Eaves et al. "Line-sum-symmetric scalings of square nonnegative matrices". In: *Mathematical Programming Studies*. Vol. 25. Springer, 1985, pp. 124–141. https://doi.org/10.1007/BF02592021.
6. Matthias Ehrgott. *Multicriteria Optimization*. 2nd ed. Springer, 2005. ISBN: 978-3-540-21398-7. https://doi.org/10.1007/3-540-27659-9.
7. J.W. Forrester. *Principles of Systems*. System dynamics series. Productivity Press, 1990. ISBN: 978-09-15299-87-4. https://books.google.co.id/books?id=2p6UrgEACAAJ.
8. Roger A. Horn and Charles R. Johnson. *Matrix Analysis*. 2nd. Cambridge: Cambridge University Press, 2012. https://doi.org/10.1017/CBO9781139020411.
9. Edwin T. Jaynes. "Information Theory and Statistical Mechanics". In: *Physical Review* 106.4 (1957), pp. 620–630. https://doi.org/10.1103/PhysRev.106.620.
10. John G. Kemeny and J. Laurie Snell. *Finite Markov Chains*. New York: Springer, 1976. https://link.springer.com/book/9780387901923.
11. László Lovász. "Random walks on graphs: A survey". In: vol. 2. Budapest: János Bolyai Mathematical Society, 1993, pp. 1–46. https://www.cs.cmu.edu/~15859n/RelatedWork/random-walks-on-graphs.pdf.
12. J. F. Nash. "Non-Cooperative Games". In: *Annals of Mathematics* 54.2 (1951), pp. 286–295. https://doi.org/10.2307/1969529.
13. John F. Nash. "Equilibrium Points in N-Person Games". In: *Proceedings of the National Academy of Sciences* 36.1 (1950), pp. 48–49. https://doi.org/10.1073/pnas.36.1.48.
14. J. R. Norris. *Markov Chains*. Cambridge Series in Statistical and Probabilistic Mathematics. Cambridge: Cambridge University Press, 1997. https://doi.org/10.1017/CBO9780511810633.
15. T. Roughgarden and E. Tardos. "How bad is selfish routing?" In: *Proceedings 41st Annual Symposium on Foundations of Computer Science*. 2000, pp. 93–102. https://doi.org/10.1109/SFCS.2000.892069.
16. Eugene Seneta. *Non-negative Matrices and Markov Chains*. Rev. ed. New York: Springer, 2006. https://doi.org/10.1007/0-387-32792-4. https://link.springer.com/book/10.1007/0-387-32792-4.
17. Joseph M. Sussman. *Perspectives on Intelligent Transportation Systems (ITS)*. Springer, 2005. ISBN: 978-0-387-23257-7. https://doi.org/10.1007/b101063.
18. K. Teknomo. "Ideal Flow Based on Random Walk on Directed Graph". In: *Proceedings of the 9th International Collaboration Symposium on Information, Production and Systems*. Kitakyushu: Waseda University, 2015.
19. Kardi Teknomo. "Premagic and Ideal Flow Matrices". In: *arXiv preprint arXiv:1706.08856* (2017). arXiv: 1706.08856 [cs.DM]. https://arxiv.org/pdf/1706.08856.
20. Kardi Teknomo, R. Gardon, and C. Saloma. "Ideal Flow Traffic Analysis: A Case Study on a Campus Road Network". In: *Philippine Journal of Science* 148.1 (2019), pp. 51–62. https://archium.ateneo.edu/discs-faculty-pubs/271/.
21. Kardi Teknomo and R. W. Gardon. "Traffic assignment based on parsimonious data: The ideal flow network". In: 2019 IEEE Intelligent Transportation Systems Conference (ITSC). 2019, pp. 1393–1398. https://doi.org/10.1109/ITSC.2019.8917426. https://ieeexplore.ieee.org/document/8917426.
22. J. G. Wardrop. "Some theoretical aspects of road traffic research". In: Proceedings of the Institution of Civil Engineers 1.3 (1952), pp. 325–362. https://doi.org/10.1680/iicep.1952.11362. https://www.researchgate.net/publication/227555985_Wardrop_Equilibria.
23. A. G. Wilson. "A statistical theory of spatial distribution models". In: Transportation Research 1.3 (1967), pp. 253–269. https://doi.org/10.1016/0041-1647(67)90035-4.

Chapter 2
Classical Paradigm of Transportation Modeling

2.1 The Four-Step Model: An Aggregate, Sequential Framework

Before we can appreciate the paradigm shift offered by the Ideal Flow Network, we must first build a rigorous understanding of the classical theory that has been the bedrock of transportation planning for more than half a century. The dominant approach, known as the *four-step travel demand model*, is a comprehensive and logically structured framework designed to forecast travel patterns based on land use, demographics, and the characteristics of the transportation system. It represents a monumental intellectual effort to bring scientific order to the complex and often chaotic dynamics of urban mobility. This section provides a grounded explanation of this traditional paradigm.

The practice of modern transportation planning is inextricably linked to the development of quantitative models designed to forecast future travel patterns. At the heart of this history lies the classical four-step model (FSM), also known as the Urban Transportation Planning (UTP) process [24]. This framework was born of necessity in the mid-twentieth century, a period of rapid suburbanization and massive public investment in urban highway systems. Seminal studies, such as the Detroit Metropolitan Area Traffic Study [20] and the Chicago Area Transportation Study (CATS) in the 1950s [9], pioneered this sequential modeling approach to provide a rational and systematic method for predicting the consequences of large-scale infrastructure projects. The FSM provided planners with a structured methodology to answer critical questions:

- How many trips will be made?
- Where will they go?
- By what mode of transport?
- And which routes will they take?

K. Teknomo, *Ideal Flow Network for Transportation Research*,
Lecture Notes in Intelligent Transportation and Infrastructure,
https://doi.org/10.1007/978-981-95-4727-2_2

By breaking down the complex phenomenon of travel demand into a manageable sequence of four steps, it established a paradigm that would dominate the field for decades. The FSM simplifies the complexity of individual travel decisions into a series of four distinct, sequential calculations performed at a macroscopic level. The fundamental spatial unit is the Traffic Analysis Zone (TAZ). All travel is assumed to originate from or be destined for the centroid of a TAZ. The model's structure follows a logical progression, as illustrated below. Each step takes the output of the previous one as its primary input. The four steps are as follow.

1. *Trip Generation:* Forecasts the total number of trips originating from (productions) and destined for (attractions) each zone.
2. *Trip Distribution:* Links the productions and attractions to create an Origin-Destination (OD) matrix of trips between every pair of zones.
3. *Mode Choice:* Splits the OD matrix into several matrices, one for each mode of transport (e.g., private car, public transit).
4. *Traffic Assignment:* Assigns the vehicle-trip OD matrix to the highway network to predict traffic flows on individual links.

While the classical four-step model (FSM) remains a foundational element of transportation planning, the field has evolved significantly. The limitations of the classical paradigm have driven the development of more sophisticated modeling approaches.

- *Activity-Based Models (ABMs):* A fundamental shift occurred with ABMs. The core premise is that the demand for travel is *derived* from the need to participate in activities. ABMs simulate the daily activity patterns of individual agents, capturing complex behaviors like trip chaining (e.g., home-to-work-to-shop-to-home) that are invisible to the FSM.
- *Data-Driven Approaches:* Spurred by the proliferation of mobile technology, data-driven approaches leverage vast new sources of information, such as GPS traces and smart card transactions. These methods use machine learning techniques to identify patterns directly from observed data, often to supplement or replace the costly survey components of traditional frameworks.

A critical feature is the role of travel time. The travel times used as an input to Trip Distribution and Mode Choice should be consistent with the congested travel times that result from the final Traffic Assignment step. This is often achieved by implementing a feedback loop, where the entire process is iterated until input and output travel times converge.

We will begin by formalizing how transportation networks are represented in conventional models, defining the essential concepts of Traffic Analysis Zones (TAZs) and link attributes. We will then undertake an in-depth exploration of the four-step cycle itself, detailing the models and assumptions that govern each stage of the process, from the generation of trips to their final assignment onto the road network. We provide a critically informed exposition of the classical Four-Step Model (FSM) in transportation planning. The theoretical core centers on Traffic Assignment, introducing Wardrop's equilibrium principles [34]. The

Frank-Wolfe algorithm [14] is a cornerstone of transportation science, providing the fundamental iterative method for solving the user equilibrium traffic assignment problem formulated with convex link cost functions [17, 28].

A modern complementary method, the Ideal Traffic Assignment framework which based on Ideal Flow Network (IFN), is then briefly presented to overcome key data limitations of the FSM. The chapter concludes with a synthesis and critical reflection on the classical paradigm. The purpose is not merely to describe this process, but to deconstruct its machinery, providing the reader with the necessary background to critically evaluate its methods, strengths, and limitations.

The fundamental unit of a trip in the conventional four-step transportation model is the *person trip*. This unit is the basis for the initial stages of the model before mode choice is determined.

The distinction between a person trip and other units is essential. The *person trip* represents a single, one-way movement by an individual from an origin to a destination. The initial steps of the model, specifically *Trip Generation* and *Trip Distribution*, operate on this unit. Trip Generation forecasts the total number of person trips produced in and attracted to each zone based on demographic and socioeconomic data. The output of Trip Generation is a set of productions and attractions, often represented as zonal vectors (P_i) and (A_j), which serve as the inputs for the subsequent Trip Distribution step The output of Trip Distribution is a matrix of person trips, often called as Origin-Destination (OD) matrix denoted as D_{od}, which quantifies the demand for travel between a given origin o and destination d. A *person trip* is a measure of human movement. For example, a single car containing three people constitutes three person trips but only one vehicular trip. A *vehicular trip* is a measure of vehicle movement. It is a single, one-way journey by a vehicle, regardless of the number of occupants. The vehicular trip is derived from the person trip data during the *Mode Choice* step of the model, when person trips are allocated to specific modes of transportation. The four-step model begins with the person as the core unit of travel. This is a deliberate design choice that allows the model to capture the behavior of individuals before assigning them to a specific mode or route. Traffic assignment is the fourth and final step in the conventional four-step transportation model. This stage determines the specific routes that travel will take through the transportation network. It translates the demand for trips, as represented by the origin-destination matrix, into a pattern of traffic flows on the network's links. The process is foundational for evaluating network performance and identifying future infrastructure needs.

The Four-Step Model translates socioeconomic inputs into network flows through a sequential pipeline with feedback.

Trip ends are computed as *productions* P_i and *attractions* A_j for each zone i, j, also denoted O_i and D_j in some sources. Balancing enforces $\sum_i P_i = \sum_j A_j$ so total productions equal total attractions before distribution.

Step 1: Trip Generation This step forecasts daily or period trip ends by purpose using regression or cross-classification, applies reasonableness checks and caps, and then balances totals. The outputs are zonal vectors (P_i) and (A_j) for the modeled period and purpose.

Step 2: Trip Distribution This step allocates the balanced ends into an OD matrix by penalizing long or costly separations via a friction function.

Definition 2.1 (Doubly-Constrained Gravity (Overview Form)) Given balanced (P_i) and (A_j) and impedances $c_{ij} > 0$, the trips are $T_{ij} = a_i P_i b_j A_j F(c_{ij})$ with $a_i, b_j > 0$ chosen so that $\sum_j T_{ij} = P_i$ and $\sum_i T_{ij} = A_j$.

Iterative proportional fitting (IPF) [1] provides the balancing multipliers and preserves positivity. IPF is a matrix adjustment technique [11] used to estimate OD matrices from marginal totals, preserving row and column constraints.

Step 3: Mode Choice This step splits each T_{ij} across available modes using random utility theory with interpretable time and cost coefficients. The concept of utility [35], representing the satisfaction or preference an individual associates with a choice, is the behavioral foundation of modern mode choice models

Definition 2.2 (Multinomial Logit (Overview Form)) With systematic utilities U_{ijm} and scale $\mu > 0$, the probabilities are $p_{ijm} = \exp(\mu U_{ijm}) \big/ \sum_{k\in\mathcal{M}} \exp(\mu U_{ijk})$, and trips by mode are $T_{ijm} = T_{ij} p_{ijm}$.

Alternative-specific constants capture average preference not explained by measured attributes, and probabilities across modes sum to one. Ben-Akiva and Lerman [3] introduced ASCs in discrete choice models to capture unobserved utility components, critical for mode and route choice in transport modeling.

Step 4: Traffic Assignment This step loads the mode-specific OD matrices onto the network using flow-dependent link costs and the User Equilibrium principle, producing path choices and link volumes.

Dial [12] introduced the concept of network augmentation to model capacity constraints and route choices, forming the basis for augmented network representations in traffic assignment. Peeta and Ziliaskopoulos [25] developed dynamic traffic assignment models that preserve flow balance across time and space.

At equilibrium no traveler can reduce generalized cost by unilaterally changing routes between the same OD pair.

Impedance links the steps through cost and time.

Bureau of Public Roads [7] introduced the BPR function to model travel time as a function of volume-to-capacity ratio, forming the basis for link performance modeling in traffic assignment.

Definition 2.3 (BPR Link Travel Time) For link (i, j) with free-flow time t_{ij}^0 and capacity c_{ij}, the congested travel time is $t_{ij} = t_{ij}^0 \left[1 + \gamma \left(\frac{f_{ij}}{c_{ij}}\right)^\eta\right]$ with calibrated $\gamma, \eta > 0$.

The convex rise in t_{ij} with flow f_{ij} captures increasing delay under congestion and supports stable equilibrium solutions.

Generalized cost combines time and money into a single metric for distribution and choice.

Definition 2.4 (Generalized Cost) A common form is $c_{ij} = a_1\, t_{ij}(f_{ij}) + a_2\, d_{ij} + (\text{toll})_{ij}$ where t_{ij} is congested time, d_{ij} is length, and a_1, a_2 map units to a consistent utility or monetary scale.

Weights reflect values of time and penalties for waiting, access, or transfers as appropriate to the purpose and market segments.

In practice the model may iterate a feedback loop: assign flows to get updated skims, redistribute with new impedances, recompute mode splits, and reassign until changes in OD, costs, and assignment gaps fall below tolerances.

This overview sets the stage for detailed derivations, calibration steps, and numerical examples presented in the following sections.

2.2 Step 1: Trip Generation

Trip Generation answers how many person trips are produced by each origin zone and attracted to each destination zone for the travel purpose under study. First, we partition the study area into N traffic analysis zones (TAZ) indexed by $i, j \in \{1, \ldots, N\}$. This step answers: "How much travel is there?" It estimates the total number of trips that begin or end in each TAZ. It does so by estimating the total number of trips that begin or end in each TAZ within a given time period (typically a 24-hour day), without yet considering where these trips might go. These trip ends are classified as either *Productions* (P), which are typically associated with the home end of a trip, or *Attractions* (A), associated with the non-home end (e.g., workplace, shop). Productions are typically tied to where people live, and attractions are tied to where activities happen, such as jobs, schools, and shops. Let $P_i \geq 0$ be trip productions from zone i and $A_j \geq 0$ be trip attractions to zone j with $\sum_i P_i = \sum_j A_j$ after balancing.

Definition 2.5 (Trip Productions and Attractions) For each zone i, P_i and A_i summarize trip ends by purpose consistent with the adopted household or employment based generator.

The models are usually stratified by trip purpose, such as Home-Based Work (HBW), Home-Based Other (HBO), and Non-Home-Based (NHB).

2.2.1 Mathematical Formulation and Examples

Two common approaches are regression and cross-classification, which can be used alone or together.

2.2.2 Regression Models For Productions And Attractions

A regression model relates the expected number of trips in a zone to observable attributes of that zone. Coefficients are estimated from survey or count data and should be interpreted with their units so results are plausible. Typical predictors include population, numbers of workers, cars, jobs by sector, and income or land-use intensity.

This method establishes a statistical relationship between trips and a zone's socioeconomic characteristics.

Definition 2.6 (Trip Generation Regression Models) The trip productions for zone i, P_i, and attractions for zone j, A_j, are estimated using linear regression equations of the form:

$$P_i = \beta_0 + \sum_{k=1}^{K} \beta_k X_{ik} + \epsilon_i$$

$$A_j = \gamma_0 + \sum_{l=1}^{L} \gamma_l Y_{jl} + \eta_j$$

where X_{ik} and Y_{jl} are explanatory variables (e.g., population, employment), and β_k and γ_l are calibrated coefficients.

Example 2.1 (Regression for Trip Productions and Attractions) The agency specifies home-based work (HBW) productions and attractions as linear functions.

$$P_i = 100 + 0.8 \times (\text{Population}_i) + 0.01 \times (\text{MedianIncome}_i)$$

$$A_j = 50 + 1.2 \times (\text{RetailEmployment}_j) + 0.9 \times (\text{ServiceEmployment}_j)$$

For a zone with a population of 8000 and median income of \$45,000, the productions would be: $P_2 = 100 + 0.8(8000) + 0.01(45000) = 100 + 6400 + 450 = 6,950$ trips. For a zone with 1000 retail jobs and 1500 service jobs, the attractions would be: $A_2 = 50 + 1.2(1000) + 0.9(1500) = 50 + 1200 + 1350 = 2,600$ trips.

The coefficient 0.8 means eight tenths of a trip per additional resident, so adding 100 people adds roughly 80 trips, holding other attributes fixed. The 0.01 in front of income means every \$100 increase raises productions by about one trip, which can be sensible if higher incomes correlate with more commuting or escort trips.

Always check units and reasonableness to avoid absurd outputs.

Example 2.2 (Sanity Checks and Nonnegativity) If a regression predicts $P_i = -120$ for a very small zone, cap it at zero since productions cannot be negative:

$$P_i^{\text{adj}} = \max\{0,\ P_i\}.$$

If a single outlier zone has unusually high income or employment and the formula produces extremely large trips, compare to observed counts or survey expansions for consistency.

2.2.2.1 Cross-Classification

Cross-classification avoids assuming a single global slope by using observed trip rates for household categories such as size, workers, income, and car ownership. This method avoids the assumption of linearity by using observed trip rates for different household categories.

Definition 2.7 (Cross-Classification Model) Let N_{ic} be the number of households of category c in zone i, and let R_c be the average daily trip rate for that category from survey expansion.

Then the production model is

$$P_i = \sum_{c=1}^{C} N_{ic}\, R_c.$$

1. Households in the study area are classified into a multi-dimensional matrix of categories based on key characteristics (e.g., household size, number of workers, income level, car ownership).
2. From household travel survey data, the average trip production rate is calculated for each category of household.
3. The total trip productions for a zone i are then calculated by summing the trips produced by all households within that zone:

$$P_i = \sum_{c=1}^{C} N_{ic} \times R_c$$

where N_{ic} is the number of households of category c in zone i, and R_c is the average trip rate for a household in category c.

Example 2.3 (Cross-Classification) A zone contains 150 medium-income households with at least one car with trip rate $R_{\text{car+}} = 5.0$ trips per day and 20 medium-income households with no car with trip rate $R_{\text{no-car}} = 2.5$ person trips per day.

Then Productions = $P_i = 150 \times 5.0 + 20 \times 2.5 = 750 + 50 = 800$ person trips per day.

Example 2.4 (Cross-Classification with Multiple Categories) Suppose categories are based on household size $\{S \in \{1, 2, 3+\}\}$ and car ownership $\{\text{Car} = 0, 1+\}$, producing six cells with rates from the survey.

If zone i has $\{N_{S=1,\text{Car}=0} = 60,\ N_{S=1,\text{Car}=1+} = 90,\ N_{S=2,\text{Car}=0} = 30,\ N_{S=2,\text{Car}=1+} = 110,\ N_{S=3+,\text{Car}=0} = 20,\ N_{S=3+,\text{Car}=1+} = 140\}$ and rates $\{2.2, 4.0, 3.1, 5.3, 3.8, 6.4\}$ respectively, then

$$P_i = 60 \times 2.2 + 90 \times 4.0 + 30 \times 3.1 + 110 \times 5.3 + 20 \times 3.8 + 140 \times 6.4$$

$$= 132 + 360 + 93 + 583 + 76 + 896 = 2140 \text{ person trips per day.}$$

This reflects heterogeneity across household types without a single linear slope.

Cross-classification is like tallying how many of each type live in a zone and giving each type its own average trip rate. It is powerful when behavior differs a lot across categories, such as between zero-car and multi-car households.

2.2.2.2 Balancing Productions and Attractions

In a consistent system the sum of productions equals the sum of attractions for the modeled purpose:

$$\sum_i P_i = \sum_j A_j.$$

Because production and attraction models are often developed independently, their totals rarely match exactly. The usual practice is to choose the more reliable side (often productions) as the control total and scale the other side proportionally.

Lemma 2.1 (Scalar Balancing) *If $P = \sum_i P_i$ and $A = \sum_j A_j$ are finite and positive and $P \neq A$, then multiplying each attraction by P/A yields balanced totals while preserving each zone's share:*

$$A'_j = \frac{P}{A} A_j, \quad \sum_j A'_j = P.$$

An analogous rule holds if attractions are treated as the control total.

Proof Summing gives $\sum_j A'_j = (P/A)\sum_j A_j = P$, and for any j_1, j_2 the ratio $A'_{j_1}/A'_{j_2} = A_{j_1}/A_{j_2}$ so relative weights are preserved. □

Balancing is a turn the volume knob step. Balancing reconciles small total mismatches without changing each zone's relative magnitude.

It raises or lowers the whole attraction vector together so totals match, but it does not change which zones are big or small relative to each other.

This example introduces the concept of a balancing factor to align total productions and attractions by scaling attraction values proportionally.

Example 2.5 (Balancing) Suppose total calculated productions are 28,800, and total attractions are 21,490. Balancing Factor = Total Productions / Total Attractions $= 28,800/21,490 \approx 1.340$. Each zonal attraction value A_j is multiplied by this factor to get the new balanced attraction A'_j, ensuring the totals match.

When production totals are fixed, this method adjusts each attraction value using a global factor to preserve relative shares while matching totals.

Example 2.6 (Balancing by Scaling Attractions) Suppose total calculated productions are 28,800 and total attractions are 21,490.

The balancing factor is 28,800/21,490 $\approx$ 1.340.

Set $A'_j = 1.340 \times A_j$ for every zone j.

Then $\sum_j A'_j = 28{,}800$ and zonal attraction shares are unchanged.

If attraction data is deemed more reliable, this approach scales production values to match attraction totals, maintaining zonal proportions.

Example 2.7 (Balancing by Scaling Productions) If attractions are more reliable, scale productions instead by A/P.

For $P = 30{,}500$ and $A = 29{,}000$, use $P'_i = (29{,}000/30{,}500)\ P_i \approx 0.951\ P_i$ so $\sum_i P'_i = 29{,}000$.

Example 2.8 (Balancing Trip Generation) Demonstrates how to reconcile mismatched production and attraction totals by uniformly scaling the production vector.

Given $P = (120, 80, 100)$ and $A = (110, 90, 130)$ we have $P = 300$ and $A = 330$.

Balance productions by $A/P = 1.1$ to obtain $\tilde{P} = 1.1 \times P = (132, 88, 110)$ so totals match.

Example 2.9 (Balancing Trip Generation 2 Variables) Uses a linear combination of households and employment to generate productions, then applies a global scaling factor to match attractions while preserving zonal structure.

Let HH is the number of households in the zone and EMP is the number of employments. Let $P_i = \alpha\,\mathrm{HH}_i + \beta\,\mathrm{EMP}_i$ with $(\alpha, \beta) = (1.2, 0.5)$.

Compute base P from zone attributes, compare totals to A, and scale the chosen side by the global factor so that $\sum_i P_i = \sum_j A_j$, preserving the generator-implied proportions across zones.

2.2.3 Practical Guidance and Common Pitfalls

As practical guide, always check units and reasonableness for every coefficient in regression models.

Income and employment variables often require rescaling (e.g., thousands of dollars, hundreds of jobs) to avoid numerical instability. Enforce nonnegativity [4] and optionally apply upper caps if extreme outliers produce implausibly large trip ends. In cross-classification, ensure survey expansion weights and category definitions match our census or synthetic population so rates transfer correctly. Always balance productions and attractions before passing results to Trip Distribution.

Example 2.10 (A Small System) Three zones have household and job data and are assigned rates for two household categories: zero-car and one-or-more car.

1. computes P_i via cross-classification and A_j via a regression on sectoral jobs.
2. caps negatives at zero and rescales overly large predictions if needed after comparing with observed cordon totals.
3. balances to ensure $\sum_i P_i = \sum_j A_j$.

The balanced (P_i, A_j) pair is now ready for Trip Distribution with the gravity model.

2.3 Step 2: Trip Distribution

This step connects the productions and attractions to form an Origin-Destination (OD) matrix, T_{ij}, which quantifies person trips from each origin i to each destination j. The most common method is the *Gravity Model.*

2.3.1 Mathematical Formulation: The Gravity Model

Gravity models [36, 37] estimate trip distribution based on origin-destination masses and impedance functions. Constraints ensure row/column totals match observed data. The gravity model allocates trips between zones based on trip ends and travel impedance, in direct analogy to physical gravity. It is conceptually analogous to Newton's law of universal gravitation: the interaction (number of trips) between two zones is directly proportional to their size (productions and attractions) and inversely proportional to the distance (or travel impedance) between them.

Let $T_{ij} \geq 0$ denote trips from origin i to destination j.

Let productions P_i and attractions A_j be balanced totals from Trip Generation so that $\sum_j T_{ij} = P_i$ and $\sum_i T_{ij} = A_j$ after balancing.

Let $c_{ij} > 0$ denote generalized impedance between zones, such as travel time or generalized cost.

Let $F(c_{ij}) > 0$ denote the friction factor that decreases with impedance, and let $K_{ij} > 0$ be an optional socioeconomic adjustment.

Definition 2.8 (Singly-Constrained Gravity) Given origin totals (P_i), the number of trips from origin zone i to destination zone j, T_{ij}, is given by:

$$T_{ij} = P_i \frac{A_j F(c_{ij}) K_{ij}}{\sum_{k=1}^{N} A_k F(c_{ik}) K_{ik}}$$

so that each origin i distributes P_i across destinations proportionally to $A_j F(c_{ij}) K_{ij}$ and satisfies $\sum_j T_{ij} = P_i$. where:

- T_{ij} is the number of trips from origin i to destination j.
- P_i is the total trip productions from zone i (from Step 1).
- A_j is the total trip attractions to zone j (from Step 1).
- c_{ij} is the generalized cost or impedance of travel between i and j.
- $F(c_{ij})$ is the *friction factor function*.
- K_{ij} is a socioeconomic adjustment factor.
- N is the total number of zones.

The formula is just attractions times a distance penalty, normalized to 100%. Bigger destinations get more trips, but higher impedance pushes trips away.

Example 2.11 (Singly Constrained) Consider one origin i and three destinations $j = 1, 2, 3$ with $P_i = 100$ trips. Let attractions be $(A_1, A_2, A_3) = (120, 80, 60)$. Let impedances be $(c_{i1}, c_{i2}, c_{i3}) = (10, 20, 30)$ minutes. Use $F(c) = e^{-\beta c}$ with $\beta = 0.05$ and set $K_{ij} = 1$. Compute weights $w_j = A_j F(c_{ij}) = (120e^{-0.5},\ 80e^{-1.0},\ 60e^{-1.5}) \approx (72.79,\ 29.43,\ 13.39)$. Denominator $D = \sum_j w_j \approx 115.61$. Then shares are $(w_1/D, w_2/D, w_3/D) \approx (0.630,\ 0.255,\ 0.116)$. Therefore $T_{i1} \approx 63.0$, $T_{i2} \approx 25.5$, and $T_{i3} \approx 11.6$, which sum to 100 as required.

Definition 2.9 (Doubly-Constrained Gravity) Given productions (P_i) and attractions (A_j), the doubly-constrained gravity model sets

$$T_{ij} = a_i\, P_i\, b_j\, A_j\, F(c_{ij})\, K_{ij}$$

with positive multipliers (a_i) and (b_j) chosen so that both margins hold, i.e., $\sum_j T_{ij} = P_i$ and $\sum_i T_{ij} = A_j$.

Example 2.12 (Doubly Constrained with Exponential Friction) Let $P = (100, 100)$ and $A = (120, 80)$. Let the impedance matrix be $c = \begin{bmatrix} 10 & 20 \\ 20 & 10 \end{bmatrix}$ minutes with $F(c) = e^{-\beta c}$ and $\beta = 0.1$ and $K_{ij} = 1$. Compute base weights $W_{ij} = P_i A_j F(c_{ij}) = \begin{bmatrix} 100 \times 120e^{-1} & 100 \times 80e^{-2} \\ 100 \times 120e^{-2} & 100 \times 80e^{-1} \end{bmatrix} \approx \begin{bmatrix} 441.2 & 108.1 \\ 162.2 & 294.3 \end{bmatrix}$.

We seek $T_{ij} = a_i b_j W_{ij}$ that matches both margins. Initialize $b = (1, 1)$. Row scaling gives $a_1 = 100/(441.2+108.1) = 0.1865$ and $a_2 = 100/(162.2+294.3) = 0.2218$. Column scaling then gives $b_1 = 120/(a_1 \times 441.2+a_2 \times 162.2) \approx 1.018$ and $b_2 = 80/(a_1 \times 108.1 + a_2 \times 294.3) \approx 0.974$. Repeating the two scalings once more yields $a_1 \approx 0.1860$, $a_2 \approx 0.2224$, $b_1 \approx 1.020$, and $b_2 \approx 0.970$ to two decimals. The balanced matrix is then

$$T = \begin{bmatrix} a_1 b_1 W_{11} & a_1 b_2 W_{12} \\ a_2 b_1 W_{21} & a_2 b_2 W_{22} \end{bmatrix} \approx \begin{bmatrix} 76.0 & 24.0 \\ 44.0 & 56.0 \end{bmatrix},$$

which sums by rows to (100, 100) and by columns to (120, 80) up to rounding.

The value of a rescales each origin row to hit P_i and b rescales each destination column to hit A_j. Alternating these two simple scalings converges to the unique matrix that hits both sets of totals while respecting the friction weights.

Lemma 2.2 (IPF Feasibility) *If $P_i > 0$, $A_j > 0$, and $F(c_{ij}) K_{ij} > 0$ for all (i, j), then there exist positive multipliers (a_i) and (b_j) such that the matrix $T_{ij} = a_i b_j P_i A_j F(c_{ij}) K_{ij}$ satisfies both margin constraints.*

Proof Set $W_{ij} = P_i A_j F(c_{ij}) K_{ij}$, which is strictly positive. The biproportional form $T_{ij} = a_i b_j W_{ij}$ has a feasible solution by iterative proportional fitting, which alternately rescales rows and columns and preserves positivity at each step. Standard IPF results guarantee convergence to a matrix meeting both sets of margins. □

Example 2.13 (IPF Feasibility) Use the data from the previous 2×2 example but replace the last column by a very small attraction $A_2 = 5$ to stress the method. Even with A_2 tiny, W_{12} and W_{22} remain positive, and the first row scaling computes finite a_1, a_2 because the denominators are positive sums. The first column scaling computes finite b_1, b_2 because the row-scaled column sums are positive. Thus feasible positive multipliers exist despite an extreme attraction split, illustrating Lemma 2.2.

Theorem 2.1 (Uniqueness of Balanced Solution) *Under the conditions of Lemma 2.2, the balanced matrix T that satisfies both sets of margins is unique.*

Proof When all W_{ij} are strictly positive, IPF minimizes the strictly convex Kullback-Leibler divergence [15, 16] (see Sect. 5.2.2) over the set of matrices with the prescribed margins. A strictly convex objective on a convex feasible set admits a unique minimizer, hence a unique T. □

Gravity grows with P_i and A_j and shrinks with impedance via $F(c_{ij})$, and IPF enforces both sets of totals.

Example 2.14 (Uniqueness Observed from Different Starts) Run IPF on the 2×2 example twice: once starting with $b = (1, 1)$ and once with a deliberately skewed start $b = (2, 0.5)$. Both runs converge numerically to the same balanced matrix

$T \approx \begin{bmatrix} 76 & 24 \\ 44 & 56 \end{bmatrix}$ to rounding. This confirms in practice that the doubly-constrained gravity solution is unique for a given (P, A, W) triple.

Example 2.15 (Doubly Constrained with Calibration Target) Let $P = (140, 120, 140)$ and $A = (150, 120, 130)$.

Let $c = \begin{bmatrix} 5 & 10 & 15 \\ 10 & 5 & 15 \\ 15 & 10 & 5 \end{bmatrix}$ minutes and $F_\beta(c) = e^{-\beta c}$.

Choose β so that the resulting average impedance $\bar{c}(T) = \left(\sum_{ij} T_{ij} c_{ij}\right) / \left(\sum_{ij} T_{ij}\right)$ equals 9.0 minutes.

For a trial $\beta = 0.08$, build $W_{ij} = P_i A_j e^{-0.08 c_{ij}}$ and run IPF to get a balanced $T(\beta)$.

Compute $\bar{c}(T(\beta))$ and adjust β up or down using bisection since $\bar{c}$ decreases in β, repeating until the target is met within tolerance.

The final T has higher diagonal entries due to lower impedances while exactly matching the P and A margins.

Calibration dials the friction through β until the model's average trip length matches the observed one. IPF keeps the totals right while the friction controls the near versus far balance.

2.4 Step 3: Mode Choice

After establishing the total number of trips between each OD pair (T_{ij}), the mode choice step divides these trips among the available modes of transport. This model converts systematic utilities into probabilities and then into mode-specific flows.

2.4.1 Mathematical Formulation: The Multinomial Logit (MNL) Model

The Multinomial Logit (MNL) model calculates the probability that a traveler will choose a particular mode based on the relative utility of all available alternatives modes.

First, for a trip between zones i and j, OD pair (i, j) and each available mode $m \in \mathcal{M}$, define a systematic utility U_{ijm}. A common linear-in-parameters utility is

$$U_{ijm} = \sum_{k=1}^{K} \beta_k X_{ijmk},$$

where X_{ijmk} are mode attributes such as in-vehicle travel time (IVTT), out-of-vehicle time (OVTT), and monetary cost, and β_k are calibrated coefficients with interpretable units.

IVTT [3] refers to the time spent traveling inside a vehicle, often used in mode choice models and generalized cost formulations. Out-of-Vehicle Time (OVTT) refers to the portion of a transit journey spent not riding inside the vehicle. It typically includes: walking time to and from stops or stations, waiting time at stops or terminals, transfer time between vehicles or modes. OVTT [18] is a critical factor in mode choice modeling because travelers often perceive it as more burdensome than in-vehicle time. Reducing OVTT, through better stop placement, real-time information, or seamless transfers, can significantly improve the attractiveness of public transport.

Let $\mu > 0$ be a scale parameter (often absorbed into β_k in practice).

The probability that a randomly chosen traveler will select mode m is then given by:

$$P_{ij}(m) = \frac{e^{U_{ijm}}}{\sum_{k \in M_{ij}} e^{U_{ijk}}}$$

The number of trips by mode m is then simply the mode-specific trips $T_{ijm} = T_{ij} \times P_{ij}(m)$.

Definition 2.10 (Multinomial Logit) For (i, j) and modes $m \in \mathcal{M}$ with utilities U_{ijm}, the choice probabilities are $p_{ijm} = \exp(\mu U_{ijm}) / \sum_k \exp(\mu U_{ijk})$ and $T_{ijm} = T_{ij} p_{ijm}$.

Example 2.16 (Binary MNL with IVTT, OVTT, Cost) Consider an OD pair with $T_{ij} = 2000$ trips and two modes, Car and Bus.

Let the utilities be $U_m = \text{ASC}_m - 0.02\,\text{IVTT} - 0.05\,\text{OVTT} - 0.10\,\text{Cost}$ with $\mu = 1$.

Interpretation is per unit: minutes for IVTT and OVTT, currency units for Cost, and ASC_m is a mode-specific constant for unobserved preference.

Suppose Car has $(\text{ASC} = 0,\ \text{IVTT} = 20,\ \text{OVTT} = 5,\ \text{Cost} = 4.00)$ and Bus has $(\text{ASC} = -0.5,\ \text{IVTT} = 35,\ \text{OVTT} = 15,\ \text{Cost} = 1.50)$.

Compute $U_{\text{car}} = 0 - 0.02 \times 20 - 0.05 \times 5 - 0.10 \times 4.00 = -1.05$.

Compute $U_{\text{bus}} = -0.5 - 0.02 \times 35 - 0.05 \times 15 - 0.10 \times 1.50 = -2.10$.

Exponentiate: $\exp(U_{\text{car}}) \approx 0.3499$ and $\exp(U_{\text{bus}}) \approx 0.1225$.

Normalize: $p_{\text{car}} = \frac{0.3499}{0.3499+0.1225} \approx 0.741$ and $p_{\text{bus}} \approx 0.259$.

Split trips: $T_{ij,\text{car}} \approx 2000 \times 0.741 = 1482$ and $T_{ij,\text{bus}} \approx 518$.

We multiply attributes by their coefficients to get a utility score for each mode. Exponentials turn those scores into positive weights. Dividing by the sum converts weights into shares that add to one.

Lemma 2.3 (Probability System) *For any* (i, j), *the MNL probabilities satisfy* $p_{ijm} \in (0, 1)$ *and* $\sum_{m \in \mathcal{M}} p_{ijm} = 1$.

Proof Exponentials are strictly positive. Dividing a positive weight by the positive sum of all weights produces a number between 0 and 1. Summing the numerators and dividing by the same sum yields 1. □

Example 2.17 (Three Modes Sum to One) Let modes be Car, Bus, Rail with the same coefficient set as above and $\mu = 1$. Attributes are Car (ASC $= 0$, IVTT $=$ 22, OVTT $= 5$, Cost $= 3.50$), Bus (ASC $= -0.5$, IVTT $= 40$, OVTT $=$ 10, Cost $= 1.80$), Rail (ASC $= -0.2$, IVTT $= 28$, OVTT $= 8$, Cost $=$ 2.50). Compute $U_{\text{car}} = -1.04$, $U_{\text{bus}} = -1.98$, and $U_{\text{rail}} = -1.41$. Exponentials are $(0.353,\ 0.138,\ 0.244)$ which sum to 0.735. Normalize to probabilities $(0.480,\ 0.188,\ 0.332)$ which sum to $0.480+0.188+0.332 = 1.000$ up to rounding.

A property of the Multinomial Logit Model (MNL) stating [19] that the relative odds between two alternatives are unaffected by the presence of a third. Often criticized for unrealistic substitution patterns.

Theorem 2.2 (Independence of Irrelevant Alternatives (IIA)) *For modes m and n, the odds ratio satisfies* $\frac{p_{ijm}}{p_{ijn}} = \exp\{\mu(U_{ijm} - U_{ijn})\}$ *and does not depend on any other alternative in* $\mathcal{M}$.

Proof Divide p_{ijm} by p_{ijn} and note that the common denominator cancels. The ratio equals $\exp\{\mu(U_{ijm} - U_{ijn})\}$, which excludes all other alternatives. □

MNL translates utilities into shares, but the IIA property can overstate substitution between similar modes.

Example 2.18 (Odds Unchanged When a Third Mode Improves) Re-use the three-mode utilities above where $U_{\text{car}} = -1.04$ and $U_{\text{bus}} = -1.98$. The odds predicted by IIA are $\exp(U_{\text{car}} - U_{\text{bus}}) = \exp(0.94) \approx 2.56$. From the computed probabilities we had $p_{\text{car}}/p_{\text{bus}} = 0.480/0.188 \approx 2.55$ which matches rounding error. Now improve Rail by reducing OVTT from 8 to 3 minutes so U_{rail} increases by 0.25 to -1.16. Recompute exponentials: $(0.353,\ 0.138,\ 0.313)$ and sum 0.804. New probabilities are $(0.439,\ 0.172,\ 0.389)$ for Car, Bus, Rail respectively. The odds $p_{\text{car}}/p_{\text{bus}} = 0.439/0.172 \approx 2.55$ remain the same within rounding, exactly as IIA predicts.

Example 2.19 (Binary Mode Choice (Car vs. Bus)) Let IVTT is in-vehicle-travel-time and OVTT is out-of-vehicle-travel-time and ASC is an alternative specific constant (mode constant). For an OD pair with 2000 total trips, assume the following utility function:

$$U_m = \text{ASC}_m - 0.02(\text{IVTT}) - 0.05(\text{OVTT}) - 0.10(\text{Cost})$$

Attribute	Car	Bus
ASC	0.0	-0.5
IVTT (min)	20	35
OVTT (min)	5	15
Cost ($)	4.00	1.50

1. *Calculate Utilities:*
   $U_{car} = 0.0 - 0.02(20) - 0.05(5) - 0.10(4.00) = -1.05$
   $U_{bus} = -0.5 - 0.02(35) - 0.05(15) - 0.10(1.50) = -2.10$
2. *Calculate Probabilities:*
   $e^{U_{car}} = e^{-1.05} \approx 0.3499$
   $e^{U_{bus}} = e^{-2.10} \approx 0.1225$
   Sum = $0.3499 + 0.1225 = 0.4724$
   Prob(car) = $0.3499/0.4724 \approx 74.1\%$
   Prob(bus) = $0.1225/0.4724 \approx 25.9\%$
3. *Split Trips:*
   Car trips = $2,000 \times 0.741 = 1,482$
   Bus trips = $2000 \times 0.259 = 518$

2.4.2 Comprehensive Three-Mode Planning Example

This example shows how to go from utilities to mode splits and then how to test a service change. We use the three-mode attributes from Lemma Example with $T_{ij} =$ 2000 trips.

Example 2.20 (Base Split and Trips By Mode) With probabilities $(p_{\text{car}}, p_{\text{bus}}, p_{\text{rail}}) =$ $(0.480, 0.188, 0.332)$, compute

$$T_{ij,\text{car}} = 2000 \times 0.480 = 960, \quad T_{ij,\text{bus}}$$

$$= 2000 \times 0.188 = 376, \quad T_{ij,\text{rail}} = 2000 \times 0.332 = 664.$$

These sums equal 2000 up to rounding and define the base mode-split OD for this pair.

Example 2.21 (Service Improvement Scenario: Bus Priority and Stop Consolidation) Suppose a project reduces Bus IVTT from 40 to 38 minutes, reduces OVTT from 10 to 6 minutes via stop consolidation, and reduces Cost from $1.80 to $1.50 via fare policy.

The new Bus utility is $V'_{\text{bus}} = -0.5 - 0.02 \times 38 - 0.05 \times 6 - 0.10 \times 1.5 = -1.71$.
Car and Rail remain $U_{\text{car}} = -1.04$ and $U_{\text{rail}} = -1.41$.
Exponentials are $(0.353,\ 0.181,\ 0.244)$ with sum 0.778.

New probabilities are $(p'_{\text{car}}, p'_{\text{bus}}, p'_{\text{rail}}) = (0.453, 0.232, 0.314)$.

Trips become (906, 464, 628) for Car, Bus, Rail respectively.

Car loses about 54 trips, Bus gains about 88 trips, and Rail losses about 36 trips, reflecting relative improvements.

A few minutes off waiting and walking time matter a lot because OVTT typically has a higher penalty than IVTT. Even modest fare reductions can shift shares when modes are otherwise close. This is exactly the sort of what-if analysis MNL is designed for.

2.4.3 Using Mode Choice in Planning Workflows

After computing p_{ijm}, produce $T_{ijm} = T_{ij} p_{ijm}$ for all OD pairs and modes. Feed each mode-specific OD matrix into the corresponding assignment model (e.g., highway User Equilibrium for Car and Bus-in-traffic, transit path-building for Rail or Bus Rapid Transit) to obtain flows and loads by facility. Compare base and project scenarios on corridor flows, screen line volumes, person-throughput, emissions proxies, and reliability metrics. Because utilities have interpretable coefficients, we can transparently explain which attributes (time, cost, frequency) drive the changes, and then test mitigations or design variations.

Calibrate coefficients with maximum likelihood on survey choices, check signs (time and cost should be negative), and verify elasticities are reasonable.

Segment the market (e.g., peak vs off-peak, captive vs choice riders) when IIA across close substitutes is questionable, or use nested or mixed logit if appropriate in our modeling framework.

Always ensure units are consistent (minutes and currency) and keep track of the scale parameter μ if we do not absorb it into β_k.

When we change a service attribute in a scenario, recompute utilities, probabilities, and T_{ijm} before reassigning flows.

2.4.4 The Foundational Data Problem: Origin-Destination Matrix Estimation

The entire four-step process is built upon a single, critical input: the Origin-Destination (OD) matrix. Yet, this matrix is almost never directly observable, creating a fundamental data problem. The process of estimating an OD matrix from link counts is a classic example of an *ill-posed* or *underdetermined problem*. The number of unknowns (the flows in each of the $N \times N$ cells of the OD matrix) is vastly greater than the number of independent observations (the traffic counts on a limited number of links).

2.5 Step 4: Traffic Assignment: Principles of Network Equilibrium

After the first three steps of the conventional model have been used to generate an Origin-Destination (O-D) matrix of vehicle trips, the final and most computationally intensive task is to assign this demand to the network. The final step, traffic assignment [17], allocates the vehicle-trip OD matrix to the routes of the transportation network. This leads to the concept of network equilibrium.

John Glen Wardrop [34] articulated two distinct principles that formalize how traffic might distribute itself across a network.

- *Wardrop's First Principle—User Equilibrium:* "The journey times on all the routes actually used are equal, and less than those which would be experienced by a single vehicle on any unused route."
  This is a User Equilibrium (UE), a stable state where no single traveler can improve their travel time by unilaterally changing their route.
- *Wardrop's Second Principle—System Optimum:* This principle defines a flow pattern that minimizes the Total System Travel Time (TSTT). This state is known as the System Optimum (SO).

The core challenge of traffic assignment is to simulate how a multitude of individual, non-cooperative agents (drivers) will choose their routes through a congested network. The governing principle for this process is known as *User Equilibrium (UE).* This concept, rooted in the principles of microeconomics and game theory, provides a powerful and behaviorally intuitive description of the stable state that emerges from the "selfish routing" of individual travelers. This section provides a comprehensive and mathematically rigorous exposition of the User Equilibrium principle. We will begin by formally defining UE through the foundational principles of Wardrop and its modern interpretation as a Nash Equilibrium [22, 23]. We will then translate this behavioral concept into a precise mathematical language, introducing the *variational inequality* as its most general and robust formulation. The Variational Inequality formulation provides a more general and powerful mathematical framework for expressing and solving network equilibrium problems [28], capable of handling more complex cost structures than the equivalent optimization approach. Subsequently, we will deconstruct the iterative algorithms, most notably the *Frank-Wolfe algorithm,* that are used to compute the equilibrium solution, illustrating their mechanics with a detailed numerical example. We will discuss the crucial theoretical questions of the existence and uniqueness of UE solutions, establishing the conditions under which a stable and predictable outcome is guaranteed.

The User Equilibrium principle provides a descriptive model of a transportation network. Sheffi [27] and Daganzo [10] authored the definitive text on urban transportation networks, formalizing user equilibrium and system optimum principles central to traffic assignment. It predicts the state that emerges from the uncoordinated, self-interested actions of individual travelers. In contrast, the concept of *System Optimum (SO)* provides a normative benchmark. It does not describe what drivers *will* do, but rather prescribes the flow pattern that is best for the system as a whole. It answers the question: If a central authority with complete control over the network could direct every vehicle, what traffic assignment would minimize the total time spent by all travelers combined? The System Optimum serves as a crucial theoretical ideal. It represents the maximum possible efficiency of a given network for a given demand. By comparing the performance of the User Equilibrium state to this ideal, we can quantify the loss of efficiency caused by selfish routing. This section defines the System Optimum principle, presents its formal mathematical structure as a convex optimization problem, explains the algorithms used to find it, and analyzes the fundamental divergence between individual and collective optimality.

2.5.1 The Mathematical Formulation of Traffic Equilibrium

2.5.1.1 Network Representation in Conventional Modeling

The first step in any quantitative analysis of a transportation system is to create a formal mathematical representation of its physical structure.

A network is represented as a directed graph $\mathcal{G} = (\mathcal{V}, \mathcal{E})$. Travel time on a link a is a function of the flow f_a on that link, $t_a(f_a)$. A widely used specification is the Bureau of Public Roads (BPR) function defined in Eq. (2.3). While sharing the same graph-theoretic foundations as the Ideal Traffic Assignment (ITA) model, the conventional paradigm like FSM includes specific constructs designed to handle its unique demand-driven approach, most notably the concept of the Traffic Analysis Zone.

2.5.1.2 The Network Graph and Traffic Analysis Zones (TAZ)

At its core, the transportation system is represented as a *directed graph*, $\mathcal{G} = (\mathcal{V}, \mathcal{E})$, where $\mathcal{V}$ is the set of nodes and $\mathcal{E}$ is the set of links. In this context, nodes in the set $\mathcal{V}$ represent intersections, and links in the set $\mathcal{E}$ represent the one-way road segments connecting them. However, the traditional model operates at a macroscopic scale. It is computationally infeasible to model trips from every individual household to every individual place of work. To make the problem tractable, the study area is partitioned into a set of discrete geographic sub-areas known as *Traffic Analysis Zones (TAZs)*. A TAZ is the fundamental unit of analysis for travel demand.

All travel is modeled as originating from and being destined for these zones, rather than from specific points on the network. To integrate TAZs into the network graph, two special components are introduced:

- *Centroids:* Each TAZ is represented by a single, abstract node called a *centroid*. This centroid is a conceptual point representing the center of all travel activity within the zone. These centroids are added to the set of nodes $\mathcal{V}$.
- *Centroid Connectors:* These are fictitious links that connect a TAZ's centroid to the real road network. They represent the local street system within a zone, modeling the process of vehicles beginning a trip on local roads before entering the main arterial network, and vice versa at the destination.

This aggregation of demand into TAZs is a foundational simplification of the traditional paradigm. It transforms an intractable many-to-many problem at the individual level into a manageable zone-to-zone problem, but it does so at the cost of spatial precision, an issue that has significant implications for the model's accuracy.

In the conventional model, each link $(i, j) \in \mathcal{E}$ is endowed with a set of attributes that describe its physical and operational characteristics. These attributes are crucial for modeling how the network performs under a given traffic load. The most important are:

- *Capacity* (C): The capacity of a link $c_{ij}$is a measure of its maximum traffic-carrying ability, typically expressed in vehicles per hour (vph). It is a function of the road's physical design, including the number of lanes, lane width, and intersection design.
- *Free-Flow Travel Time* (T^0): This is the time required to traverse a link under ideal, uncongested conditions. It is calculated as the link's length divided by the free-flow speed, which is the maximum legal or safe speed.
- *Travel Time* (T): This is a mathematical function that defines the actual, congested travel time on a link, t_{ij}, as a function of the traffic flow (volume), f_{ij}, assigned to it. As flow increases and approaches the link's capacity, travel time increases non-linearly. The most common form is the Bureau of Public Roads (BPR) function in Eq. (2.3). The existence of this function, which relates travel time to flow, is the mechanism that allows the model to simulate the effects of congestion.

2.5.2 Defining User Equilibrium (Wardrop's First Principle or Nash Equilibrium)

The concept of User Equilibrium was first formally articulated by John Glen Wardrop in 1952 [34]. His work provided the foundational behavioral axioms for traffic assignment that remain the standard to this day.

Axiom 2.1 (Wardrop's First Principle) *The journey times on all the routes actually used are equal and less than those which would be experienced by a single vehicle on any unused route.*

This principle describes a stable state of network flow based on the collective outcome of individual, self-interested behavior. It is a descriptive model that aims to predict the flow pattern that will likely emerge in a real-world network where travelers are free to choose their own paths. The principle is built on two core behavioral assumptions:

1. Every traveler seeks to minimize their own personal travel time (or, more generally, their travel cost).
2. Every traveler has accurate knowledge of the travel times on all available routes.

Under these assumptions, the network will settle into an equilibrium state defined by two precise conditions for every Origin-Destination pair:

1. *Condition of Equal Cost:* For a given O-D pair, the travel cost on *all paths that carry flow* is identical.
2. *Condition of Minimal Cost:* The travel cost on any *unused path* between the O-D pair is greater than or equal to the travel cost on the used paths.

This state is an equilibrium because, once it is reached, no single driver has any incentive to unilaterally change their route.

Imagine a group of drivers all trying to get from one point to another, and they have multiple routes to choose from. Each route has a travel time that increases as more cars use it (due to congestion). *User Equilibrium* is the stable state where no driver can switch to a different route to save time. Everyone is doing the best they can for themselves. If a driver saw that taking a different route would get them to their destination faster, they would switch. The system isn't at equilibrium until all drivers on a route are satisfied that their current path is the fastest available to them, given what everyone else is doing. This is why it's called a "User Equilibrium". It is based on individual (user) choices, not on what is best for the entire system (like minimizing total travel time).

2.5.2.1 Equivalence to Nash Equilibrium

The state described by Wardrop's First Principle is a direct application of the concept of a *Nash Equilibrium* from non-cooperative game theory [22, 23].

Definition 2.11 (Nash Equilibrium) A Nash Equilibrium is a state in a strategic game where each player has chosen a strategy, and no player can benefit by changing their strategy while the other players keep their strategies unchanged.

In the context of traffic assignment, the "game" is defined as:

- *Players:* The individual drivers traveling between a specific O-D pair.
- *Strategies:* The set of available paths between that O-D pair.
- *Payoff:* The negative of the travel cost on the chosen path.

A User Equilibrium is a Nash Equilibrium because, at the equilibrium flow pattern, no individual driver can reduce their travel cost by unilaterally changing their route.

Example 2.22 (User Equilibrium) Consider a simple network with one origin (O) and one destination (D), connected by two alternative routes. Let the total travel demand be 4000 vehicles per hour. The link performance functions (in minutes) are:

- Route 1: $t_1(f_1) = 10 + 2f_1$
- Route 2: $t_2(f_2) = 15 + f_2$

with the flow conservation constraint: $f_1 + f_2 = 4$ (with flows in thousands). The system reaches equilibrium when the travel times are equal: $10+2f_1 = 15+(4-f_1)$, which solves to $f_1 = 3$ and $f_2 = 1$. Travel time $t_1 = 10 + 2(3) = 16$ minutes and $t_2 = 15+1 = 16$ minutes. The equilibrium travel time is 16 minutes on both routes.

2.5.3 *Mathematical Formulation: Variational Inequality*

This subsection is needed because the Variational Inequality (VI) expresses User Equilibrium (UE) in a single, general condition that does not depend on any particular algorithm, and it extends naturally to networks and cost models where a simple objective may not exist. Nagurney and Zhang [21] introduced a unified framework for modeling traffic equilibria using variational inequalities.

It allows us to prove existence and uniqueness under clear assumptions, to compare UE with System Optimum, and to design and certify algorithms like Frank-Wolfe or MSA using a common gap measure.

Definition 2.12 (User Equilibrium as a Variational Inequality) Let $\mathcal{X}$ be the set of feasible link flows that satisfy nonnegativity and flow conservation for the given OD demands. A feasible link-flow vector $f^* \in \mathcal{X}$ is a User Equilibrium if

$$\sum_{a\in A} t_a(f_a^*)\left(f_a - f_a^*\right) \geq 0 \quad \text{for all } x \in \mathcal{X},$$

where $t_a(\cdot)$ is the flow-dependent cost on link a.

The formula 2.12 is just a mathematical way to check this idea of "no better options". Let's break it down: f_a^* is the current flow of cars on all the roads in the network. This is the flow we are checking to see if it's an equilibrium. Variable f_a is any other possible flow of cars on the roads. Notation $t_a(f_a^*)$ is the travel time on each road, based on the current flow f_a^*. The difference $\left(f_a - f_a^*\right)$ represents a "change" or "shift" in flow from the current state f_a^* to the new state f_a. The formula says that if we multiply the current travel times $t_a(f_a^*)$ by this potential change in flow $\left(f_a - f_a^*\right)$, the result must be greater than or equal to zero. It means if we try to shift some traffic (f_a) away from its equilibrium path (f_a^*), we will not find a "better" path. A "better" path would mean that the new flow shift ($(f_a - f_a^*)$) would be away from the more congested, high-cost links and toward the less congested, lower-cost links. This would make the dot product negative. Because the dot product is *non-negative*, it confirms that there's no way to shift flow to get a better outcome. Every potential shift either makes things worse or has no effect on total travel costs.

At UE, the current cost vector $t(f^*)$ makes every feasible direction $x - f^*$ non-improving. If there were a feasible way to push flow that strictly reduced the dot product with costs, some travelers could lower their costs by rerouting, so the flow would not be an equilibrium.

The following Example 2.23 with the two parallel links demonstrates this perfectly. At equilibrium, the travel times are equal, so any shift in flow results in a dot product of zero. This means there's no way to improve the travel time by rerouting traffic, confirming the equilibrium.

Example 2.23 (Variational Inequality on Two Parallel Links) Total demand is $Q = 8$ between one origin and one destination.

Costs are $t_1(f_1) = 6 + 0.5f_1$ and $t_2(f_2) = 7 + 0.2f_2$ with $f_1 + f_2 = 8$.

Equalizing used-route costs gives $6 + 0.5f_1^* = 7 + 0.2(8 - f_1^*)$, so $0.7f_1^* = 2.6$ and $f_1^* = 26/7 \approx 3.714$, $f_2^* = 8 - f_1^* \approx 4.286$.

Then $t_1(f_1^*) = t_2(f_2^*) = 7.857\overline{1}$.

For any feasible x we have $f_2 - f_2^* = -(f_1 - f_1^*)$, so

$$\sum_a t_a(f_a^*)\,(f_a - f_a^*) = (t_1^* - t_2^*)\,(f_1 - f_1^*) = 0 \times (f_1 - f_1^*) = 0 \geq 0,$$

which satisfies the Variational Inequality (VI) condition in Definition 2.12.

Beckmann et al. [2] introduced a convex potential function to model user equilibrium in traffic networks, foundational for optimization.

Proposition 2.1 (Beckmann Transformation (Convex Potential)) *If each link cost $t_a(\cdot)$ is continuous and nondecreasing and separable by link, then any UE solves the convex optimization*

$$\min_{x \in \mathcal{X}} \ \Phi(x) = \sum_{a \in A} \int_0^{f_a} t_a(w)\,dw.$$

Proof The gradient of Φ is $t(x)$, and KKT optimality for minimizing Φ over $\mathcal{X}$ is equivalent to the Variational Inequality in Definition 2.12. □

Think of $\Phi(x)$ as the area under each link's cost curve up to the current flow, summed across links.

At the minimum of this area, there is no first-order descent direction, which is the Variational Inequality condition.

Example 2.24 (Evaluating $\Phi(x)$ and Locating the Minimum) Use the parallel-link data above with $f_1 = a$ and $f_2 = 8 - a$.

The potential is

$$\Phi(a) = \int_0^a (6 + 0.5w)\, dw + \int_0^{8-a} (7 + 0.2w)\, dw = 62.4 - 2.6a + 0.35a^2.$$

Differentiate to get $\Phi'(a) = -2.6 + 0.7a$ and set to zero to obtain $a^* = 2.6/0.7 = 26/7 \approx 3.714$, which matches the Variational Inequality solution.

A second derivative $\Phi''(a) = 0.7 > 0$ confirms a unique minimum.

Theorem 2.3 (UE-Potential Equivalence [2]) *Under separable, nondecreasing link costs $t_a(\cdot)$, a feasible flow is a UE if and only if it minimizes $\Phi(x) = \sum_a \int_0^{f_a} t_a(w)\, dw$ over $\mathcal{X}$.*

Proof The Variational Inequality in Definition 2.12 is the first-order optimality condition for minimizing Φ; conversely, any minimizer satisfies the Variational Inequality. □

Example 2.25 (UE vs System Optimum On the Same Network) The System Optimum minimizes $\sum_a f_a\, t_a(f_a)$ subject to feasibility.

For the parallel-link case, with $f_1 = a$ and $f_2 = 8 - a$, the SO objective is

$$S(a) = a\,(6 + 0.5a) + (8 - a)\,\big(7 + 0.2(8 - a)\big) = 68.8 - 4.2a + 0.7a^2.$$

Set $S'(a) = -4.2 + 1.4a = 0$ to get $a_{\text{SO}} = 3$ and $f_2 = 5$.

Compute totals: $S(3) = 62.5$ and $S(26/7) \approx 62.92$, so UE is slightly less efficient than SO.

Interpretation: selfish routing equalizes used-route costs (UE), while SO balances marginal delays to reduce the system-wide sum of travel times.

Corollary 2.1 (Existence and Uniqueness of UE Link Flows) *If each $t_a(\cdot)$ is continuous and nondecreasing, a UE exists. If each $t_a(\cdot)$ is strictly increasing, the UE link-flow vector is unique.*

Proof Existence follows because Φ is convex and continuous on a nonempty compact feasible set. Strict monotonicity makes Φ strictly convex in x, which implies a unique minimizer. □

Example 2.26 (Why Strict Monotonicity Implies a Unique Split) Let $t_1(x) = 6 + 0.5x$ and $t_2(x) = 7 + 0.2x$ as above.

The equality $t_1(a) = t_2(8 - a)$ becomes $6 + 0.5a = 7 + 0.2(8 - a)$ which simplifies to a strictly increasing left-hand side minus right-hand side equal to zero with a single root.

Graphically, the two cost curves cross once, so there is exactly one a that equalizes costs, hence one UE link-flow vector.

2.5.4 Algorithms for UE Assignment

We need algorithms because costs depend on flows and flows depend on costs, so UE cannot be written in closed form on general networks. The Variational Inequality gives us a principled gap to monitor and convexity gives us descent directions, which motivates the Frank-Wolfe (FW) method and the Method of Successive Averages (MSA).

Definition 2.13 (Frank-Wolfe Direction and Line Search) At iterate f^k, compute link costs $t(f^k)$, run all-or-nothing loading on shortest paths to obtain $\bar{f}^k$, set $d^k = \bar{f}^k - f^k$, and choose $\alpha_k \in [0, 1]$ to minimize $\Phi(f^k + \alpha_k d^k)$.

Sheffi [27] formalized the all-or-nothing assignment method as a baseline for traffic assignment, used in iterative algorithms for estimating traffic flows.

Example 2.27 (Three FW Steps on Parallel Links) Start from $f^0 = (0, 0)$ so $t_1 = 6$ and $t_2 = 7$ and all $Q = 8$ goes to link 1, i.e., $\bar{f}^0 = (8, 0)$ and $d^0 = (8, 0)$.

Line search minimizes $\Phi(\alpha \times 8,\ 0)$ over $\alpha \in [0, 1]$, yielding $\alpha_0 = 1$ and $f^1 = (8, 0)$.

Now $t_1 = 10$ and $t_2 = 7$, so $\bar{f}^1 = (0, 8)$ and $d^1 = (-8, 8)$.

Minimize $\Phi(8 - 8\alpha,\ 8\alpha)$ over α to get $\alpha_1 \approx 0.464$, giving $f^2 \approx (4.286, 3.714)$.

Repeat to obtain f^3 essentially equal to the analytic UE $(3.714, 4.286)$ within rounding.

Interpretation: FW zigzags between all-or-nothing solutions with exact line searches that monotonically reduce Φ.

Definition 2.14 (Method of Successive Averages (MSA) [27]) At iterate f^k, compute the all-or-nothing flow $\bar{f}^k$ under $t(f^k)$ and update $f^{k+1} = (1 - \alpha_k)f^k + \alpha_k \bar{f}^k$ with $\alpha_k = 1/(k + 1)$.

Example 2.28 (MSA On Parallel Links with Diminishing Steps) Start at $f^0 = (0, 0)$.

Iteration 0: $\bar{f}^0 = (8, 0)$ so $f^1 = (1) \times \bar{f}^0 = (8, 0)$.

Iteration 1: costs $(10, 7)$ give $\bar{f}^1 = (0, 8)$ and with $\alpha_1 = 1/2$ we set $f^2 = \frac{1}{2}(8, 0) + \frac{1}{2}(0, 8) = (4, 4)$.

Iteration 2: costs $(8, 7.8)$ give $\bar{f}^2 = (0, 8)$ and with $\alpha_2 = 1/3$ we set $f^3 = \frac{2}{3}(4, 4) + \frac{1}{3}(0, 8) = (2.667, 5.333)$.

Continuing yields a sublinear but steady approach to $(3.714, 4.286)$ as the Variational Inequality (VI) gap shrinks (see Definition 2.12).

Interpretation: MSA trades optimal line searches for simplicity and robustness using averaging steps that get smaller over time.

2.5.5 Algorithms for Solving User Equilibrium

The most famous method is the *Frank-Wolfe algorithm*, an iterative method that solves the non-linear UE problem by repeatedly solving a linear subproblem (an All-or-Nothing assignment). We need this subsection because equilibrium is defined implicitly through flows and costs, so we cannot write the UE solution in closed form on realistic networks. They also provide diagnostics (gaps) so we know when to stop with confidence. Algorithms turn the definition into a reproducible, convergent procedure with measurable progress at each iteration.

2.5.5.1 The Frank-Wolfe Algorithm (Method of Convex Combinations)

The *Frank-Wolfe algorithm* is a classic and widely used method for solving the UE problem. It is an iterative descent algorithm that, at each step, finds a linear approximation of the objective function and moves in that direction. The Frank-Wolfe (FW) method solves UE by repeatedly building a 'best' all-or-nothing direction under current costs and then moving along that direction by an optimal step length.

FW alternates between *route everyone on current shortest paths* and *how far should we move toward that loading.* Because the Beckmann potential is convex under separable nondecreasing costs, each step reduces the potential until the solution flattens at equilibrium.

Frank-Wolfe Algorithm

Step 0: Initialization Set the iteration counter $m = 0$. Compute an All-or-Nothing (AON) assignment using free-flow costs to obtain the initial link flows $\mathbf{f}^{(0)}$.

Step 1: Update Link Costs Evaluate the current link costs based on the flow:

$$\mathbf{t}^{(m)} = \mathbf{t}\left(\mathbf{f}^{(m)}\right)$$

Step 2: Direction Finding Perform an AON assignment using the updated costs $\mathbf{t}^{(m)}$ to generate auxiliary flows $\bar{\mathbf{f}}^{(m)}$. Compute the descent direction:

$$\mathbf{d}^{(m)} = \bar{\mathbf{f}}^{(m)} - \mathbf{f}^{(m)}$$

Step 3: Line Search Determine the optimal step size $\lambda_m \in [0, 1]$ that minimizes the Beckmann potential along the segment:

$$\mathbf{f}^{(m)} + \lambda\,\mathbf{d}^{(m)}$$

Step 4: Flow Update Update the link flows using the computed step size:

$$\mathbf{f}^{(m+1)} = (1 - \lambda_m)\mathbf{f}^{(m)} + \lambda_m \bar{\mathbf{f}}^{(m)}$$

Step 5: Convergence Check Evaluate the convergence criterion (e.g., relative Frank-Wolfe gap). If the criterion is below a specified tolerance, terminate the algorithm. Otherwise, increment the iteration counter $m \leftarrow m + 1$ and repeat from Step 1.

Example 2.29 (FW On Two Parallel Links With Exact Line Search) Total demand is $Q = 8$ between one OD pair.

Costs are $t_1(f_1) = 6 + 0.5 f_1$ and $t_2(f_2) = 7 + 0.2 f_2$.

Step 0. Free-flow costs are $(6, 7)$ so AON sends all 8 to link 1 giving $f^{(0)} = (8, 0)$.

Step 1. Costs at $f^{(0)}$ are $t_1(8) = 10$ and $t_2(0) = 7$, giving $(10, 7)$.

Step 2. AON under $(10, 7)$ sends all 8 to link 2 (cheaper), so $\bar{f}^{(0)} = (0, 8)$ and direction $d^{(0)} = (-8, 8)$.

Step 3. Line search minimizes $\Phi(8 - 8\lambda,\ 8\lambda)$. The objective function is:

$$\begin{aligned}
\Phi(\lambda) &= \int_0^{8(1-\lambda)} (6 + 0.5w)\, dw + \int_0^{8\lambda} (7 + 0.2w)\, dw \\
&= \Big[6w + 0.25w^2\Big]_0^{8-8\lambda} + \Big[7w + 0.1w^2\Big]_0^{8\lambda} \\
&= (64 - 24\lambda + 22.4\lambda^2).
\end{aligned}$$

Differentiate to get $\Phi'(\lambda) = -24 + 44.8\lambda = 0$.
Solving for λ:

$$\lambda_0 = \frac{24}{44.8} = \frac{15}{28} \approx 0.5357.$$

Since $0.5357 \in [0, 1]$, this is the optimal step size.

Step 4. Update $f^{(1)} = (8, 0) + 0.5357(-8, 8) \approx (3.714,\ 4.286)$.

Step 5. Check costs: $t_1(3.714) \approx 7.857$ and $t_2(4.286) \approx 7.857$. The costs are equal, so we have reached the analytic UE.

Interpretation. While standard FW often zigzags, with linear cost functions on parallel links, the exact line search locates the equilibrium in a single valid step.

Example 2.30 (A Simple Convergence Gap for FW) Let the FW gap be $G^{(m)} = \sum_a t_a\big(f_a^{(m)}\big)\big(\bar{f}_a^{(m)} - f_a^{(m)}\big)$.

At $f^{(0)} = (8, 0)$ with costs $(10, 7)$ and $\bar{f}^{(0)} = (0, 8)$ we get $G^{(0)} = 10 \times (0 - 8) + 7 \times (8 - 0) = -80 + 56 = -24$.

The absolute value indicates a large potential descent direction.

At $f^{(1)} \approx (4.286, 3.714)$ under updated costs the gap magnitude is close to zero, indicating near-equilibrium.

2.5.5.2 The Method of Successive Averages (MSA)

The *Method of Successive Averages (MSA)* [27] is a simpler alternative to Frank-Wolfe [14] that uses a predetermined step size, $\lambda_m = \frac{1}{m+1}$, sacrificing per-iteration optimality for computational speed. It is often used in practice for large-scale networks. MSA computes the same AON direction as FW but uses a predetermined shrinking step $\lambda_m = \frac{1}{m+1}$ instead of an optimal line search.

MSA takes smaller and smaller bites toward equilibrium. It is simple, memory-light, and robust on large networks, though it typically needs more iterations than FW.

Example 2.31 (MSA on the Same Two Parallel Links) Start with $f^{(0)} = (0, 0)$.

Iteration $m = 0$: AON gives $\bar{f}^{(0)} = (8, 0)$ and $\lambda_0 = 1$, so $f^{(1)} = (8, 0)$.

Iteration $m = 1$: costs $(10, 7)$ give $\bar{f}^{(1)} = (0, 8)$ and $\lambda_1 = 1/2$, so $f^{(2)} = \frac{1}{2}(8, 0) + \frac{1}{2}(0, 8) = (4, 4)$.

Iteration $m = 2$: costs $(8, 7.8)$ give $\bar{f}^{(2)} = (0, 8)$ and $\lambda_2 = 1/3$, so $f^{(3)} = \frac{2}{3}(4, 4) + \frac{1}{3}(0, 8) = (2.667, 5.333)$.

Continuing yields $(3.714, 4.286)$ within tolerance while the gap shrinks steadily.

Interpretation. MSA trades per-iteration optimality for simplicity and scalability.

2.5.6 *Existence and Uniqueness of UE Solutions*

We need these results to justify that the algorithms are converging to something that actually exists and to understand when the answer is unique. Existence prevents 'chasing a mirage,' and uniqueness tells us when link volumes are determinate given the inputs. For the User Equilibrium model to be a reliable tool, we must be confident that a solution exists and that it is unique.

Theorem 2.4 (Existence of UE) *A User Equilibrium flow pattern exists if the feasible set $\mathcal{X}$ of link flows is nonempty and compact and each link cost $t_a(f_a)$ is continuous in f_a.*

Compactness means flows live in a closed and bounded box defined by OD totals and conservation, and continuity means costs do not jump discontinuously. Then the convex Beckmann potential attains a minimum on $\mathcal{X}$, which is the UE.

Example 2.32 (Checking User Equilibrium Existence) Two parallel links carry $Q = 8$ with $t_1(f_1) = 6 + 0.5f_1$ and $t_2(f_2) = 7 + 0.2f_2$. Feasible flows satisfy $f_1 \in [0, 8]$ and $f_2 = 8 - f_1 \in [0, 8]$, which is a compact segment. Both costs are continuous in their flows. Therefore a UE exists and, as computed earlier, equals $(3.714, 4.286)$.

This condition is met by virtually all realistic link performance functions. The question of *uniqueness* is more subtle.

Theorem 2.5 (Uniqueness of Link Flows) *If link costs are continuous and strictly increasing in their own flows, then the UE link-flow vector is unique.*

Strictly increasing costs make the Beckmann potential strictly convex. A strictly convex function has only one minimizer, so the UE link flows cannot differ. The strict monotonicity condition intuitively means that if we increase the flow on a set of links, the average cost on that set of links must also increase. Standard BPR functions in Eq. (2.3) are strictly monotone. Therefore, for most standard traffic assignment problems, the equilibrium link flow pattern is unique. However, even when the link flows are unique, the *path flows are generally not unique*. For example, on a network with two identical parallel routes, the unique equilibrium of link flows will have the traffic split evenly, but there are infinitely many ways to assign individual path flows to achieve this link flow. The traditional UE model can reliably predict the total volume of traffic on a road segment, but it cannot predict the specific path choice of any individual driver.

Example 2.33 (A Unique Split with Strictly Increasing Costs) With $Q = 8$, $t_1(f_1) = 6 + 0.5f_1$ and $t_2(f_2) = 7 + 0.2f_2$ are strictly increasing. Equalizing used-link costs gives $f_1^* = 26/7$ and $f_2^* = 8 - 26/7$, which is the only solution. If we change t_2 to a flat cost $t_2(f_2) = 7$ over $[0, 8]$, any $f_1 \in [2, 8]$ with $f_2 = 8 - f_1$ equalizes costs at 7, and link flows are no longer unique. This contrast illustrates why strict monotonicity is essential for uniqueness.

Even when link flows are unique, path flows are generally not, because many path decompositions can yield the same link totals. This matters for interpreting route-level outputs but not for link-level volumes used in planning.

Example 2.34 (Path-Flow Non-Uniqueness with Identical Parallel Routes) Consider two identical parallel paths between the same OD with $Q = 10$ and the same link cost function on each path segment. The UE link flows are $f_1 = 5$ and $f_2 = 5$. One valid path decomposition is $(f^{(A)}, f^{(B)}) = (5, 5)$. Another is $(6, 4)$ combined with a $+1$ unit loop on path B and a -1 unit loop on path A that cancels on links, still yielding $f_1 = 5$ and $f_2 = 5$. Infinitely many decompositions map to the same link totals, so path flows are not identified even though link flows are.

2.5.7 *Defining System Optimum (Wardrop's Second Principle)*

System Optimum (SO) is a state of a transportation network where the total travel time for all users is minimized. This objective represents a global, or societal, goal. To achieve SO, a central authority would have to direct each driver onto a specific route, even if that route is not the fastest for that individual, to ensure that the collective sum of all travel times is as low as possible. This state is sometimes referred to as Wardrop's Second Principle, as it contrasts with the individual-centric User Equilibrium principle. SO does not reflect individual user behavior; it is a normative state that demonstrates the maximum potential efficiency of the network.

The System Optimum is founded on the second of Wardrop's two principles, which shifts the objective from individual cost minimization to collective welfare maximization.

Axiom 2.2 (Wardrop's Second Principle) *The average journey time is a minimum.*

This principle is more precisely stated as: the traffic pattern is assigned such that the *total system travel time* is minimized. This represents a global efficiency objective.

Pareto Optimality is a state in which it is impossible to make any single individual better off without making at least one other individual worse off. This concept is fundamental to welfare economics.

A situation is considered Pareto Optimal [13] if there is no possible "Pareto improvement," which is a change that benefits at least one person without harming anyone A system operating at a System Optimum state is also Pareto Optimal. The reasoning is as follows: The total travel time is at its minimum. By definition, the system has achieved the lowest possible aggregate travel cost. No individual can be made better off without another being worse off. If a driver were to switch routes to reduce their own travel time, this action would necessarily increase the total travel time for the system. Since the system's total travel time is already at its minimum, any increase in that total must be caused by an increase in travel time for other users. Therefore, a shift from SO is not a Pareto improvement. It makes one user better off (shorter travel time) but makes others worse off (longer travel times), which violates the condition of Pareto Optimality.

Thus, every System Optimum state is a Pareto Optimal state, but not every Pareto Optimal state is a System Optimum. The set of all Pareto Optimal flows is much larger and includes a wide range of distributions that are not necessarily the most efficient for the system as a whole. The SO state is one of *Pareto Optimality* for the system of travelers; it is impossible to reduce one traveler's journey time without increasing the total time spent by others. The conceptual model for System Optimum is that of a benevolent "system manager" or a central traffic controller. This authority has perfect information about the network's cost functions and the O-D demand matrix. Its goal is to assign routes to all vehicles in a way that minimizes the sum of all their travel times. This is a cooperative routing scenario, where individual preferences are subordinated to the collective good. The resulting flow pattern is often counter-intuitive from an individual's perspective. To achieve the SO, some drivers may be required to take routes that are longer for them personally than other available options. This sacrifice is necessary to prevent their presence on a shorter route from imposing disproportionately large delays on a greater number of other travelers. Because of this, the SO state is generally not a stable equilibrium in the absence of control; individual drivers would have an incentive to deviate from their assigned "system-optimal" path to find a faster "user-optimal" one.

2.5.8 *Mathematical Formulation: Convex Optimization*

The System Optimum problem is naturally formulated as a convex optimization problem [6]. The goal is to find the feasible flow pattern that minimizes a specific objective function: the total system travel time.

Definition 2.15 (Total System Travel Time) The Total System Travel Time (TSTT), denoted by $Z(\mathbf{f})$, is the sum, over all links in the network, of the flow on each link multiplied by the travel time on that link.

$$Z(\mathbf{f}) = \sum_{a \in \mathcal{E}} f_a \times t_a(f_a) \tag{2.1}$$

where f_a is the flow on link a and $t_a(f_a)$ is the travel time on link a as a function of its flow.

The SO problem is then to find the flow vector $\mathbf{f}^{SO}$ that solves:

$$\min_{\mathbf{f} \in \mathcal{F}} Z(\mathbf{f}) = \sum_{a \in \mathcal{E}} f_a \times t_a(f_a) \tag{2.2}$$

where $\mathcal{F}$ is the set of all feasible flow patterns that satisfy the O-D demand and non-negativity constraints. If the link travel time functions $t_a(f_a)$ are non-decreasing, the objective function $Z(\mathbf{f})$ is convex, which guarantees that a global minimum exists.

2.5.8.1 The Concept of Marginal Cost

A profound insight into the SO problem comes from analyzing the conditions at the optimum. The key is to understand the cost from the perspective of the system manager. When a system manager considers adding one additional vehicle to a link, they account not only for the travel time that this new vehicle will experience, but also for the additional delay this vehicle imposes on *all other existing vehicles* on that link. This total system-level cost is the *marginal cost* of travel.

Definition 2.16 (Marginal Cost [2]) The marginal cost of travel on a link a, denoted $t_a^*(f_a)$, is the derivative of the total travel time on that link, $C_a(f_a) = f_a \times t_a(f_a)$, with respect to the flow on that link.

$$t_a^*(f_a) = \frac{d(f_a \times t_a(f_a))}{df_a} \tag{2.3}$$

Using the product rule for differentiation, this becomes:

$$t_a^*(f_a) = t_a(f_a) + f_a \frac{dt_a(f_a)}{df_a} \tag{2.4}$$

The marginal cost consists of two components:

1. $t_a(f_a)$: The average cost experienced by the additional driver (this is the cost that drivers consider in the UE model).
2. $f_a \frac{dt_a}{df_a}$: The *externality cost*. This is the additional delay imposed by the new vehicle on all f_a existing vehicles on the link.

This externality is precisely what self-interested drivers ignore. The following fundamental theorem connects the SO and UE problems.

Theorem 2.6 (SO as a UE with Marginal Costs) *The System Optimum flow pattern is equivalent to the User Equilibrium flow pattern on a modified network where the cost on each link is replaced by its marginal cost function,* $t_a^*(f_a)$.

This theorem is powerful because it means that the conditions for a System Optimum are that the *marginal costs* on all used paths between an O-D pair are equal and minimal.

2.5.9 Finding the System Optimum

Theorem 2.6 provides a direct and elegant method for computing the SO solution. We can use the exact same iterative algorithms developed for the User Equilibrium problem, such as the Frank-Wolfe algorithm, with one critical modification.

Finding System Optimum with Frank-Wolfe
This algorithm follows the same structure as the User Equilibrium (UE) Frank-Wolfe algorithm described in Algorithm 2.5.5.1, with a key modification in Step 2. *Step 2 (SO Direction Finding).* For each link a, compute the marginal cost based on the current flow pattern:

$$t_a^{*(m)} = t_a^* \left(f_a^{(m)} \right)$$

Using these marginal costs, perform an All-or-Nothing (AON) assignment to determine shortest paths. The resulting AON flow pattern $\mathbf{y}^{(m)}$ serves as the descent direction for the System Optimum problem. All other steps, including line search, flow update, and convergence check, remain unchanged from the UE variant. Repeated shortest-path searches based on marginal costs yield a flow pattern that minimizes total system travel time.

2.5.10 The "Price of Anarchy" in General Networks

The User Equilibrium represents the outcome of selfish behavior, while the System Optimum represents the outcome of perfect cooperation. In most realistic congested networks, these two states will not be the same, and their divergence is a central challenge in network design. The inefficiency created by uncoordinated, selfish routing can be formally quantified by a concept from game theory known as the *Price of Anarchy (PoA)* [26].

Definition 2.17 (Price of Anarchy) The Price of Anarchy is the ratio of the total system travel time at User Equilibrium to the total system travel time at System Optimum.

$$PoA = \frac{TSTT(\text{UE})}{TSTT(\text{SO})} = \frac{\sum_{a\in\mathcal{E}} f_a^{\text{UE}} \times t_a(f_a^{\text{UE}})}{\sum_{a\in\mathcal{E}} f_a^{\text{SO}} \times t_a(f_a^{\text{SO}})} \tag{2.5}$$

The PoA is always greater than or equal to 1. A value of 1 indicates that the UE and SO states coincide. In general, the PoA will be greater than 1, and its value measures the percentage loss in system-wide efficiency that results from uncoordinated behavior.

Example 2.35 (Quantifying the Price of Anarchy) Let us return to the two-route network from the previous sections.

- Demand: 4000 vph
- Cost functions: $t_1(f_1) = 10 + 2f_1$, $t_2(f_2) = 15 + f_2$ (flows in thousands)

1. *Calculate TSTT at User Equilibrium:* The UE solution is $f_1^{\text{UE}} = 3000$, $f_2^{\text{UE}} = 1000$. The travel time on both routes is 16 minutes.

$$TSTT(UE) = (3000 \times 16) + (1000 \times 16) = 64,000 \text{ vehicle-minutes.}$$

2. *Calculate the System Optimum Solution:* At SO, the marginal costs are equal.
   - $t_1^*(f_1) = 10 + 4f_1$
   - $t_2^*(f_2) = 15 + 2f_2$

   Solving $t_1^*(f_1) = t_2^*(f_2)$ with the constraint $f_1 + f_2 = 4$ gives:

$$f_1^{\text{SO}} \approx 2,167 \text{ vehicles}, \; f_2^{\text{SO}} \approx 1,833 \text{ vehicles.}$$

3. *Calculate TSTT at System Optimum:* We find the actual travel times at the SO flows:
   - $t_1(f_1^{\text{SO}}) \approx 14.33$ minutes.
   - $t_2(f_2^{\text{SO}}) \approx 16.83$ minutes.

$$TSTT(SO) \approx (2167 \times 14.33) + (1833 \times 16.83) \approx 61,917 \text{ vehicle-minutes.}$$

4. *Calculate the Price of Anarchy:*

$$PoA = \frac{64,000}{61,917} \approx 1.034$$

For this simple network, the Price of Anarchy is approximately 1.034. This means that the uncoordinated, selfish behavior of drivers results in a total system travel time that is 3.4% higher than the theoretical minimum achievable under perfect coordination.

2.5.11 An Complementary Paradigm: The Ideal Traffic Assignment (ITA)

The FSM's critical dependency on costly and often outdated OD matrices is a major weakness. Methods to estimate OD matrix from traffic count exist (see [8, 38]). The Ideal Traffic Assignment (ITA) framework utilizes Ideal Flow Network (IFN) model to represent a novel complementary technique, shifting the focus from a demand-centric view to a supply-centric one, leveraging the physical structure of the transportation network itself to infer traffic patterns.

The ITA is a parsimonious framework for predicting traffic flows from road structure alone. Mathematically, ITA is based on an IFN, which is the limiting distribution of relative flows within a strongly connected network where flow is conserved at each node. Its computation is grounded in *Markov Chains* and the *Principle of Maximum Entropy*. The core of the IFN model is a *stochastic transition probability matrix, S*, where each element represents the probability of moving from one node to another.

Combining ITA with the Four Step Model (FSM) can make the transportation planning even stronger because we can to relate socio-economic factors to total demand κ. Here are some ideas.

1. Trip Generation: Estimate the number of trips produced or attracted by each zone based on socio-economic variables. For example: Trips per household = f(population, income, car ownership, employment)
2. Trip Distribution: Determine where trips go using gravity models or destination choice models. The gravity model is:

$$T_{ij} = k \times \frac{O_i \times D_j}{d_{ij}^{\beta}}$$

   where T_{ij} is trips from zone i to j, O_i is trip origins, D_j is trip destinations, d_{ij} is distance, and β is a parameter.

3. Mode Choice: Predict the mode of travel (e.g., car, transit) using discrete choice models (e.g., logit models):

$$P(\text{car}) = \frac{e^{U_{\text{car}}}}{e^{U_{\text{car}}} + e^{U_{\text{transit}}}}$$

   where utility U depends on travel time, cost, income, etc.
4. Route Assignment: Assign vehicle trips to routes based on travel time, resulting in link flows and total κ.
   From this, κ (total vehicle flow per hour) can be expressed as:

$$\kappa = f(\text{population, car ownership, GDP, land use, travel time})$$

   Empirical models often use regression equations, such as:

$$\kappa = a \times \text{population} + b \times \text{number of cars} + c \times \text{GDP} + \epsilon$$

   where parameters a, b, c are estimated from data.

We can also use Capacity and Congestion to Infer Flow and Model calibration. Inductively, if we know the total capacity $\tilde{c}$ and the maximum congestion ξ (from observed data), we can infer the total flow κ using the *Network Flow Identity* (NFI) in Eq. (13.19):

$$\kappa = \tilde{c} \times \varpi \times \xi \tag{2.6}$$

However, this requires an estimate of ϖ, which depends on the congestion distribution. Here's how to proceed:

1. Estimate ϖ from Data: Collect data on link-level flows and capacities to compute the congestion matrix G. Then:

$$\varpi = \frac{\text{average}(\mathsf{G})}{\max(\mathsf{G})} = \frac{\kappa/\tilde{c}}{\xi}$$

   If κ is unknown, ϖ can be estimated from historical data or similar networks.
2. Infer κ: With $\tilde{c}$, ξ, and ϖ, calculate $\kappa = \tilde{c} \times \varpi \times \xi$.
3. Relate to Socio-Economic Factors: Use statistical data (e.g., from census, traffic counts) to regression κ against population, car ownership, GDP, etc. For example: We collect data for multiple cities or time periods: κ_i, population$_i$, cars$_i$, GDP$_i$. Then, we fit a model such as $\kappa = \beta_0 + \beta_1 \times \text{population} + \beta_2 \times \text{cars} + \beta_3 \times \text{GDP} + \epsilon$. This model can then be used for future predictions.

This approach allows calibration of the relationship between socio-economic factors and κ, which can be used for planning purposes. For instance, if population is projected to grow, the model can predict future κ, and then the theory can be used to determine required capacity $\tilde{c}$ or efficiency improvements ϖ to maintain acceptable congestion ξ. To see the application of this concept, refer to the comprehensive examples in Sect. 14.7.

2.6 Detailed Comparison: Traditional Traffic Models vs. Ideal Traffic Assignment

The FSM was a product of its time. Its legacy is mixed; it provided the foundational language for transportation analysis but is beset by limitations that have become more apparent as policy questions have grown more complex.

2.6.1 Limitations and Legacy of the Classical Paradigm

The FSM's major limitations can be summarized as:

- *The Aggregation Problem:* By averaging diverse characteristics within a TAZ, the FSM loses information.
- *The Trip-Based Fallacy:* The model's focus on independent trips is a fundamental misrepresentation of reality, as people often link trips into complex tours.
- *The Static View:* The FSM produces a static snapshot of an average condition, unable to represent dynamics.

A mastery of its principles is the necessary first step for anyone wishing to understand and appreciate the more advanced models, like ITA, that have been built upon its foundations and in response to its limitations.

Having established a thorough understanding of the machinery of the conventional four-step model and its equilibrium principles, we are now in a position to conduct a detailed comparative analysis. The Ideal Traffic Assignment (ITA) framework is not merely an incremental refinement of these classical methods; it represents a fundamental departure in philosophy, data requirements, and mathematical foundations. Understanding these differences is essential for appreciating the unique insights that each paradigm offers and for selecting the appropriate tool for a given analytical task. This section provides a systematic, side-by-side comparison across five key dimensions, culminating in a summary of the respective strengths and weaknesses that define their ideal domains of application.

2.6.2 Comparison with Traditional Traffic Assignment Models

The Ideal Traffic Assignment (ITA) framework offers a distinct, supply-side perspective on network flow that contrasts sharply with the demand-driven assumptions of traditional traffic assignment models. While both paradigms seek to model traffic distribution, their foundational principles, data requirements, and theoretical objectives differ significantly. Below, we delineate these differences.

- *All-or-Nothing (AON):* This is the simplest assignment method. It assumes that link travel times are fixed (typically at their free-flow values) and are unaffected by congestion. Consequently, AON assigns the entire volume for a given O-D pair to the single, shortest path.
  - *Critique:* The primary limitation of AON is its internal inconsistency. The very act of assigning all traffic to one path would create congestion, thus increasing its travel time and violating the initial assumption that it was the shortest path.
  - *Contrast with ITA:* ITA, by its nature, models congestion and distributes flow across multiple, often redundant, paths. It describes a state of equilibrium where path choices and travel times are mutually consistent.
- *User Equilibrium (UE):* UE models represent the behavioral standard in transportation science. They are based on Wardrop's First Principle [34], which posits that travelers are rational, selfish agents who will switch routes until no one can unilaterally improve their own travel time.
  - *Nature:* UE is fundamentally a *demand-driven* model. Its solution depends on an iterative process of assigning a given Origin-Destination (O-D) matrix of trips onto the network until a stable equilibrium is reached.
  - *Contrast with ITA:* ITA is a *supply-driven* model. It does not require an O-D matrix. Instead, it describes the most probable, entropically maximal flow state that the network's structure (its topology and capacities) can support, independent of any specific demand pattern.
- *System Optimum (SO):* The SO model is a normative benchmark, not a descriptive model of behavior. It calculates the flow distribution that would minimize the *total travel time* for all users in the system, as if a central planner could dictate the route for every vehicle.
  - *Purpose:* SO represents the state of maximum possible efficiency for the network as a whole. The gap between the UE and SO states is often termed the "Price of Anarchy."
  - *Contrast with ITA:* ITA is descriptive, not normative. It models the most likely state to emerge from the system's properties, not a centrally managed ideal.

2.6.2.1 The Unification of Equilibria

Traditionally, UE and SO are viewed as divergent states, reflecting the conflict between individual rationality and collective efficiency. However, this chapter's framework serves as a foundation for a powerful synthesis. As will be proven in the later chapter on Premagic Equilibrium Network Optimality (13), this apparent conflict can be resolved. We will demonstrate that under the specific topological and operational conditions described by this theory, the selfishly rational state (UE) and the collectively efficient state (SO) *converge into a single, identical state*, effectively eliminating the Price of Anarchy.

2.6.2.2 The Advantage of Parsimony

The primary advantage of the ITA framework over its traditional counterparts is its *parsimony*. By generating a plausible and profound traffic assignment from network topology and link capacities alone, it bypasses the need for extensive, and often unavailable, O-D demand data. This makes it an exceptionally powerful tool for long-range strategic analysis, the evaluation of new infrastructure designs, and planning in data-poor environments where traditional models would be inapplicable.

2.6.3 Philosophical Approaches: Demand-Centric vs. Structure-Centric

The most fundamental difference between the two paradigms lies in their core philosophy, their answer to the question, "What is the primary driver of traffic flow patterns?"

- *The Traditional Paradigm: A Demand-Centric Worldview* The conventional four-step model is unequivocally *demand-centric*. Its entire logical structure is built to serve one purpose: to take an exogenously defined quantum of travel demand, represented by the O-D matrix, and assign it to a network. In this worldview, the transportation network is a largely passive entity, a container of fixed capacity that accommodates the flows dictated by the travel needs of the population. The primary analytical question is: "Given this known demand, how will the network perform?"
- *The Ideal Traffic Assignment Paradigm: A Supply-Side Worldview* In stark contrast, the Ideal Traffic Assignment framework is a *supply-side model*. Its foundational premise is that the transportation network is an active system whose physical properties are the primary determinants of the emergent flow pattern. The philosophy posits that the network itself channels and constrains movement in an inherent, predictable way, using principles from Markov chains and information theory to find the most probable, balanced, steady-state flow

distribution that the network infrastructure can support. The primary analytical question is: "What is the natural, steady-state flow distribution that is intrinsic to this specific network structure?"

2.6.4 Data Requirements: O-D vs. Observable Counts

The contrasting philosophies lead directly to vastly different data requirements, which is the most significant practical distinction between the two paradigms.

- *Traditional Models: Data-Intensive* The conventional model is critically dependent on a complete *Origin-Destination matrix*, a reliance that is its greatest practical flaw, as O-D matrices are expensive to create, difficult to update, and often the largest source of error in transport models [5]. This data-intensive approach creates a high barrier to entry and makes the traditional model a cumbersome tool.
- *Ideal Traffic Assignment: Data-Parsimonious via Equivalence and Calibration* The ITA framework is designed to be *data-parsimonious*. It achieves this through the concepts of network equivalence and calibration.

  1. *Equivalence and Relative Flow:* The IFN model first solves for the *pattern* of flow, not its magnitude. Two IFNs are *equivalent* if their flow matrices are proportional by a positive scaling factor ($\mathsf{F}_1 = \varsigma \mathsf{F}_2$) [29]. This means the model can first determine a scale-invariant, relative flow pattern, often represented as a *cardinal network* ($\mathsf{F}_{\text{cardinal}}$) with the smallest possible integer flows [31]. This step requires only the network's capacity matrix as input.
  2. *Calibration with Sparse Data:* To convert this relative pattern into absolute flows, the model is calibrated against reality. This does not require an O-D matrix. Instead, it uses a small set of *observable link flow counts* from a handful of sensor locations [33].

  The calibration is a formal optimization problem. We seek the optimal global scaling factor, ς_{opt}, that minimizes the Sum of Squared Errors (SSE) between the scaled model flows and the observed flows for a small set of monitored links, $\mathcal{W}$.

$$\min_{\varsigma>0} \text{SSE}(\varsigma) = \sum_{(i,j)\in\mathcal{W}} (\varsigma f_{ij,\text{cardinal}} - f_{ij,\text{obs}})^2 \tag{2.7}$$

  The analytical solution for this optimal factor is [32]:

$$\varsigma_{\text{opt}} = \frac{\sum_{(i,j)\in\mathcal{W}} f_{ij,\text{cardinal}} \times f_{ij,\text{obs}}}{\sum_{(i,j)\in\mathcal{W}} (f_{ij,\text{cardinal}})^2} \tag{2.8}$$

Example 2.36 (ITA Calibration) To illustrate this powerful technique, consider a four-approach intersection with a theoretical cardinal flow network, $\mathsf{F}_{\text{cardinal}}$, derived from its geometry [29].

$$\mathsf{F}_{\text{cardinal}} = \begin{pmatrix} 0 & 20 & 10 & 5 \\ 10 & 0 & 15 & 10 \\ 5 & 10 & 0 & 20 \\ 20 & 5 & 10 & 0 \end{pmatrix} \tag{2.9}$$

Suppose a traffic survey provides two peak-hour measurements: flow from B to C is $f_{BC,\text{obs}} = 500$ veh/hr and flow from D to A is $f_{DA,\text{obs}} = 650$ veh/hr [29]. The corresponding flows in our cardinal model are $f_{BC,\text{cardinal}} = 15$ and $f_{DA,\text{cardinal}} = 20$ [29]. We can now calculate the optimal scaling factor:

$$\varsigma_{\text{opt}} = \frac{(15 \times 500) + (20 \times 650)}{15^2 + 20^2} = \frac{7500 + 13000}{225 + 400} = \frac{20500}{625} = 32.8 \tag{2.10}$$

This factor allows us to scale the entire matrix to estimate the absolute flow on all 12 links of the intersection [29]. From just two measurements, we have inferred a complete, physically consistent (premagic [30]) flow matrix.

2.6.5 Underlying Mathematical Foundations

The two paradigms are built upon entirely different branches of mathematics.

- *Traditional Models: Mathematical Programming and Optimization* The mathematical heart of traditional traffic assignment is *deterministic equilibrium-seeking*. The tools used are convex optimization and variational inequalities [21].
- *Ideal Traffic Assignment: Stochastic Processes and Information Theory* The mathematical core of Ideal Traffic Assignment is the analysis of *stochastic steady-states*. The analytical procedure is a sequential transformation of matrices, relying on Markov chain theory and linear algebra to find the unique stationary distribution vector $\boldsymbol{\pi}$ of the system's stochastic matrix S. A key theoretical result is that all equivalent networks share the exact same stochastic matrix, which is an *invariant* of the network's structure [29].

2.6.6 Behavioral Assumptions

The two paradigms embody different views on how to model traveler behavior.

- *Traditional Models: Explicit, Individualistic Rationality* The conventional model is built upon a strong, explicit behavioral assumption of the perfectly rational

"homo economicus." Each driver is assumed to have perfect information and will always choose the path that minimizes their personal travel cost.
- *Ideal Traffic Assignment: The Behavioral Stochastic Matrix (BSM)* The ITA framework models behavior at an aggregate level through the *Behavioral Stochastic Matrix* (S), which represents the equilibrium turning probabilities at each node. This BSM can be specified based on network structure (e.g., capacity) or as the end result of an iterative feedback process where choices adapt to congestion.

2.6.7 Strengths, Weaknesses, and Guidance on Application

Traditional traffic assignment and the Ideal Traffic Assignment Framework represent two distinct approaches to modeling and analyzing transportation networks. The traditional approach is fundamentally demand-centric, relying on a four-step process that begins with estimating travel demand through an Origin-Destination (O-D) matrix. Its primary strength lies in its ability to provide a comprehensive forecast of demand, making it an industry standard for detailed planning. However, this model's main weakness is its critical dependence on the O-D matrix, which is often difficult and expensive to acquire accurately.

In contrast, the Ideal Traffic Assignment Framework operates from a supply-side or structure-centric philosophy. This paradigm's core strength is its data parsimony, as it can analyze a network's structural properties a nd be calibrated even with sparse data. This makes it particularly powerful for assessing the impact of supply-side changes, such as new infrastructure, or for real-time network estimation. Its primary limitation is that it produces relative flow distributions, which require external scaling to represent absolute traffic volumes. This framework simplifies baseline behavior to focus on the inherent properties of the network's design. The detailed comparison in Table 2.1 of the core philosophies and use cases of the traditional four-step model and the Ideal Traffic Assignment Framework makes it clear that each paradigm has a distinct set of strengths and weaknesses.

2.6.7.1 Guidance on Application: When to Use Each Paradigm

- *Use Traditional Models (UE/SO) when:*
  - The primary goal is *demand forecasting* based on projected changes in population, employment, or land use.
  - A *high-quality, reliable, and recent O-D matrix is available*, or the budget exists to create one.
  - The project requires *legally or contractually defensible forecasts* using established, regulator-approved methodologies.

- *Use the Ideal Traffic Assignment Framework when:*
  - The primary goal is *strategic, supply-side analysis* of the network's structure.
  - The key question is: "How will this change in infrastructure affect the inherent efficiency and flow pattern of the network?"
  - An O-D matrix is *unavailable, outdated, or prohibitively expensive.*
  - The goal is to develop *real-time network state estimation* or "nowcasting" systems by leveraging its efficient calibration with live sensor data.

The Ideal Traffic Assignment framework can be seen as a powerful, complementary paradigm. It provides a different and often more practical lens through which to view network performance, one focused on the inherent properties of the network's structure and its response to the flow it carries.

2.7 Summary

This chapter delivered the traditional four-step travel demand model. It begins with the fundamental concept of Traffic Analysis Zones (TAZs), which simplifies the complex problem of individual travel by aggregating it into discrete geographical areas. Then we delve into the critical attributes of network links, such as capacity and free-flow travel time, and the use of travel time functions (like the BPR function) to simulate congestion. The chapter concludes by offering guidance on when to use each paradigm: traditional models for demand forecasting and legally defensible forecasts, and the Ideal Traffic Assignment framework for strategic, supply-side analysis, especially when O-D matrices are unavailable or for developing real-time network state estimation systems. This chapter thus serves as a detailed primer on the conventional approach, setting a clear baseline against which the ITA framework can be compared and appreciated.

Table 2.1 Comparison of traditional and Ideal Traffic Assignment Paradigms

Criterion	Traditional Four-Step Model	Ideal Traffic Assignment Framework
Core Philosophy	Demand-Centric	Supply-Side / Structure-Centric
Key Strength	Comprehensive demand forecasting; industry standard.	Data parsimony; powerful for structural analysis and calibration with sparse data.
Key Limitation	Critical dependence on O-D matrix.	Produces relative flows requiring scaling; simplified baseline behavior.
Typical Use Case	Analysis of demand-side policies with a known O-D matrix.	Analysis of supply-side infrastructure changes; real-time estimation.

References

1. Michael Bacharach. *Biproportional Matrices and Input-Output Change*. Cambridge, UK: Cambridge University Press, 1970. ISBN: 9780521075947.
2. M. J. Beckmann, C. B. McGuire, and C. B. Winsten. *Studies in the Economics of Transportation*. New Haven, CT: Yale University Press, 1956. https://cowles.yale.edu/sites/default/files/2022-08/specpub-beckmann-mcguire-winsten.pdf.
3. Moshe Ben-Akiva and Steven R. Lerman. *Discrete Choice Analysis: Theory and Application to Travel Demand*. MIT Press, 1985. https://mitpress.mit.edu/9780262536400/discrete-choice-analysis/.
4. Abraham Berman and Robert J. Plemmons. *Nonnegative Matrices in the Mathematical Sciences*. Philadelphia: Society for Industrial and Applied Mathematics, 1994. https://doi.org/10.1137/1.9781611971262. https://epubs.siam.org/doi/book/10.1137/1.9781611971262.
5. Piet H. L. Bovy and Eliahu Stern. *Route Choice: Wayfinding in Transport Networks*. Dordrecht: Kluwer Academic Publishers, 1990. https://doi.org/10.1007/978-94-017-0222-3.
6. Stephen Boyd and Lieven Vandenberghe. *Convex Optimization*. Cambridge University Press, 2004. https://doi.org/10.1017/CBO9780511804441.
7. Bureau of Public Roads. *Traffic Assignment Manual*. Tech. rep. U.S. Department of Commerce, Urban Planning Division, 1964.
8. Ennio Cascetta and Alessandra A. Improta. "Estimation of Travel Demand Using Traffic Counts and Other Data Sources". In: *Transportation and Network Analysis: Current Trends: Miscellanea in honor of Michael Florian*. Ed. by Michel Gendreau and Patrice Marcotte. Boston, MA: Springer US, 2002, pp. 71–94. ISBN: 978-1-4757-6871-8. https://doi.org/10.1007/978-1-4757-6871-8_5.
9. Chicago Area Transportation Study. *Chicago Area Transportation Study: Final Report. Volume I: Survey Findings*. Chicago, IL: State of Illinois, 1959.
10. C. F. Daganzo and Y. Sheffi. "On Stochastic Models of Traffic Assignment". In: *Transportation Science* 11.3 (1977), pp. 253–274. https://doi.org/10.1287/trsc.11.3.253.
11. W. Edwards Deming and Frederick F. Stephan. "On a Least Squares Adjustment of a Sampled Frequency Table When the Expected Marginal Totals Are Known". In: *Annals of Mathematical Statistics* 11.4 (1940), pp. 427–444. https://doi.org/10.1214/aoms/1177731829.
12. Robert B. Dial. "A probabilistic multipath traffic assignment model which obviates path enumeration". In: *Transportation Research* 5.2 (1971), pp. 83–111. https://doi.org/10.1016/0041-1647(71)90012-8.
13. Matthias Ehrgott. *Multicriteria Optimization*. 2nd ed. Springer, 2005. ISBN: 978-3-540-21398-7. https://doi.org/10.1007/3-540-27659-9.
14. Marguerite Frank and Philip Wolfe. "An algorithm for quadratic programming". In: *Naval Research Logistics Quarterly* 3.1-2 (1956), pp. 95–110. https://doi.org/10.1002/nav.3800030109.
15. S. Kullback. *Information Theory and Statistics*. Reprint of the 1959 edition. Dover Publications, 1997.
16. S. Kullback and R. A. Leibler. "On Information and Sufficiency". In: *The Annals of Mathematical Statistics* 22.1 (1951), pp. 79–86. https://doi.org/10.1214/aoms/1177729694.
17. L. J. LeBlanc, E. K. Morlok, and W. P. Pierskalla. "An Accurate and Efficient Approach to Solving the Road Network Equilibrium Traffic Assignment Problem". In: *Transportation Research* 9.5 (1975), pp. 309–318. https://onlinepubs.trb.org/Onlinepubs/trr/1974/491/491-002.pdf.
18. Marvin L. Manheim. *Fundamentals of Transportation Systems Analysis: Volume 1—Basic Concepts*. Cambridge, MA: MIT Press, 1979. ISBN: 9780262131292.
19. Daniel McFadden. "Conditional Logit Analysis of Qualitative Choice Behavior". In: *Frontiers in Econometrics*. Ed. by Paul Zarembka. New York, NY: Academic Press, 1974, pp. 105–142. https://eml.berkeley.edu/reprints/mcfadden/zarembka.pdf.

20. Michigan State Highway Department. *Report on the Detroit Metropolitan Area Traffic Study: Data Summary and Interpretation*. Part 1 of the Detroit Metropolitan Area Traffic Study. Lansing, MI: Speaker-Hines and Thomas, State Printers, 1955.
21. Anna Nagurney and Dingzhu Zhang. *Projected Dynamical Systems and Variational Inequalities with Applications*. Springer, 1996. https://doi.org/10.1007/978-1-4757-2494-0. https://link.springer.com/book/10.1007/978-1-4615-2301-7.
22. J. F. Nash. "Non-Cooperative Games". In: *Annals of Mathematics* 54.2 (1951), pp. 286–295. https://doi.org/10.2307/1969529.
23. John F. Nash. "Equilibrium Points in N-Person Games". In: *Proceedings of the National Academy of Sciences* 36.1 (1950), pp. 48–49. https://doi.org/10.1073/pnas.36.1.48.
24. Juan de Dios Ortúzar and Luis G. Willumsen. *Modelling Transport*. 4th. Chichester: John Wiley & Sons, 2011. https://doi.org/10.1002/9781119993308.
25. Srinivas Peeta and Athanasios K. Ziliaskopoulos. "Foundations of dynamic traffic assignment: The past, the present and the future". In: *Networks and Spatial Economics* 1.3 (2001), pp. 233–265. https://doi.org/10.1023/A:1012827724855.
26. T. Roughgarden and E. Tardos. "How bad is selfish routing?" In: *Proceedings 41st Annual Symposium on Foundations of Computer Science*. 2000, pp. 93–102. https://doi.org/10.1109/SFCS.2000.892069.
27. Yosef Sheffi. *Urban Transportation Networks: Equilibrium Analysis with Mathematical Programming Methods*. Englewood Cliffs, NJ: Prentice-Hall, 1985.
28. M.J. Smith. "The existence, uniqueness and stability of traffic equilibria". In: *Transportation Research Part B: Methodological* 13.4 (1979), pp. 295–304. ISSN: 0191-2615. https://doi.org/10.1016/0191-2615(79)90022-5. https://www.sciencedirect.com/science/article/pii/0191261579900225.
29. K. Teknomo. "Ideal Flow Based on Random Walk on Directed Graph". In: *Proceedings of the 9th International Collaboration Symposium on Information, Production and Systems*. Kitakyushu: Waseda University, 2015.
30. Kardi Teknomo. "Premagic and Ideal Flow Matrices". In: *arXiv preprint arXiv:1706.08856* (2017). arXiv: 1706.08856 [cs.DM]. https://arxiv.org/pdf/1706.08856.
31. Kardi Teknomo. *The Signatures of Ideal Flow Networks*. 2024. arXiv: 2408.06344 [cs.DM]. https://arxiv.org/abs/2408.06344.
32. Kardi Teknomo, R. Gardon, and C. Saloma. "Ideal Flow Traffic Analysis: A Case Study on a Campus Road Network". In: *Philippine Journal of Science* 148.1 (2019), pp. 51–62. https://archium.ateneo.edu/discs-faculty-pubs/271/.
33. Kardi Teknomo and R. W. Gardon. "Traffic assignment based on parsimonious data: The ideal flow network". In: *2019 IEEE Intelligent Transportation Systems Conference (ITSC)*. 2019, pp. 1393–1398. https://doi.org/10.1109/ITSC.2019.8917426. https://ieeexplore.ieee.org/document/8917426.
34. J. G. Wardrop. "Some theoretical aspects of road traffic research". In: *Proceedings of the Institution of Civil Engineers* 1.3 (1952), pp. 325–362. https://doi.org/10.1680/iicep.1952.11362. https://www.researchgate.net/publication/227555985_Wardrop_Equilibria.
35. Stanley Leon Warner. *Stochastic choice of mode in urban travel: a study in binary choice*. Evanston, IL: Northwestern University Press, 1962. ISBN: 978-0810102798.
36. A. G. Wilson. "A statistical theory of spatial distribution models". In: *Transportation Research* 1.3 (1967), pp. 253–269. https://doi.org/10.1016/0041-1647(67)90035-4.
37. Alan G. Wilson. "Entropy in urban and regional modelling: retrospect and prospect". In: *Geographical Analysis* 42.4 (2010), pp. 364–394. https://doi.org/10.1111/j.1538-4632.2010.00798.x.
38. H. J. Van Zuylen and L. G. Willumsen. "The most likely trip matrix estimated from traffic counts". In: *Transportation Research Part B: Methodological* 14.3 (1980), pp. 281–293. https://doi.org/10.1016/0191-2615(80)90008-9.

Chapter 3
Matrix Algebra for Network Analysis

3.1 Introduction

In this chapter, we will introduce the essential concepts of graph theory and matrix algebra that underpin modern network analysis. The objective is to establish a clear and logical mathematical framework, presented in a sequential manner, that enables the reader to understand how a physical network's structure gives rise to its dynamic behavior. We build this foundation from first principles. Each new concept rests upon the preceding ones.

The learning path is structured as follows. We begin with the foundational language of *graph theory* to describe network topology and connectivity. Graph theory [6, 8, 11, 15, 16] provides the structural foundation for transport networks, including connectivity, paths, cycles, and flows. We then introduce the necessary tools from *matrix algebra*, including permutation matrices, which are essential for defining key matrix properties. Subsequently, we formalize the representation of networks using *adjacency and incidence matrices*.

With these dual foundations in place, we establish the critical equivalences between physical network properties and algebraic characteristics. We demonstrate that a network is *strongly connected* if and only if its adjacency matrix is *irreducible*, and that *flow conservation* is captured by the *premagic* property. This leads to the analytical core of the chapter: the *Perron-Frobenius theorem* [12, 22]. This powerful theorem provides the mathematical guarantee for a unique equilibrium state in strongly connected systems. We conclude by synthesizing these concepts into the *Ideal Flow Network (IFN) model* and introducing the computational tools, such as the *generalized inverse*, required for its application.

The ultimate goal is to provide a rigorous, self-contained theoretical reference that illuminates the principles used throughout this book to model complex systems.

K. Teknomo, *Ideal Flow Network for Transportation Research*,
Lecture Notes in Intelligent Transportation and Infrastructure,
https://doi.org/10.1007/978-981-95-4727-2_3

3.2 Fundamentals of Graph Theory

The field of graph theory can be traced back to a problem concerning the city of Konigsberg, Prussia, in 1736. The city contained two islands connected to the mainland by seven bridges. The problem, posed by Leonhard Euler [10], asked if it was possible for a citizen to take a walk from home, cross each of the seven bridges exactly once, and return to the starting point. Euler simplified the problem by representing each landmass as a point, which he called a vertex, and each bridge as a line, which he called an edge. Using this abstraction, Euler proved that no such path existed. This foundational work established a new way of analyzing problems based on network structure.

To analyze transportation systems with mathematical rigor, we must first abstract their physical form into a precise mathematical object. The field of graph theory, provides the necessary language and tools for this essential task. Science progresses through the creation of such models, which are purposeful abstractions of reality designed to help us understand, predict, and control the systems we are interested in [5]. By representing a complex physical road network as a mathematical object called a *directed graph*, or digraph, we can strip away geographically superfluous details like road curvature and elevation. This allows us to focus on the essential structural property that governs system-wide flow: *connectivity*.

This section provides a detailed, self-contained introduction to the algebraic graph theory required for transport network analysis [4]. The objective is to build a robust and intuitive understanding of these foundational concepts from first principles [20]. We will define the core components of a network, explore the ways movement and connectivity are described, and finally, translate these physical concepts into the powerful language of linear algebra, which forms the computational backbone of modern network science [9].

3.2.1 Core Concepts: Graphs, Digraphs, Nodes, and Links

A *graph* is a diagram consisting of a set of points joined together by a set of lines [8]. The points are formally called *vertices*, and the lines are formally called *edges*. Each edge connects exactly two vertices. Vertices that are connected by an edge are called *adjacent* or neighbors. A vertex can exist on its own as an isolated point, but an edge can only exist to connect vertices. An edge that connects a vertex to itself is called a *simple loop*. A graph that contains no loops and no multiple edges between any two vertices is known as a *simple graph*. A *complete graph* is a special type of simple graph where every pair of distinct vertices is connected by an edge.

The edges in a graph are typically considered to have no direction. However, if the edges are assigned a direction, the graph is called a *directed graph*, or *digraph*. In this book, we are primarily concerned with directed graphs. To maintain clarity, we will adopt a specific terminology. For undirected graphs, we will use the terms *vertices* and *edges*. For directed graphs, we will use the terms *nodes* and *links*. The term *network* is used as a common synonym for a digraph.

Definition 3.1 (Graph and Digraph) A *graph* is a pair of two sets, $\mathcal{G} = \{\mathcal{V}, \mathcal{E}\}$, consisting of a finite set of vertices $\mathcal{V}$ and a finite set of edges $\mathcal{E}$. A *digraph* is also a pair of two sets, $\mathcal{D} = \{\mathcal{V}, \mathcal{E}\}$, where $\mathcal{V}$ is a finite set of nodes and $\mathcal{E}$ is a finite set of links, with each link defined as an ordered pair of nodes from $\mathcal{V}$.

In the context of urban transportation, this abstract definition has a direct and intuitive physical interpretation.

- *Nodes* ($\mathcal{V}$): Nodes represent key locations in the network where travel decisions are made or where flow can change direction. Most commonly, nodes represent road *intersections*. They can also be used to model other critical points, such as traffic analysis zones (TAZs), major trip generators like airports or shopping centers, or intermodal hubs where travelers can switch between, for instance, a rail system and the road network [21].
- *Links* ($\mathcal{E}$): Links represent the pathways that connect the nodes. In a road network, a link corresponds to a *one-way segment of a road* that directly connects two nodes. The directionality is a crucial feature of this definition. A link (i, j) from node i to node j is distinct from a potential link (j, i) in the opposite direction. Therefore, a standard two-way street that connects intersections A and B must be represented by *two separate directed links*: the link (A, B) and the link (B, A) [1].

This abstraction is powerful because it distills a complex physical reality into its essential topological structure. Attributes like the number of lanes, the speed limit, or the presence of a traffic signal are considered properties *of* a link, but they are not part of the fundamental definition of the network's layout itself.

To further characterize the structure at each node, we must be able to count the number of connections entering and leaving it.

Definition 3.2 (Node Degree) For any node $i \in N$ in a digraph, the *in-degree* is the number of incoming links that terminate at node i. The *out-degree* is the number of outgoing links that originate from node i [29].

In a graph, the number of edges connected to a vertex is called the *degree* of that vertex. In a digraph, this concept is refined. The *in-degree* of a node is the number of links directed into it. The *out-degree* of a node is the number of links directed away from it.

Example 3.1 (A Simple Traffic Interchange) Let us consider a simple traffic interchange represented by the three-node network shown in Fig. 3.1.

This network can be formally defined by its set of nodes and links:

- Nodes: $N = \{A, B, C\}$
- Links: $E = \{(A, B), (A, C), (B, C), (C, A)\}$

Fig. 3.1 A simple network for calculating node degrees

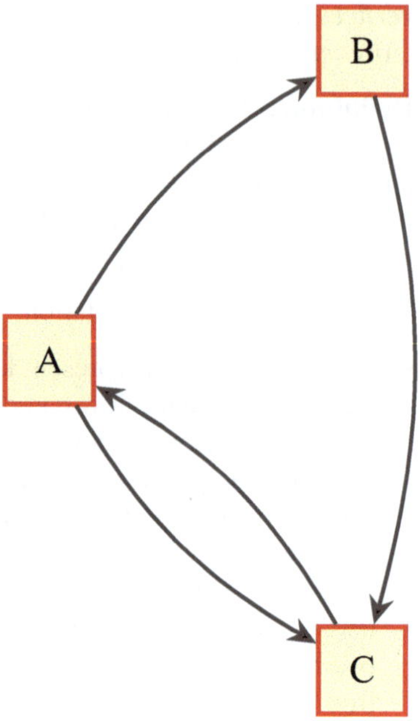

We can now calculate the degrees for each node using Definition 3.2:

- *Node A:* Has one incoming link, (C, A), and two outgoing links, (A, B) and (A, C). Therefore, in-degree(A) = 1, out-degree(A) = 2.
- *Node B:* Has one incoming link, (A, B), and one outgoing link, (B, C). Therefore, in-degree(B) = 1, out-degree(B) = 1.
- *Node C:* Has two incoming links, (A, C) and (B, C), and one outgoing link, (C, A). Therefore, in-degree(C) = 2, out-degree(C) = 1.

An important property of any directed graph is that the sum of all in-degrees must equal the sum of all out-degrees, because each link contributes exactly one to both totals. Here, the sum of in-degrees is $1 + 1 + 2 = 4$, and the sum of out-degrees is $2 + 1 + 1 = 4$. This confirms the total number of links in the network is 4 [29, 30].

Example 3.2 (Modeling a Roundabout) A roundabout provides a more complex and non-trivial example of network representation. Consider a four-entry roundabout connecting roads from the North, South, East, and West. A common way to model this is to represent each entry/exit point as a separate node on a central cycle (Fig. 3.2).

Let the external approach nodes be $N_{ext}, S_{ext}, E_{ext}, W_{ext}$ and the nodes on the roundabout circle be $N_{circ}, S_{circ}, E_{circ}, W_{circ}$.

- *Nodes:* $N = \{N_{ext}, S_{ext}, E_{ext}, W_{ext}, N_{circ}, S_{circ}, E_{circ}, W_{circ}\}$
- *Links* (E):

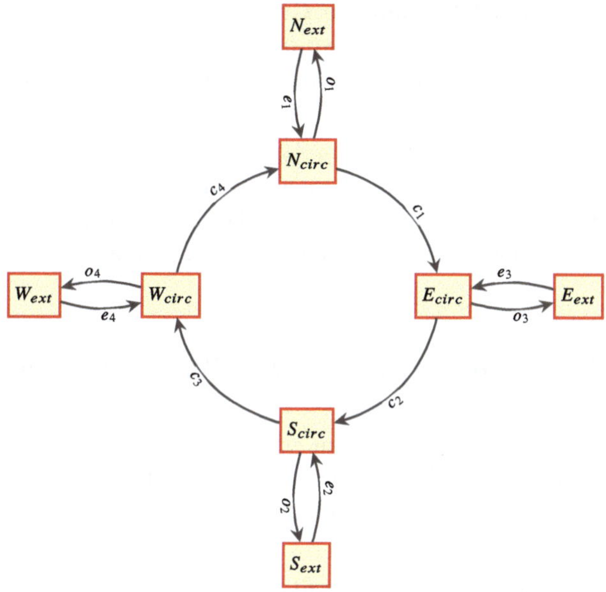

Fig. 3.2 A graph representation of a four-entry roundabout

- *Entry links:* (N_{ext}, N_{circ}), (S_{ext}, S_{circ}), (E_{ext}, E_{circ}), (W_{ext}, W_{circ})
- *Circulatory links (one-way):*
  - (N_{circ}, E_{circ})
  - (E_{circ}, S_{circ})
  - (S_{circ}, W_{circ})
  - (W_{circ}, N_{circ})
- *Exit links:* (N_{circ}, N_{ext}), (S_{circ}, S_{ext}), (E_{circ}, E_{ext}), (W_{circ}, W_{ext})

Let's analyze the degree of one of the circulatory nodes, for example, N_{circ}:

- *in-degree*(N_{circ}): It has two incoming links: one from the external road (N_{ext}, N_{circ}) and one from the preceding node on the roundabout (W_{circ}, N_{circ}). So, its in-degree is 2.
- *out-degree*(N_{circ}): It has two outgoing links: one to exit the roundabout (N_{circ}, N_{ext}) and one to continue circulating (N_{circ}, E_{circ}). So, its out-degree is 2.

This example illustrates how a physically complex intersection can be decomposed into a set of simple nodes and one-way links, which perfectly preserves the allowed movements. The graph representation makes the allowed paths explicit and amenable to mathematical analysis. Note that this roundabout network is also an *Eulerian circuit* 3.9 because in each node, the in-degree is equal to the out-degree.

Definition 3.3 (Complete Graph) A complete graph is a family of undirected graphs in which every two distinct vertices are joined by exactly one edge. The complete graph with n nodes is symbolized by K_n. Each vertex of a complete graph has a degree of $n - 1$. A complete graph K_n has $\frac{1}{2}n(n - 1)$ edges.

Definition 3.4 (Clique) A *clique* is a complete subgraph. Given an undirected graph $\mathcal{G} = (\mathcal{V}, \mathcal{E})$, a clique is a subset of nodes in the graph such that every two distinct nodes in the clique are adjacent. A *maximal clique* is the largest clique such that no additional adjacent node can be included.

3.2.2 Connectivity: Walks, Paths, and Strong Connectivity

An agent can move from one node to another within a network, but only by traversing the links. The sequence of links that an agent follows defines its movement. The properties of these sequences determine the connectivity of the network.

Definition 3.5 (Walk, Trail, Path, and Cycle) The following terms describe sequences of movement within a graph or digraph.

- A *walk* is a sequence of links that an agent can traverse, where nodes and links may be repeated. The number of links in the sequence is the *length* of the walk.
- A *trail* is a walk in which all links are unique. Nodes may be repeated.
- A *path*, also known as a *trajectory*, is a walk in which all links and all nodes are unique.
- A *closed walk* is a walk that starts and ends at the same node.
- A *closed trail* is a closed walk in which all links are unique.
- A *cycle*, also known as a *circuit*, is a closed walk in which all nodes and links are unique, except for the start and end node, which is the same.

Example 3.3 (Walk, Trail, and Path) Consider the network shown in Fig. 3.3.

- An agent performing a *walk* might traverse the sequence of links $ab \to bc \to ce \to eb \to bd \to df \to fb \to bc$. This walk has a length of eight. The link bc is used twice, node b is visited three times, and node c is visited twice.
- An agent performing a *trail* might traverse the sequence of links $ab \to bc \to ce \to eb \to bd$. This trail uses five unique links but passes through node b twice.
- An agent performing a *path* from node a to node f might traverse the sequence of links $ab \to bc \to ce \to eg \to gf$. All nodes and links in this sequence are unique.

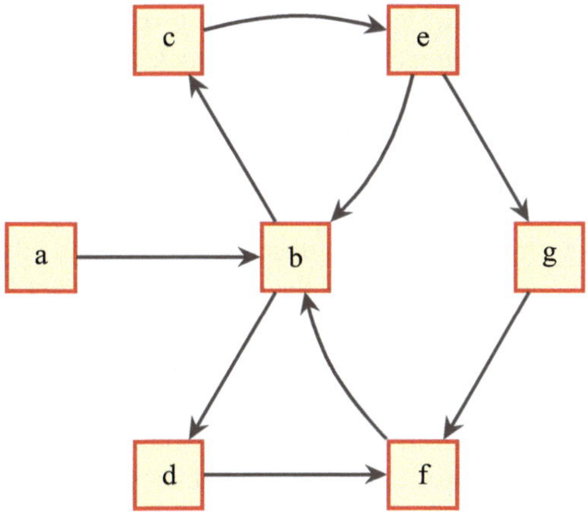

Fig. 3.3 A network illustrating different types of movement

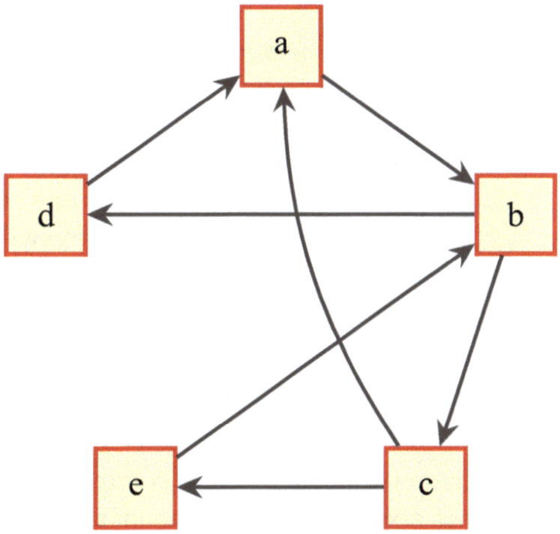

Fig. 3.4 A network illustrating closed movements

Example 3.4 (Closed Trail and Cycle) Consider the network shown in Fig. 3.4.

- An agent performing a *closed trail* might traverse the sequence $ab \to bc \to ce \to eb \to bd \to da$. This sequence uses six distinct links but passes through nodes a and b twice.
- An agent performing a *cycle* might traverse the sequence $ab \to bc \to ca$. This sequence returns to the starting node while using unique nodes and links.

The overall structure of a directed network can be classified based on its connectivity. This classification is fundamental because, as we will see, it corresponds directly to the algebraic properties of the network's matrix representation. A directed network falls into one of two primary categories. The first category is *strongly connected*, where every node is reachable from every other node. This property is equivalent to its adjacency matrix being *irreducible* [18]. The second category includes all networks that are not strongly connected. These networks are equivalent to their adjacency matrix being *reducible*, and they can be further subdivided into *weakly connected* and *unconnected* networks. Weakly connected networks are often characterized by the presence of terminal structures, such as sources and sinks, which can be individual nodes or entire components.

Definition 3.6 (Connectivity Types)

- A *subgraph* is a subset of vertices and edges taken from a graph.
- A *sub-network* is the equivalent concept for a digraph.
- A *component* is a special subgraph where it is possible to move from any vertex to any other vertex within that subgraph.
- A *strongly connected component* (SCC) is a sub-network where it is possible to move from any node to any other node within that sub-network.
- A network is *connected* if its underlying undirected graph is connected. This means that if we ignore the direction of the links, there is a path between any two nodes.
- A network is *unconnected* if it consists of two or more components between which there are no links.
- A network is *weakly connected* if its underlying undirected graph is connected, but the digraph itself is not strongly connected. This implies that while the network is whole, there is at least one pair of nodes (u, v) such that one cannot get from u to v by following the directed links.
- A network is *strongly connected* if there is a directed path from every node to every other node in the network.

Example 3.5 (Identifying Connectivity) Consider the three networks in Fig. 3.5.

(a) This network is *strongly connected*. From any node, an agent can reach any other node. Its adjacency matrix would be irreducible.

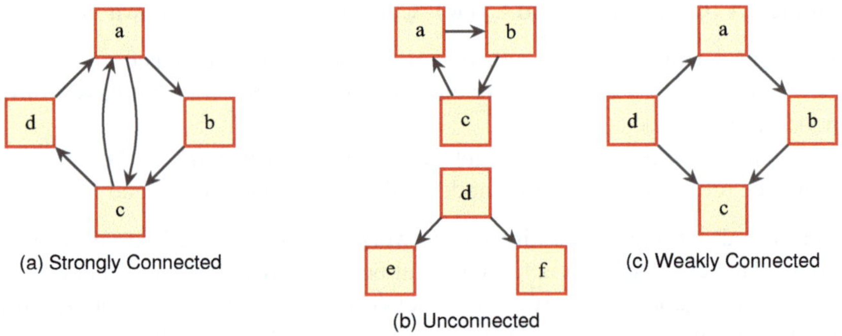

Fig. 3.5 Examples of (**a**) a strongly connected network, (**b**) an unconnected network, and (**c**) a weakly connected network

(b) This network is *unconnected*. It consists of two separate components. Its adjacency matrix would be reducible.
(c) This network is *weakly connected*. The underlying graph is connected, but it is not strongly connected. It contains a source component and a sink component. Its adjacency matrix would be reducible.

Definition 3.7 (Source and Sink Structures)

- A *source node* is a node with an in-degree of zero. An agent can only leave a source node.
- A *sink node* is a node with an out-degree of zero. Once an agent enters a sink node, it cannot leave.
- A *source component* is a strongly connected component where once an agent leaves any node within it, the agent can never return to any node in that component.
- A *sink component* is a strongly connected component that, once entered, an agent can never leave.

Example 3.6 (Identifying Sink and Source Structures)

- In the network Fig. 3.6, node d is a *sink node* because there is inflow to the node but no outflow from it. An agent entering node d is trapped.
- In the network Fig. 3.7, node d is a *source node* because an agent can only exit from it. There is no inflow to the node.
- The network Fig. 3.8 contains a *sink component*. Nodes $\{b, c, d\}$ form a strongly connected component. Once an agent from outside (e.g., from node e) enters this component, it is trapped and can never leave.
- Conversely, the network Fig. 3.9 shows a *source component*. Nodes $\{a, b, e\}$ form a strongly connected component. Once an agent leaves this component (e.g., by moving to node c), it can never return.

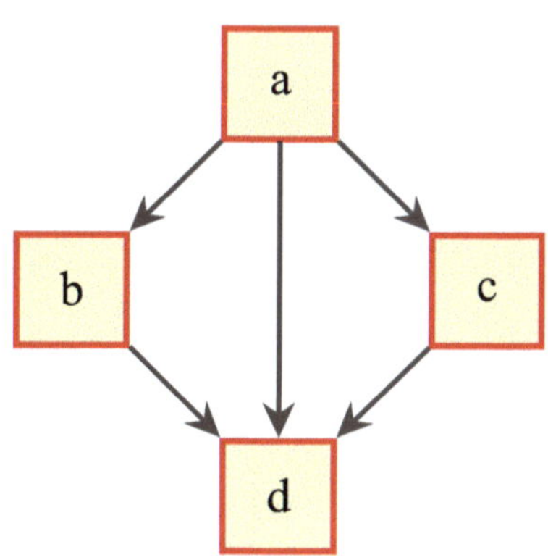

Fig. 3.6 A network with a source node (a) and a sink node (d)

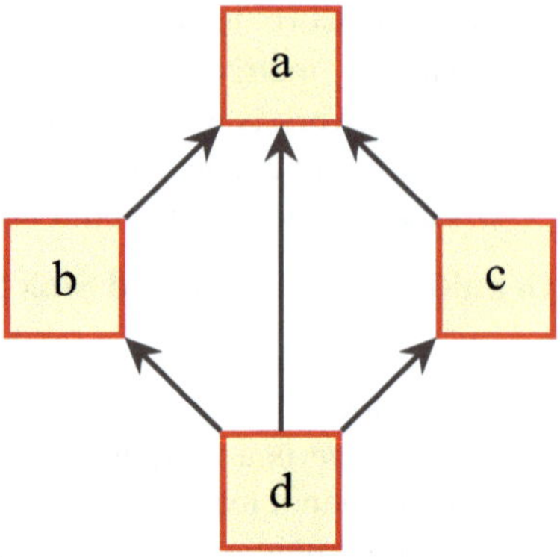

Fig. 3.7 A network with a source node (d) and a sink node (a)

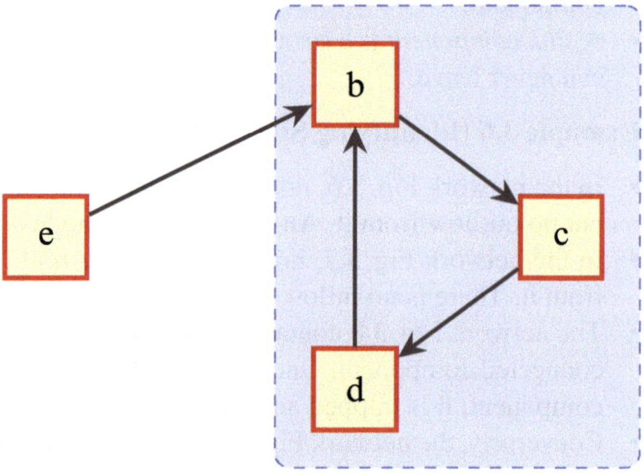

Fig. 3.8 A network with a source node (e) and a sink component (b, c, d)

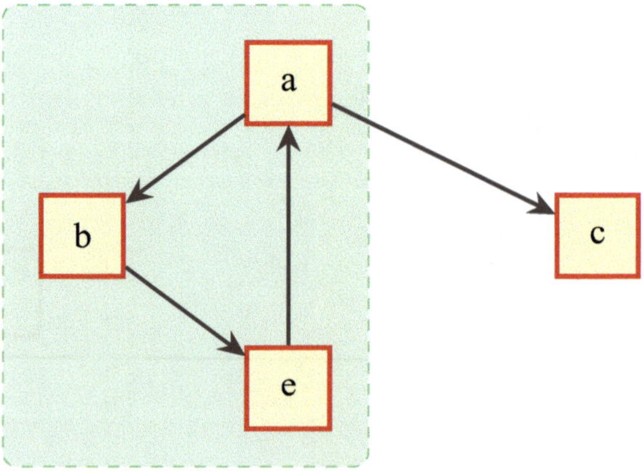

Fig. 3.9 A network with a source component (a, b, e) and a sink node (c)

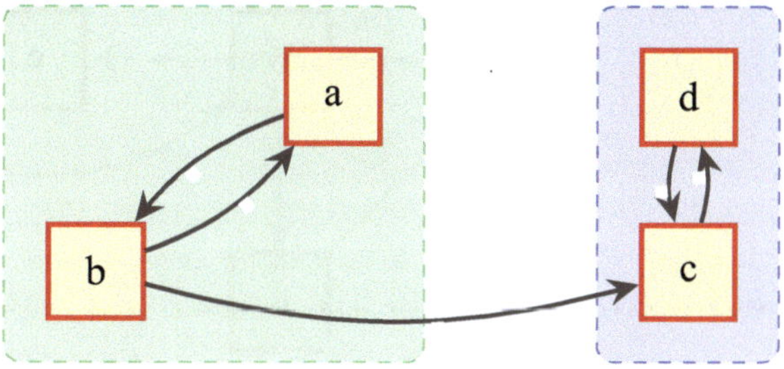

Fig. 3.10 A weakly connected network with a source component and a sink component

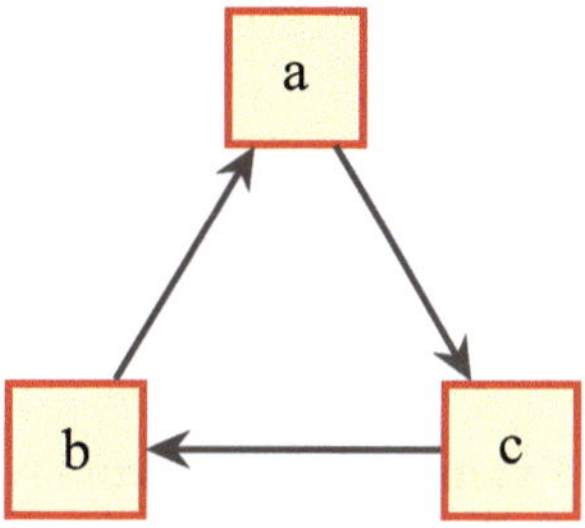

Fig. 3.11 A minimally strong network

Example 3.7 (Sink Component and Source Component) This is an example of weakly connected network. It has a source component and a sink component and one-way bottleneck between different regions of the graph. Consider the system in Fig. 3.10, representing two city districts connected by a one-way bridge.

In this network Fig. 3.10, it is possible to travel from District A (nodes a, b) to District B (nodes c, d) via the link (b, c). However, it is impossible to travel back. No path exists from node c or d to node a or b. Such a system cannot achieve a stable, circulating equilibrium. Any traffic flowing from District A to District B becomes permanently trapped there; over time, flow would drain from A and accumulate indefinitely in B, violating the principle of a steady state [23]. Therefore, such a network cannot be analyzed as a single integrated system using the Ideal Flow Network. It must be broken down into its constituent *strong components*.

Definition 3.8 (Minimally Strong Network) A *minimally strong network* is a strongly connected network with the property that if any single link were to be removed, the network would no longer be strongly connected.

Example 3.8 (Minimally Strong Network) A simple directed cycle is a common example of a *minimally strong network*. Consider the three-node network in Fig. 3.11. The network is strongly connected because there is a path from any node to any other node (e.g., $a \to b \to c \to a$). However, if any single link is removed, such as the link $b \to c$, it is no longer possible to get from node b to node c, and the network ceases to be strongly connected.

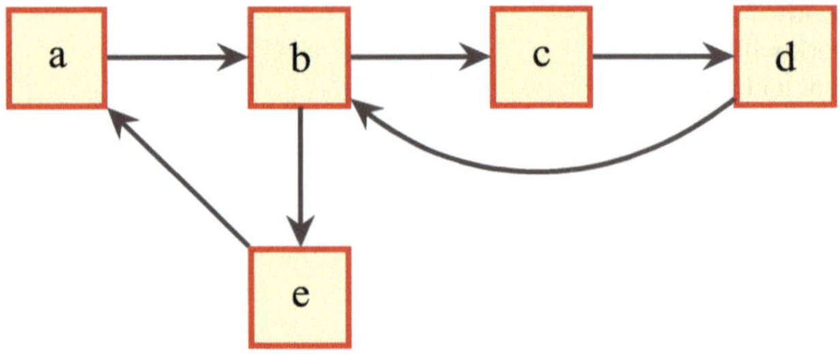

Fig. 3.12 An Eulerian network where in-degree equals out-degree for all nodes

Definition 3.9 (Eulerian Network) A network is called *Eulerian* if it is possible to find a closed trail that traverses every link in the network exactly once. A key property of an Eulerian network is that for every node, the in-degree is equal to the out-degree.

Example 3.9 (Eulerian Network) Consider the network in Fig. 3.12. This network is *Eulerian* because an agent can traverse every link exactly once in a closed trail, for instance, the trail $a \to b \to c \to d \to b \to e \to a$. A necessary condition for a network to be Eulerian is that for every node, the in-degree must equal the out-degree. We can verify this property for the example network:

- Node a: in-degree=1, out-degree=1
- Node b: in-degree=2, out-degree=2
- Node c: in-degree=1, out-degree=1
- Node d: in-degree=1, out-degree=1
- Node e: in-degree=1, out-degree=1

Since the condition holds for all nodes, an Eulerian circuit exists.

3.3 Foundational Matrix Algebra

The previous section established the language of graph theory to describe the topology of transportation networks. To analyze these structures quantitatively, to calculate flows, verify conservation laws, and find equilibrium states, we must translate this topological language into the powerful and precise language of linear algebra. Matrix algebra provides the computational engine for network science. It enables us to represent complex networks and the flows within them as structured arrays of numbers that can be manipulated through a well-defined set of rules. To analyze the structure and dynamics of networks, we must first establish the language of matrix algebra. Matrices and vectors provide a powerful framework for representing network topology and the flows that move through it.

This section provides a thorough review of the fundamental concepts of linear algebra that are indispensable for understanding the Ideal Traffic Assignment model. Our objective is to build a solid foundation, moving from basic definitions to the more advanced concepts required for steady-state analysis. We will begin by defining the core objects of our analysis, vectors and matrices, and the operations we can perform on them. We will then explore how systems of linear equations are used to solve for unknown quantities in a network, particularly the equilibrium distribution of flow. Finally, we will provide an intuitive introduction to the concepts of eigenvalues and eigenvectors, explaining their profound connection to the stability and long-term behavior of dynamic systems like traffic networks. Throughout, we will ground these abstract mathematical ideas in concrete examples related to transportation analysis.

In this text, we denote matrices with bold uppercase letters, such as A, and vectors with bold lowercase letters, such as $\mathbf{v}$.

3.3.1 Basic Matrix and Vector Types

At the heart of linear algebra are two fundamental objects: the vector and the matrix. In network science, they are used to store and manipulate all essential information about the system, from its physical layout to the volume of traffic on its links.

Definition 3.10 (Scalar, Vector, Matrix)

- A *scalar* is simply a single number, such as the total flow in a network, κ, or a parameter like α.
- A *vector* is an ordered list of numbers, typically arranged in a column. An n-dimensional column vector $\mathbf{v}$ has the form:

$$\mathbf{v} = \begin{pmatrix} v_1 \\ v_2 \\ \vdots \\ v_n \end{pmatrix} \tag{3.1}$$

- The transpose of a column vector, denoted $\mathbf{v}^T$, is a *row vector*: $\mathbf{v}^T = \begin{pmatrix} v_1 & v_2 & \cdots & v_n \end{pmatrix}$. In network analysis, vectors are often used to store attributes associated with each node, such as the total flow entering or leaving each intersection.
- A *matrix* is a rectangular array of numbers arranged in rows and columns. An $m \times n$ matrix M has m rows and n columns. The element in the i-th row and j-th column is denoted M_{ij}. In our work, we are primarily concerned with *square matrices*, where the number of rows equals the number of columns ($n \times n$). These are used to represent relationships between the nodes of a network, such as connectivity or flow volumes.

Example 3.10 (Transpose Matrix) The transpose of a matrix M, denoted M^T, is formed by interchanging its rows and columns. The element at position (i, j) in M becomes the element at position (j, i) in M^T. In a network context, the transpose operation has a clear physical meaning: it reverses the direction of all specified relationships. For our example flow matrix F, the transpose is:

$$\mathsf{F}^T = \begin{pmatrix} 0 & 40 & 10 & 0 \\ 50 & 0 & 60 & 0 \\ 20 & 0 & 0 & 70 \\ 0 & 30 & 0 & 0 \end{pmatrix} \tag{3.2}$$

Here, the entry $[\mathsf{F}^T]_{12} = 40$ now represents the flow from node 2 to node 1, which was the original entry F_{21}.

We begin by defining the most basic matrix and vector types that form the basis of our analysis.

Definition 3.11 (Basic Matrix and Vector Types) Let n be a positive integer.

- A *square matrix* is a matrix with an equal number of rows and columns ($n \times n$). Adjacency and flow matrices in network models are typically square.
- A *non-negative matrix* $\mathsf{M} = [m_{ij}]$ is a matrix where all entries are non-negative, i.e., $m_{ij} \geq 0$ for all i, j.
- A *positive matrix* is a matrix where all entries are strictly positive, i.e., $m_{ij} > 0$ for all i, j.
- The *one vector* $\mathbf{j}$ is a column vector where all entries are one. It is useful for summing the rows of a matrix M via the product $\mathsf{M}\mathbf{j}$.
- The *identity matrix* I is a square matrix with ones on the main diagonal and zeros elsewhere. For any compatible matrix A, we have $\mathsf{AI} = \mathsf{IA} = \mathsf{A}$.

Example 3.11 (Non-Negative and Positive Matrices) A matrix A is non-negative if all of its elements are either zero or positive numbers. For instance, M_1 is a non-negative matrix.

$$\mathsf{M}_1 = \begin{pmatrix} 0 & 1 & 0 & 0 \\ 0 & 0 & 7 & 0 \\ 1 & 1 & 0 & 4 \\ 0 & 4 & 0 & 0 \end{pmatrix}$$

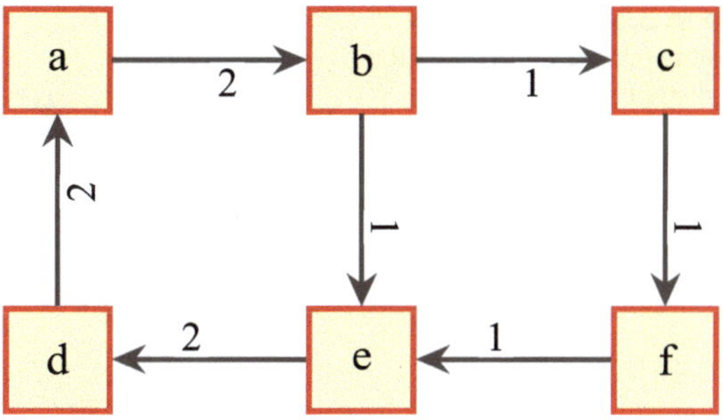

Fig. 3.13 A traffic flow network for the Avaloria district

A matrix is positive if all of its elements are strictly positive numbers.

$$\mathsf{A} = \begin{pmatrix} 1 & 3 & 2/3 \\ 9 & 7 & 8 \\ 4 & 6 & 5 \end{pmatrix}$$

Example 3.12 (Traffic Flow Matrix) Consider a small traffic network in the Avaloria district with six nodes, as shown in Fig. 3.13.

The relative flow of vehicles per hour between these locations can be represented by a flow matrix F. This is a square, non-negative matrix. The entry $f_{ab} = 2$ indicates that 2 vehicles per hour travel from location a to location b.

$$\mathsf{F} = \begin{pmatrix} 0 & 2 & 0 & 0 & 0 & 0 \\ 0 & 0 & 1 & 0 & 1 & 0 \\ 0 & 0 & 0 & 0 & 0 & 1 \\ 2 & 0 & 0 & 0 & 0 & 0 \\ 0 & 0 & 0 & 2 & 0 & 0 \\ 0 & 0 & 0 & 0 & 1 & 0 \end{pmatrix} \tag{3.3}$$

Example 3.13 (Non-Negative Flow Matrices) The flow matrices representing traffic in a network are non-negative, as flow cannot be negative. Consider the flow matrix F for the Brindenton district road network shown in Fig. 3.14.

$$\mathsf{F} = \begin{pmatrix} 0 & 5 & 0 & 0 & 0 & 0 \\ 0 & 0 & 2 & 3 & 0 & 0 \\ 5 & 0 & 0 & 0 & 0 & 0 \\ 0 & 0 & 0 & 0 & 7 & 0 \\ 0 & 0 & 0 & 0 & 0 & 7 \\ 0 & 0 & 3 & 4 & 0 & 0 \end{pmatrix}$$

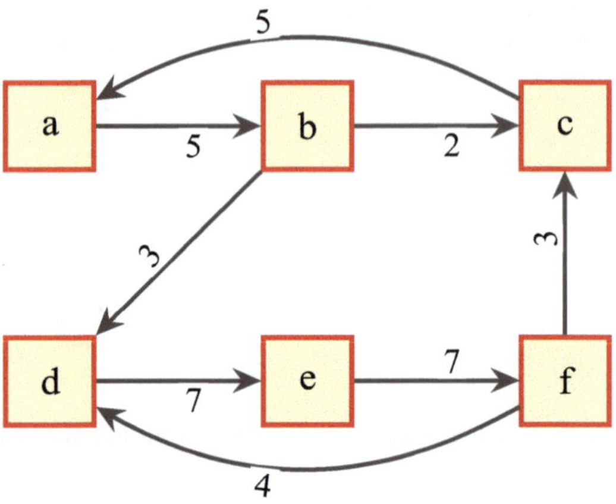

Fig. 3.14 A traffic flow network for the Brindenton district

3.3.2 Fundamental Matrix Operations

The following matrix operations are used throughout this book.

3.3.2.1 Binarized Matrix

A matrix can be *binarized* by converting all of its positive entries to one and all other entries (zero or negative) to zero. This operation is denoted by a beret symbol, $\breve{\mathsf{F}}$, or by the logical operation $(\mathsf{F} > 0)$. The binarized version of a flow matrix is its *adjacency matrix*.

Example 3.14 (Binarizing a Flow Matrix) Consider the flow matrix for the Brindenton district network.

$$\mathsf{F} = \begin{pmatrix} 0 & 5 & 0 & 0 & 0 & 0 \\ 0 & 0 & 2 & 3 & 0 & 0 \\ 5 & 0 & 0 & 0 & 0 & 0 \\ 0 & 0 & 0 & 0 & 7 & 0 \\ 0 & 0 & 0 & 0 & 0 & 7 \\ 0 & 0 & 3 & 4 & 0 & 0 \end{pmatrix}$$

The binarized version of this matrix, $\breve{\mathsf{F}}$, is its adjacency matrix.

$$\breve{\mathsf{F}} = (\mathsf{F} > 0) = \begin{pmatrix} 0 & 1 & 0 & 0 & 0 & 0 \\ 0 & 0 & 1 & 1 & 0 & 0 \\ 1 & 0 & 0 & 0 & 0 & 0 \\ 0 & 0 & 0 & 0 & 1 & 0 \\ 0 & 0 & 0 & 0 & 0 & 1 \\ 0 & 0 & 1 & 1 & 0 & 0 \end{pmatrix}$$

3.3.2.2 Matrix Multiplication

The *multiplication* of two matrices A and B produces a matrix $\mathsf{C} = \mathsf{AB}$. The entry c_{ij} is the vector inner product of the i-th row of A and the j-th column of B, i.e., $c_{ij} = \sum_{k=1}^{m} a_{ik}b_{kj}$. Multiplication is defined only if the number of columns in A equals the number of rows in B.

Example 3.15 (Matrix Multiplication) Given $\mathsf{A} = \begin{pmatrix} 5 & 3 \\ 2 & 1 \end{pmatrix}$ and $\mathsf{B} = \begin{pmatrix} 2 & 4 \\ 7 & 2 \end{pmatrix}$, the product is:

$$\mathsf{C} = \mathsf{AB} = \begin{pmatrix} (5 \times 2) + (3 \times 7) & (5 \times 4) + (3 \times 2) \\ (2 \times 2) + (1 \times 7) & (2 \times 4) + (1 \times 2) \end{pmatrix} = \begin{pmatrix} 31 & 26 \\ 11 & 10 \end{pmatrix}$$

Example 3.16 (Matrix-Vector Multiplication) The product of an $m \times n$ matrix M and an $n \times 1$ column vector $\mathbf{v}$ is an $m \times 1$ column vector. This operation is fundamental for aggregating information. A particularly important use case is multiplication by the one vector, $\mathbf{j}$. The product $\mathsf{M}\mathbf{j}$ results in a column vector where each entry is the sum of the corresponding row of M. In a flow matrix, this gives the *total outflow* from each node.

Let's calculate the total outflow for our example matrix F:

$$\text{Outflow} = \mathsf{F}\mathbf{j} = \begin{pmatrix} 0 & 50 & 20 & 0 \\ 40 & 0 & 0 & 30 \\ 10 & 60 & 0 & 0 \\ 0 & 0 & 70 & 0 \end{pmatrix} \begin{pmatrix} 1 \\ 1 \\ 1 \\ 1 \end{pmatrix} = \begin{pmatrix} 70 \\ 70 \\ 70 \\ 70 \end{pmatrix}$$

This tells us that the total flow leaving node 1 is 70 veh/hr, the total flow leaving node 2 is 70 veh/hr, and so on.

Similarly, the product $\mathsf{M}^T\mathbf{j}$ gives the column sums of M. In a flow matrix, this represents the *total inflow* to each node.

$$\text{Inflow} = \mathsf{F}^T\mathbf{j} = \begin{pmatrix} 0 & 40 & 10 & 0 \\ 50 & 0 & 60 & 0 \\ 20 & 0 & 0 & 70 \\ 0 & 30 & 0 & 0 \end{pmatrix} \begin{pmatrix} 1 \\ 1 \\ 1 \\ 1 \end{pmatrix} = \begin{pmatrix} 50 \\ 110 \\ 90 \\ 30 \end{pmatrix}$$

This calculation reveals that the inflow to node 1 is 50 veh/hr, while its outflow is 70 veh/hr. This network is not in a steady state; flow is not conserved.

3.3.2.3 Hadamard Product (Element-wise Product)

The *Hadamard product*, or element-wise product, is a direct element-by-element multiplication of two matrices of the same size [17]. If $\mathsf{C} = \mathsf{A} \circ \mathsf{B}$, then $c_{ij} = a_{ij}b_{ij}$.

Example 3.17 (Hadamard Product) Using the same matrices as in the previous example:

$$\mathsf{A} \circ \mathsf{B} = \begin{pmatrix} 5 \times 2 & 3 \times 4 \\ 2 \times 7 & 1 \times 2 \end{pmatrix} = \begin{pmatrix} 10 & 12 \\ 14 & 2 \end{pmatrix}$$

3.3.2.4 Hadamard Division

The *Hadamard division* [17] is the element-wise inverse operation. If $\mathsf{C} = \mathsf{A} \oslash \mathsf{B}$, then $c_{ij} = a_{ij}/b_{ij}$, with the special treatment that division by zero results in zero to preserve the matrix pattern.

Example 3.18 (Hadamard Division) Given $\mathsf{A} = \begin{pmatrix} 5 & 7 & 8 \\ 6 & 2 & 0 \end{pmatrix}$ and $\mathsf{B} = \begin{pmatrix} 2 & 0 & 2 \\ 3 & 2 & 5 \end{pmatrix}$, the Hadamard division is:

$$\mathsf{A} \oslash \mathsf{B} = \begin{pmatrix} 5/2 & 7/0 & 8/2 \\ 6/3 & 2/2 & 0/5 \end{pmatrix} = \begin{pmatrix} 2.5 & 0 & 4 \\ 2 & 1 & 0 \end{pmatrix}$$

Example 3.19 (Calculating the Congestion Matrix) A key performance indicator for a transportation network is the level of congestion on each link, often defined as the ratio of flow to capacity. Let's assume we have a flow matrix F and a corresponding capacity matrix C. The *congestion matrix*, G, is calculated using Hadamard division: $\mathsf{G} = \mathsf{F} \oslash \mathsf{C}$.

3.3.2.5 Matrix Partition and Concatenation

A matrix can be partitioned into smaller sub-matrices, or blocks. Conversely, matrices can be joined together, or *concatenated*, either horizontally or vertically, to form a larger *augmented matrix*. For horizontal concatenation, the matrices must have the same number of rows. For vertical concatenation, they must have the same number of columns.

Example 3.20 (Matrix Partitioning) The matrix A can be partitioned in several ways:

$$\mathsf{A} = \begin{pmatrix} 1 & 2 & 7 \\ 4 & 3 & 8 \\ 9 & 8 & 5 \end{pmatrix} = \left(\begin{array}{cc|c} 1 & 2 & 7 \\ \hline 4 & 3 & 8 \\ 9 & 8 & 5 \end{array}\right) = \begin{pmatrix} \mathsf{A}_{11} & \mathsf{A}_{12} \\ \mathsf{A}_{21} & \mathsf{A}_{22} \end{pmatrix}$$

Example 3.21 (Matrix Partitioning and Concatenation) Given $\mathsf{A} = \begin{pmatrix} 3 & 5 & 5 \\ 7 & 4 & 0 \\ 0 & 1 & 8 \end{pmatrix}$ and $\mathsf{B} = \begin{pmatrix} 3 & 7 & 7 \\ 7 & 7 & 7 \\ 4 & 2 & 4 \end{pmatrix}$, the horizontal concatenation is:

$$[\mathsf{A}|\mathsf{B}] = \left(\begin{array}{ccc|ccc} 3 & 5 & 5 & 3 & 7 & 7 \\ 7 & 4 & 0 & 7 & 7 & 7 \\ 0 & 1 & 8 & 4 & 2 & 4 \end{array}\right)$$

3.3.3 The Permutation Matrix

A *permutation* of an n-element set is a linear reordering of the elements of that set.

A permutation can be represented as a function σ that maps each element's original position to its new position.

Example 3.22 (Permutation) For example, $\sigma(1) = 2, \sigma(2) = 3, \sigma(3) = 1$ means the first element moves to the second position, the second to the third, and the third to the first. This is often written as:

$$\sigma = \begin{pmatrix} 1 & 2 & 3 \\ 2 & 3 & 1 \end{pmatrix}$$

In matrix algebra, a permutation is represented by a permutation matrix, which is used to reorder the rows or columns of another matrix.

Definition 3.12 (Permutation Matrix) A *permutation matrix* P is a square (0, 1)-matrix that has exactly one unit entry in each row and each column, and zeros elsewhere. Multiplying a matrix A by P from the left (PA) permutes the rows of A. Multiplying from the right (AP) permutes the columns of A. Every permutation matrix is an *orthogonal matrix*, meaning its inverse is equal to its transpose: $\mathsf{P}^{-1} = \mathsf{P}^T$.

Example 3.23 (Permutation and Matrix Multiplication) Consider the permutation σ that maps (1, 2, 3) to (3, 1, 2). This corresponds to the permutation matrix P.

$$\mathsf{P} = \begin{pmatrix} 0 & 0 & 1 \\ 1 & 0 & 0 \\ 0 & 1 & 0 \end{pmatrix} \quad \text{and} \quad \mathsf{P}^T = \begin{pmatrix} 0 & 1 & 0 \\ 0 & 0 & 1 \\ 1 & 0 & 0 \end{pmatrix}$$

Let $\mathsf{A} = \begin{pmatrix} 1 & 4 & 7 \\ 2 & 5 & 8 \\ 3 & 6 & 9 \end{pmatrix}$. The following operations demonstrate the effect of P and its transpose on A: Multiplying from the left permutes the rows of A:

$$\mathsf{PA} = \begin{pmatrix} 0 & 0 & 1 \\ 1 & 0 & 0 \\ 0 & 1 & 0 \end{pmatrix} \begin{pmatrix} 1 & 4 & 7 \\ 2 & 5 & 8 \\ 3 & 6 & 9 \end{pmatrix} = \begin{pmatrix} 3 & 6 & 9 \\ 1 & 4 & 7 \\ 2 & 5 & 8 \end{pmatrix}$$

Multiplying from the right permutes the columns of A:

$$\mathsf{AP} = \begin{pmatrix} 1 & 4 & 7 \\ 2 & 5 & 8 \\ 3 & 6 & 9 \end{pmatrix} \begin{pmatrix} 0 & 0 & 1 \\ 1 & 0 & 0 \\ 0 & 1 & 0 \end{pmatrix} = \begin{pmatrix} 4 & 7 & 1 \\ 5 & 8 & 2 \\ 6 & 9 & 3 \end{pmatrix}$$

Thus,

- $\mathsf{PA} = \begin{pmatrix} 3 & 6 & 9 \\ 1 & 4 & 7 \\ 2 & 5 & 8 \end{pmatrix}$ (Rows of A are permuted by σ).
- $\mathsf{AP} = \begin{pmatrix} 4 & 7 & 1 \\ 5 & 8 & 2 \\ 6 & 9 & 3 \end{pmatrix}$ (Columns of A are permuted by σ).
- $\mathsf{P}^T\mathsf{A} = \begin{pmatrix} 2 & 5 & 8 \\ 3 & 6 & 9 \\ 1 & 4 & 7 \end{pmatrix}$ (Rows of A are permuted by σ^{-1}).
- $\mathsf{P}^T\mathsf{AP} = \begin{pmatrix} 5 & 8 & 2 \\ 6 & 9 & 3 \\ 4 & 7 & 1 \end{pmatrix}$ (Rows and columns are permuted).

3.3.4 The Generalized Inverse

The standard matrix inverse A^{-1} is defined only for non-singular square matrices. However, many real-world problems involve matrices that are rectangular or singular. To solve linear systems in these cases, we require the concept of a generalized inverse.

Definition 3.13 (Generalized Inverse) For any $m \times n$ matrix A, a *generalized inverse* (or *pseudo-inverse*), denoted A^{-}, is any matrix that satisfies the condition:

$$\mathsf{AA}^{-}\mathsf{A} = \mathsf{A} \tag{3.4}$$

In general, the generalized inverse is not unique.

3.3.4.1 The Moore-Penrose Inverse

To ensure a unique solution with desirable properties, we use a specific type of generalized inverse called the Moore-Penrose inverse, which is unique for any given matrix.

Definition 3.14 (Moore-Penrose Inverse) The *Moore-Penrose inverse* of a matrix A, denoted A^+, is the unique matrix that satisfies the following four conditions:

1. $\mathsf{AA}^+\mathsf{A} = \mathsf{A}$
2. $\mathsf{A}^+\mathsf{AA}^+ = \mathsf{A}^+$
3. $(\mathsf{AA}^+)^T = \mathsf{AA}^+$ (The product is Hermitian)
4. $(\mathsf{A}^+\mathsf{A})^T = \mathsf{A}^+\mathsf{A}$ (The product is Hermitian)

The first two conditions define a reflexive generalized inverse. The four conditions together guarantee uniqueness. The Moore-Penrose inverse is typically computed using Singular Value Decomposition (SVD), as discussed in Sect. 3.7.1.2.

3.3.4.2 Solving Linear Systems

Many fundamental problems in science and engineering can be expressed as a system of linear equations. This is particularly true in network analysis, where we often need to find a set of unknown quantities (like flows) that simultaneously satisfy a set of constraints (like conservation laws).

The primary use of the generalized inverse is to find a meaningful solution to the linear system $\mathsf{Ax} = \mathbf{b}$. In software like Matlab, the backslash operator "ïs used to solve this system, representing the operation $\mathbf{x} = \mathsf{A}^+\mathbf{b}$. The specific formula for A^+ depends on the dimensions of A.

- *Overdetermined Systems* ($m > n$): If there are more equations than unknowns, a unique solution may not exist. The generalized inverse provides the *least-squares solution*, which minimizes the error norm $||\mathsf{Ax} - \mathbf{b}||$. The *left generalized inverse* is used:

$$\mathsf{A}^+ = (\mathsf{A}^T\mathsf{A})^{-1}\mathsf{A}^T \tag{3.5}$$

- *Underdetermined Systems* ($m < n$): If there are fewer equations than unknowns, there are infinitely many solutions. The generalized inverse provides one of these solutions. The *right generalized inverse* is used:

$$\mathsf{A}^+ = \mathsf{A}^T(\mathsf{AA}^T)^{-1} \tag{3.6}$$

Note that numerical software typically uses SVD [14, 28] to compute A^+, which yields the unique minimum-norm solution for underdetermined systems.

Example 3.24 (Overdetermined System, $m > n$**)** Solve the linear system $\mathsf{A}\mathsf{x} = \mathbf{b}$ given:

$$\mathsf{A} = \begin{pmatrix} 1 & 5 \\ 2 & 1 \\ 3 & -3 \end{pmatrix}, \quad \mathbf{b} = \begin{pmatrix} 1 \\ 8 \\ 2 \end{pmatrix}$$

Since the matrix is 3×2, we use the left generalized inverse.

$$\mathsf{A}^T\mathsf{A} = \begin{pmatrix} 1 & 2 & 3 \\ 5 & 1 & -3 \end{pmatrix} \begin{pmatrix} 1 & 5 \\ 2 & 1 \\ 3 & -3 \end{pmatrix} = \begin{pmatrix} 14 & -2 \\ -2 & 35 \end{pmatrix}$$

$$(\mathsf{A}^T\mathsf{A})^{-1} = \frac{1}{(14)(35) - (-2)(-2)} \begin{pmatrix} 35 & 2 \\ 2 & 14 \end{pmatrix} = \frac{1}{486} \begin{pmatrix} 35 & 2 \\ 2 & 14 \end{pmatrix}$$

$$\mathsf{A}^+ = (\mathsf{A}^T\mathsf{A})^{-1}\mathsf{A}^T = \frac{1}{486} \begin{pmatrix} 35 & 2 \\ 2 & 14 \end{pmatrix} \begin{pmatrix} 1 & 2 & 3 \\ 5 & 1 & -3 \end{pmatrix} = \frac{1}{486} \begin{pmatrix} 45 & 72 & 99 \\ 72 & 18 & -36 \end{pmatrix}$$
$$= \frac{1}{54} \begin{pmatrix} 5 & 8 & 11 \\ 8 & 2 & -4 \end{pmatrix}$$

The least-squares solution is:

$$\mathbf{x} = \mathsf{A}^+\mathbf{b} = \frac{1}{54} \begin{pmatrix} 5 & 8 & 11 \\ 8 & 2 & -4 \end{pmatrix} \begin{pmatrix} 1 \\ 8 \\ 2 \end{pmatrix} = \frac{1}{54} \begin{pmatrix} 5 + 64 + 22 \\ 8 + 16 - 8 \end{pmatrix} = \frac{1}{54} \begin{pmatrix} 91 \\ 16 \end{pmatrix}$$

As expected for an inconsistent system, the calculated solution $\mathbf{x}$ is the least-squares solution, which minimizes the error norm $||\mathsf{A}\mathsf{x} - \mathbf{b}||$, but does not provide an exact solution.

Example 3.25 (Underdetermined System, $m < n$**)** Solve the linear system $\mathsf{A}\mathsf{x} = \mathbf{b}$ given:

$$\mathsf{A} = \begin{pmatrix} 1 & 2 & 3 \\ 5 & 1 & -3 \end{pmatrix}, \quad \mathbf{b} = \begin{pmatrix} 1 \\ 8 \end{pmatrix}$$

Since the matrix is 2×3, we use the right generalized inverse.

$$\mathsf{A}\mathsf{A}^T = \begin{pmatrix} 1 & 2 & 3 \\ 5 & 1 & -3 \end{pmatrix} \begin{pmatrix} 1 & 5 \\ 2 & 1 \\ 3 & -3 \end{pmatrix} = \begin{pmatrix} 14 & -2 \\ -2 & 35 \end{pmatrix}$$

$$(\mathsf{A}\mathsf{A}^T)^{-1} = \frac{1}{486} \begin{pmatrix} 35 & 2 \\ 2 & 14 \end{pmatrix}$$

$$\mathsf{A}^+ = \mathsf{A}^T(\mathsf{A}\mathsf{A}^T)^{-1} = \begin{pmatrix} 1 & 5 \\ 2 & 1 \\ 3 & -3 \end{pmatrix} \frac{1}{486} \begin{pmatrix} 35 & 2 \\ 2 & 14 \end{pmatrix} = \frac{1}{486} \begin{pmatrix} 45 & 72 \\ 72 & 18 \\ 99 & -36 \end{pmatrix} = \frac{1}{54} \begin{pmatrix} 5 & 8 \\ 8 & 2 \\ 11 & -4 \end{pmatrix}$$

One possible solution is:

$$\mathbf{x} = \mathsf{A}^+\mathbf{b} = \frac{1}{54} \begin{pmatrix} 5 & 8 \\ 8 & 2 \\ 11 & -4 \end{pmatrix} \begin{pmatrix} 1 \\ 8 \end{pmatrix} = \frac{1}{54} \begin{pmatrix} 69 \\ 24 \\ -21 \end{pmatrix} = \frac{1}{18} \begin{pmatrix} 23 \\ 8 \\ -7 \end{pmatrix}$$

Testing this solution: $\mathsf{A}\mathsf{x} = \frac{1}{18} \begin{pmatrix} 1 & 2 & 3 \\ 5 & 1 & -3 \end{pmatrix} \begin{pmatrix} 23 \\ 8 \\ -7 \end{pmatrix} = \frac{1}{18} \begin{pmatrix} 18 \\ 144 \end{pmatrix} = \begin{pmatrix} 1 \\ 8 \end{pmatrix} = \mathbf{b}$. The solution is correct.

3.4 Algebraic Representations of a Network

A graph or digraph can be represented numerically using matrices. Two of the most common and useful representations are the adjacency matrix and the incidence matrix. These matrices translate the physical topology of a network into an algebraic object that can be manipulated and analyzed.

3.4.1 The Adjacency Matrix

The *adjacency matrix* represents the direct connections between the nodes of a network. Two nodes are considered adjacent if there is a link connecting them. The adjacency matrix is a square matrix where both the rows and columns represent the nodes of the network, typically in a fixed order $(v_1, v_2, \ldots, v_n)$.

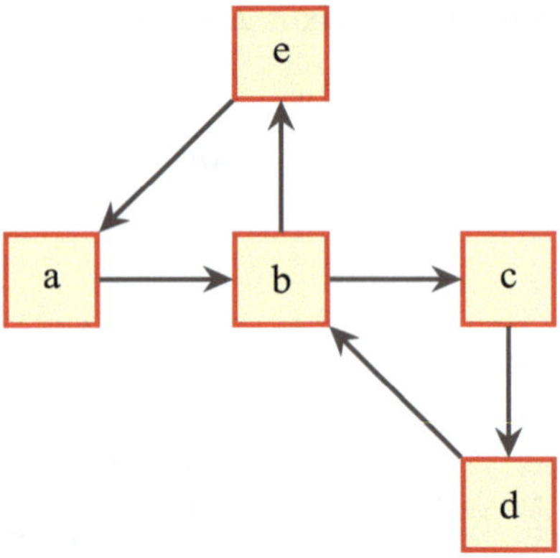

Fig. 3.15 A simple five-node network

Definition 3.15 (Adjacency Matrix) The adjacency matrix $\mathsf{A} = [a_{ij}]$ of a digraph is defined as:

$$a_{ij} = \begin{cases} 1 & \text{if there is a link from node } v_i \text{ to node } v_j \\ 0 & \text{otherwise} \end{cases}$$

For a weighted network, the entry a_{ij} can represent the weight or capacity of the link. As established in Sect. 3.3.2.1, the (0, 1) adjacency matrix is the binarized version of a network's flow or weight matrix. A network can also be represented as an *adjacency list*, which, for each node, lists the nodes it connects to.

Cormen et al. [7] describe adjacency list representations in graph algorithms, essential for efficient storage and traversal in transport network modeling.

Example 3.26 (Representing a Network) Consider the network below. Its adjacency list is $\{a : \{b : 1\}, b : \{c : 1, e : 1\}, c : \{d : 1\}, d : \{b : 1\}, e : \{a : 1\}\}$. Its corresponding adjacency matrix A is (Fig. 3.15):

$$\mathsf{A} = \begin{pmatrix} 0\,1\,0\,0\,0 \\ 0\,0\,1\,0\,1 \\ 0\,0\,0\,1\,0 \\ 0\,1\,0\,0\,0 \\ 1\,0\,0\,0\,0 \end{pmatrix}$$

The adjacency matrix is a complete and unambiguous map of the network's direct, one-step connections. A '1' at position (i, j) indicates that it is possible to travel directly from node i to node j. Let's construct the adjacency matrices for two of our previous examples.

Example 3.27 (Interchange)

- Nodes are ordered (A, B, C).
- Links are (A,B), (A,C), (B,C), (C,A).
- The adjacency matrix $\mathsf{A}_{\text{interchange}}$ is:

$$\mathsf{A}_{\text{interchange}} = \begin{pmatrix} 0 & 1 & 1 \\ 0 & 0 & 1 \\ 1 & 0 & 0 \end{pmatrix} \tag{3.7}$$

Example 3.28 (Weakly Connected System)

- Nodes are ordered $(1, 2, 3, 4)$.
- Links are (1,2), (2,1), (2,3), (3,4), (4,3).
- The adjacency matrix A_{weak} is:

$$\mathsf{A}_{\text{weak}} = \begin{pmatrix} 0 & 1 & 0 & 0 \\ 1 & 0 & 1 & 0 \\ 0 & 0 & 0 & 1 \\ 0 & 0 & 1 & 0 \end{pmatrix} \tag{3.8}$$

The adjacency matrix is more than just a static map. A remarkable theorem reveals that its powers encode information about longer movements through the network.

3.4.1.1 Properties of the Adjacency Matrix

The structure of the adjacency matrix reveals important properties of the network.

- *Symmetry:* For an undirected graph, the adjacency matrix is always symmetric ($\mathsf{A} = \mathsf{A}^T$), since an edge (v_i, v_j) implies an edge (v_j, v_i).
- *Self-Loops:* The diagonal entries a_{ii} are non-zero if and only if the network contains self-loops. For a simple graph, the diagonal is all zeros.
- *Degree:* For an undirected graph, the degree of a vertex v_i is the sum of its corresponding row or column: $\deg(v_i) = \sum_j a_{ij}$. For a digraph, the *in-degree* of node v_j is the sum of column j, $\sum_i a_{ij}$, and the *out-degree* of node v_i is the sum of row i, $\sum_j a_{ij}$.
- *Source and Sink Nodes:* A row of all zeros indicates a *sink node* (zero out-degree). A column of all zeros indicates a *source node* (zero in-degree).
- *Number of Walks:* The entry (i, j) of the matrix A^k gives the total number of distinct walks of length k from node v_i to node v_j.

3.4.2 The Incidence Matrix

The *incidence matrix* provides an alternative representation of a network, describing the relationship between nodes and links instead of between nodes and nodes. The rows of an incidence matrix represent the nodes, and the columns represent the links.

Biggs [3, 4] introduced algebraic graph theory tools including incidence matrices, which are essential for representing node-edge relationships in directed transport networks.

While the adjacency matrix is a common representation of network topology, for the formulation of flow conservation in optimization problems, the *node-link incidence matrix* is often more direct.

Definition 3.16 (Node-Link Incidence Matrix) For a directed graph $G = (N, E)$ with $n = |N|$ nodes and $m = |E|$ links, the node-link incidence matrix, M, is an $n \times m$ matrix where each row corresponds to a node and each column corresponds to a link. The entries of the matrix are defined as:

$$\mathsf{M}_{ik} = \begin{cases} +1 & \text{if node } i \text{ is the origin of link } k \\ -1 & \text{if node } i \text{ is the destination of link } k \\ 0 & \text{otherwise} \end{cases} \tag{3.9}$$

The incidence matrix $\mathsf{B} = [b_{ij}]$ of a digraph is defined as:

$$b_{ij} = \begin{cases} +1 & \text{if link } e_j \text{ leaves node } v_i \\ -1 & \text{if link } e_j \text{ enters node } v_i \\ 0 & \text{otherwise} \end{cases}$$

For an undirected graph, both entries corresponding to an edge are $+1$.

Example 3.29 (Constructing an Incidence Matrix) Consider a simple network with 4 nodes and 5 links as shown below.

- Nodes: $N = \{1, 2, 3, 4\}$
- Links: $E = \{e_1, e_2, e_3, e_4, e_5\} = \{(1, 2), (1, 3), (2, 3), (3, 4), (4, 1)\}$

The corresponding 4×5 node-link incidence matrix M is:

$$\mathsf{M} = \begin{pmatrix} +1 & +1 & 0 & 0 & -1 \\ -1 & 0 & +1 & 0 & 0 \\ 0 & -1 & -1 & +1 & 0 \\ 0 & 0 & 0 & -1 & +1 \end{pmatrix} \tag{3.10}$$

The primary utility of this matrix is in expressing flow conservation. If we define a vector of link flows $\mathbf{f} = (f_1, f_2, f_3, f_4, f_5)^T$, the product $\mathsf{M}\mathbf{f}$ results in a vector where each element is the net flow (outflow—inflow) at the corresponding node.

3.4.2.1 Relationship Between Adjacency and Incidence Matrices

An adjacency matrix can be converted into an incidence matrix. The number of links in the network, which corresponds to the number of columns in the incidence matrix, is equal to the number of non-zero entries in the adjacency matrix.

Example 3.30 (Converting Adjacency to Incidence Matrix) Consider the adjacency matrix from Example 3.26.

$$\mathsf{A} = \begin{pmatrix} 0\,1\,0\,0\,0 \\ 0\,0\,1\,0\,1 \\ 0\,0\,0\,1\,0 \\ 0\,1\,0\,0\,0 \\ 1\,0\,0\,0\,0 \end{pmatrix}$$

This matrix has 5 non-zero entries, so the network has 5 links. The incidence matrix B will have 5 rows (for the nodes) and 5 columns (for the links ab, bc, be, cd, db, ea). For each non-zero entry $a_{ij} = 1$ in A, we create a column in B for the link (i, j). In that column, we place a $+1$ at row i and a -1 at row j. Following this procedure for all non-zero entries yields the incidence matrix:

$$\mathsf{B} = \begin{array}{c} \\ a \\ b \\ c \\ d \\ e \end{array} \begin{array}{c} \begin{array}{cccccc} ab & bc & be & cd & db & ea \end{array} \\ \begin{pmatrix} +1 & 0 & 0 & 0 & 0 & -1 \\ -1 & -1 & -1 & 0 & +1 & 0 \\ 0 & +1 & 0 & -1 & 0 & 0 \\ 0 & 0 & 0 & +1 & -1 & 0 \\ 0 & 0 & +1 & 0 & 0 & +1 \end{pmatrix} \end{array}$$

Conversely, one could construct the adjacency matrix from the incidence matrix. Each column of B defines a single link. For a column representing link (i, j), we would place a 1 in the entry a_{ij} of the adjacency matrix.

3.4.2.2 Properties of the Incidence Matrix

- *Column Sum:* For a digraph, the sum of each column in its incidence matrix is always zero, as the $+1$ and -1 cancel out. For an undirected graph, the sum of each column is always two.

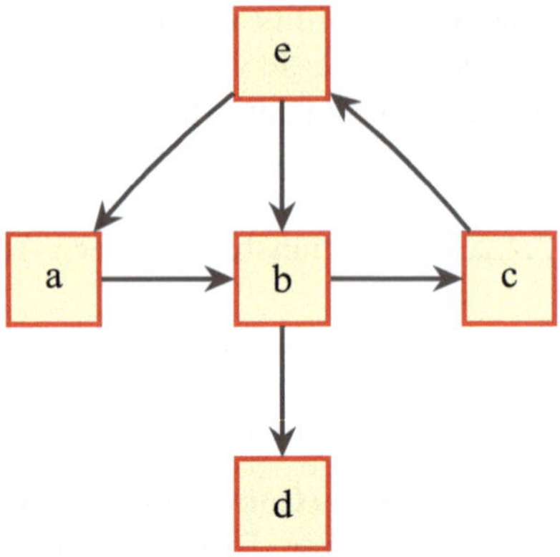

Fig. 3.16 Network for Example 3.31

- *Degree:* The number of positive entries (+ 1) in a row gives the out-degree of the corresponding node. The number of negative entries (− 1) in a row gives the in-degree.

Example 3.31 (Network to Matrix Representation 1) Represent the network shown in Fig. 3.16 using an adjacency list and an adjacency matrix.

The adjacency list is:

$$\{a : \{b : 1\}, b : \{c : 1, d : 1\}, c : \{e : 1\}, d : \{\}, e : \{a : 1, b : 1\}\}.$$

Observe that node d has no out-links, and therefore its list is empty. This indicates that node d is a sink node.

The corresponding adjacency matrix is:

$$\mathsf{A} = \begin{array}{c} \\ a \\ b \\ c \\ d \\ e \end{array} \begin{array}{c} \begin{array}{ccccc} a & b & c & d & e \end{array} \\ \begin{pmatrix} 0 & 1 & 0 & 0 & 0 \\ 0 & 0 & 1 & 1 & 0 \\ 0 & 0 & 0 & 0 & 1 \\ 0 & 0 & 0 & 0 & 0 \\ 1 & 1 & 0 & 0 & 0 \end{pmatrix} \end{array}$$

Notice that the fourth row of the matrix is all zeros, which confirms that node d is a sink node.

Example 3.32 (Network to Matrix Representation 2) Represent the network shown in Fig. 3.17 using an adjacency list and an adjacency matrix.

The adjacency list is:

$$\{a : \{b : 1\}, b : \{c : 1\}, c : \{e : 1\}, d : \{b : 1, c : 1\}, e : \{a : 1, b : 1\}\}.$$

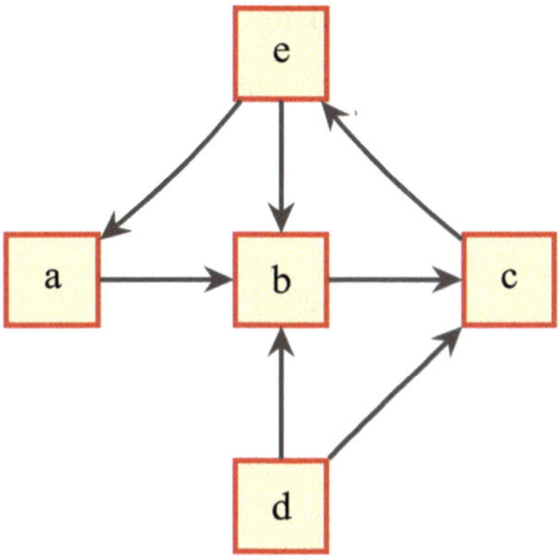

Fig. 3.17 Network for Example 3.32

The corresponding adjacency matrix is:

$$\mathsf{A} = \begin{matrix} & a & b & c & d & e \\ a & 0 & 1 & 0 & 0 & 0 \\ b & 0 & 0 & 1 & 0 & 0 \\ c & 0 & 0 & 0 & 0 & 1 \\ d & 0 & 1 & 1 & 0 & 0 \\ e & 1 & 1 & 0 & 0 & 0 \end{matrix}$$

Notice that the fourth column of the matrix is all zeros, which indicates that node d is a source node.

Example 3.33 (Matrix to Network Representation 1) Given the adjacency matrix $\mathsf{A} = [0, 1, 0, 0, 0; 0, 0, 1, 1, 1; 0, 0, 0, 1, 0; 0, 0, 0, 0, 0; 1, 0, 0, 0, 0]$, derive the adjacency list and draw the corresponding network.

The adjacency list is:

$$\{a : \{b : 1\}, b : \{c : 1, d : 1, e : 1\}, c : \{d : 1\}, d : \{\}, e : \{a : 1\}\}.$$

The network is drawn in Fig. 3.18. Observe that node d is a sink node, corresponding to the all-zero fourth row in the matrix.

Example 3.34 (Matrix to Network Representation 2) Given the adjacency matrix $\mathsf{A} = [0, 1, 0, 0, 0; 0, 0, 1, 0, 1; 0, 0, 0, 1, 0; 0, 0, 1, 0, 0; 1, 0, 0, 0, 0]$, derive the adjacency list and draw the corresponding network.

The adjacency list is:

$$\{a : \{b : 1\}, b : \{c : 1, e : 1\}, c : \{d : 1\}, d : \{c : 1\}, e : \{a : 1\}\}.$$

The network is drawn in Fig. 3.19.

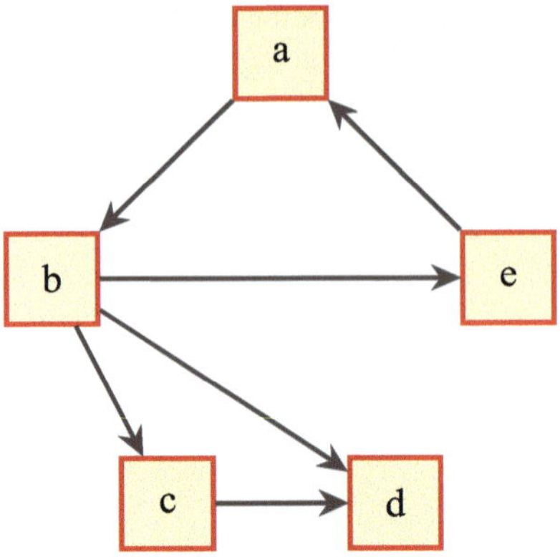

Fig. 3.18 Network for Example 3.33

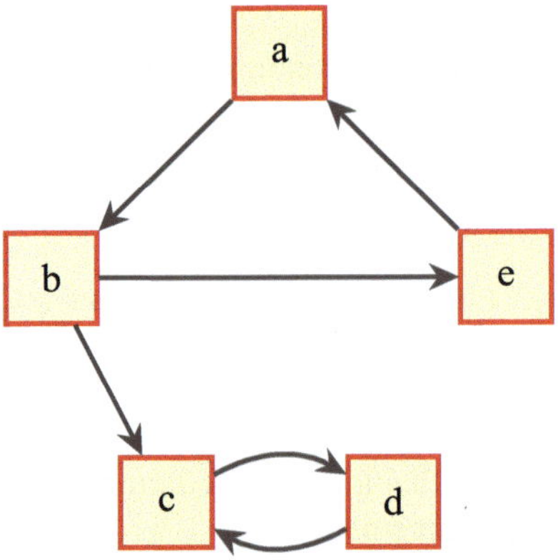

Fig. 3.19 Network for Example 3.34

Example 3.35 (Identifying Components 1) Identify the components in the network shown in Fig. 3.20.

Nodes $\{a, b, e\}$ and $\{b, c, d\}$ form strongly connected components. Because all nodes in the network are mutually reachable through these interconnected components, the entire network is strongly connected.

Example 3.36 (Identifying Components 2) Identify the components in the network shown in Fig. 3.21.

This network has two strongly connected components: $\{a, b, e\}$ and $\{c, d\}$. The component $\{a, b, e\}$ is a *source component* because once an agent leaves it (via the link $b \to c$), it can never return. The component $\{c, d\}$ is a *sink component* because once an agent enters it, it is trapped within that component and can never leave. The network as a whole is weakly connected.

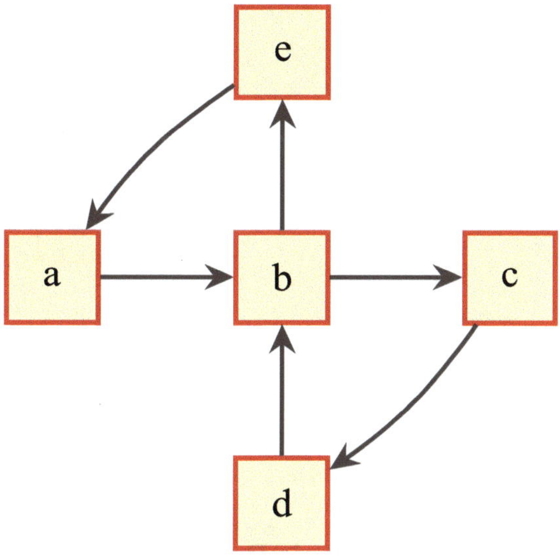

Fig. 3.20 Network for Example 3.35

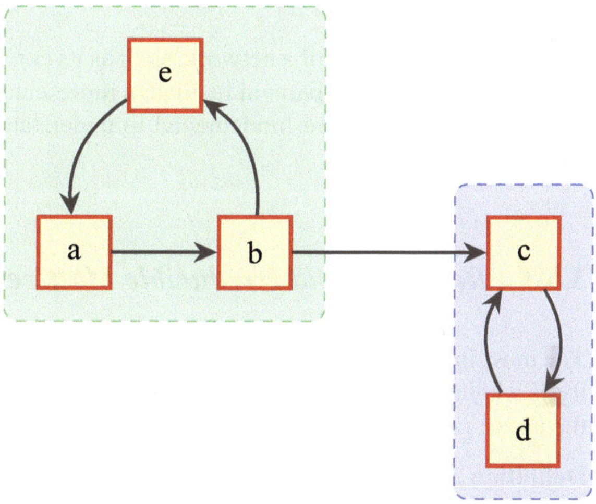

Fig. 3.21 Network for Example 3.36

Example 3.37 (Identifying a Sink Component) Identify the sink component in the network shown in Fig. 3.22.

Nodes $\{b, c, d\}$ form a strongly connected component. Node e is a source node. An agent can leave node e but cannot return. Once an agent enters the component $\{b, c, d\}$, it will be trapped there forever. Thus, the component $\{b, c, d\}$ is a sink component.

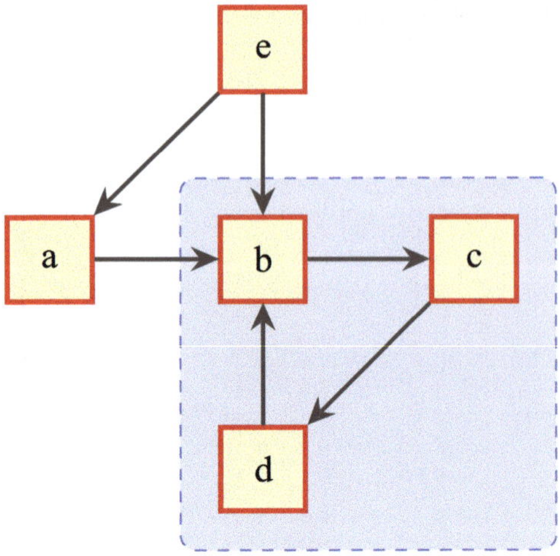

Fig. 3.22 Network for Example 3.37

3.5 Key Algebraic Properties of Network Matrices

The physical properties of a network, such as its connectivity and flow patterns, have direct algebraic counterparts in its matrix representation. This section defines these key properties, which are fundamental to understanding how a network's structure dictates its behavior.

3.5.1 Reducible and Irreducible Matrices

The most fundamental algebraic property, which corresponds to network connectivity, is whether a matrix is reducible or irreducible. This distinction is defined through the use of permutation matrices.

Definition 3.17 (Congruence) A matrix A is *congruent* to a matrix B if there exists a permutation matrix P such that $\mathsf{B} = \mathsf{P}^T\mathsf{AP}$. This operation corresponds to applying the same reordering to both the rows and columns of A, which is equivalent to relabeling the nodes of the corresponding network.

Definition 3.18 (Reducible Matrix) A square matrix A is *reducible* (or decomposable) if it is congruent to a block upper-triangular matrix. That is, there exists a permutation matrix P such that:

$$\mathsf{P}^T\mathsf{AP} = \begin{pmatrix} \mathsf{B}_{11} & \mathsf{B}_{12} \\ 0 & \mathsf{B}_{22} \end{pmatrix} \tag{3.11}$$

where B_{11} and B_{22} are square submatrices and $\mathsf{0}$ is a zero matrix. A matrix that is not reducible is called *irreducible*. The network associated with a reducible matrix is either unconnected or weakly connected. A matrix with a row of all zeros or a column of all zeros is always reducible.

Example 3.38 (Testing for Reducibility 1) Test if the following matrix is reducible.

$$\mathsf{A} = \begin{pmatrix} 1 & 3 & 2 \\ 0 & 7 & 0 \\ 4 & 6 & 5 \end{pmatrix}$$

We can find a permutation matrix P that transforms A into the required block form. Let $\sigma = (1, 3, 2)$, which corresponds to swapping the second and third rows/columns.

$$\mathsf{P} = \begin{pmatrix} 1 & 0 & 0 \\ 0 & 0 & 1 \\ 0 & 1 & 0 \end{pmatrix}$$

Applying the congruence transformation gives:

$$\mathsf{P}^T\mathsf{AP} = \begin{pmatrix} 1 & 0 & 0 \\ 0 & 0 & 1 \\ 0 & 1 & 0 \end{pmatrix} \begin{pmatrix} 1 & 3 & 2 \\ 0 & 7 & 0 \\ 4 & 6 & 5 \end{pmatrix} \begin{pmatrix} 1 & 0 & 0 \\ 0 & 0 & 1 \\ 0 & 1 & 0 \end{pmatrix} = \left(\begin{array}{cc|c} 1 & 2 & 3 \\ 4 & 5 & 6 \\ \hline 0 & 0 & 7 \end{array}\right)$$

Since A is congruent to a block upper-triangular matrix, it is reducible.

Example 3.39 (Testing for Reducibility 2) The following matrices are reducible by inspection.

$$\mathsf{M}_1 = \begin{pmatrix} 1 & 3 & 2 \\ 4 & 5 & 6 \\ 0 & 0 & 0 \end{pmatrix} \quad , \quad \mathsf{M}_2 = \begin{pmatrix} 1 & 0 & 4 \\ 2 & 0 & 5 \\ 3 & 0 & 6 \end{pmatrix}$$

Matrix M_1 is reducible because it has a row of all zeros. Matrix M_2 is reducible because it has a column of all zeros.

The most important connection for our purposes is that between the physical property of strong connectivity and an algebraic property of the adjacency matrix.

An *irreducible matrix* is a square matrix that is not reducible. This means there is no permutation matrix P that can transform it into the block upper-triangular form shown in Eq. (3.11). The network associated with an irreducible matrix is always strongly connected.

Definition 3.19 (Irreducible Matrix) A square matrix M is *irreducible* if it is not reducible. A matrix is *reducible* if there exists a permutation of its rows and columns that can transform it into a block upper-triangular form [2, 19]:

$$\mathsf{P}^T\mathsf{MP} = \begin{pmatrix} \mathsf{X} & \mathsf{Y} \\ 0 & \mathsf{Z} \end{pmatrix}$$

where P is a permutation matrix, X and Z are square matrices, and $\mathsf{0}$ is a matrix of zeros.

The block of zeros in a reducible matrix is the algebraic signature of a network that is not strongly connected. It signifies a partition of the network where flow can pass from one part to another, but not back [2]. This leads to the fundamental theorem linking the two concepts.

Theorem 3.1 (Connectivity and Irreducibility) *A directed graph D is strongly connected if and only if its adjacency matrix* A *is irreducible [2, 13].*

This theorem is the cornerstone of our analytical framework. It allows us to use the powerful theorems of linear algebra that apply to irreducible matrices, most notably the Perron-Frobenius theorem [12, 22], to prove the existence of a unique, stable equilibrium flow distribution in any realistic, fully integrated transportation network [24, 26]. This bridge between the physical world of roads and intersections and the abstract world of irreducible matrices is what makes a scientific, predictive model of network flow possible.

A common method to test for irreducibility involves matrix powers. A square non-negative matrix A is called *primitive* if there exists a positive integer k such that A^k is a positive matrix (i.e., all its entries are greater than zero). For an $n \times n$ adjacency matrix A, a sufficient condition for irreducibility is that the matrix $\mathsf{R} = (\mathsf{A} + \mathsf{I})^{n-1}$ is positive.

Example 3.40 (Testing for Irreducibility) The network in Fig. 3.15 is strongly connected. Determine if its adjacency matrix is irreducible. The adjacency matrix is:

$$\mathsf{A} = \begin{pmatrix} 0 & 1 & 0 & 0 & 0 \\ 0 & 0 & 1 & 0 & 1 \\ 0 & 0 & 0 & 1 & 0 \\ 0 & 1 & 0 & 0 & 0 \\ 1 & 0 & 0 & 0 & 0 \end{pmatrix}$$

To test for irreducibility, we compute $\mathsf{R} = (\mathsf{A} + \mathsf{I})^{n-1}$ for $n = 5$.

$$\mathsf{A} + \mathsf{I} = \begin{pmatrix} 1 & 1 & 0 & 0 & 0 \\ 0 & 1 & 1 & 0 & 1 \\ 0 & 0 & 1 & 1 & 0 \\ 0 & 1 & 0 & 1 & 0 \\ 1 & 0 & 0 & 0 & 1 \end{pmatrix}$$

$$\mathsf{R} = (\mathsf{A} + \mathsf{I})^4 = \begin{pmatrix} 5 & 6 & 6 & 4 & 6 \\ 6 & 9 & 6 & 6 & 6 \\ 1 & 6 & 5 & 5 & 4 \\ 4 & 6 & 6 & 5 & 6 \\ 5 & 6 & 4 & 1 & 5 \end{pmatrix}$$

Since all entries of the resulting matrix R are positive, the matrix A is irreducible. This confirms the algebraic counterpart to the network's property of being strongly connected.

Example 3.41 (Testing a Reducible Matrix) The network in Fig. 3.16 is weakly connected. Its adjacency matrix is:

$$\mathsf{A} = \begin{pmatrix} 0 & 1 & 0 & 0 & 0 \\ 0 & 0 & 1 & 1 & 0 \\ 0 & 0 & 0 & 0 & 1 \\ 0 & 0 & 0 & 0 & 0 \\ 1 & 1 & 0 & 0 & 0 \end{pmatrix}$$

We test for irreducibility with $n = 5$:

$$\mathsf{R} = (\mathsf{A} + \mathsf{I})^4 = \begin{pmatrix} 5 & 5 & 6 & 10 & 6 \\ 6 & 5 & 5 & 11 & 5 \\ 0 & 0 & 1 & 0 & 4 \\ 0 & 0 & 0 & 1 & 0 \\ 5 & 6 & 4 & 5 & 5 \end{pmatrix}$$

Because matrix R contains zero entries, we conclude that matrix A is not irreducible; it is reducible. This aligns with the network being weakly connected. We can also observe that the fourth row of A is all zeros, which immediately implies it is reducible.

3.5.1.1 Properties Related to Irreducibility and Connectivity

The irreducibility of an adjacency matrix is not just an abstract algebraic property; it directly relates to the number of ways one can travel within the corresponding network.

Theorem 3.2 (Walks and Powers of A) *Let A be the adjacency matrix of a digraph. The entry at position (i, j) of the matrix power A^k, denoted $[\mathsf{A}^k]_{ij}$, is equal to the number of distinct walks of length k from node i to node j [4, 13].*

Proof We use induction on the walk length k. *Base Case* ($k = 1$): By definition, $[\mathsf{A}^1]_{ij} = \mathsf{A}_{ij}$, which is 1 if a link (a walk of length 1) exists from i to j, and 0 otherwise. The theorem holds for $k = 1$ [25]. *Inductive Step:* Assume the theorem holds for length $k \geq 1$. A walk of length $k+1$ from i to j consists of a walk of length k from i to some intermediate node m, followed by a single link from m to j. The total number of such walks is the sum over all possible intermediate nodes m of the number of k-length walks from i to m multiplied by the existence of a link from m to j. By the inductive hypothesis, this is $\sum_{m=1}^{n}[\mathsf{A}^k]_{im} \cdot \mathsf{A}_{mj}$. This summation is precisely the definition of the matrix product $(\mathsf{A}^k\mathsf{A})_{ij}$, which equals $[\mathsf{A}^{k+1}]_{ij}$ [17, 25]. Thus, the theorem holds for all $k \geq 1$. □

This property allows us to investigate the overall connectivity of a network by summing the powers of its adjacency matrix.

Example 3.42 (Counting Walks with the Adjacency Matrix) Let's apply this theorem to our interchange example. We already have its adjacency matrix, $\mathsf{A}_{\text{interchange}}$. Let's find the number of walks of length 2 and 3.

- *Walks of length 2:* We compute A^2:

$$\mathsf{A}^2 = \begin{pmatrix} 0 & 1 & 1 \\ 0 & 0 & 1 \\ 1 & 0 & 0 \end{pmatrix} \begin{pmatrix} 0 & 1 & 1 \\ 0 & 0 & 1 \\ 1 & 0 & 0 \end{pmatrix} = \begin{pmatrix} 1 & 0 & 1 \\ 1 & 0 & 0 \\ 0 & 1 & 1 \end{pmatrix} \tag{3.12}$$

  Interpretation: The entry $[\mathsf{A}^2]_{1,1} = 1$ tells us there is exactly one walk of length 2 from node A back to itself. By inspecting the graph, we can see this walk is the cycle $A \to C \to A$ [29]. The entry $[\mathsf{A}^2]_{1,3} = 1$ indicates one walk of length 2 from A to C, which is $A \to B \to C$.
- *Walks of length 3:* We compute $\mathsf{A}^3 = \mathsf{A}^2 \cdot \mathsf{A}$:

$$\mathsf{A}^3 = \begin{pmatrix} 1 & 0 & 1 \\ 1 & 0 & 0 \\ 0 & 1 & 1 \end{pmatrix} \begin{pmatrix} 0 & 1 & 1 \\ 0 & 0 & 1 \\ 1 & 0 & 0 \end{pmatrix} = \begin{pmatrix} 1 & 1 & 1 \\ 0 & 1 & 1 \\ 1 & 0 & 1 \end{pmatrix} \tag{3.13}$$

  Interpretation: The entry $[\mathsf{A}^3]_{2,3} = 1$ tells us there is one walk of length 3 from node B to node C. We can trace this walk on the graph: $B \to C \to A \to C$. This powerful property connects the abstract algebra of matrices directly to the physical process of movement in the network.

Definition 3.20 (Transitive Closure Matrix) A graph is *connected* if there is a path between any two of its nodes. The connectivity can be analyzed using the *transitive closure matrix*, T, defined as:

$$\mathsf{T} = \mathsf{A} + \mathsf{A}^2 + \cdots + \mathsf{A}^{n-1}$$

The binarized version of this matrix, $\check{\mathsf{T}}$, indicates whether a path of any length up to $n-1$ exists between two nodes.

The structure of the transitive closure matrix reveals important connectivity details. A zero in the (i, j) entry of $\check{\mathsf{T}}$ means there is no path from node i to node j. A block of non-zero entries can indicate a connected component.

Example 3.43 (Transitive Closure of an Unconnected Network) Consider the following adjacency matrix A for an unconnected digraph with $n = 4$.

$$\mathsf{A} = \begin{pmatrix} 0 & 1 & 0 & 0 \\ 0 & 0 & 1 & 0 \\ 1 & 1 & 0 & 0 \\ 0 & 0 & 0 & 0 \end{pmatrix}$$

The transitive closure matrix T is:

$$\mathsf{T} = \mathsf{A} + \mathsf{A}^2 + \mathsf{A}^3 = \begin{pmatrix} 0 & 1 & 0 & 0 \\ 0 & 0 & 1 & 0 \\ 1 & 1 & 0 & 0 \\ 0 & 0 & 0 & 0 \end{pmatrix} + \begin{pmatrix} 0 & 0 & 1 & 0 \\ 1 & 1 & 0 & 0 \\ 0 & 1 & 1 & 0 \\ 0 & 0 & 0 & 0 \end{pmatrix} + \begin{pmatrix} 1 & 1 & 0 & 0 \\ 0 & 1 & 1 & 0 \\ 1 & 1 & 1 & 0 \\ 0 & 0 & 0 & 0 \end{pmatrix} = \begin{pmatrix} 1 & 2 & 1 & 0 \\ 1 & 2 & 2 & 0 \\ 2 & 3 & 2 & 0 \\ 0 & 0 & 0 & 0 \end{pmatrix}$$

The binarized matrix is:

$$\check{\mathsf{T}} = \begin{pmatrix} 1 & 1 & 1 & 0 \\ 1 & 1 & 1 & 0 \\ 1 & 1 & 1 & 0 \\ 0 & 0 & 0 & 0 \end{pmatrix}$$

The zero row and column for node d indicate it is an isolated component. The block of ones for nodes $\{a, b, c\}$ indicates they form a separate connected component.

A closely related concept is the reachability matrix, which includes paths of length zero (i.e., a node can reach itself).

Definition 3.21 (Reachability Matrix) Node v_j is *reachable* from node v_i if there is a path from v_i to v_j. The *reachability matrix* $\mathsf{R} = [r_{ij}]$ is the binarized version of $(\mathsf{I} + \mathsf{T})$.

$$r_{ij} = \begin{cases} 1 & \text{if node } v_j \text{ is reachable from node } v_i \\ 0 & \text{otherwise} \end{cases}$$

An all-ones reachability matrix indicates a strongly connected network, which is the definition of an irreducible matrix.

3.6 Equivalence of Network and Matrix Properties

The preceding sections have established the distinct languages of graph theory and matrix algebra. We now formalize the direct correspondence between these two domains. A network's most important physical characteristic, its connectivity, is perfectly mirrored by an algebraic property of its adjacency matrix.

3.6.1 Strong Connectivity and Matrix Irreducibility

A transport network is functional only if it is possible to travel between any two points. This property is called strong connectivity. In linear algebra, this concept is captured by the irreducibility of the network's adjacency matrix.

Theorem 3.3 (Irreducibility and Strong Connectivity) *A directed graph* $\mathcal{D}$ *is* strongly connected *if and only if its adjacency matrix* A *is* irreducible.

Proof First, assume the adjacency matrix A is reducible. By definition, there exists a permutation matrix P that transforms A into a block upper-triangular form:

$$\mathsf{P}^T\mathsf{A}\mathsf{P} = \begin{pmatrix} \mathsf{B}_{11} & \mathsf{B}_{12} \\ 0 & \mathsf{B}_{22} \end{pmatrix}$$

The permutation P partitions the set of nodes $\mathcal{V}$ into two disjoint subsets, $\mathcal{V}_1$ and $\mathcal{V}_2$, corresponding to the blocks B_{11} and B_{22}, respectively. The zero block 0 in the lower-left position signifies that there are no links originating from any node in $\mathcal{V}_2$ and terminating at any node in $\mathcal{V}_1$. Therefore, it is impossible to find a path from a node in $\mathcal{V}_2$ to a node in $\mathcal{V}_1$. This violates the definition of a strongly connected network. Thus, if A is reducible, the network is not strongly connected.

Conversely, assume the network $\mathcal{D}$ is not strongly connected. This means the set of nodes $\mathcal{V}$ can be partitioned into two non-empty sets, $\mathcal{V}_1$ and $\mathcal{V}_2$, such that there is no path from any node in $\mathcal{V}_2$ to any node in $\mathcal{V}_1$. If we order the nodes such that all nodes from $\mathcal{V}_1$ come first, followed by all nodes from $\mathcal{V}_2$, the adjacency matrix A will take the block upper-triangular form. This reordering is equivalent to applying a permutation matrix P. The resulting matrix has the form of Eq. (3.11), which by definition is reducible. Thus, if the network is not strongly connected, its adjacency matrix A is reducible.

Combining these two implications, we conclude that a network is strongly connected if and only if its adjacency matrix is irreducible. □

Example 3.44 (Connectivity and Reducibility) Consider the networks in Fig. 3.23.

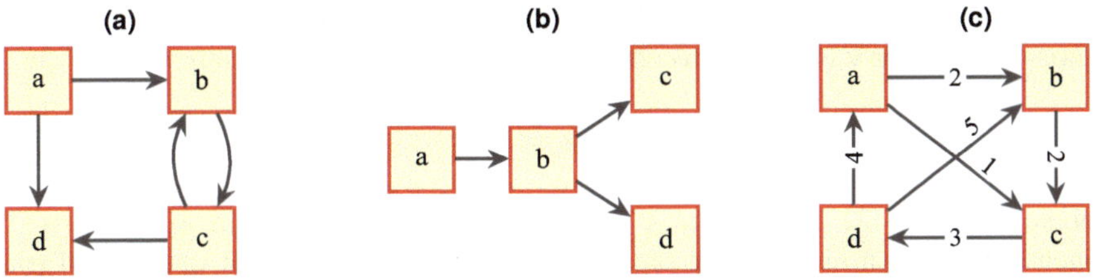

Fig. 3.23 A set of three networks illustrating different types of connectivity

- *Network (a)* is weakly connected. Node a is a source and node d is a sink. It is not possible to travel from d to a. Therefore, its adjacency matrix is reducible.
- *Network (b)* is weakly connected. It has one source node and two sink nodes. It is not strongly connected, so its adjacency matrix is reducible.
- *Network (c)* is strongly connected. From any node, it is possible to reach any other node. Therefore, its adjacency matrix is irreducible. This network is also an example of an Ideal Flow Network (IFN), as the flow is conserved at each node.

3.7 Equilibrium Analysis of Networks

Having established the equivalence between a network's physical connectivity and the algebraic irreducibility of its matrix representation, we can now analyze its long-term behavior. In any strongly connected network, a unique and stable equilibrium flow distribution can exist. The mathematical tools of eigenvalue and eigenvector analysis, culminating in the powerful Perron-Frobenius theorem, provide the foundation for identifying and understanding this equilibrium state.

3.7.1 Eigenvalues and Eigenvectors

The concepts of eigenvalues and eigenvectors are central to linear algebra and reveal the intrinsic properties of a matrix transformation. In network analysis, they allow us to find the stable states of a system.

Intuitively, the core idea of Eigenvector is vectors that don't change direction. For any given square matrix A, there exist special vectors called *eigenvectors*. When the transformation A is applied to an eigenvector $\mathbf{v}$, the resulting vector points in the *exact same direction* as the original vector $\mathbf{v}$. The transformation only scales the vector by a factor called the *eigenvalue*, λ.

Definition 3.22 (Eigenvalue and Eigenvector) For a square matrix A of size $n \times n$, a non-zero vector $\mathbf{v}$ is an *eigenvector* of A if it satisfies the equation:

$$\mathsf{A}\mathbf{v} = \lambda \mathbf{v} \tag{3.14}$$

for some scalar λ. The scalar λ is called the *eigenvalue* corresponding to the eigenvector $\mathbf{v}$. An eigenvector represents a direction that is left unchanged (only scaled) by the transformation of matrix A. The eigenvalue λ is the factor by which the eigenvector is scaled. To find the eigenvalues, we solve the characteristic equation $\det(\mathsf{A} - \lambda \mathsf{I}) = 0$.

Example 3.45 (Calculating Eigenvalues and Eigenvectors) Find the eigenvalues and corresponding eigenvectors for the matrix:

$$\mathsf{A} = \begin{pmatrix} 1 & 2 \\ 4 & 3 \end{pmatrix}$$

First, we solve the characteristic equation:

$$\begin{aligned} \det(\mathsf{A} - \lambda \mathsf{I}) &= \det \begin{pmatrix} 1-\lambda & 2 \\ 4 & 3-\lambda \end{pmatrix} \\ &= (1-\lambda)(3-\lambda) - (2)(4) \\ &= 3 - \lambda - 3\lambda + \lambda^2 - 8 \\ &= \lambda^2 - 4\lambda - 5 = 0 \end{aligned}$$

Factoring the quadratic equation gives $(\lambda - 5)(\lambda + 1) = 0$. The eigenvalues are $\lambda_1 = 5$ and $\lambda_2 = -1$.

To find the eigenvector for $\lambda_1 = 5$, we solve $(\mathsf{A} - 5\mathsf{I})\mathbf{v}_1 = \mathbf{0}$:

$$\begin{pmatrix} 1-5 & 2 \\ 4 & 3-5 \end{pmatrix} \begin{pmatrix} v_{11} \\ v_{12} \end{pmatrix} = \begin{pmatrix} -4 & 2 \\ 4 & -2 \end{pmatrix} \begin{pmatrix} v_{11} \\ v_{12} \end{pmatrix} = \begin{pmatrix} 0 \\ 0 \end{pmatrix}$$

This gives the equation $-4v_{11} + 2v_{12} = 0$, which simplifies to $v_{12} = 2v_{11}$. The eigenvector is any vector of the form $k \begin{pmatrix} 1 \\ 2 \end{pmatrix}$. We can choose $\mathbf{v}_1 = \begin{pmatrix} 1 \\ 2 \end{pmatrix}$.

To find the eigenvector for $\lambda_2 = -1$, we solve $(\mathsf{A} - (-1)\mathsf{I})\mathbf{v}_2 = \mathbf{0}$:

$$\begin{pmatrix} 1-(-1) & 2 \\ 4 & 3-(-1) \end{pmatrix} \begin{pmatrix} v_{21} \\ v_{22} \end{pmatrix} = \begin{pmatrix} 2 & 2 \\ 4 & 4 \end{pmatrix} \begin{pmatrix} v_{21} \\ v_{22} \end{pmatrix} = \begin{pmatrix} 0 \\ 0 \end{pmatrix}$$

This gives the equation $2v_{21} + 2v_{22} = 0$, which simplifies to $v_{21} = -v_{22}$. The eigenvector is any vector of the form $k \begin{pmatrix} 1 \\ -1 \end{pmatrix}$. We can choose $\mathbf{v}_2 = \begin{pmatrix} 1 \\ -1 \end{pmatrix}$.

3.7.1.1 Properties of Eigenvalues

- The sum of the eigenvalues of a matrix is equal to its trace (the sum of its diagonal elements): $\sum_{i=1}^{n} \lambda_i = \text{tr}(\mathsf{A})$.
- The product of the eigenvalues of a matrix is equal to its determinant: $\prod_{i=1}^{n} \lambda_i = \det(\mathsf{A})$.
- If λ is an eigenvalue of A, then λ^k is an eigenvalue of A^k for any positive integer k.
- The eigenvalues of a symmetric matrix are always real numbers.

This concept of Eigenvalues and Eigenvectors is directly applicable to traffic networks modeled as Markov chains. Let the stochastic matrix S represent the transformation during one time step. The stationary distribution is the distribution that does not change over time. In the language of eigenvectors, this means the scaling factor must be 1. The stationary distribution $\boldsymbol{\pi}^T$ is the specific left eigenvector of the stochastic matrix S that corresponds to an eigenvalue of $\lambda = 1$:

$$\boldsymbol{\pi}^T \mathsf{S} = 1 \cdot \boldsymbol{\pi}^T = \boldsymbol{\pi}^T \tag{3.15}$$

When the network's flow distribution reaches this state, it becomes stable. A critical question is whether such an eigenvector with $\lambda = 1$ always exists and is unique and physically meaningful. The answer is provided by the *Perron-Frobenius Theorem* [12, 22] 3.7.2. Let A be an irreducible, non-negative square matrix. Then:

1. There exists a unique largest real eigenvalue λ_{max}, which is positive. This is called the *Perron root*.
2. The corresponding eigenvector (both left and right) is real, has strictly positive components, and is unique up to scaling. This is the *Perron vector*.

When this Perron-Frobenius theorem [12] is applied to an irreducible stochastic matrix S [24] (which corresponds to a strongly connected network), the implications are profound: the Perron root is guaranteed to be exactly 1, and therefore a unique, positive, and physically meaningful stationary distribution $\boldsymbol{\pi}^T$ is guaranteed to exist. This powerful result is the mathematical bedrock that guarantees the existence of a unique, stable equilibrium in any fully integrated transportation network.

3.7.1.2 Singular Value Decomposition (SVD)

While eigenvalue decomposition is defined only for square matrices, a related and more general factorization called *Singular Value Decomposition* [14] (SVD) exists for any rectangular matrix. SVD is a fundamental tool for computation, particularly for finding the generalized inverse of a matrix.

Definition 3.23 (Singular Value Decomposition [14, 28]) For any real $m \times n$ matrix A, the SVD is a factorization of the form:

$$\mathsf{A} = \mathsf{U}\Sigma\mathsf{V}^T \tag{3.16}$$

where:

- U is an $m \times m$ orthogonal matrix. Its columns are the eigenvectors of AA^T and are called the *left singular vectors*.
- V is an $n \times n$ orthogonal matrix. Its columns are the eigenvectors of $\mathsf{A}^T\mathsf{A}$ and are called the *right singular vectors*.
- Σ is an $m \times n$ diagonal matrix. Its diagonal entries σ_i are the *singular values* of A.

The singular values are the square roots of the non-zero eigenvalues of both AA^T and $\mathsf{A}^T\mathsf{A}$. They are arranged in descending order, $\sigma_1 \geq \sigma_2 \geq \cdots \geq 0$.

The deep connection between SVD and eigenvalue decomposition is evident: the singular vectors of A are the eigenvectors of the related symmetric matrices AA^T and $\mathsf{A}^T\mathsf{A}$. This decomposition is particularly useful for computing the Moore-Penrose generalized inverse, A^+, which is essential for solving linear systems that may not have a unique solution.

Example 3.46 (SVD and the Moore-Penrose Inverse) The Moore-Penrose inverse A^+ can be computed directly from the SVD of A:

$$\mathsf{A}^+ = \mathsf{V}\Sigma^+\mathsf{U}^T$$

where Σ^+ is formed by taking the reciprocal of each non-zero singular value in Σ and then transposing the resulting matrix. For instance, consider the linear equation $\mathsf{Ax} = \mathbf{b}$ where A is a 2×3 matrix:

$$\mathsf{A} = \begin{pmatrix} 1 & 2 & 3 \\ 5 & 1 & -3 \end{pmatrix}, \quad \mathbf{b} = \begin{pmatrix} 1 \\ 8 \end{pmatrix}$$

This system has infinitely many solutions. The SVD provides a robust method to find the unique solution with the minimum Euclidean norm, which is given by $\mathbf{x} = \mathsf{A}^+\mathbf{b}$. Numerical software like Matlab uses SVD to compute this solution. For this system, the minimum norm solution is $\mathbf{x} = (1.5, 0, -1/6)^T$. We can verify that this is a valid solution:

$$\mathsf{Ax} = \begin{pmatrix} 1 & 2 & 3 \\ 5 & 1 & -3 \end{pmatrix} \begin{pmatrix} 1.5 \\ 0 \\ -1/6 \end{pmatrix} = \begin{pmatrix} 1.5 + 0 - 0.5 \\ 7.5 + 0 + 0.5 \end{pmatrix} = \begin{pmatrix} 1 \\ 8 \end{pmatrix} = \mathbf{b}$$

3.7.2 The Perron-Frobenius Theorem

The Perron-Frobenius theorem is a cornerstone of matrix theory and network analysis. It provides the mathematical guarantee for the existence of a unique, stable equilibrium in any system represented by an irreducible non-negative matrix, such as a strongly connected network. The theorem was originally developed by Oskar Perron for positive matrices and was later generalized by Georg Frobenius to irreducible non-negative matrices [12, 22].

Theorem 3.4 (Perron-Frobenius for Irreducible Matrices) *Let* A *be an* $n \times n$ *irreducible non-negative matrix. Then the following statements are true:*

1. *The* spectral radius $\rho(\mathsf{A}) = \max_i |\lambda_i|$ *is a simple, positive eigenvalue of* A. *This dominant eigenvalue is called the* Perron root.
2. *The corresponding right eigenvector* $\mathbf{r}$ *and left eigenvector* $\mathbf{l}^T$ *are unique up to a positive scaling factor.*
3. *These unique eigenvectors, called the* Perron vectors, *can be chosen to have strictly positive entries.*

The power of this theorem lies in its guarantee of a unique, positive equilibrium state for a wide class of systems.

Corollary 3.1 (Uniqueness and Positivity) *For any irreducible non-negative matrix, there exists a unique largest real eigenvalue which is positive and simple. Its corresponding eigenvector can be chosen so that all of its components are positive. No other eigenvalue has a corresponding non-negative eigenvector.*

Corollary 3.2 (Application to Stochastic Matrices) *If* S *is an irreducible row-stochastic matrix, its Perron root is exactly 1. The corresponding right Perron vector is the one vector,* $\mathbf{j}$.

Example 3.47 (Perron-Frobenius Verification) Consider the irreducible matrix $\mathsf{A} = \begin{pmatrix} 0 & 1 \\ 1/2 & 1/2 \end{pmatrix}$. We find its eigenvalues by solving the characteristic equation $\det(\mathsf{A} - \lambda \mathsf{I}) = 0$:

$$\begin{aligned} (0-\lambda)(1/2-\lambda) - (1)(1/2) &= 0 \\ -\frac{1}{2}\lambda + \lambda^2 - \frac{1}{2} &= 0 \\ 2\lambda^2 - \lambda - 1 &= 0 \\ (2\lambda + 1)(\lambda - 1) &= 0 \end{aligned}$$

The eigenvalues are $\lambda_1 = 1$ and $\lambda_2 = -1/2$. The spectral radius is $\rho(\mathsf{A}) = \max(|1|, |-1/2|) = 1$. This is the Perron root, and it is positive and simple, as the theorem predicts.

The eigenvector for $\lambda_1 = 1$ is found by solving $(\mathsf{A} - 1\mathsf{I})\mathbf{v} = \mathbf{0}$:

$$\begin{pmatrix} -1 & 1 \\ 1/2 & -1/2 \end{pmatrix} \begin{pmatrix} v_1 \\ v_2 \end{pmatrix} = \begin{pmatrix} 0 \\ 0 \end{pmatrix}$$

This yields $-v_1 + v_2 = 0$, or $v_1 = v_2$. The Perron vector can be chosen as $\mathbf{v} = (1, 1)^T$, which has strictly positive entries, confirming the theorem.

Example 3.48 (Three-Node Cycle Network) Consider a simple three-node cycle, which is a minimally strong network. Its adjacency matrix is:

$$\mathsf{A} = \begin{pmatrix} 0 & 1 & 0 \\ 0 & 0 & 1 \\ 1 & 0 & 0 \end{pmatrix}$$

Since the network is strongly connected, the matrix A is irreducible. By inspection, we can see that the row sums are all equal to 1. Therefore, this is a row-stochastic matrix. According to Corollary 3.2, its Perron root must be $\lambda = 1$. The corresponding right Perron vector is the one vector, $\mathbf{v} = (1, 1, 1)^T$. We can verify this:

$$\mathsf{A}\mathbf{v} = \begin{pmatrix} 0 & 1 & 0 \\ 0 & 0 & 1 \\ 1 & 0 & 0 \end{pmatrix} \begin{pmatrix} 1 \\ 1 \\ 1 \end{pmatrix} = \begin{pmatrix} 1 \\ 1 \\ 1 \end{pmatrix} = 1 \cdot \mathbf{v}$$

The results are consistent with the Perron-Frobenius theorem.

3.7.3 Stochastic Matrices and the Stationary Distribution

The Perron-Frobenius theorem finds its most direct application in network analysis through the use of stochastic matrices. These matrices represent the transition probabilities of a random walk on a network, which is equivalent to a Markov chain.

Definition 3.24 (Stochastic Matrix)

- A non-negative square matrix S is called a *row-stochastic matrix* if all of its row sums are equal to one. That is, $\mathsf{S}\mathbf{j} = \mathbf{j}$.
- A non-negative square matrix T is called a *column-stochastic matrix* if all of its column sums are equal to one. That is, $\mathsf{T}^T\mathbf{j} = \mathbf{j}$.
- A non-negative square matrix Ω is called a *doubly stochastic matrix* if all of its row sums and all of its column sums are equal to one.

Any flow matrix F can be converted into a row-stochastic matrix S by dividing each element by its corresponding row sum: $s_{ij} = f_{ij}/\sum_k f_{ik}$. The entry s_{ij} represents the probability of moving from node i to node j in a single step.

For a system represented by an irreducible row-stochastic matrix, the Perron-Frobenius theorem guarantees a unique equilibrium state. This state is described by the stationary distribution vector.

Example 3.49 (Doubly Stochastic Matrix) Consider the following matrix Ω.

$$\Omega = \begin{pmatrix} 0.5 & 0.25 & 0.25 \\ 0.25 & 0.5 & 0.25 \\ 0.25 & 0.25 & 0.5 \end{pmatrix}$$

The sum of each row is $0.5 + 0.25 + 0.25 = 1$. The sum of each column is also 1. Therefore, Ω is a doubly stochastic matrix. For any irreducible doubly stochastic matrix, the stationary distribution is uniform: $\boldsymbol{\pi}^T = (1/n, 1/n, \ldots, 1/n)$.

Theorem 3.5 (Stationary Distribution) *If* S *is an* $n \times n$ *irreducible row-stochastic matrix, then there exists a unique row vector* $\boldsymbol{\pi}^T$ *such that:*

$$\boldsymbol{\pi}^T \mathsf{S} = \boldsymbol{\pi}^T \quad \textit{and} \quad \sum_{i=1}^{n} \pi_i = 1 \tag{3.17}$$

The vector $\boldsymbol{\pi}^T$ *is the* stationary distribution vector. *It is the unique left Perron vector of* S *corresponding to the Perron root* $\lambda = 1$, *normalized so that its elements sum to one. Each element* π_i *represents the long-term proportion of time spent at node i in a random walk, or equivalently, the proportion of total flow passing through node i at equilibrium.*

Example 3.50 (Calculating the Stationary Distribution) Consider the irreducible adjacency matrix $\mathsf{A} = \begin{pmatrix} 0 & 1 \\ 1 & 1 \end{pmatrix}$. The row sums are 1 and 2. The corresponding row-stochastic matrix S is:

$$\mathsf{S} = \begin{pmatrix} 0/1 & 1/1 \\ 1/2 & 1/2 \end{pmatrix} = \begin{pmatrix} 0 & 1 \\ 1/2 & 1/2 \end{pmatrix}$$

We need to find $\boldsymbol{\pi}^T = (\pi_1, \pi_2)$ such that $\boldsymbol{\pi}^T \mathsf{S} = \boldsymbol{\pi}^T$. This gives the system of equations:

$$\frac{1}{2}\pi_2 = \pi_1$$

$$\pi_1 + \frac{1}{2}\pi_2 = \pi_2$$

Both equations simplify to $\pi_2 = 2\pi_1$. Using the normalization condition $\pi_1 + \pi_2 = 1$, we have:

$$\pi_1 + 2\pi_1 = 1 \implies 3\pi_1 = 1 \implies \pi_1 = 1/3$$

Therefore, $\pi_2 = 2/3$. The stationary distribution is $\boldsymbol{\pi}^T = (1/3, 2/3)$. This means that in the long run, agents will spend twice as much time at node 2 as at node 1.

Example 3.51 (Stationary Distribution for a Three-Node Network) Consider the network represented by the stochastic matrix:

$$\mathsf{S} = \begin{pmatrix} 0.5 & 0.5 & 0 \\ 0 & 0.5 & 0.5 \\ 1 & 0 & 0 \end{pmatrix}$$

We solve for $\boldsymbol{\pi}^T = (\pi_1, \pi_2, \pi_3)$ in the system $\boldsymbol{\pi}^T\mathsf{S} = \boldsymbol{\pi}^T$:

$$\begin{aligned} 0.5\pi_1 + \pi_3 &= \pi_1 \\ 0.5\pi_1 + 0.5\pi_2 &= \pi_2 \\ 0.5\pi_2 &= \pi_3 \end{aligned}$$

From the second equation, $0.5\pi_1 = 0.5\pi_2 \implies \pi_1 = \pi_2$. From the third equation, $\pi_3 = 0.5\pi_1$. Using the normalization condition $\pi_1 + \pi_2 + \pi_3 = 1$:

$$\pi_1 + \pi_1 + 0.5\pi_1 = 1 \implies 2.5\pi_1 = 1 \implies \pi_1 = 1/2.5 = 0.4$$

Thus, $\pi_2 = 0.4$ and $\pi_3 = 0.5 \times 0.4 = 0.2$. The stationary distribution is $\boldsymbol{\pi}^T = (0.4, 0.4, 0.2)$.

Example 3.52 (Stationary Distribution of a Four-Node Network) Consider the strongly connected network from Fig. 3.23c. Let its flow matrix F be:

$$\mathsf{F} = \begin{pmatrix} 0 & 2 & 1 & 0 \\ 0 & 0 & 0 & 2 \\ 0 & 0 & 0 & 3 \\ 4 & 0 & 0 & 0 \end{pmatrix}$$

The row sums are $(3, 2, 3, 4)^T$. The corresponding row-stochastic matrix S is:

$$\mathsf{S} = \begin{pmatrix} 0 & 2/3 & 1/3 & 0 \\ 0 & 0 & 0 & 1 \\ 0 & 0 & 0 & 1 \\ 1 & 0 & 0 & 0 \end{pmatrix}$$

We solve the system $\boldsymbol{\pi}^T \mathsf{S} = \boldsymbol{\pi}^T$ with the condition $\sum \pi_i = 1$.

$$\pi_4 = \pi_1$$
$$(2/3)\pi_1 = \pi_2$$
$$(1/3)\pi_1 = \pi_3$$
$$\pi_2 + \pi_3 = \pi_4$$

Substituting the first three equations into the fourth gives $(2/3)\pi_1 + (1/3)\pi_1 = \pi_1$, which simplifies to $\pi_1 = \pi_1$. This confirms consistency. Now, using the normalization condition:

$$\pi_1 + \pi_2 + \pi_3 + \pi_4 = 1$$
$$\pi_1 + (2/3)\pi_1 + (1/3)\pi_1 + \pi_1 = 1$$
$$3\pi_1 = 1 \implies \pi_1 = 1/3$$

From this, we find $\pi_2 = 2/9$, $\pi_3 = 1/9$, and $\pi_4 = 1/3$. The stationary distribution is $\boldsymbol{\pi}^T = (1/3, 2/9, 1/9, 1/3)$, or approximately $(0.333, 0.222, 0.111, 0.333)$.

Example 3.53 (Stationary Distribution of a Five-Node Network) Consider the irreducible adjacency matrix from Example 3.40.

$$\mathsf{A} = \begin{pmatrix} 0 & 1 & 0 & 0 & 0 \\ 0 & 0 & 1 & 0 & 1 \\ 0 & 0 & 0 & 1 & 0 \\ 0 & 1 & 0 & 0 & 0 \\ 1 & 0 & 0 & 0 & 0 \end{pmatrix}$$

The row sums are $(1, 2, 1, 1, 1)^T$. The row-stochastic matrix S is:

$$\mathsf{S} = \begin{pmatrix} 0 & 1 & 0 & 0 & 0 \\ 0 & 0 & 1/2 & 0 & 1/2 \\ 0 & 0 & 0 & 1 & 0 \\ 0 & 1 & 0 & 0 & 0 \\ 1 & 0 & 0 & 0 & 0 \end{pmatrix}$$

Solving the system $\boldsymbol{\pi}^T \mathsf{S} = \boldsymbol{\pi}^T$ and $\sum \pi_i = 1$ yields the equations:

$$\pi_5 = \pi_1$$
$$\pi_1 + \pi_4 = \pi_2$$

$$(1/2)\pi_2 = \pi_3$$
$$\pi_3 = \pi_4$$
$$(1/2)\pi_2 = \pi_5$$

From these equations, we can derive the relationships: $\pi_1 = \pi_5$, $\pi_3 = \pi_4$, and $\pi_3 = (1/2)\pi_2$. Substituting into the second equation: $\pi_1 + (1/2)\pi_2 = \pi_2 \implies \pi_1 = (1/2)\pi_2 \implies \pi_2 = 2\pi_1$. Now we have all nodes in terms of π_1: $\pi_2 = 2\pi_1$, $\pi_3 = \pi_1$, $\pi_4 = \pi_1$, $\pi_5 = \pi_1$. Using the normalization condition:

$$\pi_1 + 2\pi_1 + \pi_1 + \pi_1 + \pi_1 = 1$$
$$6\pi_1 = 1$$
$$\pi_1 = \frac{1}{6}$$

The stationary distribution is $\boldsymbol{\pi}^T = (\frac{1}{6}, \frac{2}{6}, \frac{1}{6}, \frac{1}{6}, \frac{1}{6})$, or $(\frac{1}{6}, \frac{1}{3}, \frac{1}{6}, \frac{1}{6}, \frac{1}{6})$.

3.8 Synthesis and Computation

The concepts developed in the previous sections, strong connectivity, matrix irreducibility, flow conservation, and the premagic property, converge in the definition of an Ideal Flow Network. This model provides a complete theoretical framework for analyzing stable, conservative flow systems.

3.8.1 *The Ideal Flow Network (IFN) Model*

An Ideal Flow Network (IFN) represents a system in perfect equilibrium. Physically, it is a network where agents can move between any two nodes and where the flow is conserved at every node. Algebraically, these two conditions correspond to the irreducibility and premagic properties of the network's flow matrix.

Definition 3.25 (Ideal Flow Matrix) An *Ideal Flow Matrix* is a non-negative square matrix that is both *premagic* and *irreducible*. An *Ideal Flow Network* is a network whose corresponding flow matrix is an Ideal Flow Matrix. This implies that the network must be strongly connected and exhibit flow conservation at every node. Furthermore, for a flow matrix to be a valid representation of a given network structure, all links in the network must correspond to positive (non-zero) entries in the flow matrix.

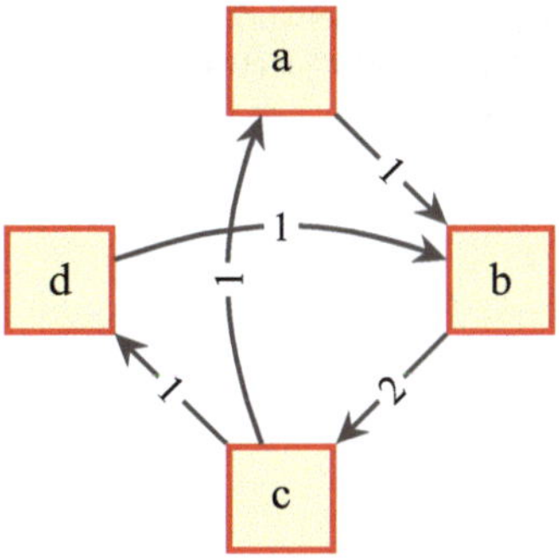

Fig. 3.24 A network that is not a valid Ideal Flow Network (IFN) due to a zero flow value

Example 3.54 (A Non-IFN due to Reducibility) Identify if the following premagic matrix is an Ideal Flow Matrix.

$$\mathsf{F} = \begin{pmatrix} 0\,2\,0\,0\,0\,0 \\ 2\,0\,0\,0\,0\,0 \\ 0\,0\,0\,3\,0\,0 \\ 0\,0\,3\,0\,0\,0 \\ 0\,0\,0\,0\,0\,4 \\ 0\,0\,0\,0\,4\,0 \end{pmatrix}$$

The matrix is premagic, as the row sums $(2, 2, 3, 3, 4, 4)^T$ equal the column sums. However, the network corresponding to this matrix consists of three separate, unconnected components ($\{a, b\}$, $\{c, d\}$, $\{e, f\}$). An unconnected network is not strongly connected, which means its adjacency matrix is reducible. Since the matrix is reducible, it is not an Ideal Flow Matrix.

Example 3.55 (A Non-IFN due to Zero Flow on a Link) Identify if the given premagic matrix F is an Ideal Flow Matrix for the network shown in Fig. 3.24.

$$\mathsf{F} = \begin{pmatrix} 0\,1\,0\,0 \\ 0\,0\,2\,0 \\ 1\,0\,0\,1 \\ 0\,1\,0\,0 \end{pmatrix}$$

The matrix is irreducible (the network is strongly connected) and premagic (row sums and column sums are $(1, 2, 2, 1)^T$). However, the network structure includes a link from node c to node b, but the corresponding matrix entry f_{cb} is 0. An Ideal Flow Network requires all existing links to carry positive flow. Therefore, while the matrix itself has the properties of an Ideal Flow Matrix, it is not a valid IFN representation for this specific network.

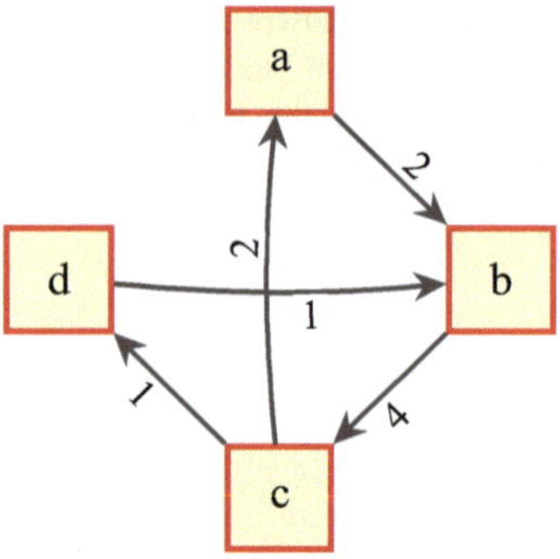

Fig. 3.25 A network that is a valid Ideal Flow Network (IFN)

Example 3.56 (A Valid Ideal Flow Network) Identify if the given premagic matrix F is an Ideal Flow Matrix for the network shown in Fig. 3.25.

$$\mathsf{F} = \begin{pmatrix} 0 & 2 & 0 & 0 \\ 0 & 0 & 4 & 0 \\ 2 & 0 & 0 & 1 \\ 0 & 1 & 0 & 0 \end{pmatrix}$$

- *Non-negative:* All entries are ≥ 0.
- *Irreducible:* The network is strongly connected, so the matrix is irreducible.
- *Premagic:* The row sums are $(2, 4, 3, 1)^T$. The column sums are also $(2, 4, 3, 1)^T$. The matrix is premagic.
- *Positive Flow:* All links shown in the network diagram have a corresponding positive flow value in the matrix.

Since all conditions are met, this is a valid Ideal Flow Network.

Graph theory and linear algebra are closely related fields, and the analysis of networks relies on translating concepts from one domain to the other. Table 3.1 summarizes the key equivalences that form the foundation of algebraic graph theory.

Table 3.1 Equivalence of Terminology between Network Theory and Matrix Algebra.

Network Property (Graph Theory)	Matrix Property (Linear Algebra)
Strongly Connected	Irreducible
Weakly Connected or Unconnected	Reducible
Flow is Conserved at each node	Premagic (Line-Sum Symmetric)
Ideal Flow Network (IFN)	Ideal Flow Matrix

3.8.2 Steady Matrices

This section introduces special classes of matrices whose non-zero entries follow a regular, or "steady," pattern. These concepts are important for the analysis of more complex network phenomena, such as congestion. This is a core concept from the Ideal Flow Network model provided by Teknomo [27].

3.8.2.1 Row-Steady, Column-Steady, and Steady Matrices

The concept of steadiness is defined relative to a given zero-one pattern.

Definition 3.26 (Pattern and Steady Matrices)

- A (0,1) non-negative matrix P is called a *pattern matrix*. It defines the locations of the non-zero entries for a class of matrices.
- A *row-steady matrix* R is a non-negative matrix that has the same pattern as P and has equal values for all non-zero entries within each row.
- A *column-steady matrix* C is a non-negative matrix that has the same pattern as P and has equal values for all non-zero entries within each column.
- A *steady matrix* B is a non-negative matrix that has the same pattern as P and has the same value for all of its non-zero entries. Thus, $\mathsf{B} = k\mathsf{P}$ for some scalar $k > 0$.

Example 3.57 (Types of Steady Matrices) Given the pattern matrix P, the matrices R, C, and B are examples of row-steady, column-steady, and steady matrices, respectively.

$$\mathsf{P} = \begin{pmatrix} 0 & 1 & 1 & 0 \\ 1 & 0 & 1 & 0 \\ 1 & 1 & 0 & 1 \\ 0 & 0 & 1 & 0 \end{pmatrix}, \quad \mathsf{R} = \begin{pmatrix} 0 & 2 & 2 & 0 \\ 3 & 0 & 3 & 0 \\ 7 & 7 & 0 & 7 \\ 0 & 0 & 9 & 0 \end{pmatrix}$$

$$\mathsf{C} = \begin{pmatrix} 0 & 2 & 5 & 0 \\ 1 & 0 & 5 & 0 \\ 1 & 2 & 0 & 4 \\ 0 & 0 & 5 & 0 \end{pmatrix}, \quad \mathsf{B} = \begin{pmatrix} 0 & 3 & 3 & 0 \\ 3 & 0 & 3 & 0 \\ 3 & 3 & 0 & 3 \\ 0 & 0 & 3 & 0 \end{pmatrix}$$

Proposition 3.1 (Generating Premagic Steady Matrices) *Given a pattern matrix* P*, a premagic steady matrix* $\mathsf{B} = k\mathsf{P}$ *exists if and only if the pattern matrix* P *is itself either premagic or symmetric.*

Proof If P is premagic or symmetric, then it satisfies $\mathsf{P}\mathbf{j} = \mathsf{P}^T\mathbf{j}$. Multiplying by a scalar k preserves this property, so $\mathsf{B} = k\mathsf{P}$ is also premagic. Conversely, assume a premagic steady matrix $\mathsf{B} = k\mathsf{P}$ exists. Then $\mathsf{B}\mathbf{j} = \mathsf{B}^T\mathbf{j}$, which implies $k\mathsf{P}\mathbf{j} = k\mathsf{P}^T\mathbf{j}$. Since $k > 0$, it follows that $\mathsf{P}\mathbf{j} = \mathsf{P}^T\mathbf{j}$, meaning P must be premagic. □

3.8.2.2 Generating Premagic Steady Matrices

While a premagic steady matrix does not exist for every pattern, it is always possible to construct premagic row-steady and premagic column-steady matrices for any irreducible pattern.

Proposition 3.2 (Existence of Premagic Row-Steady Matrix) *Given an irreducible pattern matrix* P*, there exists a premagic row-steady matrix associated with that pattern.*

Proof Let κ be a desired total flow scaling factor. To construct the premagic row-steady matrix, begin by forming the row-stochastic matrix $\mathsf{S} = \mathsf{P} \oslash (\mathsf{P}\mathbf{j}\mathbf{j}^T)$, where $\mathbf{j}$ is the all-ones vector. Next, solve for the stationary distribution $\boldsymbol{\pi}^T$ of S by solving the augmented system $[\mathsf{S}^T - \mathsf{I};\ \mathbf{j}^T]\boldsymbol{\pi} = [\mathbf{0};\ \kappa]$. Then construct the premagic flow matrix as $\mathsf{F} = \text{diag}(\boldsymbol{\pi})\mathsf{S}$. Finally, define the premagic row-steady matrix as $\mathsf{R} = \mathsf{F} \oslash \mathsf{P}$. For each row i, the non-zero entries of R are equal to $\pi_i/(\sum_j p_{ij})$.

The matrix F is premagic because $\boldsymbol{\pi}^T$ is the stationary distribution of S. Since P is irreducible, S inherits this property, ensuring the existence of a unique positive $\boldsymbol{\pi}$. The Hadamard division by P preserves the pattern and guarantees that R is row-steady. □

Proposition 3.3 (Existence of Premagic Column-Steady Matrix) *Given an irreducible pattern matrix* P*, there exists a premagic column-steady matrix associated with that pattern.*

Proof The construction mirrors the row-steady case, but operates on the transpose of the pattern matrix. Let κ be a desired total flow scaling factor. Begin by forming the row-stochastic matrix $\mathsf{S} = \mathsf{P}^T \oslash (\mathsf{P}^T\mathbf{j}\mathbf{j}^T)$, where $\mathbf{j}$ is the all-ones vector. Solve for the stationary distribution $\boldsymbol{\pi}^T$ of S using the system $[\mathsf{S}^T - \mathsf{I};\ \mathbf{j}^T]\boldsymbol{\pi} = [\mathbf{0};\ \kappa]$. Construct the premagic flow matrix as $\mathsf{F} = \text{diag}(\boldsymbol{\pi})\mathsf{S}$, and define the premagic column-steady matrix as $\mathsf{C} = (\mathsf{F} \oslash \mathsf{P}^T)^T$.

This construction ensures that F is premagic, since $\boldsymbol{\pi}^T$ is the stationary distribution of the irreducible matrix S. The Hadamard division by P^T, followed by transposition, guarantees that C is column-steady and conforms to the original pattern. □

3.9 Summary

This chapter has provided a sequential review of the mathematical foundations required for network analysis. We began by establishing the fundamental language of *graph theory* to describe network topology, defining concepts such as nodes, links, paths, and the critical property of strong connectivity. We then developed the necessary tools from *matrix algebra*, including basic operations and the permutation matrices used to define reducibility.

The core of this chapter was the formal bridge built between these two domains. We demonstrated how a network's structure is encoded algebraically in its *adjacency and incidence matrices*. This led to the central equivalences: a network is *strongly connected* if and only if its adjacency matrix is *irreducible*, and a network conserves flow at each node if and only if its flow matrix is *premagic*.

With this foundation, we introduced the analytical engine of network equilibrium analysis: the *Perron-Frobenius theorem*. This theorem guarantees that any irreducible, non-negative matrix has a unique, positive dominant eigenvalue (the Perron root) and a corresponding positive eigenvector. For the row-stochastic matrices that represent network transitions, this theorem ensures the existence of a unique *stationary distribution vector*, which defines the proportional equilibrium flow across all nodes. Finally, we synthesized these concepts into the *Ideal Flow Network (IFN) model* and introduced the computational tools, such as the *generalized inverse*, required for its practical application.

References

1. Jørgen Bang-Jensen and Gregory Z. Gutin. *Digraphs: Theory, Algorithms and Applications*. London: Springer, 2008. https://doi.org/10.1007/978-1-84800-998-1. https://link.springer.com/book/10.1007/978-1-84800-998-1.
2. Abraham Berman and Robert J. Plemmons. *Nonnegative Matrices in the Mathematical Sciences*. Philadelphia: Society for Industrial and Applied Mathematics, 1994. https://doi.org/10.1137/1.9781611971262. https://epubs.siam.org/doi/book/10.1137/1.9781611971262.
3. N. L. Biggs, E. K. Lloyd, and R. J. Wilson. *Graph Theory, 1736–1936*. Oxford University Press, 1986.
4. Norman Biggs. *Algebraic Graph Theory*. 2nd. Cambridge University Press, 1993. https://doi.org/10.1017/CBO9780511608704.
5. George E. P. Box. "Science and Statistics". In: *Journal of the American Statistical Association* 71.356 (1976), pp. 791–799. https://doi.org/10.1080/01621459.1976.10480949.
6. Gary Chartrand et al. *Graphs & Digraphs*. 7th ed. Boca Raton, FL: Chapman and Hall/CRC, 2024. ISBN: 9781032606989. https://doi.org/10.1201/9781032606989.
7. T. H. Cormen et al. *Introduction to Algorithms*. 3rd. MIT Press, 2009.
8. Reinhard Diestel. *Graph Theory*. 5th. Berlin, Heidelberg: Springer, 2017. https://doi.org/10.1007/978-3-662-53622-3.
9. Ernesto Estrada. *The Structure of Complex Networks: Theory and Applications*. Oxford: Oxford University Press, 2012. https://doi.org/10.1093/acprof:oso/9780199591756.001.0001.
10. Leonhard Euler. "Solutio problematis ad geometriam situs pertinentis". In: *Commentarii academiae scientiarum Petropolitanae* 8 (1741). Presented in 1736, pp. 128–140.
11. L. R. Foulds. *Graph Theory Applications*. Springer-Verlag, 1992. https://doi.org/10.1007/978-1-4612-0933-1.
12. G. Frobenius. "Ueber Matrizen aus nicht negativen Elementen". In: *Sitzungsberichte der Königlich Preußischen Akademie der Wissenschaften* (1912), pp. 456–477.
13. Chris Godsil and Gordon Royle. *Algebraic Graph Theory*. New York: Springer, 2001. https://doi.org/10.1007/978-1-4613-0163-9.
14. Gene H. Golub and Charles F. Van Loan. *Matrix Computations*. 4th ed. Johns Hopkins University Press, 2013. ISBN: 9781421407944.
15. F. Harary. *Graph Theory*. Addison-Wesley, 1969.

16. F. Harary and R. Z. Norman. “Some properties of line digraphs”. In: *Rendiconti del Circolo Matematico di Palermo* 9.2 (1960), pp. 161–168. https://doi.org/10.1007/BF02854556.
17. Roger A. Horn and Charles R. Johnson. *Matrix Analysis*. 2nd. Cambridge: Cambridge University Press, 2012. https://doi.org/10.1017/CBO9781139020411.
18. John G. Kemeny and J. Laurie Snell. *Finite Markov Chains*. New York: Springer, 1976. https://link.springer.com/book/9780387901923.
19. Carl D. Meyer. *Matrix Analysis and Applied Linear Algebra*. Philadelphia: Society for Industrial and Applied Mathematics, 2000. https://doi.org/10.1137/1.9780898719512.
20. Mark E. J. Newman. *Networks: An Introduction*. Oxford: Oxford University Press, 2010. https://doi.org/10.1093/acprof:oso/9780199206650.001.0001. https://academic.oup.com/book/27303.
21. Juan de Dios Ortúzar and Luis G. Willumsen. *Modelling Transport*. 4th. Chichester: John Wiley & Sons, 2011. https://doi.org/10.1002/9781119993308
22. Oskar Perron. “Zur Theorie der Matrices”. In: *Mathematische Annalen* 64.2 (1907), pp. 248–263. https://doi.org/10.1007/BF01449896.
23. Pranab Kumar Sen and Julio M. Singer. “Large Sample Methods in Statistics”. In: *Encyclopedia of Statistical Sciences* (2006). https://doi.org/10.1002/0471667196.ess1333.pub2.
24. Eugene Seneta. *Non-negative Matrices and Markov Chains*. Rev. ed. New York: Springer, 2006. https://doi.org/10.1007/0-387-32792-4 https://link.springer.com/book/10.1007/0-387-32792-4.
25. Richard P. Stanley. *Enumerative Combinatorics*. 2nd. Vol. 1. Cambridge: Cambridge University Press, 2011. https://doi.org/10.1017/CBO9781139058520.
26. K. Teknomo. “Ideal Flow Based on Random Walk on Directed Graph”. In: *Proceedings of the 9th International Collaboration Symposium on Information, Production and Systems*. Kitakyushu: Waseda University, 2015.
27. Kardi Teknomo. *Introduction to Ideal Flow Networks*. 2022.
28. Lloyd N. Trefethen and David Bau III. *Numerical Linear Algebra*. Philadelphia, PA: Society for Industrial and Applied Mathematics, 1997. https://doi.org/10.1137/1.9780898719574. eprint: https://epubs.siam.org/doi/pdf/10.1137/1.9780898719574. https://epubs.siam.org/doi/abs/10.1137/1.9780898719574.
29. Douglas B. West. *Introduction to Graph Theory*. 2nd. Upper Saddle River, NJ: Prentice Hall, 2001.
30. Robin J. Wilson. *Introduction to Graph Theory*. 4th. Harlow, England: Prentice Hall, 1996.

Chapter 4
Introduction to Markov Chains

4.1 Introduction

This chapter presents a comprehensive exploration of Markov Chains, a class of stochastic processes fundamental to fields such as transportation, traffic engineering, and urban transportation planning. It begins with foundational concepts, including the Markov property and transition matrices, and progressively introduces advanced topics such as stationary distributions, convergence behavior, and spectral analysis.

Formal propositions and theorems are complemented by intuitive explanations to ensure conceptual clarity for a broad audience. Practical examples illustrate how Markov Chains can effectively model real-world systems, including vehicle movements, traffic congestion, and land use dynamics.

By the end of the chapter, the theoretical framework and applied methodologies provided will support the use of Markov Chains in analyzing and designing complex systems.

4.2 Definitions and Axioms of Markov Chains

Markov Chain is a mathematical model used to describe a system that transitions from one state to another. The key feature is that the future state of the system depends only on the current state and not on how it got there (this is the memoryless property).

Seneta [9] provides a rigorous treatment of non-negative matrices and Markov chains, including stochastic matrices which underpin transition probabilities in ideal flow networks.

K. Teknomo, *Ideal Flow Network for Transportation Research*,
Lecture Notes in Intelligent Transportation and Infrastructure,
https://doi.org/10.1007/978-981-95-4727-2_4

Definition 4.1 (Markov Chain) A *Markov chain* is a sequence of random variables $\{X_t\}_{t=0}^{\infty}$ with the property that the future state of the system depends only on the present state and not on the past states. Formally, for any $t \geq 0$ and states $x_0, x_1, \ldots, x_t, x_{t+1}$,

$$P(X_{t+1} = x_{t+1} \mid X_t = x_t, X_{t-1} = x_{t-1}, \ldots, X_0 = x_0)$$
$$= P(X_{t+1} = x_{t+1} \mid X_t = x_t). \tag{4.1}$$

Think of the state space as a collection of all possible states the system can be in. For example, if we're modeling the weather, the state space could be "sunny," "cloudy," and "rainy."

Definition 4.2 (State Space) The *state space* of a Markov chain is the set of all possible states that the process can be in. It is denoted as S, where $S = \{s_1, s_2, \ldots, s_n\}$ for a finite state space, or $S = \{s_1, s_2, \ldots\}$ for an infinite state space.

Transition Probability describes the likelihood of moving from one state to another. If we are in state i, the transition probability tells us how likely we are to move to state j in the next time step.

Transition Matrix gives a compact way to store all the transition probabilities between states. Each row corresponds to a current state, and each column corresponds to a potential next state.

Definition 4.3 (Transition Probability) The *transition probability* P_{ij} is the probability that the system moves from state i to state j in one time step, i.e.,

$$P_{ij} = P(X_{t+1} = s_j \mid X_t = s_i). \tag{4.2}$$

Definition 4.4 (Transition Probability Matrix) The *transition probability matrix* S of a Markov chain is a square matrix whose element s_{ij} represents the transition probability from state s_i to state s_j. That is,

$$\mathsf{S} = \begin{pmatrix} s_{11} & s_{12} & \cdots & s_{1n} \\ s_{21} & s_{22} & \cdots & s_{2n} \\ \vdots & \vdots & \ddots & \vdots \\ s_{n1} & s_{n2} & \cdots & s_{nn} \end{pmatrix}.$$

We need to distinguish Homogeneous vs Inhomogeneous Markov Chain. If the system's behavior is time-independent (the transition probabilities do not change over time), it is homogeneous. Otherwise, if the probabilities change over time, it is inhomogeneous.

Definition 4.5 (Homogeneous Markov Chain) A Markov chain is said to be *homogeneous* if the transition probabilities do not depend on time. In other words, the transition probability matrix S remains constant over time, i.e.,

$$P(X_{t+1} = s_j \mid X_t = s_i) = P(X_{t+1} = s_j \mid X_t = s_i) = s_{ij}, \quad \forall t. \tag{4.3}$$

Definition 4.6 (Inhomogeneous Markov Chain) An *inhomogeneous* Markov chain has transition probabilities that depend on time, i.e.,

$$P(X_{t+1} = s_j \mid X_t = s_i, t) = s_{ij}(t), \quad \text{where } s_{ij}(t) \neq s_{ij}(t'). \tag{4.4}$$

The transition Matrix of Markov Chain must be stochastic. In some special cases, the transition Matrix can be doubly stochastic.

Definition 4.7 (Stochastic Matrix [6]) A matrix S is said to be *stochastic* if it satisfies the following two conditions:

$$s_{ij} \geq 0, \quad \forall i, j \quad \text{and} \quad \sum_j s_{ij} = 1, \quad \forall i. \tag{4.5}$$

This means that all elements are non-negative, and the sum of the elements in each row is equal to one.

Definition 4.8 (Doubly Stochastic Matrix) A matrix S is *doubly stochastic* if it satisfies the following conditions:

$$s_{ij} \geq 0, \quad \forall i, j \quad \text{and} \quad \sum_j s_{ij} = 1, \quad \forall i \quad \text{and} \quad \sum_i s_{ij} = 1, \quad \forall j. \tag{4.6}$$

Example 4.1 Consider a Markov chain with two states $\{s_1, s_2\}$. The transition probability matrix is:

$$\mathsf{S} = \begin{pmatrix} 0.6 & 0.4 \\ 0.3 & 0.7 \end{pmatrix}.$$

This means that if the system is in state s_1, there is a 60% chance of staying in s_1 and a 40% chance of transitioning to s_2. If the system is in state s_2, there is a 30% chance of transitioning to s_1 and a 70% chance of staying in s_2.

Example 4.2 Let's now compute the steady-state (or stationary) distribution for the following Markov chain, represented by the transition probability matrix:

$$\mathsf{S} = \begin{pmatrix} 0.2 & 0.8 \\ 0.7 & 0.3 \end{pmatrix}.$$

The steady-state distribution $\boldsymbol{\pi}^T = (\pi_1\ \pi_2)$ satisfies the equation:

$$\boldsymbol{\pi}^T \mathsf{S} = \boldsymbol{\pi}^T,$$

which translates into the system of equations:

$$\begin{cases} 0.2\pi_1 + 0.7\pi_2 = \pi_1 \\ 0.8\pi_1 + 0.3\pi_2 = \pi_2 \end{cases}.$$

These two equations are linearly dependent. Simplifying the first equation gives:

$$0.7\pi_2 = \pi_1 - 0.2\pi_1$$

$$0.7\pi_2 = 0.8\pi_1$$

To find a unique solution, this relation is combined with the normalization constraint, $\pi_1 + \pi_2 = 1$. The normalization constraint comes from knowledge that the total probability must sum to 1.

We can solve this system by substitution. From the normalization constraint, we have $\pi_2 = 1 - \pi_1$. Substituting this into the simplified equation:

$$0.8\pi_1 = 0.7(1 - \pi_1)$$

$$0.8\pi_1 = 0.7 - 0.7\pi_1$$

$$0.8\pi_1 + 0.7\pi_1 = 0.7$$

$$1.5\pi_1 = 0.7$$

Solving for π_1 yields:

$$\pi_1 = \frac{0.7}{1.5} = \frac{7}{15}$$

Subsequently, we find π_2:

$$\pi_2 = 1 - \pi_1 = 1 - \frac{7}{15} = \frac{8}{15}$$

Thus, the stationary distribution is $\pi^T = \left(\frac{7}{15}\ \frac{8}{15}\right)$. The approximate values of the steady-state probabilities are:

$$\pi_1 = \frac{7}{15} \approx 0.4667, \quad \pi_2 = \frac{8}{15} \approx 0.5333$$

4.3 Transition Probability Matrix

The transition probability matrix encapsulates all the transition probabilities between states.

Definition 4.9 (Transition Probability Matrix) The *transition probability matrix* S of a Markov chain is a square matrix where the element s_{ij} represents the probability of transitioning from state s_i to state s_j in one time step. That is,

$$s_{ij} = P(X_{t+1} = s_j \mid X_t = s_i). \tag{4.7}$$

The matrix S is stochastic, meaning each row sums to 1: $\sum_j s_{ij} = 1, \quad \forall i$.

Proposition 4.1 (Stochastic Matrix Property) *The transition probability matrix* S *is always a stochastic matrix, meaning that:*

$$s_{ij} \geq 0, \quad \forall i, j \quad \textit{and} \quad \sum_j s_{ij} = 1, \quad \forall i.$$

Proof By definition, s_{ij} is the probability of transitioning from state s_i to state s_j, which must be non-negative. Since S represents all possible transitions from state s_i, the sum of the probabilities of transitioning from state s_i to all other states must be 1:

$$\sum_j s_{ij} = 1.$$

Thus, S is a stochastic matrix. □

Example 4.3 (Markov Chain) Consider a Markov chain with 3 states $\{s_1, s_2, s_3\}$ and the following transition probability matrix S:

$$\mathsf{S} = \begin{pmatrix} 0.5 & 0.3 & 0.2 \\ 0.2 & 0.6 & 0.2 \\ 0.1 & 0.4 & 0.5 \end{pmatrix}.$$

This matrix satisfies the stochastic matrix property because all entries are non-negative and the sum of each row is 1.

$$0.5 + 0.3 + 0.2 = 1, \quad 0.2 + 0.6 + 0.2 = 1, \quad 0.1 + 0.4 + 0.5 = 1.$$

Thus, it is a valid transition probability matrix.

4.4 Computing Powers of the Transition Matrix

The n-step transition probabilities help analyze the system's behavior over multiple time steps.

Definition 4.10 (n-Step Transition Probability) The *n-step transition probability* $P_{ij}^{(n)}$ is the probability of transitioning from state s_i to state s_j in n steps. t is given by the (i, j)-th entry of the matrix S^n, where S^n denotes the n-th power of the transition probability matrix:

$$P_{ij}^{(n)} = (\mathsf{S}^n)_{ij}. \tag{4.8}$$

Proposition 4.2 (n-Step Transition Probability Matrix) *The n-step transition probability matrix* S^n *can be computed by multiplying the transition probability matrix* S *by itself n times:*

$$\mathsf{S}^n = \mathsf{S} \times \mathsf{S} \times \cdots \times \mathsf{S}, \quad n \textit{ times}. \tag{4.9}$$

4.5 State Behavior and Classifications

In this section, we will define and explore the different types of states that can exist in a Markov chain, including recurrent, transient, absorbing, and other classifications. We will also introduce essential concepts like irreducibility and periodicity.

4.5.1 State Classification

Recurrent · and transient states describe the long-term behavior of the system. Recurrent states are guaranteed to be revisited, while transient states may never be revisited.

Definition 4.11 (Recurrent State) A state s_i in a Markov chain is called *recurrent* if, starting from s_i, there is a probability of 1 that the chain will eventually return to s_i. Formally, state s_i is recurrent if:

$$P(\text{ever return to } s_i \mid \text{starting from } s_i) = 1. \tag{4.10}$$

Definition 4.12 (Transient State) A state s_i is called *transient* if, starting from s_i, there is a probability less than 1 that the chain will eventually return to s_i. Formally, state s_i is transient if:

$$P(\text{ever return to } s_i \mid \text{starting from } s_i) < 1. \tag{4.11}$$

Proposition 4.3 (Classification of States) *A state in a Markov chain is either recurrent or transient, but not both. A Markov chain that contains only recurrent states is said to be* recurrent, *while a chain containing transient states is classified as* transient.

Proof By definition, a recurrent state will always return to itself with probability 1, while a transient state has a non-zero probability of never returning. Hence, a state can only belong to one of these categories, and thus the classification is mutually exclusive. □

4.5.2 *Absorbing and Non-Absorbing States*

Absorbing states are those from which the system cannot escape once entered.

Definition 4.13 (Absorbing State) An *absorbing state* is a state s_i such that once the process enters s_i, it cannot leave. That is, for an absorbing state:

$$P(\text{stay in } s_i \mid X_t = s_i) = 1. \tag{4.12}$$

In other words, $s_{ii} = 1$ and $s_{ij} = 0$ for all $j \neq i$.

Definition 4.14 (Non-Absorbing State) A *non-absorbing state* is a state that has a non-zero probability of transitioning to at least one other state. That is, for a non-absorbing state s_i, there exists a $j \neq i$ such that $s_{ij} > 0$.

Proposition 4.4 (Absorbing Chain) *A Markov chain is said to be* absorbing *if it has at least one absorbing state and it is possible to reach an absorbing state from every non-absorbing state in the chain. In an absorbing chain, once an absorbing state is reached, the process remains there forever.*

Proof If a Markov chain has an absorbing state, it means that once the system enters that state, it cannot leave. Therefore, the process "gets stuck" in that state. For the chain to be absorbing, every state must either lead to an absorbing state or be an absorbing state itself. □

Grinstead and Snell [3, 4] provide a rigorous treatment of absorbing states in Markov chains.

4.5.3 Irreducibility and Periodicity

Irreducibility and periodicity define the structure of the chain, with irreducibility ensuring that all states can be reached, and periodicity describing the cyclic behavior of the system.

Definition 4.15 (Irreducible Chain [5]) A Markov chain is said to be *irreducible* if it is possible to reach any state from any other state, i.e., for any pair of states s_i and s_j, there exists some time step n such that the (i, j)-th entry of S^n is positive:

$$P_{ij}^{(n)} > 0 \quad \text{for some } n. \tag{4.13}$$

Definition 4.16 (Periodic Chain [5]) A Markov chain is *periodic* if there exists a $d > 1$ such that the chain can only return to a state after multiples of d time steps. The period d of a state s_i is defined as the greatest common divisor (gcd) of the set of time steps at which the chain can return to s_i. A chain is periodic if all states share the same period.

Definition 4.17 (Aperiodic Chain) A Markov chain is *aperiodic* if all states have period 1, meaning the chain can return to a state at irregular time steps. Formally, the period of every state in an aperiodic chain is 1.

Norris [7] provides a formal definition and analysis of aperiodic Markov chains, which ensure convergence to stationary distributions in stochastic flow models.

Proposition 4.5 (Irreducibility and Periodicity) *For an irreducible Markov chain, the chain is either aperiodic or periodic. If a chain is irreducible and aperiodic, then the long-term behavior of the chain will converge to a unique stationary distribution independent of the initial state.*

Proof If a Markov chain is irreducible, it means that it is possible to reach any state from any other state. If the chain is aperiodic, there is no fixed time period for revisiting a state, and thus the chain will eventually reach a steady-state distribution. If the chain is periodic, it will oscillate between states and not converge to a steady-state distribution. □

Example 4.4 (Absorbing State) Consider a Markov chain with the following transition probability matrix:

$$\mathsf{S} = \begin{pmatrix} 1 & 0 \\ 0 & 1 \end{pmatrix}.$$

This is a 2-state absorbing Markov chain where both states are absorbing. Once the chain enters state s_1, it stays there forever, and the same applies to state s_2. Thus, this is an absorbing chain.

Example 4.5 (Irreducibility and Periodicity) Consider the following transition probability matrix:

$$\mathsf{S} = \begin{pmatrix} 0 & 1 & 0 \\ 0 & 0 & 1 \\ 1 & 0 & 0 \end{pmatrix}.$$

This matrix represents a cyclic Markov chain with 3 states. Starting from any state, the chain will move cyclically through the states, i.e., $s_1 \to s_2 \to s_3 \to s_1$, and so on. The chain is irreducible because all states can reach any other state. However, it is periodic because it has a period of 3, meaning it takes exactly 3 steps to return to the starting state.

4.6 Steady-State Analysis

In this section, we will explore the steady-state behavior of a Markov chain. Specifically, we will discuss the concept of the stationary distribution, prove its existence under certain conditions, and outline methods to compute it.

4.6.1 *Stationary Distribution*

The *stationary distribution* of a Markov chain describes the long-term behavior of the system. It provides the probabilities of being in each state once the system has reached equilibrium, where the distribution no longer changes from one transition to the next. This *steady-state* behavior reflects how the system behaves after many transitions, regardless of its initial state, provided such a distribution exists and is unique.

Definition 4.18 (Stationary (Steady-State) Distribution) A *stationary distribution* $\boldsymbol{\pi}^T = (\pi_1, \pi_2, \ldots, \pi_n)$ is a probability vector that satisfies the equation:

$$\boldsymbol{\pi}^T \mathsf{S} = \boldsymbol{\pi}^T, \quad \sum_{i=1}^{n} \pi_i = 1, \quad \pi_i \geq 0 \; \forall i. \tag{4.14}$$

This means that if the system's state probabilities are described by $\boldsymbol{\pi}$ at time t, they will remain unchanged at time $t + 1$, indicating equilibrium in the long run. In many contexts, this stationary distribution is also referred to as the *steady-state distribution*.

This Eq. (4.14) defines a distribution that is invariant under the transition process, meaning if the system's state probabilities are described by $\boldsymbol{\pi}^T$, they will remain so after one transition.

Proposition 4.6 (Existence and Uniqueness [9]) *If the Markov chain is irreducible and positive recurrent (or if the transition probability matrix* S *is regular), a unique stationary distribution exists. The stationary distribution is independent of the initial distribution.*

Proof Irreducibility ensures all states communicate, and positive recurrence guarantees return times are finite. Under these conditions, the Markov chain settles into a unique stationary distribution. If S is regular, then some power S^k has strictly positive entries, implying irreducibility and convergence. The existence and uniqueness of the stationary distribution are guaranteed by the Perron-Frobenius theorem [1, 8] explained in Sect. 3.7.2. □

Example 4.6 (Steady-State Calculation) Consider a Markov chain with two states $\{s_1, s_2\}$ and the following transition probability matrix:

$$\mathsf{S} = \begin{pmatrix} 0.6 & 0.4 \\ 0.3 & 0.7 \end{pmatrix}.$$

To find the steady-state distribution $\boldsymbol{\pi}^T = (\pi_1, \pi_2)$, we solve the system:

$$\pi_1 = 0.6\pi_1 + 0.3\pi_2, \quad \pi_2 = 0.4\pi_1 + 0.7\pi_2.$$

We also know that $\pi_1 + \pi_2 = 1$. Substituting $\pi_2 = 1 - \pi_1$ into the first equation:

$$\pi_1 = 0.6\pi_1 + 0.3(1 - \pi_1),$$

$$\pi_1 = 0.6\pi_1 + 0.3 - 0.3\pi_1,$$

$$\pi_1 = 0.3\pi_1 + 0.3,$$

$$0.7\pi_1 = 0.3,$$

$$\pi_1 = \frac{3}{7} \approx 0.4286, \quad \pi_2 = 1 - \pi_1 = \frac{4}{7} \approx 0.5714.$$

Thus, the steady-state distribution is $\pi_1 = 0.4286$ and $\pi_2 = 0.5714$.

Example 4.7 (Steady-State Calculation) Consider a Markov chain with 3 states $\{s_1, s_2, s_3\}$ and the following transition probability matrix:

$$\mathsf{S} = \begin{pmatrix} 0.2 & 0.6 & 0.2 \\ 0.4 & 0.4 & 0.2 \\ 0.3 & 0.3 & 0.4 \end{pmatrix}.$$

We want to find the steady-state distribution $\boldsymbol{\pi} = (\pi_1, \pi_2, \pi_3)$. This gives the system of equations:

$$\pi_1 = 0.2\pi_1 + 0.4\pi_2 + 0.3\pi_3,$$

$$\pi_2 = 0.6\pi_1 + 0.4\pi_2 + 0.3\pi_3,$$

$$\pi_3 = 0.2\pi_1 + 0.2\pi_2 + 0.4\pi_3.$$

Also, $\pi_1 + \pi_2 + \pi_3 = 1$. Solving this system (either by substitution or matrix methods) gives:

$$\pi_1 \approx 0.3125, \quad \pi_2 \approx 0.4375, \quad \pi_3 \approx 0.2500.$$

Thus, the steady-state distribution is $\pi_1 = 0.3125$, $\pi_2 = 0.4375$, and $\pi_3 = 0.2500$.

4.6.2 *Computing the Stationary Distribution*

Computation of the stationary distribution involves solving a system of linear equations, which can be done either analytically or numerically.

To compute the stationary distribution $\boldsymbol{\pi}$, we need to solve the system of equations:

$$\boldsymbol{\pi}^T \mathsf{S} = \boldsymbol{\pi}^T.$$

which can be rewritten as:

$$\begin{pmatrix}\pi_1 & \pi_2 & \dots & \pi_n\end{pmatrix} \begin{pmatrix} s_{11} & s_{12} & \dots & s_{1n} \\ s_{21} & s_{22} & \dots & s_{2n} \\ \vdots & \vdots & \ddots & \vdots \\ s_{n1} & s_{n2} & \dots & s_{nn} \end{pmatrix} = \begin{pmatrix}\pi_1 & \pi_2 & \dots & \pi_n\end{pmatrix}.$$

This system of equations can be solved by eliminating one of the variables (since $\sum_i \pi_i = 1$) and solving the resulting linear system.

Example 4.8 (2-State Markov Chain Stationary Distribution) Consider the following transition probability matrix for a 2-state Markov chain:

$$\mathsf{S} = \begin{pmatrix} 0.4 & 0.6 \\ 0.5 & 0.5 \end{pmatrix}.$$

To find the stationary distribution $\boldsymbol{\pi}^T = (\pi_1, \pi_2)$, we solve the system:

$$\pi_1 = 0.4\pi_1 + 0.5\pi_2, \quad \pi_2 = 0.6\pi_1 + 0.5\pi_2,$$

with the constraint $\pi_1 + \pi_2 = 1$.

Substitute $\pi_2 = 1 - \pi_1$ into the first equation:

$$\pi_1 = 0.4\pi_1 + 0.5(1 - \pi_1),$$

$$\pi_1 = 0.4\pi_1 + 0.5 - 0.5\pi_1,$$

$$\pi_1 = -0.1\pi_1 + 0.5,$$

$$1.1\pi_1 = 0.5,$$

$$\pi_1 = \frac{5}{11} \approx 0.4545, \quad \pi_2 = 1 - \pi_1 = \frac{6}{11} \approx 0.5455.$$

Thus, the stationary distribution is $\pi_1 \approx 0.4545$ and $\pi_2 \approx 0.5455$.

Example 4.9 (2-State Markov Chain Stationary Distribution Matrix Form) Solve the previous Example 4.8 using matrix form.

Answer:

Consider the transition probability matrix for a 2-state Markov chain.

$$\mathsf{S} = \begin{pmatrix} 0.4 & 0.6 \\ 0.5 & 0.5 \end{pmatrix}$$

The goal is to find the stationary distribution vector $\boldsymbol{\pi} = \begin{pmatrix} \pi_1 \\ \pi_2 \end{pmatrix}$.

1. *Formulate the Homogeneous System:*
   First, we construct the matrix $(\mathsf{S}^T - \mathsf{I})$.

$$\mathsf{S}^T - \mathsf{I} = \begin{pmatrix} 0.4 & 0.5 \\ 0.6 & 0.5 \end{pmatrix} - \begin{pmatrix} 1 & 0 \\ 0 & 1 \end{pmatrix} = \begin{pmatrix} -0.6 & 0.5 \\ 0.6 & -0.5 \end{pmatrix}$$

   The system of equations is $(\mathsf{S}^T - \mathsf{I})\boldsymbol{\pi} = \mathsf{0}$, which is:

$$\begin{pmatrix} -0.6 & 0.5 \\ 0.6 & -0.5 \end{pmatrix} \begin{pmatrix} \pi_1 \\ \pi_2 \end{pmatrix} = \begin{pmatrix} 0 \\ 0 \end{pmatrix}$$

   The two equations, $-0.6\pi_1 + 0.5\pi_2 = 0$ and $0.6\pi_1 - 0.5\pi_2 = 0$, are linearly dependent.

2. *Augment the System with the Normalization Constraint:*
   We replace the homogeneous equation with the normalization constraint $\pi_1 + \pi_2 = 1$. This creates a new linear system $\mathsf{A}'\boldsymbol{\pi} = \mathsf{b}'$.

$$\mathsf{A}' = \begin{pmatrix} -0.6 & 0.5 \\ 1 & 1 \end{pmatrix}, \quad \mathsf{b}' = \begin{pmatrix} 0 \\ 1 \end{pmatrix}$$

   The system to solve is:

$$\begin{pmatrix} -0.6 & 0.5 \\ 1 & 1 \end{pmatrix} \begin{pmatrix} \pi_1 \\ \pi_2 \end{pmatrix} = \begin{pmatrix} 0 \\ 1 \end{pmatrix}$$

3. *Solve the System by Matrix Inversion:*
   The solution is given by $\boldsymbol{\pi} = (\mathsf{A}')^{-1}\mathsf{b}'$. We first find the inverse of A'.
   The determinant of A' is:

$$\det(\mathsf{A}') = (-0.6)(1) - (0.5)(1) = -1.1$$

   The inverse matrix $(\mathsf{A}')^{-1}$ is:

$$(\mathsf{A}')^{-1} = \frac{1}{-1.1} \begin{pmatrix} 1 & -0.5 \\ -1 & -0.6 \end{pmatrix} = \begin{pmatrix} \frac{-1}{1.1} & \frac{0.5}{1.1} \\ \frac{1}{1.1} & \frac{0.6}{1.1} \end{pmatrix}$$

   Now, we can find $\boldsymbol{\pi}$:

$$\boldsymbol{\pi} = \begin{pmatrix} \pi_1 \\ \pi_2 \end{pmatrix} = \begin{pmatrix} \frac{-1}{1.1} & \frac{0.5}{1.1} \\ \frac{1}{1.1} & \frac{0.6}{1.1} \end{pmatrix} \begin{pmatrix} 0 \\ 1 \end{pmatrix} = \begin{pmatrix} (\frac{0.5}{1.1}) \\ (\frac{0.6}{1.1}) \end{pmatrix} = \frac{1}{11} \begin{pmatrix} 5 \\ 6 \end{pmatrix}$$

   The stationary distribution is $\pi_1 = \frac{5}{11} \approx 0.4545$ and $\pi_2 = \frac{6}{11} \approx 0.5455$.

Example 4.10 (Stationary Distribution) Consider the following transition probability matrix for a 3-state Markov chain:

$$\mathsf{S} = \begin{pmatrix} 0.5 & 0.2 & 0.3 \\ 0.3 & 0.5 & 0.2 \\ 0.4 & 0.4 & 0.2 \end{pmatrix}.$$

We want to find the stationary distribution $\boldsymbol{\pi}^T = (\pi_1, \pi_2, \pi_3)$. The system of equations is:

$$\pi_1 = 0.5\pi_1 + 0.3\pi_2 + 0.4\pi_3,$$

$$\pi_2 = 0.2\pi_1 + 0.5\pi_2 + 0.4\pi_3,$$

$$\pi_3 = 0.3\pi_1 + 0.2\pi_2 + 0.2\pi_3.$$

We also have the constraint $\pi_1 + \pi_2 + \pi_3 = 1$.
This system can be solved using substitution. Here we are using matrix method.

1. *Formulate the Homogeneous System:*
   First, we construct the matrix $(\mathsf{S}^T - \mathsf{I})$.

$$\mathsf{S}^T - \mathsf{I} = \begin{pmatrix} 0.5 & 0.3 & 0.4 \\ 0.2 & 0.5 & 0.4 \\ 0.3 & 0.2 & 0.2 \end{pmatrix} - \begin{pmatrix} 1 & 0 & 0 \\ 0 & 1 & 0 \\ 0 & 0 & 1 \end{pmatrix} = \begin{pmatrix} -0.5 & 0.3 & 0.4 \\ 0.2 & -0.5 & 0.4 \\ 0.3 & 0.2 & -0.8 \end{pmatrix}$$

   This gives the system $(\mathsf{S}^T - \mathsf{I})\boldsymbol{\pi} = \mathbf{0}$, which has linearly dependent rows.
2. *Augment the System with the Normalization Constraint:*
   We replace the third row with the constraint $\pi_1 + \pi_2 + \pi_3 = 1$. This yields the system $\mathsf{A}'\boldsymbol{\pi} = \mathsf{b}'$.

$$\mathsf{A}' = \begin{pmatrix} -0.5 & 0.3 & 0.4 \\ 0.2 & -0.5 & 0.4 \\ 1 & 1 & 1 \end{pmatrix}, \quad \mathsf{b}' = \begin{pmatrix} 0 \\ 0 \\ 1 \end{pmatrix}$$

3. *Solve the System by Matrix Inversion:*
   The solution is $\boldsymbol{\pi} = (\mathsf{A}')^{-1}\mathsf{b}'$. We first find the inverse of A'.
   The determinant of A' is:

$$\det(\mathsf{A}') = -0.5(-0.5 - 0.4) - 0.3(0.2 - 0.4) + 0.4(0.2 - (-0.5))$$

$$= -0.5(-0.9) - 0.3(-0.2) + 0.4(0.7) = 0.45 + 0.06 + 0.28 = 0.79$$

   The adjugate matrix, $\operatorname{adj}(\mathsf{A}')$, is the transpose of the cofactor matrix:

$$\operatorname{adj}(\mathsf{A}') = \begin{pmatrix} -0.9 & 0.1 & 0.32 \\ 0.2 & -0.9 & 0.28 \\ 0.7 & 0.8 & 0.19 \end{pmatrix}$$

   The inverse is:

$$(\mathsf{A}')^{-1} = \frac{1}{\det(\mathsf{A}')}\operatorname{adj}(\mathsf{A}') = \frac{1}{0.79}\begin{pmatrix} -0.9 & 0.1 & 0.32 \\ 0.2 & -0.9 & 0.28 \\ 0.7 & 0.8 & 0.19 \end{pmatrix}$$

   Finally, we calculate $\boldsymbol{\pi}$:

$$\boldsymbol{\pi} = \frac{1}{0.79}\begin{pmatrix} -0.9 & 0.1 & 0.32 \\ 0.2 & -0.9 & 0.28 \\ 0.7 & 0.8 & 0.19 \end{pmatrix}\begin{pmatrix} 0 \\ 0 \\ 1 \end{pmatrix} = \frac{1}{0.79}\begin{pmatrix} 0.32 \\ 0.28 \\ 0.19 \end{pmatrix} = \begin{pmatrix} 32/79 \\ 28/79 \\ 19/79 \end{pmatrix}$$

The stationary distribution is:

$$\pi_1 = \frac{32}{79} \approx 0.4051, \quad \pi_2 = \frac{28}{79} \approx 0.3544, \quad \pi_3 = \frac{19}{79} \approx 0.2405$$

4.6.3 Limiting Distribution vs. Stationary Distribution

The limiting distribution describes the behavior of the system as time tends to infinity and, in the case of irreducible and aperiodic chains, is the same as the stationary distribution.

Definition 4.19 (Limiting Distribution) The *limiting distribution* of a Markov chain is the distribution to which the system converges as $t \to \infty$. If the chain is irreducible and aperiodic, the limiting distribution is the same as the stationary distribution. Formally, the limiting distribution $\boldsymbol{\pi}$ satisfies:

$$\lim_{t \to \infty} P(X_t = s_i) = \pi_i. \tag{4.15}$$

Proposition 4.7 (Limiting Distribution) *If a Markov chain is irreducible and aperiodic, the limiting distribution exists and is unique. The limiting distribution is the same as the stationary distribution.*

Proof For irreducible and aperiodic chains, the system eventually forgets its initial state, and the probability distribution converges to the stationary distribution. The existence and uniqueness of the limiting distribution follow from the Perron-Frobenius theorem. □

4.7 Markov Chain Theorems and Properties

In this section, we will explore the most important theorems and properties of Markov chains. These will provide deep insights into the long-term behavior and mathematical structure of Markov Chains.

4.7.1 Fundamental Theorem of Markov Chains

The Fundamental Theorem of Markov chains tells us that the long-term behavior is determined by the stationary distribution.

Theorem 4.1 (Fundamental Theorem of Markov Chains) *Let* S *be the transition probability matrix of a Markov chain with state space* $S = \{s_1, s_2, \ldots, s_n\}$*. Then, the long-term behavior of the Markov chain depends only on the stationary distribution* $\boldsymbol{\pi}$*, which satisfies* $\boldsymbol{\pi}^T\mathsf{S} = \boldsymbol{\pi}^T$*. In the long run, an irreducible and aperiodic Markov chain will always converge to a unique stationary distribution, independent of the initial state.*

Proof The transition probability matrix S governs the evolution of the system. For an irreducible chain, every state can reach any other state eventually, meaning the system is strongly-connected. For an aperiodic chain, the system does not cycle in a fixed pattern. This implies that the system "forgets" its initial state after a sufficient number of steps. Over time, the probability distribution across states reaches a steady state, where the probabilities no longer change. This steady state is the stationary distribution, and it is unique for irreducible and aperiodic chains because the chain's long-term behavior is not dependent on where it started. □

This Theorem 4.1 follows from the fact that for an irreducible and aperiodic Markov chain, the transition probabilities eventually converge to a unique stationary distribution. The Markov chain "forgets" its initial state and its behavior becomes independent of the starting point. The chain reaches equilibrium at the stationary distribution $\boldsymbol{\pi}$. Think of it as if the system's "memory" of its starting point fades away, and it reaches a state where the probabilities of being in each state are constant over time. This is the equilibrium point, and it is independent of the initial state.

4.7.2 Kolmogorov-Chapman Equation

The Kolmogorov-Chapman equation helps compute transition probabilities over multiple steps. This equation is particularly useful when we want to compute the probability of transitioning from one state to another in a given number of steps n.

Definition 4.20 (Kolmogorov-Chapman Equation [2, 5]) The Kolmogorov-Chapman equation provides a way to compute n-step transition probabilities. It states that the n-step transition probability from state s_i to state s_j can be decomposed into a sum over all possible intermediate states s_k that the system could go through in the m-step and $n-m$-step intervals:

$$P^{(n)} * ij = \sum *k \in SP^{(m)} * ikP^{(n-m)} * kj \tag{4.16}$$

for any m such that $0 < m < n$. This recursive relation is useful for computing transition probabilities over multiple time steps.

This recursive formulation essentially breaks down the multi-step transition into two parts:

1. The probability of moving from s_i to some intermediate state s_k after m steps.
2. The probability of moving from s_k to s_j after the remaining $n - m$ steps.

The recursive nature of this equation helps simplify the computation of transition probabilities for complex systems.

Proposition 4.8 (Kolmogorov-Chapman Equation Property) *For any time steps m and n, the n-step transition probability $P_{ij}^{(m+n)}$ can be computed as:*

$$P_{ij}^{(m+n)} = \sum_k P_{ik}^{(m)} P_{kj}^{(n)}. \tag{4.17}$$

This equation ensures that the transition probability over $m + n$ steps can be broken down into a product of probabilities over m and n steps.

Proof The Kolmogorov-Chapman equation follows directly from the law of total probability. The idea is to condition on the intermediate states at time m and then apply the transition probabilities for both steps m and n. □

4.7.3 Spectral Analysis of Markov Chains

Spectral analysis of the transition probability matrix provides insights into the eigenvalues and their impact on the chain's behavior.

Definition 4.21 (Eigenvalue Decomposition of Transition Matrix) For a given transition probability matrix S, the eigenvalue decomposition provides the following relation:

$$\mathsf{S}\mathsf{v}_i = \lambda_i \mathsf{v}_i, \tag{4.18}$$

where λ_i is the eigenvalue associated with the eigenvector v_i. For a stochastic matrix, the largest eigenvalue is always 1, and the corresponding eigenvector is the stationary distribution $\boldsymbol{\pi}$.

Proposition 4.9 (Spectral Radius) *The spectral radius $\rho(\mathsf{S})$ of the transition probability matrix S is the largest absolute value of its eigenvalues. For a stochastic matrix, the spectral radius is always equal to 1:*

$$\rho(\mathsf{S}) = 1. \tag{4.19}$$

This property is important because it indicates that the Markov chain cannot "explode" in the long run; it always remains bounded.

Proof For a stochastic matrix, the sum of each row is 1, which implies that one eigenvalue is 1. The other eigenvalues must have absolute values less than or equal to 1. Hence, the spectral radius $\rho(\mathsf{S})$ is equal to 1. □

4.7.4 Convergence and Long-Term Behavior

The Convergence Theorem guarantees that irreducible and aperiodic Markov chains converge to a stationary distribution in the long run.

Theorem 4.2 (Convergence to Stationary Distribution) *For an irreducible and aperiodic Markov chain, the distribution of states $\boldsymbol{\pi}^{(t)}$ at time t converges to the stationary distribution $\boldsymbol{\pi}$ as $t \to \infty$:*

$$\lim_{t\to\infty} \boldsymbol{\pi}^{(t)} = \boldsymbol{\pi}. \tag{4.20}$$

This means that regardless of the initial state, the Markov chain will eventually converge to a unique stationary distribution.

Proof This result follows from the fact that, for irreducible and aperiodic chains, the transition probability matrix S^t approaches a rank-1 matrix as $t \to \infty$, where each row is the stationary distribution $\boldsymbol{\pi}$. Hence, the system reaches equilibrium in the long run. □

Example 4.11 (Convergence) Consider a Markov chain with the following transition probability matrix:

$$\mathsf{S} = \begin{pmatrix} 0.5 & 0.5 \\ 0.2 & 0.8 \end{pmatrix}.$$

We want to find the limiting distribution of this Markov chain. The stationary distribution $\boldsymbol{\pi} = (\pi_1, \pi_2)$ satisfies:

$$\pi_1 = 0.5\pi_1 + 0.2\pi_2, \quad \pi_2 = 0.5\pi_1 + 0.8\pi_2,$$

with the constraint $\pi_1 + \pi_2 = 1$.

Substitute $\pi_2 = 1 - \pi_1$ into the first equation:

$$\pi_1 = 0.5\pi_1 + 0.2(1 - \pi_1),$$

$$\pi_1 = 0.5\pi_1 + 0.2 - 0.2\pi_1,$$

$$\pi_1 = 0.3\pi_1 + 0.2,$$

$$0.7\pi_1 = 0.2,$$

$$\pi_1 = \frac{2}{7} \approx 0.2857, \quad \pi_2 = 1 - \pi_1 = \frac{5}{7} \approx 0.7143.$$

Thus, the limiting (and stationary) distribution is $\pi_1 \approx 0.2857$ and $\pi_2 \approx 0.7143$.

4.8 Applications of Markov Chains

In this section, we will explore some practical applications of Markov Chains. These examples will help demonstrate how Markov Chains are used to model real-world systems in various fields

4.8.1 Markov Chains in Transportation Research

In transportation research, Markov Chains are often used to model vehicle movements, traffic flows, and pedestrian behavior, where the state of the system changes probabilistically over time. Markov Chains help model the flow of vehicles through a road network by representing intersections as states. Each state's transition probabilities indicate how likely it is for vehicles to move from one intersection to another in a given time period. In traffic engineering, Markov Chains are often used to model the state of traffic at different locations on a highway or within a city. The evolution of traffic congestion, allowing traffic engineers to predict peak congestion times and evaluate traffic management strategies. These models can help understand traffic congestion, the likelihood of accidents, or the flow of vehicles at intersections, providing insights into congestion and optimization of traffic systems.

Example 4.12 (Modeling Vehicle Movements in Traffic Networks) In traffic engineering, Markov Chains can be used to model the movement of vehicles between different intersections or nodes in a road network. Each intersection is considered a state, and the transition probabilities between states represent the likelihood of a vehicle moving from one intersection to another in a given time interval.

Let the state space $S = \{s_1, s_2, \ldots, s_n\}$ represent the set of intersections in a transportation network. The transition probability matrix S then models the probability of moving from one intersection to another over a fixed time period, such as 1 minute. Consider a simple network with three intersections. The states are:

$$S = \{s_1 : \text{Intersection A}, s_2 : \text{Intersection B}, s_3 : \text{Intersection C}\}.$$

The transition probability matrix might look like this:

$$\mathsf{S} = \begin{pmatrix} 0.6 & 0.3 & 0.1 \\ 0.2 & 0.5 & 0.3 \\ 0.1 & 0.4 & 0.5 \end{pmatrix}.$$

This matrix represents the probability that a vehicle will move from one intersection to another during each time step. For example, from Intersection A (s_1), there is a 60% chance it will stay at s_1, a 30% chance it will move to s_2, and a 10% chance it will move to s_3. Similarly, the transition probability matrix defines the movement probabilities for the other intersections.

Using this Markov Chain model, transportation planners can analyze traffic flow patterns evolve over time, and predict congestion at various intersections, helping to optimize traffic light timings, road capacities, and routing strategies.

Example 4.13 (Modeling Traffic Congestion) Consider a simple system where the traffic flow can be in one of three states: low congestion, moderate congestion, and high congestion. Markov Chains can model how the traffic congestion at an intersection evolves over time.

Let s_1 represent low congestion, s_2 represent moderate congestion, and s_3 represent high congestion. The transition probability matrix for this system could be as follows:

$$\mathsf{S} = \begin{pmatrix} 0.7 & 0.2 & 0.1 \\ 0.3 & 0.5 & 0.2 \\ 0.1 & 0.4 & 0.5 \end{pmatrix}.$$

This transition probability matrix represents the probabilities of moving between different levels of congestion. For example, if the system is in a state of low congestion (s_1), there is a 70% chance it will remain in low congestion, a 20% chance it will transition to moderate congestion, and a 10% chance it will transition to high congestion.

By modeling the congestion dynamics using Markov Chains, traffic engineers can predict when congestion is likely to occur, the impact of changes in traffic volume, and the effectiveness of congestion mitigation measures such as adjusting traffic light timings or implementing alternate routes or plan for infrastructure improvements.

4.8.2 Markov Chains in Urban Transportation Planning

In urban transportation planning, Markov Chains are used to model population dynamics, land use changes, the movement of populations, and the development of infrastructure over time. The probabilistic nature of Markov Chains allows planners to simulate various urban growth scenarios and evaluate their potential impacts and predict the long-term development of cities, aiding in effective urban transportation planning and infrastructure development.

Example 4.14 (Population Growth Model) Consider a simple population growth model where the population can be in one of three states: low, medium, or high population. The transitions between these states are governed by probabilities, which depend on factors such as birth rates and death rates.

The transition probability matrix might look like:

$$\mathsf{S} = \begin{pmatrix} 0.8 & 0.1 & 0.1 \\ 0.3 & 0.5 & 0.2 \\ 0.1 & 0.4 & 0.5 \end{pmatrix}.$$

In this model, if the population is in the low state, it has an 80% chance of staying in the low state, a 10% chance of moving to the medium state, and a 10% chance of moving to the high state.

By analyzing this Markov Chain, we can predict the long-term behavior of the population and plan conservation efforts accordingly.

Example 4.15 (Modeling Land Use Changes) Urban planners can use Markov Chains to model the changes in land use within a city over time. By examining the transition probabilities between different land-use types, planners can predict how an urban area will evolve, assisting with zoning decisions, infrastructure investment, and sustainability planning. Consider a city with four types of land use: residential, commercial, industrial, and agricultural. The state space $S = \{s_1, s_2, s_3, s_4\}$ represents these land use types:

$$s_1 : \text{Residential}, \quad s_2 : \text{Commercial}, \quad s_3 : \text{Industrial}, \quad s_4 : \text{Agricultural}.$$

The transition probability matrix S describes the probabilities of land transitioning from one use type to another. For example, land that is currently agricultural may have a higher probability of transitioning to residential or commercial land due to urbanization.

A possible transition probability matrix could look like this:

$$\mathsf{S} = \begin{pmatrix} 0.5 & 0.3 & 0.1 & 0.1 \\ 0.2 & 0.5 & 0.2 & 0.1 \\ 0.1 & 0.2 & 0.6 & 0.1 \\ 0.1 & 0.2 & 0.1 & 0.6 \end{pmatrix}.$$

This matrix represents the likelihood of transitioning from one land use type to another over time. For example, there is a 50% chance that residential land will remain residential, a 30% chance it will transition to commercial, and so on.

By analyzing the Markov Chain model, urban planners can predict future land use patterns, the need for infrastructure development, and potential zoning changes, helping to design cities with efficient land-use patterns and sustainable growth strategies..

4.8.3 Markov Chains in Queueing Theory

Markov Chains are fundamental tools in modeling queueing systems and analyzing their performance. In queueing theory, continuous-time Markov chains (CTMCs) are often used to represent systems where entities (e.g., customers or data packets) arrive, wait, and are served according to probabilistic rules.

Example 4.16 (M/M/1 Queue) The M/M/1 queue is a classic model consisting of:

- A single server,
- Exponentially distributed interarrival times with rate λ,
- Exponentially distributed service times with rate μ,
- An infinite buffer (queue capacity).

The system is modeled as a continuous-time Markov Chain (CTMC) where each state $n \in \mathbb{N}_0$ represents the number of customers in the system. Transitions occur due to arrivals (rate λ) and departures (rate μ), with the following dynamics:

- From state n to $n+1$: arrival at rate λ,
- From state n to $n-1$: service completion at rate μ (only if $n \geq 1$).

The infinitesimal generator matrix Q for the M/M/1 queue is:

$$\mathsf{Q} = \begin{pmatrix} -\lambda & \lambda & 0 & 0 & \dots \\ \mu & -(\lambda+\mu) & \lambda & 0 & \dots \\ 0 & \mu & -(\lambda+\mu) & \lambda & \dots \\ 0 & 0 & \mu & -(\lambda+\mu) & \dots \\ \vdots & \vdots & \vdots & \vdots & \ddots \end{pmatrix}$$

This matrix governs the time evolution of the system's state probabilities. Solving the steady-state equations $\mathsf{Q}\boldsymbol{\pi} = \mathsf{0}$ with normalization $\sum_{n=0}^{\infty} \pi_n = 1$ yields performance metrics such as:

- π_0: probability of an empty system,
- $\mathbb{E}[N]$: expected number of customers in the system,
- $\mathbb{E}[W]$: expected waiting time (via Little's Law).

4.9 Summary

Markov Chains serve as a mathematical lens for understanding, predicting, and optimizing dynamic systems, especially within transportation and urban contexts. This chapter not only formalizes the underlying stochastic principles but also connects them to tangible phenomena such as traffic flow and land use evolution. By bridging rigorous theory with applied modeling, it reveals how probabilistic transitions and equilibrium behavior underpin strategic decision-making in complex environments.

References

1. G. Frobenius. "Ueber Matrizen aus nicht negativen Elementen". In: *Sitzungsberichte der Königlich Preußischen Akademie der Wissenschaften* (1912), pp. 456–477.
2. G. R. Grimmett and D. R. Stirzaker. *Probability and Random Processes*. 3rd. Oxford University Press, 2001.
3. Charles M. Grinstead and J. Laurie Snell. *Introduction to Probability*. American Mathematical Society, 1997. https://doi.org/10.1090/mbk/004. https://math.dartmouth.edu/~prob/prob/prob.pdf.
4. Charles M. Grinstead and J. Laurie Snell. *Introduction to Probability*. 2nd Revised. Providence, RI: American Mathematical Society, 2012. https://open.umn.edu/opentextbooks/textbooks/21.
5. John G. Kemeny and J. Laurie Snell. *Finite Markov Chains*. New York: Springer, 1976. https://link.springer.com/book/9780387901923.
6. Carl D. Meyer. *Matrix Analysis and Applied Linear Algebra*. Philadelphia: Society for Industrial and Applied Mathematics, 2000. https://doi.org/10.1137/1.9780898719512.
7. J. R. Norris. *Markov Chains*. Cambridge Series in Statistical and Probabilistic Mathematics. Cambridge: Cambridge University Press, 1997. https://doi.org/10.1017/CBO9780511810633.
8. Oskar Perron. "Zur Theorie der Matrices". In: *Mathematische Annalen* 64.2 (1907), pp. 248–263. https://doi.org/10.1007/BF01449896.
9. Eugene Seneta. *Non-negative Matrices and Markov Chains*. Rev. ed. New York: Springer, 2006. https://doi.org/10.1007/0-387-32792-4. https://link.springer.com/book/10.1007/0-387-32792-4.

Chapter 5
The Principle of Maximum Entropy in IFN Estimation

5.1 Introduction

The analysis of complex systems, such as urban transportation networks, requires robust theoretical frameworks that can manage uncertainty and incomplete information. Information theory, and specifically the concept of entropy, provides such a framework. Originally developed by Claude Shannon [15], entropy quantifies the average level of *information* or *uncertainty* inherent in a variable's possible outcomes.

This chapter introduces the foundational concepts of entropy and the Principle of Maximum Entropy, emphasizing their role in rational inference under conditions of incomplete data. These principles are then extended to network analysis, with formal definitions of node entropy and network entropy. The formulation of an Ideal Flow Network is subsequently presented as a maximum entropy optimization problem, accompanied by illustrative examples demonstrating its computational implementation.

5.2 The Concepts of Entropy: From Physics to Information

The concept of entropy originates in the domain of physics, where it served as a measure of disorder within a thermodynamic system. This foundational idea was developed by Clausius and later formalized by Boltzmann and Gibbs. The Gibbs formulation, in particular, introduced the connection between entropy and probability.

K. Teknomo, *Ideal Flow Network for Transportation Research*,
Lecture Notes in Intelligent Transportation and Infrastructure,
https://doi.org/10.1007/978-981-95-4727-2_5

To apply the Principle of Maximum Entropy, we must first quantify the concept of uncertainty. We develop the mathematical foundations in a structured manner. We first introduce Shannon entropy for a general probabilistic system. Then, we extend this concept to a transport network.

5.2.1 Shannon's Information Entropy

The entropy formulation we employ in this text is based on the work of Claude Shannon [15], who adapted these principles for information theory. Information entropy, or Shannon entropy, quantifies the average uncertainty associated with a random variable. It quantifies the average amount of information, or surprise, associated with the outcome of a random process.

Definition 5.1 (Shannon Entropy [3]) Let a system have a set of n mutually exclusive states. Let the probability of the system being in state i be p_i. Shannon entropy H^S for a discrete probability distribution with n outcomes $(p_1, p_2, \ldots, p_n)$ is defined as:

$$H^S(p_1, \ldots, p_n) = -\sum_{i=1}^{n} p_i \ln p_i \tag{5.1}$$

This formula is applicable under the constraint that the probabilities must satisfy $\sum_{i=1}^{n} p_i = 1$ and $p_i \geq 0$. By convention, we define $0 \ln(0) = 0$, as a certain event (probability 1 or 0) provides no surprise.

Entropy is measured in different units depending on the logarithmic base used. When the natural logarithm ($\ln$) is applied, entropy is expressed in *nats*. When the base-2 logarithm ($\log_2$) is used, the unit becomes *bits*.

The brilliance of Shannon's formulation lies in its consistency and uniqueness, which are grounded in a set of foundational axioms. Entropy is the unique, consistent measure of uncertainty that satisfies three key properties: continuity, monotonicity for a uniform distribution, and a composition law for compound events.

Axiom 5.1 (The Axioms of Entropy) *The function of entropy, H^S, is the only function that satisfies the following three axioms:*

1. *Continuity: H^S is a continuous function of the probabilities $p_1, \ldots, p_n$. A small change in probability results in a small change in entropy.*
2. *Monotonicity: If all probabilities are equal, $p_i = \frac{1}{n}$, then H^S is a monotonically increasing function of n. This means that a system with more equally likely outcomes has greater uncertainty.*
3. *Composition Law: The total entropy of a system is invariant to how its events are grouped. If we partition a set of n events into r groups, the total entropy is the sum of the entropy of the groups and the weighted average of the entropies within each group.*

Proposition 1.1:

Example 5.1 (Composition Law) For a simple illustration of the Composition Law, consider a probability distribution with three elements (p_1, p_2, p_3). If we group the last two elements together, the entropy is given by:

$$H^S(p_1, p_2, p_3) = H^S(p_1, p_2 + p_3) + (p_2 + p_3)H^S\left(\frac{p_2}{p_2 + p_3}, \frac{p_3}{p_2 + p_3}\right)$$

This principle ensures that the measure of uncertainty is consistent regardless of the level of detail at which we choose to analyze the system.

In simpler terms, entropy is a formal way of measuring our uncertainty about the outcome of an event. If we have a perfectly fair coin, our uncertainty is at its maximum. If we know a coin is heavily biased to land on heads, our uncertainty is low. Entropy provides a number to quantify this intuition. It is zero when an outcome is certain (probability is 1) and zero when an outcome is impossible (probability is 0).

The entropy is at its maximum when all possible outcomes (states) are equally likely. that is, $p_i = 1/n$ for all i. This corresponds to the state of maximum uncertainty, representing a state of maximum doubt or unpredictability. Conversely, the entropy is minimized (with a value of 0) when one state is certain to occur, that is, $p_k = 1$ for some state k. This corresponds to a state of complete certainty.

Example 5.2 (Route Choice at an Intersection) Consider a traffic intersection in the city of Port Sterling. A vehicle arrives from the south and can proceed in one of three directions of the possible outgoing routes:

- Route 1: turn left (L),
- Route 2: go straight (S), or
- Route 3: turn right (R).

First, consider a scenario where we have no traffic data. Without any traffic data, we can only assume each choice is equally likely. The traffic is distributed equally among the three routes. The probability for each route is $p_1 = p_2 = p_3 = 1/3$. The entropy, calculated using Eq. (5.1), is: $H^S = -\sum_{i=1}^{3} \frac{1}{3} \ln\left(\frac{1}{3}\right) = -3\left(\frac{1}{3} \ln \frac{1}{3}\right) = -\ln \frac{1}{3} = \ln(3) \approx 1.0986$. This is the maximum possible entropy for a three-state system, reflecting our maximal uncertainty about the choice of route.

Second, suppose sensors provide data indicating the probabilities. Consider a scenario where most traffic takes Route 1. The probabilities are $p_1 = 0.8$, $p_2 = 0.1$, and $p_3 = 0.1$. The entropy is: $H^S = -(0.8 \ln 0.8 + 0.1 \ln 0.1 + 0.1 \ln 0.1) \approx -(0.8 \times -0.2231 + 0.2 \times -2.3026) \approx 0.6390$. The entropy is lower than in the first scenario. The observed data, being non-uniform, has reduced the uncertainty and thus lowered the entropy. This reflects our increased certainty that a vehicle will choose Route 1.

Entropy has emerged as a powerful analytical and computational tool in transportation science, particularly in modeling flow distributions, network resilience, and optimization under uncertainty. The following studies exemplify diverse applications of entropy-based methods in transport contexts. In his thesis, Bhat [1] proposed a dynamic entropy-minimizing framework for transportation planning, emphasizing feedback and steady-state equilibrium. Shi et al. [16] extended entropic optimal transport theory to general graph structures via the EOFT-Sinkhorn algorithm. Their formulation incorporates flow balance and capacity constraints, enabling scalable computation of entropy-regularized transport plans on complex networks. Nutz [13] developed a rigorous framework for entropic optimal transport, including duality theory and Sinkhorn algorithms. These foundations support entropy-based modeling of transportation flows, offering tractable solutions for high-dimensional and stochastic routing problems. Zeng et al. [25] proposed an entropy-integrated centrality metric to identify critical nodes in multimodal public transport systems. By quantifying uncertainty and diversity in passenger flows, their method enhances network resilience assessment and informs targeted interventions for flow optimization.

Wang and Holguín-Veras [20] developed a tour-based entropy maximization framework for commercial vehicle flows, validating equilibrium under aggregate constraints, ideal for IFN calibration in freight systems. Lopez-Ospina et al. [10] developed an entropy-maximizing freight tour synthesis model using fuzzy logic and cost variability, offering a robust framework for routing under uncertainty in transport networks.

5.2.2 *Joint, Conditional, and Relative Entropy*

To analyze interactions within a network, we must extend the concept of entropy to multiple variables. Let X and Y be two discrete random variables. Let $p(x_i, y_j)$ be their joint probability. Joint entropy quantifies the uncertainty of a pair of random variables.

Definition 5.2 (Joint and Conditional Entropy [3]) The *joint entropy* $h(X, Y)$ is the entropy of the joint distribution $p(x_i, y_j)$:

$$h(X, Y) = -\sum_i \sum_j p(x_i, y_j) \ln p(x_i, y_j) \tag{5.2}$$

The *conditional entropy* $h(Y|X)$ measures the remaining uncertainty in Y given that X is known:

$$h(Y|X) = -\sum_i \sum_j p(x_i, y_j) \ln p(y_j|x_i) = \sum_i p(x_i) h(Y|X = x_i) \tag{5.3}$$

These quantities are related by a fundamental chain rule.

Proposition 5.1 (Chain Rule for Entropy) *The entropy of a pair of random variables is the sum of the entropy of one variable and the conditional entropy of the other, given the first.*

$$h(X, Y) = h(X) + h(Y|X) \tag{5.4}$$

Proof We start from the definition of joint entropy, Eq. (5.2), and use the definition of conditional probability, $p(x_i, y_j) = p(y_j|x_i)p(x_i)$.

$$\begin{aligned}
h(X, Y) &= -\sum_{i,j} p(x_i, y_j) \ln p(x_i, y_j) \\
&= -\sum_{i,j} p(x_i, y_j) \ln \big(p(y_j|x_i)p(x_i)\big) \\
&= -\sum_{i,j} p(x_i, y_j) \big(\ln p(y_j|x_i) + \ln p(x_i)\big) \\
&= -\sum_{i,j} p(x_i, y_j) \ln p(y_j|x_i) - \sum_i \left(\sum_j p(x_i, y_j)\right) \ln p(x_i) \\
&= h(Y|X) - \sum_i p(x_i) \ln p(x_i) \\
&= h(Y|X) + h(X)
\end{aligned}$$

This completes the proof. □

An important consequence is that knowledge of one variable can only reduce or maintain the uncertainty about another. It can never increase it.

Corollary 5.1 *Conditioning reduces entropy: $h(Y|X) \leq h(Y)$. Equality holds if and only if X and Y are independent.*

Proof This is a direct result of the non-negativity of mutual information, which we define in Definition 5.4. □

It is often necessary to compare two probability distributions over the same variable.

Definition 5.3 (Relative Entropy and Cross-Entropy) Let p and q be two probability distributions on a set of events $\{x_1, \ldots, x_n\}$. The *relative entropy*, also known as *Kullback-Leibler (KL) divergence* [7, 8], from q to p is:

$$d_{KL}(\mathsf{p}||\mathsf{q}) = \sum_{i=1}^{n} p_i \ln\left(\frac{p_i}{q_i}\right) \tag{5.5}$$

The *cross-entropy* between p and q is:

$$h(\mathsf{p}, \mathsf{q}) = -\sum_{i=1}^{n} p_i \ln(q_i) \tag{5.6}$$

These two quantities are related by the identity $h(\mathsf{p}, \mathsf{q}) = h(\mathsf{p}) + d_{KL}(\mathsf{p}||\mathsf{q})$. The KL divergence measures the inefficiency of assuming the distribution is q when the true distribution is p. It is not a true metric as it is not symmetric.

In the application side, Courtain et al. [2] proposed a relative entropy-regularized optimal transport model on directed graphs. Rubinstein and Kroese [14] introduced the Cross-Entropy method for rare-event simulation and combinatorial optimization, widely applied in transport model calibration and flow inference.

Theorem 5.1 (Gibbs' Inequality) *The relative entropy is non-negative:*

$$d_{KL}(\mathsf{p}||\mathsf{q}) \geq 0 \tag{5.7}$$

with equality if and only if $\mathsf{p} = \mathsf{q}$.

Proof We use the logarithmic inequality $\ln(x) \leq x - 1$ for all $x > 0$, with equality only at $x = 1$.

$$\begin{aligned} -d_{KL}(\mathsf{p}||\mathsf{q}) &= -\sum_i p_i \ln\left(\frac{p_i}{q_i}\right) = \sum_i p_i \ln\left(\frac{q_i}{p_i}\right) \\ &\leq \sum_i p_i \left(\frac{q_i}{p_i} - 1\right) \\ &= \sum_i (q_i - p_i) = \sum_i q_i - \sum_i p_i = 1 - 1 = 0 \end{aligned}$$

Thus, $d_{KL}(\mathsf{p}||\mathsf{q}) \geq 0$. Equality holds when $q_i/p_i = 1$ for all i, which means $\mathsf{p} = \mathsf{q}$.
□

A direct consequence of Gibbs' inequality is that $h(\mathsf{p}) \leq h(\mathsf{p}, \mathsf{q})$. The true entropy of a distribution is its minimum possible cross-entropy. Finally, we can quantify the information shared between two variables.

Mutual information quantifies the amount of information shared between two random variables. In transport modeling, it's used to assess dependency between origin-destination patterns or flow distributions. Cover and Thomas [3] provide a comprehensive foundation for entropy measures, Shannon, mutual information, and relative entropy, used in flow inference and network uncertainty modeling.

Definition 5.4 (Mutual Information) *Mutual information*, denoted by $I(X; Y)$, measures the reduction in uncertainty about X due to the knowledge of Y:

$$I(X; Y) = h(X) - h(X|Y) \tag{5.8}$$

It is symmetric, so $I(X; Y) = I(Y; X)$. Using the chain rule, it can also be expressed as:

$$I(X; Y) = h(X) + h(Y) - h(X, Y) \tag{5.9}$$

Mutual information can also be written as the KL divergence between the joint distribution and the product of the marginals: $I(X; Y) = d_{KL}(p(x, y)||p(x)p(y))$. If X and Y are independent, then $p(x, y) = p(x)p(y)$ and $I(X; Y) = 0$.

5.2.3 *Maximum Entropy*

The most important property of entropy for our application is its behavior at a maximum.

Proposition 5.2 *The entropy of a probability distribution $H^S(p_1, \ldots, p_n)$ is maximized when all probabilities are equal, i.e., $p_i = \frac{1}{n}$ for all $i = 1, \ldots, n$.*

Proof To find the maximum of H^S, we use the method of Lagrange multipliers [11], subject to the constraint that $\sum_{i=1}^{n} p_i = 1$. We form the Lagrangian function:

$$L(p_1, \ldots, p_n, \lambda) = H^S + \lambda g(p_1, \ldots, p_n) = -\sum_{i=1}^{n} p_i \ln p_i + \lambda \left(\sum_{i=1}^{n} p_i - 1 \right)$$

We set the gradient of the Lagrangian with respect to each p_i to zero:

$$\frac{\partial L}{\partial p_i} = -(\ln p_i + 1) + \lambda = 0$$

This implies $\ln p_i = \lambda - 1$, so $p_i = e^{\lambda - 1}$. Since this value is independent of the index i, all probabilities must be equal: $p_1 = p_2 = \cdots = p_n$. Substituting this into our constraint $\sum_{i=1}^{n} p_i = 1$, we get $n \cdot p_i = 1$, which gives us the result $p_i = \frac{1}{n}$. □

The practical meaning of this theorem is profound. When we have no information about a system other than the possible outcomes, the most rational assumption to make is that each outcome is equally likely. This is the state of maximum uncertainty. We can also define the entropy ratio, $R = \frac{H^S}{\max H^S}$, which normalizes the entropy of any distribution relative to its maximum possible value.

For a tangible demonstration, consider a physical space divided into smaller areas. The areas of these divisions can be modeled as probabilities, provided the total area is unity. As we introduce more, more uniform divisions, the entropy increases.

Example 5.3 (Entropy in Rectangle) Consider a sequence of rectangles, all with a total area of 1.

- Rectangle (a) has one area. It has no divisions; there is only one possible outcome, so the entropy is $H^S = 1 \ln 1 = 0$.
- Rectangle (b) is divided into two equal areas, giving it the maximum possible entropy for two outcomes. The probabilities are $p_1 = 0.5$ and $p_2 = 0.5$. The entropy is $H^S = -0.5 \ln 0.5 - 0.5 \ln 0.5 = 0.6931$. This is the maximum possible entropy for two outcomes, so the entropy ratio is $R = 1$.
- Rectangle (c) has three parts with probabilities $p_1 = 0.5, p_2 = 0.25, p_3 = 0.25$. The entropy is $H^S = 1.0397$. The maximum entropy for three outcomes is $H_{max} = -\sum_{i=1}^{3} \frac{1}{3} \ln \frac{1}{3} = 1.0986$. The entropy ratio is $R = \frac{1.0397}{1.0986} = 0.9464$.
- Rectangle (d) has five parts with probabilities $p_1 = 0.5, p_2 = 0.25, p_3 = 1/12, p_4 = 1/12, p_5 = 1/12$. The entropy is $H^S = 1.3144$. The maximum for five outcomes is $H^S_{max} = -\sum_{i=1}^{5} \frac{1}{5} \ln \frac{1}{5} = 1.6094$. The entropy ratio is $R = \frac{1.3144}{1.6094} = 0.8167$.

When rectangles are further subdivided, with their entropy values increasing as the divisions become more numerous, though not perfectly uniform. This illustrates how entropy is a direct measure of the diversity and uniformity of a distribution. The entropy is highest when the parts are equal, and the entropy ratio falls as the distribution becomes less uniform.

The connection between entropy and information is also direct. The amount of information conveyed by an event with probability p is defined as $I(p) = \log_2(1/p)$. The entropy is simply the expected value of this information. For a series of binary decisions in a decision tree, the probability of reaching a leaf node is related to the path taken, and the total uncertainty is captured by the entropy. This principle is fundamental to understanding how information is encoded and transmitted.

5.3 The Principle of Maximum Entropy

The principle of maximum entropy is a profound method for making inferences from incomplete information. It provides a formal, mathematical framework for the common-sense notion known as the Principle of Insufficient Reason. This principle, stated by Jaynes [5], dictates that the probability distribution that best represents a system with incomplete information is the one with the largest entropy, subject to the available constraints. In essence, it tells us to choose the most unbiased, least-presumptuous distribution that is consistent with all of our known data.

5.3.1 *Principle of Insufficient Reason*

When faced with a problem with multiple possible solutions and a lack of information to distinguish between them, the most rational approach is to assume the simplest, most uniform distribution. This is precisely what maximum entropy formalizes.

Proposition 5.3 *Given a set of probabilities with a single constraint (the sum of probabilities equals 1), the maximum entropy distribution is the uniform distribution.*

Proof This was proven in Proposition 5.2. The solution to the constrained optimization problem $\max(-\sum p_i \ln p_i)$ with constraint $\sum p_i = 1$ is $p_i = \frac{1}{n}$. □

This proposition provides the mathematical justification for the Principle of Insufficient Reason. For example, if we have five variables whose total value is 150, with no other information, the maximum entropy assumption leads to a uniform distribution where each variable is 30. We do not arbitrarily assign values like 10, 20, 30, 40, and 50, because that would implicitly assume a linear progression, which is not supported by our given information. Maximum entropy forces us to be honest about our lack of knowledge.

5.3.2 *Solving Incomplete Problems*

The power of maximum entropy becomes evident when dealing with more complex constraints that defy simple common-sense assumptions.

A classic example is a series of coin flips.

Example 5.4 (Biased Coins with Incomplete Information) Consider a scenario with two biased coins. The probabilities of heads for each coin, p_1 and p_2, are unknown. We are only given one piece of information from an experiment: the event of flipping two heads (HH) occurred with a probability of 0.5.

Without maximum entropy, we are stuck. There are infinitely many combinations of p_1 and p_2 that satisfy $p_1 p_2 = 0.5$. We cannot determine a unique solution from this single constraint.

The maximum entropy method provides a principled way forward. We maximize the total entropy of the system:

$$H^S = -p_1 \ln p_1 - (1 - p_1)\ln(1 - p_1) - p_2 \ln p_2 - (1 - p_2)\ln(1 - p_2)$$

Subject to the constraint:

$$p_1 p_2 = 0.5$$

Solving this optimization problem reveals that the most unbiased and probable solution is when $p_1 = p_2 = \sqrt{0.5} \approx 0.707$. The maximum entropy principle has, in effect, added a hidden assumption, that the coins are equally biased, because it leads to the most uniform distribution consistent with the data.

This approach becomes even more powerful when we introduce multiple, seemingly conflicting constraints, such as knowledge of both $p_1 p_2 = 0.5$ and $(1 - p_1)p_2 = 0.25$. Maximum entropy can still find a unique solution that satisfies all constraints while remaining as unbiased as possible.

5.3.3 MaxEnt and Queuing Efficiency

A compelling real-world application of this principle is in managing a queuing system.

Example 5.5 (Maximizing Queuing Efficiency) Imagine a queuing system with 30 customers and three servers with different service times: $r_1 = 1$, $r_2 = 2$, and $r_3 = 3$ minutes. Our goal is to distribute the customers (N_1, N_2, N_3) to achieve the most efficient system. A common, but flawed, metric for efficiency is average waiting time. The invariant principle in queuing theory tells us that the average waiting time is independent of the distribution of customers among the servers.

A better measure is to minimize the maximum queue time for any single server. This measure directly addresses customer satisfaction by reducing the longest wait. Maximum efficiency, therefore, corresponds to a scenario where the total queue time for each server is as uniform as possible. This is a direct parallel to the maximum entropy principle.

To formulate this, we define the probability of a customer being served by server i based on its total queue time $T_i = N_i r_i$. We then maximize the entropy of this probability distribution: Find N_1, N_2, N_3 to

$$\max_{N_i} H^S = -\sum_{i=1}^{3} p_i \ln p_i$$

Subject to the constraints:

$$p_i = \frac{N_i r_i}{\sum_j N_j r_j} \quad \text{and} \quad \sum_{i=1}^{3} N_i = 30$$

This optimization leads to the solution where all queue times are equal: $N_1 r_1 = N_2 r_2 = N_3 r_3$. This shows that the principle of maximum entropy naturally finds the most efficient configuration, distributing the workload to create the fairest and most uniform experience for customers.

5.3.4 MaxEnt for Solving Inverse Problems

The maximum entropy framework can be generalized to solve a wide range of inverse problems, where we must determine an unknown input from a known output. This includes solving systems of linear equations and performing linear regression. The key is to separate the unknown parameter vector $\mathbf{q}$ into two components: a probability vector $\mathbf{p}$ and a feature vector $\mathbf{f}$, such that $\mathbf{q} = \mathbf{p} \circ \mathbf{f}$. The problem is then to maximize the entropy of $\mathbf{p}$ subject to the linear constraints imposed by the known data. This provides a robust and principled method for finding a unique and unbiased solution even when the problem is underdetermined.

5.3.5 The Gravity Model and Maximum Entropy

The estimation of a trip distribution matrix, also known as an Origin-Destination (O-D) matrix, is a cornerstone of classical transportation planning. The fundamental challenge is to determine the number of trips, f_{ij}, from each origin zone i to each destination zone j when only aggregate data is available. The Principle of Maximum Entropy provides the theoretical justification for the widely used doubly-constrained gravity model by showing that it represents the most probable and unbiased distribution consistent with the known data. this is a classic problem is estimating the trip distribution (O-D) matrix F from aggregate data [4, 23]. Wilson [21–23] introduced entropy-maximizing models for spatial interaction under information-theoretic constraints. Xie et al. [24] applied maximum entropy methods to estimate subnetwork origin-destination matrices, offering a scalable approach for ideal flow calibration in reduced transport networks.

The problem is to estimate the trip matrix, represented by the matrix $\mathsf{F} = [f_{ij}]$, given the following aggregate information:

- The total number of trips originating from each zone i, denoted by the vector $\mathsf{O} = [O_i]$.
- The total number of trips terminating at each zone j, denoted by the vector $\mathsf{D} = [D_j]$.
- A known matrix of travel costs, $\mathsf{C} = [c_{ij}]$, where c_{ij} is the cost of a trip from zone i to zone j. This cost can represent travel time, monetary expense, or a combination thereof.
- A known total system-wide travel cost, C_{total}.

Without the principle of maximum entropy, this problem is intractable. We have far more unknown variables (the individual f_{ij} values) than we have equations (the total origin and destination trips). Any attempt to fill in the matrix without an underlying principle would introduce arbitrary bias. Maximum entropy solves this by formalizing the least-presumptuous assumption: we should choose the matrix that is consistent with all of our known data, but is otherwise as random and uniform as possible.

The problem is to find the matrix of trips F that maximizes the system entropy. The entropy is defined over the probability distribution of trips, $p_{ij} = f_{ij}/\kappa$, where $\kappa = \sum_{i,j} f_{ij}$ is the total number of trips in the system. The optimization formulation is as follows:

Objective Function: Maximize the system entropy, H^S.

$$\max_{\mathbf{F}} \quad H^S = -\sum_{i=1}^{n}\sum_{j=1}^{m}\left(\frac{f_{ij}}{\kappa}\right)\ln\left(\frac{f_{ij}}{\kappa}\right)$$

Constraints: The maximization is subject to three crucial sets of constraints that embody the known data.

1. *Origin trip constraints* : The sum of trips leaving each origin zone i must equal the known total number of trips originating from that zone, O_i. This is a set of n linear constraints, one for each origin zone.

$$\sum_{j=1}^{m} f_{ij} = O_i \quad \forall i = 1, \ldots, n$$

   This constraint is often referred to as flow conservation at the origin. It guarantees that our estimated trips for a given zone sum up to the observed total.
2. *Destination trip constraints*: The sum of trips arriving at each destination zone j must equal the known total number of trips terminating at that zone, D_j. This is a set of m linear constraints, one for each destination zone.

$$\sum_{i=1}^{n} f_{ij} = D_j \quad \forall j = 1, \ldots, m$$

   This is the corresponding flow conservation constraint for destinations, ensuring that our model's outputs match the observed trip arrivals.
3. *Total System Cost Constraint*: The sum of all trips, weighted by their respective costs, must equal the known total system-wide travel cost, C_{total}. This is a single, powerful linear constraint.

$$\sum_{i=1}^{n}\sum_{j=1}^{m} f_{ij}c_{ij} = C_{\text{total}}$$

   This constraint is arguably the most important. It introduces a behavioral component into the model, acknowledging that travelers' choices are not purely random; they are influenced by travel costs. This constraint ensures the model's output is consistent with the observed overall travel cost behavior of the system.

The solution to this constrained maximization problem yields a specific, elegant mathematical form for the trip distribution matrix. The resulting maximum entropy distribution for the flow is the well-known doubly-constrained gravity model.

$$f_{ij} = A_i O_i B_j D_j \exp(-\lambda c_{ij})$$

Here, the terms A_i and B_j are balancing factors (derived from the Lagrange multipliers of the origin and destination constraints) that ensure the row and column sums of the final matrix precisely match the known O_i and D_j totals. The parameter λ (derived from the Lagrange multiplier of the cost constraint) is a deterrence parameter that controls the sensitivity of travelers to cost. This shows that a cornerstone of traditional transportation planning is not merely an empirical formula, but a direct and rigorous consequence of maximizing entropy given a set of real-world constraints.

5.4 Applying Entropy to Networks

Having established the foundational principles of entropy and maximum entropy, we can now apply these concepts to analyze and characterize a network. This requires a shift in perspective, moving from general probability distributions to the specific, microscopic behavior of a network's nodes and links.

5.4.1 Defining Network-Specific Metrics

A network is composed of a set of nodes and directed links. The flow across these links defines the state of the network. We can analyze this state from both a global, network-wide perspective and a local, node-centric point of view.

5.4.1.1 Global and Local Entropy Measures

Shannon network entropy (H^S) is a macroscopic metric that treats the entire network as a single probability distribution of links.

Definition 5.5 (Shannon Network Entropy (H^S)) The Shannon network entropy (H^S) is a metric that would treat the entire network as a single system where the probability distribution is defined over all links. The probability of a flow unit being on any given link is $p_{ij} = f_{ij}/\kappa$, where f_{ij} is the flow on a link and κ is the total network flow. The formula of Shannon Network Entropy is:

$$H^S = -\sum_i \sum_j p_{ij} \ln p_{ij} \tag{5.10}$$

H^S provides a comprehensive, high-level view of flow dispersion. It measures the uncertainty associated with a flow unit's location within the network. It quantifies the predictability of flow distribution. A high value for H^S indicates a diffuse, more dispersed and less predictable flow pattern across the network's links. Conversely, a low value of Shannon network entropy suggests that traffic flow is concentrated on a few links (such as a few major arteries), making the network's state more predictable. The probability term, $p_{ij} = \frac{f_{ij}}{\kappa}$, is defined as the ratio of the flow on a specific link (f_{ij}) to the total flow in the system (κ). This transforms the raw flow values into a normalized probability distribution. The formula, $H^S = -\sum_i \sum_j p_{ij} \ln p_{ij}$, is the standard mathematical expression for Shannon entropy as we have defined earlier in Definition 5.1. It sums the information content of each link's probability, weighted by that probability itself. The negative sign ensures that the result is a positive value.

For simplicity, we can index the links from 1 to E and let the probability of flow on link k be p_k.

Proposition 5.4 (Maximum Shannon Network Entropy) *The Shannon network entropy (H^S) of any network is maximized if and only if the flow is uniformly distributed across all its links. The maximum value is:*

$$H^S_{max} = \ln(E) \tag{5.11}$$

where E is the total number of links in the network.

Proof The Shannon network entropy is defined by the probability distribution of flow across the links. Let E be the total number of links in the network, and let $p_k = f_{ij}/\kappa$ be the probability of a flow unit being on link k. The sum of these probabilities is unity, i.e., $\sum_{k=1}^{E} p_k = 1$. The entropy is given by:

$$H^S = -\sum_{k=1}^{E} p_k \ln p_k$$

To find the maximum value of H^S, we can use Gibbs' inequality. This inequality states that for any probability distribution $P = \{p_1, \ldots, p_E\}$ and another distribution $Q = \{q_1, \ldots, q_E\}$ over the same events, the following holds:

$$-\sum_{k=1}^{E} p_k \ln p_k \leq -\sum_{k=1}^{E} p_k \ln q_k$$

Equality holds if and only if $p_k = q_k$ for all k.

To find the upper bound for H^S, we choose a reference distribution Q that is uniform, where $q_k = 1/E$ for all links. Substituting this into Gibbs' inequality gives:

$$H^S \leq -\sum_{k=1}^{E} p_k \ln\left(\frac{1}{E}\right)$$

Using the property of logarithms, $\ln(1/E) = -\ln(E)$, the inequality becomes:

$$H^S \leq -\sum_{k=1}^{E} p_k(-\ln(E))$$

$$H^S \leq \ln(E)\sum_{k=1}^{E} p_k$$

Since the sum of the probabilities is 1, $\sum_{k=1}^{E} p_k = 1$, we have:

$$H^S \leq \ln(E)$$

This proves that the maximum possible value for the Shannon network entropy is $\ln(E)$. The equality holds if and only if the probability distribution P is identical to the uniform distribution Q, which means $p_k = 1/E$ for all links. This corresponds to the condition where the flow on every link is the same. □

Shannon network entropy is a comprehensive metric to describe the entire network. However, for calibrating an IFN, this global view is insufficient. The IFN model is rooted in the principle of flow conservation at each node, which implies that a network's behavior emerges from local, node-level decisions. A more suitable metric is needed to capture the uniformity of these local decisions.

To achieve this, we first define the local properties of each node, where we can analyze the states from a local, node-centric point of view.

Definition 5.6 (Node Value (π_i)) The node value, denoted by π_i, is the total amount of flow that passes through node i. For an Ideal Flow Network (IFN), this value is equal to the sum of all inflows and the sum of all outflows at node i, due to the principle of flow conservation.

Definition 5.7 (Out-link Probability (s_{ij})) The out-link probability, s_{ij}, is the probability that a unit of flow leaving node i will travel to its neighbor, node j. It is derived from the flow matrix F and the node value π_i as:

$$s_{ij} = \frac{f_{ij}}{\pi_i}$$

This forms a local probability distribution for each node, such that the sum of all out-link probabilities from a given node is unity, i.e., $\sum_j s_{ij} = 1$.

Definition 5.8 (Node Entropy (H_i)) The node entropy, H_i, is the Shannon entropy of the out-link probabilities for a single node i. It measures the uncertainty of flow distribution inherent in how that node distributes its flow to its neighbors. The formula for node entropy is:

$$H_i = -\sum_j s_{ij} \ln s_{ij}$$

By definition, a zero out-link probability ($s_{ij} = 0$) contributes zero to the entropy sum, as $\ln(0)$ is treated as zero in this context.

These local entropies must be aggregated to form a single network-level metric that reflects the importance of each node. While node entropy provides a crucial local measure, we cannot simply sum the node entropies because this would misrepresent the system. A simple average would misrepresent the system by giving equal weight to low-flow and high-flow nodes. Therefore, we use a flow-weighted average. We propose a metric that weights each node's local entropy by its significance in the total flow.

5.4.1.2 Maximum Entropy for Inferring Missing Network Data

The utility of the Maximum Entropy Principle [5, 6] is most evident when dealing with real-world problems that suffer from a lack of complete data. Calibrating an Ideal Flow Network (IFN) for a large city presents a perfect example. Gathering a complete dataset of turning movement probabilities for every intersection is economically and logistically infeasible. Instead, we typically have sparse, incomplete data from various sources. This is precisely where maximum entropy provides a powerful and principled method for inferring the missing information.

The core idea is to use all available, truthful data as constraints and then choose the most unbiased probability distribution that satisfies those constraints. This is the distribution that maximizes the entropy, making no assumptions beyond the known facts. Instead of conducting thousands of traffic counts, we are using the most truthful and unbiased inference method possible to fill in the missing data, creating a complete and consistent stochastic matrix for our entire IFN.

For any large-scale network, such as a city's road system, detailed data is a luxury. Traffic surveys, while accurate, are expensive and can only cover a tiny fraction of the network. However, we do have a wealth of macroscopic, aggregate information from sources like OpenStreetMap (road width, number of lanes, road type) and Google Maps (estimated traffic flow). This information provides vital constraints on the network's behavior. The challenge is to translate these static structural and sparse flow constraints into a complete, consistent set of microscopic turning movement probabilities for every intersection. This is a classic inverse problem.

We must formally define our objective. For a given intersection i, our task is to find the most probable and unbiased set of out-link probabilities, s_{ij}, for all outgoing links j. This collection of probabilities forms the node's distribution.

The solution lies in applying the Maximum Entropy Principle to each intersection individually. The principle tells us to maximize the uncertainty of our distribution while strictly adhering to all known data. This formalizes the idea of making the least-presumptuous assumption possible.

Optimization Formulation For a specific intersection i, we formulate the problem as: Find the out-link probabilities s_{ij} to

$$\max_{s_{ij}} \quad H_i = -\sum_{j=1}^{m_i} s_{ij} \ln s_{ij}$$

Subject to constraints:

1. Probability Sum: The sum of probabilities for all outgoing links must equal one: $\sum_{j=1}^{m_i} s_{ij} = 1$.
2. Non-negativity: All probabilities must be non-negative: $s_{ij} \geq 0$.
3. Known Data Constraints: This is where we incorporate real-world information. These constraints can take various forms, such as:
   - Known flows: If a field survey provides a specific flow value for one movement.
   - Flow relationships: If physical characteristics (e.g., road width, number of lanes) imply a proportional relationship between two movements.

This formulation allows us to bridge the gap between our sparse knowledge and a complete model.

To illustrate this method, let's explore two scenarios: a trivial case with a single constraint, and a non-trivial case with multiple, more complex constraints.

Example 5.6 (Missing Data) Consider a three-way intersection with three possible movements: left, straight, and right. Through a small field survey, we learn that the probability of a left turn (s_L) is 0.25. The probabilities for going straight (s_S) and turning right (s_R) are completely unknown.

Problem: Find the most probable and unbiased values for s_S and s_R.

Constraints:

1. $s_L = 0.25$
2. $s_L + s_S + s_R = 1$
3. $s_S, s_R \geq 0$

Maximum Entropy Solution: The problem is to maximize the entropy of the probability distribution (s_L, s_S, s_R) subject to these constraints. The most honest assumption is that the unknown probabilities are equal. The total remaining probability is $1 - s_L = 1 - 0.25 = 0.75$. This remaining probability is distributed uniformly across the two unknown movements.

$$s_S = s_R = \frac{0.75}{2} = 0.375$$

This example demonstrates the core principle: MaxEnt automatically distributes the remaining uncertainty as uniformly as possible.

Example 5.7 (Complex Constraints) Let's consider a more realistic scenario where we have multiple constraints. At a three-way intersection, we have the following information: The total flow entering the intersection (and therefore exiting) is 1000 vehicles per hour. A sensor provides a known flow of 500 veh/hr for the straight-through movement. Data from OpenStreetMap indicates that the road for the left turn has a physical capacity that is twice that of the road for the right turn. This is a crucial constraint that implies the flow and therefore the turning probability of the left turn is twice that of the right turn.

Problem: Determine the turning movement probabilities s_L, s_S, and s_R.

Constraints:

1. $F_L + F_S + F_R = 1000$ (Total flow constraint)
2. $F_S = 500$ (Known flow for straight movement)
3. $F_L = 2F_R$ (Capacity-based flow relationship)

Maximum Entropy Solution: We use these constraints to solve for the unknown flows. From the first two constraints, we know that the remaining flow to be distributed between the left and right turns is $1000 - 500 = 500$ veh/hr. Thus, $F_L + F_R = 500$. Substituting the third constraint into this equation gives us: $2F_R + F_R = 500 \implies 3F_R = 500$. Solving for F_R gives $F_R \approx 166.67$ veh/hr. We can then find $F_L = 2 \times F_R \approx 333.33$ veh/hr. Resulting Probabilities:

- $s_L = \frac{333.33}{1000} \approx 0.333$
- $s_S = \frac{500}{1000} = 0.500$
- $s_R = \frac{166.67}{1000} \approx 0.167$

This example shows that MaxEnt provides a robust, objective, and cost-effective solution for a data-starved problem. It allows us to build a complete, consistent, and logically sound stochastic matrix for the entire IFN using only a small, representative subset of real-world data, thus avoiding expensive and often infeasible field surveys.

5.4.1.3 Network Entropy: Average Node Entropy

The network entropy (H) is a metric that weights each node's local entropy (H_i) by its normalized node value, π_i/κ, reflecting the node's significance within the network. This metric is also known as the flow-weighted average of node entropies.

Definition 5.9 (Network Entropy: Average Node Entropy (H)) The network entropy, H, is a flow-based weighted average of all node entropies in the network. Each node's entropy is weighted by its normalized node value $\frac{\pi_i}{\kappa}$, where $\kappa = \sum_i \pi_i$ is the total flow in the network. The formula is:

$$H = \sum_{i=1}^{n} \frac{\pi_i}{\kappa} H_i = -\sum_{i=1}^{n} \frac{\pi_i}{\kappa} \sum_j s_{ij} \ln s_{ij} \tag{5.12}$$

This metric ensures that nodes with higher flow have a proportionally greater influence on the total network entropy.

5.4.1.4 Network Entropy Decomposition

It is critical to distinguish the Network Entropy as shown in Eq. (5.12), which is the average node entropy, from Shannon network entropy in Eq. (5.10).

While both metrics are non-negative, they measure different properties of the network. Network entropy (H) is a microscopic measure of the internal distribution of flow at each node. It captures the diversity and uniformity of choices made at every intersection. Shannon network entropy (H^S) is a macroscopic measure of the global distribution of flow across all links. Shannon network entropy is useful for a bird's-eye view but loses crucial information about the local behavior of individual nodes. For the purpose of network computation, where we want to find the most unbiased and uniform distribution of choices at every node, the flow-weighted average of node entropy (H) is the more suitable metric. The network entropy (H) is a more sensitive and appropriate metric for calibrating an IFN than the Shannon network entropy (H^S).

The IFN is defined by its ability to generate an unbiased and uniform distribution of flow choices at every node. Network entropy (H) directly measures this uniformity at the local level. It captures the diversity of choices at each intersection and aggregates this information based on flow volume. A high value of H indicates that flow is efficiently distributed across the out-links of each node, minimizing bottlenecks and maximizing local flow dispersion.

Shannon network entropy (H^S), in contrast, measures global flow dispersion. It can be high even if local distributions are highly non-uniform. For example, a network with many paths of varying length might have high H^S simply due to the large number of links, but a few nodes with concentrated outflow could still create severe local bottlenecks.

The relationship between these two metrics, $H^S = H + H_{\text{nodes}}$, as shown in Proposition 5.5 mathematically proves this distinction. Shannon entropy (H^S) is the sum of network entropy (H) and the entropy of the node flows (H_{nodes}). This decomposition shows that H isolates the portion of network uncertainty directly related to the node-level flow choices, which is the exact property we aim to optimize in an IFN. Therefore, for the purpose of defining and calibrating an IFN, network entropy (H) is the more suitable metric.

To prove the relationship $H^S = H + H_{\text{nodes}}$, we must start with the definitions of each term and use the properties of logarithms and summation. This is a fundamental result from information theory, specifically related to the decomposition of joint entropy.

Proposition 5.5 (Network Entropy Decomposition) *The Shannon network entropy (H^S) is equal to the sum of the flow-weighted average node entropy (H) and the entropy of the node flow distribution (H_{nodes}). This relationship can be expressed as:*

$$H^S = H + H_{nodes} \tag{5.13}$$

Proof Let us begins with recalling the definition of Shannon network entropy (H^S):

$$H^S = -\sum_i \sum_j p_{ij} \ln p_{ij}$$

where $p_{ij} = \frac{f_{ij}}{\kappa}$ is the probability of a flow unit being on link (i, j), and κ is the total system flow.

The probability p_{ij} can be decomposed into two components using the laws of conditional probability:

$$p_{ij} = \left(\frac{f_{ij}}{\pi_i}\right)\left(\frac{\pi_i}{\kappa}\right)$$

where $\frac{f_{ij}}{\pi_i} = s_{ij}$ is the conditional probability of traveling from node i to node j, and $\frac{\pi_i}{\kappa}$ is the marginal probability of a flow unit being at node i.

Substituting this decomposition into the equation for H^S:

$$H^S = -\sum_i \sum_j \left(\frac{f_{ij}}{\pi_i} \cdot \frac{\pi_i}{\kappa}\right) \ln \left(\frac{f_{ij}}{\pi_i} \cdot \frac{\pi_i}{\kappa}\right)$$

Using the property of logarithms, $\ln(ab) = \ln(a) + \ln(b)$, we can expand the expression:

$$H^S = -\sum_i \sum_j \left(\frac{f_{ij}}{\kappa}\right)\left[\ln\left(\frac{f_{ij}}{\pi_i}\right) + \ln\left(\frac{\pi_i}{\kappa}\right)\right]$$

Now, we separate the sum into two distinct terms:

$$H^S = -\sum_i \sum_j \left(\frac{f_{ij}}{\kappa}\right)\ln\left(\frac{f_{ij}}{\pi_i}\right) - \sum_i \sum_j \left(\frac{f_{ij}}{\kappa}\right)\ln\left(\frac{\pi_i}{\kappa}\right)$$

Let's analyze the first term:

$$-\sum_i \sum_j \left(\frac{f_{ij}}{\kappa}\right)\ln\left(\frac{f_{ij}}{\pi_i}\right) = -\sum_i \left(\frac{\pi_i}{\kappa}\right)\sum_j \left(\frac{f_{ij}}{\pi_i}\right)\ln\left(\frac{f_{ij}}{\pi_i}\right)$$

This expression is the definition of the flow-weighted average node entropy, which is our term H.

$$H = \sum_i \frac{\pi_i}{\kappa}\left[-\sum_j s_{ij}\ln s_{ij}\right] = \sum_i \frac{\pi_i}{\kappa} H_i$$

Next, we analyze the second term:

$$-\sum_i \sum_j \left(\frac{f_{ij}}{\kappa}\right)\ln\left(\frac{\pi_i}{\kappa}\right)$$

Since the term $\ln(\frac{\pi_i}{\kappa})$ is constant for a given node i, we can pull it out of the inner summation over j.

$$-\sum_i \left[\ln\left(\frac{\pi_i}{\kappa}\right)\right]\sum_j \left(\frac{f_{ij}}{\kappa}\right)$$

The inner sum, $\sum_j f_{ij}$, is equal to the total flow at node i, which is π_i. Therefore, $\sum_j \frac{f_{ij}}{\kappa} = \frac{\pi_i}{\kappa}$. Substituting this back into the expression gives:

$$-\sum_i \left(\frac{\pi_i}{\kappa}\right)\ln\left(\frac{\pi_i}{\kappa}\right)$$

This is the definition of the entropy of the node flow distribution, which is our term H_{nodes}.

Combining the results for both terms completes the proof:

$$H^S = H + H_{\text{nodes}}$$

□

5.4.1.5 Entropy Ratio

To obtain the entropy ratio of Network Entropy, recall that Network Entropy is the flow-weighted average of each node's local entropy, denoted as $H = \sum_{i=1}^{n} \frac{\pi_i}{\kappa} H_i$. The maximum possible value for this metric is a crucial reference point for network analysis.

This Proposition 5.6 establishes that the maximum value is not a single constant, but depends on the network's topology (the out-degrees of its nodes) and the flow distribution (π_i).

Proposition 5.6 (Maximum Flow-Weighted Average Node Entropy) *The flow-weighted average node entropy, H, is maximized when the outflow from each node is uniformly distributed across all its outgoing links. The maximum value is:*

$$H_{max} = \sum_{i=1}^{n} \frac{\pi_i}{\kappa} \ln(k_i) \tag{5.14}$$

where k_i is the out-degree of node i.

Proof To prove the proposition, we must establish the conditions under which the flow-weighted average node entropy is maximized and derive the corresponding maximum value. The flow-weighted average node entropy is defined as the sum of individual node entropies, weighted by the normalized node flow:

$$H = \sum_{i=1}^{n} \frac{\pi_i}{\kappa} H_i$$

where $H_i = -\sum_{j \in \text{out}(i)} s_{ij} \ln s_{ij}$ is the entropy of node i. The weights, $\frac{\pi_i}{\kappa}$, are non-negative constants for a given flow distribution.

To maximize the sum H, we must maximize each individual term H_i in the summation. The maximization of the sum is contingent on the maximization of each individual component, since the weights are positive.

The entropy of a single node, H_i, is the Shannon entropy of its out-link probability distribution $\{s_{ij}\}_{j \in \text{out}(i)}$. According to a fundamental principle of information theory, the entropy of a discrete probability distribution with k_i possible outcomes is maximized when the probabilities are uniform. For a node i with out-degree k_i, this occurs when the flow is distributed equally among all its outgoing links. The out-link probability for this condition is:

$$s_{ij} = \frac{1}{k_i} \quad \forall j \in \text{out}(i)$$

By substituting this uniform distribution into the formula for node entropy, we can derive the maximum value for a single node, denoted as $H_{i,\max}$:

$$H_{i,\max} = - \sum_{j \in \text{out}(i)} \frac{1}{k_i} \ln \left(\frac{1}{k_i} \right)$$

$$H_{i,\max} = -k_i \left(\frac{1}{k_i} \ln(k_i^{-1}) \right)$$

$$H_{i,\max} = - \ln(k_i^{-1}) = \ln(k_i)$$

This maximum value depends only on the node's out-degree.

Since the flow-weighted average node entropy, H, is a sum of these terms, its maximum value, $H_{\max}$, is obtained by substituting each node's maximum entropy, $H_{i,\max}$, into the original summation:

$$H_{\max} = \sum_{i=1}^{n} \frac{\pi_i}{\kappa} H_{i,\max}$$

$$H_{\max} = \sum_{i=1}^{n} \frac{\pi_i}{\kappa} \ln(k_i)$$

□

The maximum flow-weighted average node entropy is achieved when the outflow at every node is uniformly distributed, and its value is dependent on the flow distribution and the out-degrees of the nodes.

The entropy ratio provides a normalized, unitless measure of how close the network's actual flow distribution is to its maximal state of dispersion. It is calculated by dividing the actual network entropy by the maximum possible network entropy.

Definition 5.10 (Entropy Ratio) The entropy ratio is defined as:

$$\text{Entropy Ratio} = \frac{H}{H_{\max}} = \frac{\sum_{i=1}^{n} \frac{\pi_i}{\kappa} H_i}{\sum_{i=1}^{n} \frac{\pi_i}{\kappa} \ln(k_i)} \tag{5.15}$$

This ratio provides a measure of the network's current state of flow uniformity relative to its ideal state. It measures how close the network's node-level flow distribution is to its maximal state of uniformity. The entropy ratio is a unitless value between 0 and 1. A value of 1 indicates a perfectly uniform and maximally entropic flow distribution at every node, while a value approaching 0 indicates a highly concentrated and non-uniform flow pattern.

To make these concepts concrete, consider the following example to illustrate the difference between these two entropy measures.

Example 5.8 (Network Entropy Comparison) Here, the flows on the links are given. We will compute the node and network entropy step-by-step and contrast it with the Shannon network entropy. To compute the entropy ratio, we must first calculate the maximum possible network entropy, $H_{\max}$, using the formula:

$$H_{\max} = \sum_{i=1}^{n} \frac{\pi_i}{\kappa} \ln(k_i)$$

1. *Compute Out-link Probabilities* We first identify node out-degrees (k_i) and the node values (π_i) from the sum of outflows. For each node, we divide the outgoing link flow by the total outflow (which is the node value π_i).

   - Node A ($k_A = 2, \pi_A = 4$): $s_{AB} = \frac{3}{4}, s_{AD} = \frac{1}{4}$
   - Node B ($k_B = 2, \pi_B = 3$): $s_{BC} = \frac{2}{3}, s_{BD} = \frac{1}{3}$
   - Node C ($k_C = 1, \pi_C = 3$): $s_{CD} = \frac{3}{3} = 1$
   - Node D ($k_D = 2, \pi_D = 5$): $s_{DA} = \frac{4}{5}, s_{DE} = \frac{1}{5}$
   - Node E ($k_E = 1$,$\pi_E = 1$): $s_{EC} = \frac{1}{1} = 1$

   The total flow is the sum of all node values:

$$\kappa = \pi_A + \pi_B + \pi_C + \pi_D + \pi_E = 4 + 3 + 3 + 5 + 1 = 16$$

2. *Compute Node Entropy* We apply the Shannon formula to each node's out-link probabilities. Nodes C and E, with only one outgoing link, have zero entropy.

   - $H_A = -(\frac{3}{4} \ln \frac{3}{4} + \frac{1}{4} \ln \frac{1}{4}) \approx 0.5623$
   - $H_B = -(\frac{2}{3} \ln \frac{2}{3} + \frac{1}{3} \ln \frac{1}{3}) \approx 0.6365$
   - $H_C = -(\frac{3}{3} \ln \frac{3}{3}) = 0$
   - $H_D = -(\frac{4}{5} \ln \frac{4}{5} + \frac{1}{5} \ln \frac{1}{5}) \approx 0.5004$
   - $H_E = -(\frac{1}{1} \ln \frac{1}{1}) = 0$

3. *Compute Network Entropy (H)* We calculate the flow-weighted average of the node entropies:

$$H = \frac{\pi_A}{\kappa} H_A + \frac{\pi_B}{\kappa} H_B + \frac{\pi_C}{\kappa} H_C + \frac{\pi_D}{\kappa} H_D + \frac{\pi_E}{\kappa} H_E$$

$$H = \frac{4}{16}(0.5623) + \frac{3}{16}(0.6365) + \frac{3}{16}(0) + \frac{5}{16}(0.5004) + \frac{1}{16}(0) \approx 0.4163$$

This calculated value, $H \approx 0.4163$, reflects the average uncertainty of local flow decisions, weighted by the flow at each node.

4. *Calculate Maximum Node Entropies ($H_{i,\max}$)*:

- $H_{A,\max} = \ln(k_A) = \ln(2) \approx 0.6931$
- $H_{B,\max} = \ln(k_B) = \ln(2) \approx 0.6931$
- $H_{C,\max} = \ln(k_C) = \ln(1) = 0$
- $H_{D,\max} = \ln(k_D) = \ln(2) \approx 0.6931$
- $H_{E,\max} = \ln(k_E) = \ln(1) = 0$

5. *Compute Maximum Network Entropy ($H_{\max}$)*:

$$H_{\max} = \frac{4}{16}(0.6931) + \frac{3}{16}(0.6931) + \frac{3}{16}(0) + \frac{5}{16}(0.6931) + \frac{1}{16}(0)$$

$$H_{\max} = \frac{4+3+5+1}{16}(0.6931) = \frac{13}{16}(0.6931) \approx 0.5630$$

6. *Calculate the Entropy Ratio*: Using the calculated network entropy ($H \approx 0.4163$) from the example:

$$\text{Entropy Ratio} = \frac{H}{H_{\max}} = \frac{0.4163}{0.5630} \approx 0.74$$

This result indicates that the network's current flow distribution is approximately 74% of its maximum possible local uniformity. The low ratio, relative to 1, highlights that while the network has some level of flow dispersion, it is far from an ideal state where flow is perfectly balanced at every node. Nodes C and E, with zero entropy, are a major reason for this. This quantitative measure provides a valuable tool for network performance analysis [12].

7. *Compute Shannon Network Entropy (H^S) for comparison* First, we find the global link probabilities ($p_{ij} = f_{ij}/\kappa$) for the network.

- $p_{AB} = \frac{3}{16}$, $p_{AD} = \frac{1}{16}$
- $p_{BC} = \frac{2}{16}$, $p_{BD} = \frac{1}{16}$
- $p_{CD} = \frac{3}{16}$
- $p_{DA} = \frac{4}{16}$, $p_{DE} = \frac{1}{16}$
- $p_{EC} = \frac{1}{16}$

Now, we apply the Shannon formula to these link probabilities.

$$
\begin{aligned}
H^S = -\bigg(&\frac{3}{16}\ln\frac{3}{16} + \frac{1}{16}\ln\frac{1}{16} + \frac{2}{16}\ln\frac{2}{16} \\
&+\frac{1}{16}\ln\frac{1}{16} + \frac{3}{16}\ln\frac{3}{16} + \frac{4}{16}\ln\frac{4}{16} \\
&+\frac{1}{16}\ln\frac{1}{16} + \frac{1}{16}\ln\frac{1}{16}\bigg) \approx 1.9274
\end{aligned}
$$

The contrast between the two metrics is clear. The network entropy (H) of approximately 0.4163 is much lower than the Shannon network entropy (H^S) of approximately 1.9274. This difference is not a flaw; it is by design. The significant difference between $H \approx 0.4163$ and $H^S \approx 1.9274$ demonstrates the point. H^S is high because there are many links, each with a relatively small probability, which increases the global uncertainty. However, the low value of H reveals that there are nodes (C and E) with zero local entropy, indicating highly concentrated and non-uniform flow decisions. This local non-uniformity is a key characteristic that network entropy (H) correctly identifies, making it the superior metric for analyzing the internal behavior and local choices within an IFN. The network entropy (H) is a more sensitive measure of the internal uniformity of flow distribution. It is exactly this sensitivity to the local, node-level choices that makes it the ideal metric for calibrating an IFN.

5.5 IFN Calculation via Maximum Entropy Optimization

Having developed a nuanced understanding of entropy and its application to network metrics, we now arrive at the central thesis of this chapter: the computation of an Ideal Flow Network (IFN) can be formulated and solved as a maximum entropy optimization problem. This approach provides a principled and reproducible method for determining the most probable flow distribution that is consistent with a network's inherent physical constraints.

The application of the Ideal Flow Network in transportation planning can be viewed as a spectrum based on data availability, with the Maximum Entropy Principle (MEP) providing the rational method for inference at each level. At one end of this spectrum, when only the road network's physical topology is known, the MEP yields a uniform probability distribution over all choices at each intersection. This state corresponds to the Standard Ideal Flow, which can be thought of as a random walk through the network. As we move along the spectrum to a situation with limited, sparse data, such as traffic counts or Origin-Destination totals, each piece of information becomes a specific constraint within the MEP formulation.

Loaiza-Ganem et al. [9] introduced a novel approach using normalizing flows to construct maximum entropy distributions over network flows. Sinatra et al. [17] demonstrated that maximal-entropy random walks can be constructed with limited local information, a principle that supports decentralized modeling in ideal flow networks [18].

The principle then finds the most uniform possible flow matrix that is fully consistent with these partial measurements. Finally, in the hypothetical case of a complete dataset where all flows are known, the constraints fully determine the system. In this scenario, the MEP simply returns the observed distribution, as there is no uncertainty left to maximize; the principle acts as a logical validator, confirming that the observed state is the one that best fits the given constraints. In this way, the MEP provides a comprehensive and flexible framework for network analysis, adapting its level of inference to the quality of the available data.

5.5.1 *The Optimization Problem*

The objective of IFN computation is to find the Ideal Flow matrix F, which represents a flow distribution that is maximally unbiased. According to the Principle of Maximum Entropy, this ideal distribution is the one that maximizes the system's uncertainty, represented by the network entropy, subject to all known constraints. The constraints for an IFN are its very definition: flow must be conserved at every node, flow must be non-negative, and the total flow must be a specific value.

Theorem 5.2 (Ideal Flow Network Based on Maximum Entropy [19]) *An Ideal Flow Network, defined by its flow matrix* F*, can be computed as the solution to the following maximum entropy optimization problem:*

The optimization variables in this problem are the flows on each link, f_{ij}*. The goal is to adjust these flow values such that the total network entropy is maximized, while all network properties are strictly maintained (Table 5.1).*

5.5.2 *A Practical Example (with Excel Solver)*

The optimization problem stated above can be solved efficiently using numerical methods. A particularly accessible tool for this is the Excel Solver Add-In. This allows us to translate the formal mathematical problem into a practical computational procedure.

Example 5.9 Consider a network defined by a stochastic matrix S and a total flow $\kappa = 805$. The stochastic matrix contains the out-link probabilities s_{ij}, which are the fixed parameters of the problem.

Table 5.1 Ideal flow network based on maximum entropy

Find the flow matrix F *to maximize network entropy:*

$$\max_{\mathsf{F}} H = -\sum_{i=1}^{n} \frac{\pi_i}{\kappa} \sum_{j=1}^{n} s_{ij} \ln s_{ij} \tag{5.16}$$

Subject to constraints:

$$\pi_i = \sum_j f_{ij} \quad \text{(Node value definition)} \tag{5.17}$$

$$\sum_j f_{ij} = \sum_j f_{ji} \quad \forall i \quad \text{(Flow conservation)} \tag{5.18}$$

$$f_{ij} \geq 0 \quad \forall i, j \quad \text{(Non-negativity)} \tag{5.19}$$

$$\sum_i \sum_j f_{ij} = \kappa \quad \text{(Total flow constraint)} \tag{5.20}$$

$$\sum_j s_{ij} = \frac{f_{ij}}{\pi_i} = 1 \quad \text{(Out-link probability)} \tag{5.21}$$

$$\pi_i, \kappa \geq 0 \tag{5.22}$$

$$\mathsf{S} = \begin{bmatrix} 0 & 1/6 & 0 & 1/3 & 1/2 \\ 0 & 0 & 4/5 & 0 & 1/5 \\ 1/2 & 1/4 & 0 & 1/4 & 0 \\ 0 & 0 & 1 & 0 & 0 \\ 1/3 & 2/3 & 0 & 0 & 0 \end{bmatrix}$$

The problem is to find the node values, π_i, that satisfy the IFN constraints and maximize the network entropy.

The procedure using Excel is as follows:

1. Set up the spreadsheet: The fixed stochastic matrix S is entered. The node values π_i are set as the variables that the solver will change, initially with an arbitrary value like 1.
2. Formulate the objective function: The network entropy H is calculated from the current node values and the stochastic matrix, using the formula from Sect. 5.4.
3. Define the constraints: The constraints are entered into the solver. This includes setting the total flow to $\kappa = 805$ and enforcing the flow conservation rule for each node (the sum of inflows must equal the sum of outflows).
4. Run the solver: The solver is configured to maximize the cell containing the network entropy H by adjusting the cells containing the node values π_i.

The result of this optimization is the set of node values that maximizes the network entropy. For this example, the optimal values are determined to be: $\pi_1 = 162, \pi_2 = 165, \pi_3 = 248, \pi_4 = 116, \pi_5 = 114$. Once these values are known, the Ideal Flow matrix F is computed directly using the formula $f_{ij} = s_{ij}\pi_i$.

$$\mathsf{F} = \left(\begin{array}{c|ccccc|c} \nearrow & a & b & c & d & e & \Sigma \\ \hline a & 0 & 27 & 0 & 54 & 81 & 162 \\ b & 0 & 0 & 132 & 0 & 33 & 165 \\ c & 124 & 62 & 0 & 62 & 0 & 248 \\ d & 0 & 0 & 116 & 0 & 0 & 116 \\ e & 38 & 76 & 0 & 0 & 0 & 114 \\ \hline \Sigma & 162 & 165 & 248 & 116 & 114 & 805 \end{array}\right)$$

The Ideal Flow Network derived from this maximum entropy optimization is not merely an arbitrary solution; it is the single most probable and unbiased flow distribution consistent with the network's structure. It represents the state of maximum uncertainty, meaning that no additional assumptions about flow distribution are implicitly made.

This method transforms IFN computation from a potentially subjective process into a rigorous, objective, and repeatable one. It leverages the philosophical and mathematical power of the maximum entropy principle to provide a solution that is both theoretically sound and practically useful for understanding the inherent equilibrium state of a flow network.

5.6 Summary

The chapter deduces a powerful conclusion: the computation of an Ideal Flow Network is not an arbitrary exercise but a rigorous application of the Principle of Maximum Entropy. The ideal flow matrix represents the state of maximum uncertainty that is fully consistent with a network's structural and flow constraints. This is the most unbiased and probable flow distribution possible.

The core argument unfolds in three stages. First, the chapter establishes the foundational principle of maximum entropy as a formalized method for making rational assumptions with incomplete data. Second, it demonstrates that network-wide Shannon entropy is insufficient for this task and proposes a more nuanced metric: a flow-weighted average of node entropies. This metric captures the microscopic flow dynamics at each intersection. Finally, it combines these elements to formulate a constrained optimization problem. The solution to this problem is the IFN, proving that the network's inherent equilibrium state can be derived directly from fundamental principles of information theory.

References

1. Chandra R. Bhat. "A Dynamic, Entropy Minimizing, Equilibrium-Oriented Approach to Transportation Planning". MA thesis. Virginia Polytechnic Institute and State University, 1987. http://hdl.handle.net/10919/101162.
2. Sylvain Courtain et al. "Relative entropy-regularized optimal transport on a graph: a new algorithm and an experimental comparison". In: *International Journal of Machine Learning and Cybernetics* 14 (2023), pp. 1365–1390. https://doi.org/10.1007/s13042-022-01704-6. https://link.springer.com/article/10.1007/s13042-022-01704-6.
3. Thomas M. Cover and Joy A. Thomas. *Elements of Information Theory*. 2nd. Hoboken, NJ: Wiley-Interscience, 2006. https://doi.org/10.1002/047174882X.
4. Sven Erlander and Neil F. Stewart. *The Gravity Model in Transportation Analysis: Theory and Extensions*. Utrecht: VSP, 1990.
5. Edwin T. Jaynes. "Information Theory and Statistical Mechanics". In: *Physical Review* 106.4 (1957), pp. 620–630. https://doi.org/10.1103/PhysRev.106.620.
6. J.N. Kapur. *Maximum-entropy Models in Science and Engineering*. Wiley, 1989. ISBN: 9788122402162. https://books.google.co.id/books?id=LuNIAp3QorUC.
7. S. Kullback. *Information Theory and Statistics*. Reprint of the 1959 edition. Dover Publications, 1997.
8. S. Kullback and R. A. Leibler. "On Information and Sufficiency". In: *The Annals of Mathematical Statistics* 22.1 (1951), pp. 79–86. https://doi.org/10.1214/aoms/1177729694.
9. Gabriel Loaiza-Ganem, Yuanjun Gao, and John P. Cunningham. "Maximum Entropy Flow Networks". In: *arXiv preprint arXiv:1701.03504* (2017). https://doi.org/10.48550/arXiv.1701.03504.
10. Hector Lopez-Ospina et al. "Freight tour synthesis based on entropy maximization with fuzzy parameters including variability in transportation costs through the Markowitz model". In: *Transportation* (2025). https://doi.org/10.1007/s11116-025-10614-9.
11. David G. Luenberger. *Optimization by Vector Space Methods*. Wiley, 1997. ISBN: 9780471181170. http://sites.science.oregonstate.edu/~show/old/142_Luenberger.pdf.
12. P. Van Mieghem. *Performance Analysis of Complex Networks and Systems*. Cambridge University Press, 2014. https://doi.org/10.1017/CBO9781107415874.
13. Marcel Nutz. *Introduction to Entropic Optimal Transport*. Lecture Notes. 2022. https://www.math.columbia.edu/~mnutz/docs/EOT_lecture_notes.pdf.
14. Reuven Y. Rubinstein and Dirk P. Kroese. *The Cross-Entropy Method: A Unified Approach to Combinatorial Optimization, Monte-Carlo Simulation and Machine Learning*. Springer, 2004. https://link.springer.com/book/10.1007/978-1-4757-4321-0.
15. Claude E. Shannon. "A Mathematical Theory of Communication". In: *The Bell System Technical Journal* 27.3 (1948), pp. 379–423. https://doi.org/10.1002/j.1538-7305.1948.tb01338.x.
16. Liangliang Shi et al. "Optimal Flow Transport and its Entropic Regularization: a GPU-friendly Matrix Iterative Algorithm for Flow Balance Satisfaction". In: *The Thirteenth International Conference on Learning Representations*. 2025. https://openreview.net/forum?id=NtSlKEJ2DS.
17. Roberta Sinatra et al. "Maximal-entropy random walks in complex networks with limited information". In: *Physical Review E* 83.3 (2011), p. 030103. https://doi.org/10.1103/PhysRevE.83.030103. https://journals.aps.org/pre/abstract/10.1103/PhysRevE.83.030103.
18. K. Teknomo. "Ideal Flow Based on Random Walk on Directed Graph". In: *Proceedings of the 9th International Collaboration Symposium on Information, Production and Systems*. Kitakyushu: Waseda University, 2015.
19. Kardi Teknomo. *Introduction to Ideal Flow Networks*. 2022.
20. Qian Wang and José Holguín-Veras. "Tour-Based Entropy Maximization Formulations of Urban Commercial Vehicle Movements". In: *European Transport Conference*. 2019. https://aetransport.org/public/downloads/anlzx/3608-514ec5a965435.pdf.

21. A. G. Wilson. "A statistical theory of spatial distribution models". In: *Transportation Research* 1.3 (1967), pp. 253–269. https://doi.org/10.1016/0041-1647(67)90035-4.
22. Alan G. Wilson. "Entropy in urban and regional modelling: retrospect and prospect". In: *Geographical Analysis* 42.4 (2010), pp. 364–394. https://doi.org/10.1111/j.1538-4632.2010.00798.x.
23. Alan G. Wilson. "The use of the concept of entropy in system modelling". In: *Operational Research Quarterly* 21.2 (1970), pp. 247–265. https://doi.org/10.2307/3008472.
24. Chi Xie, Kara M. Kockelman, and S. Travis Waller. "A Maximum Entropy Method for Subnetwork Origin-Destination Trip Matrix Estimation". In: *Transportation Research Record* 2196 (2010), pp. 111–119. https://doi.org/10.3141/2196-13 (https://doi.org/10.3141/2196-12).
25. Ziqiang Zeng, Yupeng Sun, and Xinru Zhang. "Entropy-Based Node Importance Identification Method for Public Transportation Infrastructure Coupled Networks: A Case Study of Chengdu". In: *Entropy* 26.2 (2024), p. 159. https://doi.org/10.3390/e26020159. https://www.mdpi.com/1099-4300/26/2/159.

Chapter 6
Ideal Flow Network from Random Walk

6.1 Introduction

In the study of network dynamics, a central challenge is the prediction of flow patterns in the absence of comprehensive origin-destination data. This chapter presents a solution through the theoretical framework of the Ideal Flow Network (IFN), which posits that a stable flow distribution is an inherent property of a network's structure. The development proceeds by transitioning from physical intuition to mathematical formalism.

We begin with a description of the random walk [1, 5, 9], a simple model of agent movement, and demonstrate how its aggregation leads to a stable relative flow. This empirical observation is then formalized using the theory of Discrete-Time Markov Chains (DTMC), where we prove that the Ideal Flow is a direct consequence of the network's stationary distribution. Finally, we establish a profound connection between the uniform random walk and the principle of maximum entropy, demonstrating its theoretical justification. This chapter demonstrates how a complex flow pattern can be derived from a simple, local, probabilistic process, offering a powerful and data-efficient approach to network analysis.

6.2 Random Walks and Asymptotic Flow Convergence

The study of flow dynamics on a network begins with a simplified model of movement. Lovász [1, 5] provided a seminal survey on random walks in graphs, which is foundational for modeling stochastic movement in ideal flow networks.

We consider a network as a strongly connected directed graph, $\mathcal{G} = (\mathcal{V}, \mathcal{E})$. The set $\mathcal{V}$ contains n nodes, and the set $\mathcal{E}$ contains the directed links. An agent, or *random walker*, traverses the links of this graph. At each node $i \in \mathcal{V}$, an agent

K. Teknomo, *Ideal Flow Network for Transportation Research*,
Lecture Notes in Intelligent Transportation and Infrastructure,
https://doi.org/10.1007/978-981-95-4727-2_6

selects an outgoing link $(i, j) \in \mathcal{E}$ based on a specific probability distribution. The number of outgoing links from node i is its out-degree, denoted as d_i^+. In the most basic model, this probability is uniform across all outgoing links, such that the probability of moving from node i to an adjacent node j is given by $1/d_i^+$. This process is a fundamental, local rule that governs a global, collective behavior.

6.2.1 The Absolute Flow Matrix

As multiple agents move on the network, their individual trajectories can be aggregated to measure the cumulative flow. A trajectory is a sequence of visited nodes, such as $v_1 \to v_2 \to v_3 \to \cdots \to v_k$. The flow on a link (i, j) is defined as the total number of times agents traverse that specific link. This cumulative count is a function of both the number of agents (N) and the simulation time (T). We organize these counts into a square matrix called the Absolute Flow Matrix, $\mathsf{F}_{abs} \in \mathbb{Z}^{n \times n}$. The element $(\mathsf{F}_{abs})_{ij}$ is the total number of traversals from node i to node j. By its construction, F_{abs} is a non-negative integer matrix.

Proposition 6.1 (Divergence of Absolute Flow) *The elements of the Absolute Flow Matrix, F_{abs}, diverge as the number of agents or the simulation time approaches infinity. Formally, for any link $(i, j) \in \mathcal{E}$,*

$$\lim_{N,T \to \infty} (\mathsf{F}_{abs})_{ij} = \infty.$$

Proof Let ΔT be a single time step and let N be the number of agents. The total number of movements in the system over a period T is proportional to $N \cdot T$. Assuming the graph is strongly connected, every link (i, j) will be traversed with some non-zero probability. The expected number of traversals on any link (i, j) is proportional to the total number of movements, $N \cdot T$, multiplied by a positive constant representing the link's steady-state flow probability. As $N \to \infty$ or $T \to \infty$, the total number of movements approaches infinity. Therefore, the expected and observed cumulative flow on each link also approaches infinity. The absolute flow, being an unbounded measure, is therefore unsuitable for characterizing the intrinsic, stable properties of a network's topology. This limitation necessitates a method for normalization to reveal a stable flow pattern. □

6.2.2 The Principle of Asymptotic Flow Convergence

The unbounded growth of the absolute flow matrix, F_{abs}, is not useful for characterizing the inherent, stable properties of a network. This limitation can be overcome by normalizing the flow. When the cumulative flow counts are transformed into a relative measure, a remarkable and stable pattern emerges. This section formalizes the concept of relative flow and defines the Ideal Flow matrix as the asymptotic limit of this relative measure.

While the absolute flow on any link diverges, the ratio of the flow on any two links converges to a fixed value. This stability allows us to define a relative flow matrix, which represents the asymptotic flow distribution. The convergence of this distribution is guaranteed for any strongly connected network.

Theorem 6.1 (Convergence of Relative Flow Matrix) *Let $\mathcal{G} = (\mathcal{V}, \mathcal{E})$ be a strongly connected directed graph. Let $\mathsf{F}_{abs}(t)$ be the absolute flow matrix after a long simulation time t. For any scalar function of the flow, $\sigma(\mathsf{F}_{abs}(t)) > 0$, the relative flow matrix $\mathsf{F}_{rel}(t) = \mathsf{F}_{abs}(t)/\sigma(\mathsf{F}_{abs}(t))$ converges to a unique matrix as $t \to \infty$.*

Proof Let s_{ij} be the fixed probability of moving from node i to node j. The movement of agents is a series of independent trials. By the Strong Law of Large Numbers [3], the observed frequency of an event approaches its theoretical probability as the number of trials increases. Thus, the ratio of the absolute flow on link (i, j) to the total flow in the network, $\kappa(t) = \sum_{k,l}(\mathsf{F}_{abs}(t))_{kl}$, converges to the stationary flow probability of that link. Formally,

$$\lim_{t\to\infty} \frac{(\mathsf{F}_{abs}(t))_{ij}}{\kappa(t)} = \pi_i \cdot s_{ij} = \text{constant.}$$

Since this holds for every link, the matrix of these normalized values converges. Normalizing by any other flow statistic, such as the minimum non-zero flow, maintains the same ratio between link flows. For example, the ratio of two link flows converges to a constant:

$$\lim_{t\to\infty} \frac{(\mathsf{F}_{abs}(t))_{ij}}{(\mathsf{F}_{abs}(t))_{kl}} = \frac{\pi_i \cdot s_{ij}}{\pi_k \cdot s_{kl}} = \text{constant.}$$

This demonstrates that the relative flow pattern is stable and unique, regardless of the specific normalization scalar used. The relative flow matrix, F_{rel}, therefore, converges to a fixed pattern that is an intrinsic property of the network's topology. □

The proof of Theorem 6.1 leverages the Law of Large Numbers [3] to show that the ratios between link flows converge. Since the ratio of two converging sequences approaches the ratio of their limits, this implies that the relative proportions of flow across all links settle into fixed constants. The resulting matrix of converged flow ratios is unique, because the limiting probabilities, given by $\pi_i \times s_{ij}$, are themselves uniquely determined for any strongly connected network. The uniqueness of the stationary distribution $\boldsymbol{\pi}$ is guaranteed by the Perron-Frobenius theorem [2, 8], which applies to any irreducible (i.e., strongly connected) graph [4]. Importantly, the proposition asserts convergence to a unique matrix, not just a single scalar value. The proof addresses this by demonstrating that the entire matrix of flow ratios stabilizes to a unique configuration. The uniqueness holds for a given normalization factor σ. While the converged matrix depends on the choice of σ, any alternative normalization (e.g., by average, maximum or minimum flow) would yield a different, but still proportional and stable, matrix. In all cases, the ratios between link flows converge to a unique set of constants, independent of how the matrix is scaled.

6.2.3 *Definition of the Ideal Flow Matrix*

The convergent relative flow matrix can be simplified to its most fundamental integer form. This form is the Ideal Flow Matrix, which represents the simplest ratios of flow inherent to the network.

Definition 6.1 (Ideal Flow Matrix from Random Walk) The Ideal Flow matrix, F, is the unique, rational, steady-state flow distribution of a random walk on a network. It is typically scaled to produce the smallest possible non-negative integer values (called *Cardinal Network*) for each link flow, representing the minimal, conserved flow pattern inherent to the network's topology. It is formally defined as:

$$\mathsf{F} = \lim_{N \cdot T \to \infty} \frac{\mathsf{F}_{abs}}{\min((\mathsf{F}_{abs})_{ij} | (\mathsf{F}_{abs})_{ij} \neq 0)}.$$

The resulting matrix, F, is irreducible, non-negative, and premagic [11]. It is unique for a given network under the assumption of a uniform random walk.

Example 6.1 (Flow and Stochastic Matrices from Trajectories) This example demonstrates the step-by-step accumulation of flow and the calculation of flow and stochastic matrices based on a series of agent trajectories.

- Trajectories: We assign the following six trajectories to a given network.
  1. $a \to b \to c$
  2. $b \to c \to d$
  3. $c \to b \to c \to a$
  4. $c \to d \to b \to c \to a$
  5. $d \to b \to c \to b \to c \to a \to b$
  6. $a \to b \to c \to d$
- Step 1: After Trajectory 1 ($a \to b \to c$): The first trajectory adds one unit of flow to links (a, b) and (b, c). The absolute flow matrix, $\mathbf{F}_{abs,1}$, is:

$$\mathbf{F}_{abs,1} = \begin{pmatrix} 0 & 1 & 0 & 0 \\ 0 & 0 & 1 & 0 \\ 0 & 0 & 0 & 0 \\ 0 & 0 & 0 & 0 \end{pmatrix} \tag{6.1}$$

The row sums are used to normalize the matrix to obtain the stochastic matrix, $\mathbf{S}_1$. The row sums for nodes a, b, c, and d are 1, 1, 0, and 0, respectively.

$$\mathbf{S}_1 = \begin{pmatrix} 0 & 1 & 0 & 0 \\ 0 & 0 & 1 & 0 \\ 0 & 0 & 0 & 0 \\ 0 & 0 & 0 & 0 \end{pmatrix} \tag{6.2}$$

- Step 2: After Trajectory 2 ($b \to c \to d$): The second trajectory adds flow to links (b, c) and (c, d), producing the updated flow matrix, $\mathbf{F}_{abs,2}$.

$$\mathbf{F}_{abs,2} = \begin{pmatrix} 0 & 1 & 0 & 0 \\ 0 & 0 & 2 & 0 \\ 0 & 0 & 0 & 1 \\ 0 & 0 & 0 & 0 \end{pmatrix} \tag{6.3}$$

The row sums are 1, 2, 1, and 0. The stochastic matrix, $\mathbf{S}_2$, is calculated by dividing each row by its sum.

$$\mathbf{S}_2 = \begin{pmatrix} 0 & 1 & 0 & 0 \\ 0 & 0 & 1 & 0 \\ 0 & 0 & 0 & 1 \\ 0 & 0 & 0 & 0 \end{pmatrix} \tag{6.4}$$

- Step 3: After Trajectory 3 ($c \to b \to c \to a$): This trajectory adds flow to links (c, b), (b, c), and (c, a). The updated flow matrix, $\mathbf{F}_{abs,3}$, is:

$$\mathbf{F}_{abs,3} = \begin{pmatrix} 0 & 1 & 0 & 0 \\ 0 & 0 & 3 & 0 \\ 1 & 1 & 0 & 1 \\ 0 & 0 & 0 & 0 \end{pmatrix} \tag{6.5}$$

The row sums are 1, 3, 3, and 0. The stochastic matrix, $\mathbf{S}_3$, is:

$$\mathbf{S}_3 = \begin{pmatrix} 0 & 1 & 0 & 0 \\ 0 & 0 & 1 & 0 \\ 1/3 & 1/3 & 0 & 1/3 \\ 0 & 0 & 0 & 0 \end{pmatrix} \tag{6.6}$$

The stochastic matrix values are rational numbers and will converge as more trajectories are added.

Table 6.1 Stochastic Matrix Convergence from node c

Probabilities	1	2	3	4	5	6
s_{ca}	1.00	0.50	0.50	0.50	0.44	0.45
s_{cb}	0.00	0.00	0.25	0.17	0.22	0.18
s_{cd}	0.00	0.50	0.25	0.33	0.33	0.36

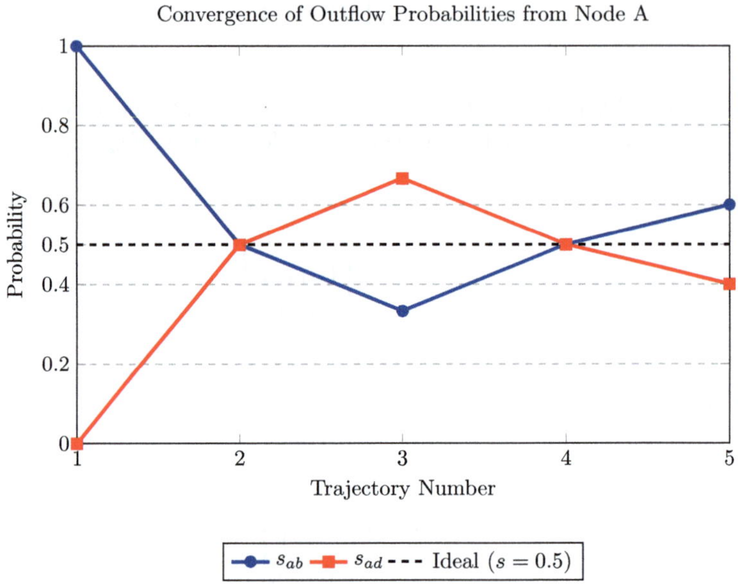

Fig. 6.1 Convergence of a stochastic network's flow distribution to a stationary state

Example 6.2 (Ideal Flow Convergence with Cyclic Trajectories) This example demonstrates how cyclic trajectories cause the flow matrix to become premagic and how the stochastic matrix converges (Fig. 6.1).

- Trajectories: We assign the following six trajectories to the network. These trajectories are explicitly cyclic, meaning they start and end at the same node.

  1. $a \to b \to c \to a$
  2. $b \to c \to d \to b$
  3. $c \to b \to c \to a \to b \to c$
  4. $c \to d \to b \to c \to a \to b \to c$
  5. $d \to b \to c \to b \to c \to a \to b \to c \to d$
  6. $a \to b \to c \to d \to b \to c \to a$

- Step 1: After Trajectory 1 ($a \to b \to c \to a$): This trajectory adds one unit of flow to links (a, b), (b, c), and (c, a). The resulting flow matrix, $\mathbf{F}_{abs,1}$, and its corresponding stochastic matrix, $\mathbf{S}_1$, are:

$$\mathbf{F}_{abs,1} = \begin{pmatrix} 0 & 1 & 0 & 0 \\ 0 & 0 & 1 & 0 \\ 1 & 0 & 0 & 0 \\ 0 & 0 & 0 & 0 \end{pmatrix} \quad \mathbf{S}_1 = \begin{pmatrix} 0 & 1 & 0 & 0 \\ 0 & 0 & 1 & 0 \\ 1 & 0 & 0 & 0 \\ 0 & 0 & 0 & 0 \end{pmatrix} \tag{6.7}$$

  This flow matrix is premagic; the row sums are (1, 1, 1, 0) and the column sums are also (1, 1, 1, 0).
- Step 3: After Trajectory 3 ($c \to b \to c \to a \to b \to c$): After three trajectories, the flow on all links is positive. The flow matrix, $\mathbf{F}_{abs,3}$, and its corresponding stochastic matrix, $\mathbf{S}_3$, are:

$$\mathbf{F}_{abs,3} = \begin{pmatrix} 0 & 2 & 0 & 0 \\ 0 & 0 & 4 & 0 \\ 2 & 1 & 0 & 1 \\ 0 & 0 & 0 & 0 \end{pmatrix} \quad \mathbf{S}_3 = \begin{pmatrix} 0 & 1 & 0 & 0 \\ 0 & 0 & 1 & 0 \\ 0.5 & 0.25 & 0 & 0.25 \\ 0 & 0 & 0 & 0 \end{pmatrix} \tag{6.8}$$

  At this stage, the network is strongly connected and the flow matrix is premagic, which are necessary conditions for an Ideal Flow Network.
- Convergence of Stochastic Matrix: The values of the stochastic matrix elements will fluctuate but converge to a stable state as more trajectories are assigned (Table 6.1). The following table and chart from the original source illustrate this convergence for the probabilities of transitions from node c.

  As seen in the table and chart, the empirical values of the stochastic matrix for node c gradually move towards a stable asymptotic value. This demonstrates that the stochastic matrix is the fundamental phenomenon behind the ideal flow.

Example 6.3 (A Cyclic Network) Consider a simple, hypothetical network with three nodes, 1, 2, and 3, connected in a cycle ($1 \to 2 \to 3 \to 1$). This network is strongly connected and symmetric. We use a uniform random walk model. We simulate a few trajectories to demonstrate flow accumulation.

- Trajectory 1: $1 \to 2 \to 3$.
- Trajectory 2: $2 \to 3 \to 1$.
- Trajectory 3: $3 \to 1 \to 2$.

After these three trajectories, the absolute flow matrix, F_{abs}, is:

$$\mathsf{F}_{abs} = \begin{pmatrix} 0 & 1 & 1 \\ 0 & 0 & 1 \\ 1 & 0 & 0 \end{pmatrix}.$$

The flow on link (1, 2) is 1, on (2, 3) is 1, and on (3, 1) is 1. The out-degree of node 1 is 2, while the out-degrees of nodes 2 and 3 are 1. The row sums and column sums are not equal yet, but the relative flows are already stable. The minimum non-zero value is 1. Normalizing by this value yields the Ideal Flow Matrix.

$$\mathsf{F} = \begin{pmatrix} 0 & 1 & 1 \\ 0 & 0 & 1 \\ 1 & 0 & 0 \end{pmatrix}.$$

With more trajectories, the flow counts would grow, but their ratios would remain constant. For instance, after three more trajectories, the flow on each link would be 2. Normalizing by the minimum flow of 2 would again yield the same matrix F, demonstrating the stability of the relative flow distribution.

6.3 A Formal Framework: Discrete-Time Markov Chains (DTMC)

The random walk provides a simple, physical metaphor for agent movement. However, to analyze its long-term behavior with precision, we must transition to a formal mathematical framework. This section establishes the connection between a random walk and a Discrete-Time Markov Chain (DTMC). This framework provides the necessary tools to prove the existence and uniqueness of the Ideal Flow.

6.3.1 Proposition: A Random Walk Is a DTMC

The dynamics of a random walk on a network can be perfectly modeled by a Discrete-Time Markov Chain. Kemeny and Snell [4] developed foundational theory on finite Markov chains. Norris [7] offers a rigorous treatment of discrete-time Markov chains, including transition matrices and convergence properties, foundational for stochastic modeling in IFNs.

This is because the random walk process satisfies the core property of a Markov chain, known as the Markov property.

Definition 6.2 (Discrete-Time Markov Chain [3, 4]) A Discrete-Time Markov Chain is a stochastic process where the conditional probability of future states depends only on the present state, not on the sequence of states that preceded it. This is often called the "memoryless" property. Formally, for a sequence of states $X_0, X_1, X_2, \ldots$, the following holds:

$$P(X_{n+1} = j | X_n = i, X_{n-1} = i_{n-1}, \ldots, X_0 = i_0) = P(X_{n+1} = j | X_n = i).$$

Proposition 6.2 (Random Walk and DTMC) *A random walk on a directed graph is a Discrete-Time Markov Chain.*

Proof Let the state of the system at time n be the node an agent occupies, denoted by X_n. The rules of a random walk state that at each node i, an agent chooses its next link (i, j) based only on a probabilistic rule for that node. The probability of choosing link (i, j), denoted s_{ij}, is independent of the path the agent took to arrive at node i. Therefore, the probability of the next state, $X_{n+1} = j$, depends only on the current state, $X_n = i$, and not on the previous states $(X_{n-1}, \ldots, X_0)$. This property directly satisfies the definition of a Discrete-Time Markov Chain. □

6.3.2 *The Transition Matrix and Irreducibility*

The dynamics of a DTMC are fully specified by its transition probability matrix. The structural properties of this matrix, dictated by the network's topology, guarantee the existence of a stable flow pattern.

Definition 6.3 (Transition Probability Matrix) The transition probability matrix, $\mathsf{S} \in \mathbb{R}^{n \times n}$, is a row-stochastic matrix where each element s_{ij} represents the probability of a transition from state (node) i to state (node) j in a single time step. The matrix is row-stochastic, meaning each row sums to one: $\sum_{j=1}^{n} s_{ij} = 1$.

For a simple random walk, the transition probability is uniform across all outgoing links. This means that for a node i with out-degree d_i^+, the probability of moving to any adjacent node j is $s_{ij} = 1/d_i^+$.

The concept of a strongly connected graph is fundamental here. A strongly connected graph means there is a path from any node to any other node. The adjacency matrix of such a graph is irreducible.

Definition 6.4 (Irreducible Matrix) A non-negative matrix M is irreducible if for every pair of indices (i, j), there exists a natural number k such that the element $(\mathsf{M}^k)_{ij}$ is positive. This means that a path of length k exists from state i to state j.

The Perron-Frobenius theorem [2, 8] is central to this analysis. It states that an irreducible, non-negative matrix has a unique positive eigenvector for its largest eigenvalue, which is 1. This theorem is the mathematical guarantee for the existence of a unique stationary distribution.

6.3.3 *The Stationary Distribution*

The stationary distribution describes the long-term, steady-state probability of an agent being at each node in the network. It is the core concept that links the Markov chain to the Ideal Flow.

Definition 6.5 (Stationary Distribution) The stationary distribution, $\boldsymbol{\pi}$, is a row vector of probabilities, $\boldsymbol{\pi}^T = (\pi_1, \pi_2, \ldots, \pi_n)$, where π_i is the long-run proportion of time a random walker spends at node i. This vector satisfies the equilibrium equation:

$$\boldsymbol{\pi}^T \mathsf{S} = \boldsymbol{\pi}^T.$$

Additionally, the elements of the vector must sum to one: $\sum_{i=1}^{n} \pi_i = 1$. The Perron-Frobenius theorem [2, 8] guarantees that for an irreducible matrix S, this unique solution exists and its elements are all positive.

This vector represents the equilibrium activity levels at each node in the network. It provides the crucial link between the network's topology and its inherent flow distribution.

Example 6.4 (Calculating a Stationary Distribution) This example demonstrates the calculation of the stationary distribution for a network with non-uniform transition probabilities. Consider a four-node network with a main artery and a side road. The transition probability matrix, S, for this system is non-symmetric:

$$\mathsf{S} = \begin{pmatrix} 0 & 0.9 & 0 & 0.1 \\ 0.2 & 0 & 0.8 & 0 \\ 0 & 0.2 & 0 & 0.8 \\ 1.0 & 0 & 0 & 0 \end{pmatrix}.$$

To find the stationary distribution, we must solve the system of linear equations given by the equilibrium equation $\boldsymbol{\pi}^T \mathsf{S} = \boldsymbol{\pi}^T$, along with the normalization constraint $\sum \pi_i = 1$. The system can be rewritten as $\boldsymbol{\pi}^T(\mathsf{S} - \mathsf{I}) = \mathsf{0}^T$. The unique solution to this system is the stationary distribution vector:

$$\boldsymbol{\pi}^T = \left(\tfrac{7}{26}\ \tfrac{15}{52}\ \tfrac{3}{13}\ \tfrac{11}{52}\right) \approx \left(0.269\ 0.288\ 0.231\ 0.212\right).$$

This result shows that in the long run, node 2 has the highest proportion of activity, followed by node 1, node 3, and finally node 4. This demonstrates how the stationary distribution, a direct result of the network's structure and probabilistic rules, reveals its long-term equilibrium activity levels.

The next section will use this stationary distribution to formally derive the Ideal Flow matrix.

6.4 The Ideal Flow as a Steady-State

The convergence of a random walk to a stable relative flow pattern is not an empirical coincidence. It is a direct consequence of the existence of a unique steady-state in the underlying Markov chain. This section formalizes this connection by proving that the Ideal Flow matrix, F, is an algebraic consequence of the stationary distribution, $\boldsymbol{\pi}$. The premagic property of F, which is central to the Ideal Flow Network theory, is a direct result of this derivation.

6.4.1 Theorem: Ideal Flow from the Stationary Distribution

The Ideal Flow matrix can be precisely defined as the product of the stationary distribution and the transition probability matrix, scaled by a total flow value. This theorem provides a powerful algebraic method to derive the flow distribution without simulation.

Theorem 6.2 (Ideal Flow from Stationary Distribution [10]) *Let* S *be the irreducible transition probability matrix of a network's Markov chain, and let* $\boldsymbol{\pi}$ *be its unique stationary distribution. The Ideal Flow matrix,* F*, with elements* $f_{ij} = \kappa \times \pi_i \times s_{ij}$*, is a non-negative, premagic matrix.*

Proof By definition, the matrix S is non-negative and the stationary distribution vector $\boldsymbol{\pi}$ is positive. The scalar κ is also positive. Thus, the elements f_{ij} are non-negative. To prove that F is premagic, we must show that the sum of each row equals the sum of its corresponding column. The total flow leaving node i (the sum of its row in F) is given by:

$$\sum_{j=1}^{n} f_{ij} = \sum_{j=1}^{n} \kappa \pi_i s_{ij} = \kappa \pi_i \sum_{j=1}^{n} s_{ij}.$$

Since S is row-stochastic, $\sum_{j=1}^{n} s_{ij} = 1$. Therefore, the total flow leaving node i is $\kappa \pi_i$. The total flow entering node j (the sum of its column in F) is given by:

$$\sum_{i=1}^{n} f_{ij} = \sum_{i=1}^{n} \kappa \pi_i s_{ij} = \kappa \sum_{i=1}^{n} \pi_i s_{ij}.$$

From the definition of the stationary distribution, we have the equilibrium equation $\boldsymbol{\pi}^T \mathsf{S} = \boldsymbol{\pi}^T$, which implies that $\sum_{i=1}^{n} \pi_i s_{ij} = \pi_j$. Therefore, the total flow entering node j is $\kappa \pi_j$. Since the total flow entering node i is $\kappa \pi_i$ and the total flow leaving is also $\kappa \pi_i$, flow is conserved at every node. This is the definition of a premagic matrix. □

6.4.2 Proposition: Rationality of the Ideal Flow

The values of the Ideal Flow matrix are always rational numbers. This is a direct consequence of its algebraic derivation from the stationary distribution.

Proposition 6.3 (Rationality of Ideal Flow [9]) *The elements of the Ideal Flow matrix,* F*, are rational numbers.*

Proof The transition probability matrix S is composed of rational entries, as the transition probabilities s_{ij} are ratios of integers (e.g., $1/d_i^+$). The stationary distribution vector $\boldsymbol{\pi}$ is the unique solution to a system of linear equations with rational coefficients. By the properties of linear algebra, the solution to such a system must also be a vector of rational numbers. The Ideal Flow matrix is defined as the product of the stationary distribution vector, the transition probability matrix, and scaled by a total flow value κ.

$$\mathsf{F} = \kappa \cdot \text{diag}(\boldsymbol{\pi}) \cdot \mathsf{S},$$

where diag($\boldsymbol{\pi}$) is a diagonal matrix with the stationary distribution $\boldsymbol{\pi}$ on its diagonal. Since all these components are rational, their product, the Ideal Flow matrix F, must also be rational. □

Example 6.5 (Deriving Ideal Flow) We continue with the non-uniform system from the previous section to derive the Ideal Flow matrix. We have the transition probability matrix S and the stationary distribution $\boldsymbol{\pi}^T$.

$$\mathsf{S} = \begin{pmatrix} 0 & 0.9 & 0 & 0.1 \\ 0.2 & 0 & 0.8 & 0 \\ 0 & 0.2 & 0 & 0.8 \\ 1.0 & 0 & 0 & 0 \end{pmatrix}, \quad \boldsymbol{\pi}^T = \begin{pmatrix} 0.269 \; 0.288 \; 0.231 \; 0.212 \end{pmatrix}.$$

We assume a total flow $\kappa = 10{,}000$ veh/hr. The Ideal Flow on each link is calculated as $f_{ij} = \kappa \times \pi_i \times s_{ij}$. For example, the flow from node 1 to node 2 is $\mathsf{F}_{12} = 10{,}000 \times \pi_1 \times s_{12} = 10{,}000 \times (0.269) \times (0.9) \approx 2423$. The complete Ideal Flow matrix is:

$$\mathsf{F} \approx \begin{pmatrix} 0 & 2423 & 0 & 269 \\ 577 & 0 & 2308 & 0 \\ 0 & 462 & 0 & 1846 \\ 2115 & 0 & 0 & 0 \end{pmatrix}.$$

The row sums of this matrix are approximately $(2692, 2885, 2308, 2115)$, and the column sums are also approximately $(2692, 2885, 2308, 2115)$, confirming the premagic property. This matrix, derived entirely from the network's structure and probabilistic rules, represents the complete, steady-state flow distribution.

6.5 The Principle of Maximum Entropy and Steady Matrices

The assumption of a uniform random walk, where all outgoing links from a node have equal probability of being chosen, requires a theoretical justification. The principle of maximum entropy provides this justification. It states that the most unbiased probability distribution, given only known constraints, is the one that maximizes information entropy. This section connects the uniform random walk to this principle and defines the class of matrices that represent this maximum entropy state.

6.5.1 *Maximization of Entropy*

The principle of maximum entropy asserts that the most probable and least structured distribution is the one with the largest entropy. In the context of a network, this means the most unbiased assumption about an agent's choices at an intersection is a uniform distribution.

Definition 6.6 (Shannon Information Entropy) The Shannon information entropy, h_i, of a set of outflow probabilities $\{s_{ij}\}$ for a node i with out-degree d_i^+ is defined as:

$$h_i = -\sum_{j=1}^{d_i^+} s_{ij} \log_2(s_{ij}).$$

Proposition 6.4 (Maximum Entropy and Uniform Distribution) *The outflow probability distribution $\{s_{ij}\}$ that maximizes the entropy h_i at node i is the uniform distribution, where each probability is $s_{ij} = 1/d_i^+$.*

Proof To maximize the entropy h_i subject to the constraint $\sum_{j=1}^{d_i^+} s_{ij} = 1$, we use the method of Lagrange multipliers [6], a method for solving constrained optimization problems. In transport modeling, it's used in equilibrium formulations and entropy maximization. The Lagrangian function is:

$$\mathcal{L}(s_{ij}, \lambda) = h_i + \lambda \left(\sum_{j=1}^{d_i^+} s_{ij} - 1 \right) = -\sum_{j=1}^{d_i^+} s_{ij} \log_2(s_{ij}) + \lambda \left(\sum_{j=1}^{d_i^+} s_{ij} - 1 \right).$$

Setting the partial derivative with respect to each s_{ij} to zero, we get:

$$\frac{\partial \mathcal{L}}{\partial s_{ij}} = -\frac{1}{\ln(2)}(1 + \ln(s_{ij})) + \lambda = 0.$$

This implies that s_{ij} is a constant for all j. Substituting this back into the constraint equation $\sum_{j=1}^{d_i^+} s_{ij} = 1$ shows that $s_{ij} = 1/d_i^+$. Thus, the uniform distribution maximizes entropy. □

This result is fundamental. It demonstrates that the uniform random walk is not an arbitrary assumption but is the most unbiased and least-structured model possible in the absence of additional information.

6.5.2 Definition of Steady Matrices

A special class of matrices, called *steady matrices*, are key to connecting maximum entropy to the Ideal Flow framework. These definitions establish a precise mathematical language for describing the structure of the transition probability matrix.

Definition 6.7 (Pattern and Steady Matrices) (See Sect. 3.8.2 for more detail)

- A (0,1) non-negative matrix P is called a *pattern matrix*. It defines the locations of the non-zero entries for a class of matrices.
- A *row-steady matrix*, R, is a non-negative matrix that has the same pattern as P and has equal values for all non-zero entries within each row.
- A *column-steady matrix*, C, is a non-negative matrix that has the same pattern as P and has equal values for all non-zero entries within each column.
- A *steady matrix*, B, is a non-negative matrix that has the same pattern as P and has the same value for all of its non-zero entries. Thus, $\mathsf{B} = k\mathsf{P}$ for some scalar $k > 0$.

6.5.3 Theorem: Maximum Entropy and the Row-Steady Matrix

This theorem provides the definitive link between the statistical principle of maximum entropy and the algebraic structure of the transition probability matrix. It shows that the Ideal Flow is a maximum entropy flow if and only if the underlying random walk is uniform.

Theorem 6.3 (Maximum Entropy and Row-Steady Matrix) *An Ideal Flow Network reaches a state of maximum entropy if and only if its transition probability matrix,* S, *is a row-steady matrix.*

Proof This is an "if and only if" proof, so we must demonstrate both directions.

- Part 1: ($\Longrightarrow$) Maximum Entropy implies Row-Steady Assume that the Ideal Flow is at a state of maximum entropy. We have proven that the maximum entropy is achieved when the outflow probabilities at each node are uniform. This means that for any node i, all transition probabilities s_{ij} for which a link (i, j)

exists are equal, specifically $s_{ij} = 1/d_i^+$. A matrix where all non-zero entries within each row are equal is, by definition, a row-steady matrix. Therefore, the transition probability matrix S is row-steady.

- Part 2: ($\Longleftarrow$) Row-Steady implies Maximum Entropy Assume that the transition probability matrix S is a row-steady matrix. By definition, this means that for each row i, all non-zero elements have the same value. Since S is a row-stochastic matrix, the sum of each row must be 1. The number of non-zero entries in row i is equal to the out-degree, d_i^+. For the row sum to be 1, the value of each non-zero entry must be $1/d_i^+$. This is the uniform probability distribution, which we have already proven to be the distribution that maximizes information entropy.

Therefore, the Ideal Flow is at a state of maximum entropy. □

6.6 Summary

This chapter has established the theoretical foundations of the Ideal Flow Network (IFN) model by analyzing the dynamics of a random walk on a strongly connected network. The following conclusions are deduced from the rigorous analysis:

1. The absolute flow matrix, F_{abs}, which counts agent traversals, increases indefinitely over time or with an increasing number of agents, making it an unstable metric for long-term analysis.
2. Normalizing this absolute flow by a statistical value, such as the minimum non-zero flow, reveals a stable, relative flow pattern that converges to an integer-valued Ideal Flow matrix, F.
3. This convergence from a random walk is formalized by a Discrete-Time Markov Chain (DTMC), where the Ideal Flow is defined by the unique stationary distribution of the network's transition probability matrix, S.
4. The assumption of a uniform random walk is not arbitrary but is the most unbiased model for an agent's movement, as it is the only distribution that maximizes the information entropy of the system.
5. This is formally proven by demonstrating that an Ideal Flow network reaches maximum entropy if and only if its transition probability matrix, S, is a row-steady matrix.

This framework provides a robust, principled, and data-efficient method for deriving a network's fundamental flow properties.

References

1. David Aldous and James Allen Fill. *Reversible Markov Chains and Random Walks on Graphs*. Available at: https://www.stat.berkeley.edu/~aldous/RWG/book.html. 2002.
2. G. Frobenius. "Ueber Matrizen aus nicht negativen Elementen". In: *Sitzungs-berichte der Königlich Preußischen Akademie der Wissenschaften* (1912), pp. 456–477.
3. G. R. Grimmett and D. R. Stirzaker. *Probability and Random Processes*. 3rd. Oxford University Press, 2001.
4. John G. Kemeny and J. Laurie Snell. *Finite Markov Chains*. New York: Springer, 1976. https://link.springer.com/book/9780387901923.
5. László Lovász. "Random walks on graphs: A survey". In: vol. 2. Budapest: János Bolyai Mathematical Society, 1993, pp. 1–46. https://www.cs.cmu.edu/~15859n/RelatedWork/random-walks-on-graphs.pdf.
6. David G. Luenberger. *Optimization by Vector Space Methods*. Wiley, 1997. ISBN: 9780471181170. http://sites.science.oregonstate.edu/~show/old/142_Luenberger.pdf.
7. J. R. Norris. *Markov Chains*. Cambridge Series in Statistical and Probabilistic Mathematics. Cambridge: Cambridge University Press, 1997. https://doi.org/10.1017/CBO9780511810633.
8. Oskar Perron. "Zur Theorie der Matrices". In: *Mathematische Annalen* 64.2 (1907), pp. 248–263. https://doi.org/10.1007/BF01449896.
9. K. Teknomo. "Ideal Flow Based on Random Walk on Directed Graph". In: *Proceedings of the 9th International Collaboration Symposium on Information, Production and Systems*. Kitakyushu: Waseda University, 2015.
10. Kardi Teknomo. "Ideal Flow of Markov Chain". In: *Discrete Mathematics, Algorithms and Applications* 10.06 (2018), p. 1850073. https://doi.org/10.1142/S1793830918500738. https://www.worldscientific.com/doi/abs/10.1142/S1793830918500738.
11. Kardi Teknomo. "Premagic and Ideal Flow Matrices". In: *arXiv preprint arXiv:1706.08856* (2017). https://arxiv.org/pdf/1706.08856.

Chapter 7
Network Preparation: Modeling Open Traffic Systems with an External Balancing Node

7.1 Introduction

The analysis of a transport system depends on its mathematical representation. We typically model a traffic system as a directed graph [1–5], $\mathcal{G} = (\mathcal{V}, \mathcal{E})$. A directed graph is a network of points, called vertices or nodes, connected by one-way paths (links or edges). The set of vertices $\mathcal{V}$ represents locations like road intersections. The set of directed edges or links $\mathcal{E}$ represents the pathways for movement between these locations. Think of a system of roads of one-way streets. The analysis of such systems, particularly for traffic assignment and congestion studies, often requires the graph to have specific structural properties.

A common requirement is that the network must be *strongly connected*. This property ensures that it is possible to travel from any node to any other node. This reflects a fully integrated system. Furthermore, the network must be *premagic* [10], which means that the total inflow equals the total outflow for every node. This condition signifies a self-contained system in a steady state. A network characterized by strong connectivity and the premagic property is called an *Ideal Flow Network* (IFN).

However, real-world urban traffic systems are sometimes open. These nodes violate the properties of strong connectivity and flow conservation. A network with a source or a sink is, by definition, weakly connected. Weak connectivity means that if we ignore the direction of the one-way roads, we can get from any point to any other point. All nodes are part of a single, continuous system, but following the one-way rules, we might not be able to return to our starting point. This presents a challenge for analytical models that require strongly connected, premagic networks.

Strong connectivity is a stricter condition. It means that we can travel from any node to any other node and back again by following the directed, one-way paths. The entire network forms a single, closed loop.

K. Teknomo, *Ideal Flow Network for Transportation Research*,
Lecture Notes in Intelligent Transportation and Infrastructure,
https://doi.org/10.1007/978-981-95-4727-2_7

To apply IFN models that require strong connectivity, we need to transform the network. We will demonstrate that any open traffic network can be augmented to meet the necessary properties for rigorous analysis. Weakly connected networks can be made into strongly connected networks by augmenting them with an external balancing "cloud" node that connects network sinks back to sources. This technique preserves the internal flow dynamics of the original system while making it compatible with powerful analytical frameworks.

This chapter provides a formal methodology to solve this problem. The learning objectives are to:

- Understand the formal definitions of network connectivity and flow conservation.
- Learn how to transform an open, weakly connected network into a closed, strongly connected, and premagic network.
- Master the concept of the external balancing node and its role in the transformation.
- Apply this transformation technique to model and analyze practical urban traffic congestion scenarios.

The following story summarizes what we will discuss in this chapter. *A transport planner faces a critical challenge: her city's traffic network was a maze of one-way streets and isolated districts, making it only weakly connected and incompatible with the Ideal Flow Network (IFN) model she desperately wanted to use. The IFN demanded a strongly connected graph, a fundamental structural requirement her data lacked. Undeterred, she rigorously partitioned her network, identifying a large residential zone acting as a source basin and a complex industrial park as a sink basin, a full component of its own. Realizing these basins were the network's disconnected entry and exit points, she introduced a single, abstract cloud node to represent the entire external environment. She then added conceptual links from all sink basin exits to the cloud and from the cloud to all source basin entries. This single, elegant theoretical maneuver instantly created a strongly connected network, successfully bridging her city's traffic gaps and allowing the IFN model to finally provide the actionable insights she needed.*

7.2 Network Preparation Decision Tree

The diagram Fig. 7.1 presents a decision tree for preparing a network for Ideal Flow Network analysis. This process ensures the network graph meets the necessary topological conditions for the Ideal Traffic Assignment framework that we discuss in detail in Chap. 9. The framework requires a strongly connected, directed graph to model traffic flow accurately.

The process begins with raw Network Data. The first step checks if the network is connected. If it is not connected, IFN analysis cannot be used. The model requires a single, unified network. If the network is connected, the next step considers if it is part of a larger environment. If the network is self-contained (not part of a

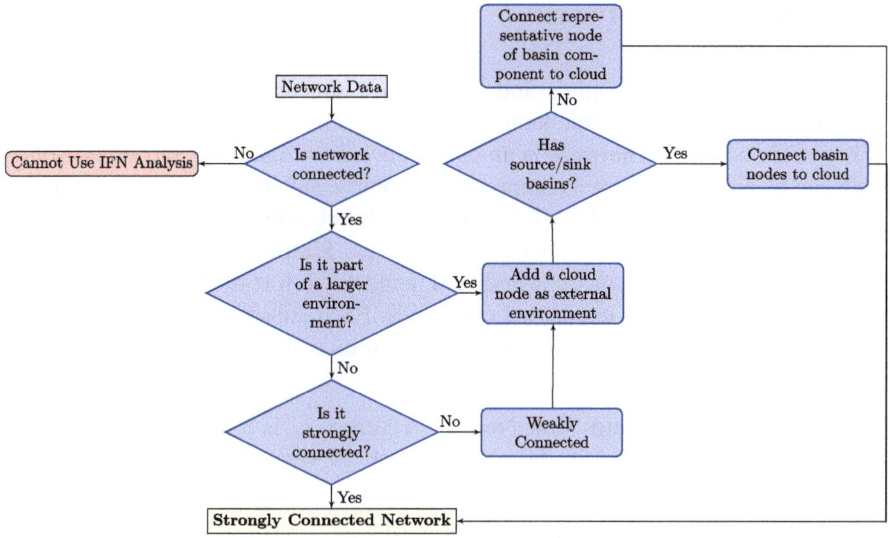

Fig. 7.1 A flowchart for preparing a network for Ideal Flow Network analysis

larger environment), the analysis proceeds to check for strong connectivity. If the network is a component of a larger system, a special cloud node is added. This node represents the external environment and acts as a single, consolidated source and sink for flows entering and leaving the network.

After the initial connectivity checks, the model assesses if the network is strongly connected. A strongly connected graph ensures that a path exists from every node to every other node. If the network is strongly connected, it is deemed a strongly connected Network and is ready for IFN analysis. If the network is only weakly connected (meaning the underlying undirected graph is connected, but the directed graph is not strongly connected), the cloud node approach is used to create a strongly connected structure.

For weakly connected networks and those with an added cloud node, the model identifies and manages source/sink basins. If basins of sources and sinks exist, all nodes within these basins are connected to the cloud node. This reconfigures the network to satisfy the strong connectivity requirement. If basins are not present, representative nodes of each basin class are connected to the cloud node. Both methods modify the graph to create a single, strongly connected component.

By following this decision tree, any initial road network can be systematically processed and transformed to meet the topological assumptions of the Ideal Traffic Assignment model [11], enabling the prediction of traffic flows from the network structure alone. The resulting prepared network is a strongly connected network that is ready for IFN analysis.

7.3 Augmented Network

In this section, the reasoning and formalities concerning the efficacy of the preceding network preparation decision tree will be presented.

A weakly connected network is an *open network* with sources and sinks. Source basin can be either a node or a component. Similarly, sink basin can be either a node or a component. The tree Fig. 7.2 visualizes the classification of directed graph structures based on their connectivity properties. Starting from general network types, it branches into unconnected and connected forms, further subdividing into weakly and strongly connected categories. Each substructure reveals nuanced topological features such as the presence of source or sink basins at node and component levels.

Definition 7.1 (Source and Sink Nodes) A vertex v_s is a *source node* if its in-degree is zero with respect to the core network. A vertex v_t is a *sink node* if its out-degree is zero with respect to the core network.

A source node is a node that only has roads leading away from it. No roads lead into a source node. A sink node is a node that only has roads leading into it. No roads lead away from a sink node.

Definition 7.2 (Strongly Connected Component (SCC)) A *component* is a special subgraph where it is possible to move from any vertex to any other vertex within that subgraph. A *strongly connected component* (SCC) is a sub-network where it is possible to move from any node to any other node within that sub-network.

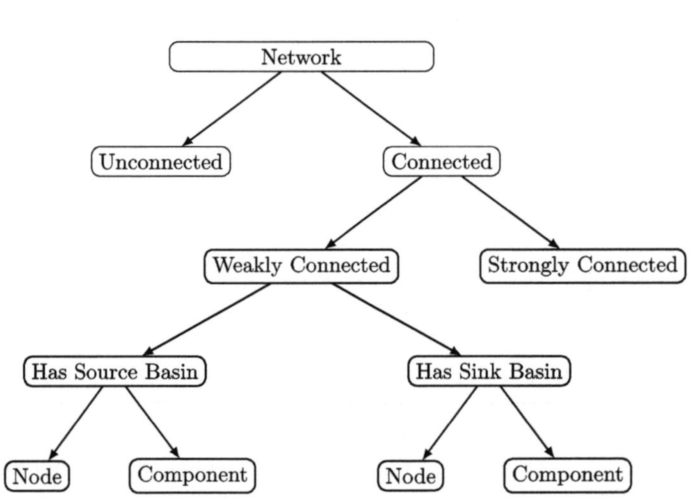

Fig. 7.2 A hierarchical classification of network connectivity types

Definition 7.3 (Source and Sink Components) A *source component* is a strongly connected component where once an agent leaves any node within it, the agent can never return to any node in that component. A *sink component* is a strongly connected component that, once entered, an agent can never leave.

The presence of sources and sinks basins (either nodes or components) makes a network weakly connected. To analyze such a network with models requiring strong connectivity, we must transform it. A weakly connected network can be transformed into an equivalent closed strongly connected network by optionally introducing only a single abstract external node (i.e., the cloud node z). The cloud node is optional because we can also connect the sink to the source basin directly without the cloud node. However, a cloud node may have another meaning that it represents the external environment outside our model.

We now ready to formalize this network preparation.

Axiom 7.1 (Network Augmentation) *An open network with sources and sinks can be transformed into an equivalent strongly connected network by introducing a single abstract external node.*

This axiom is the basis of our method. We now formalize the transformation.

Definition 7.4 (External Balancing Node) The *external balancing node*, or a *cloud node* denoted z, is an abstract vertex representing the environment outside the system.

A cloud node acts as the ultimate source for all traffic entering and the ultimate sink for all traffic leaving.

Definition 7.5 (Augmented Network [2]) Let $\mathcal{G} = (\mathcal{V}, \mathcal{E})$ be an open network. We can partition the set of all nodes $\mathcal{V}$ into a set of Maximal Strongly Connected Components (MSCCs), denoted by $\{S_1, S_2, \ldots, S_k\}$. The *source basins* are all MSCCs with no incoming edges from other MSCCs, and the *sink basins* are all MSCCs with no outgoing edges to other MSCCs. Let $\mathcal{B}_S$ be the set of all nodes belonging to source basins and $\mathcal{B}_T$ be the set of all nodes belonging to sink basins.

The *augmented network* $\mathcal{G}' = (\mathcal{V}', \mathcal{E}')$ is formed by:

1. Setting the vertex set to $\mathcal{V}' = \mathcal{V} \cup \{z\}$, where z is the external balancing node.
2. Adding a new set of edges $\mathcal{E}_{aug}$ which consists of two subsets:
   - Edges from every node in a sink basin to z: (v_t, z) for all $v_t \in \mathcal{B}_T$.
   - Edges from z to every node in a source basin: (z, v_s) for all $v_s \in \mathcal{B}_S$.

Thus, $\mathcal{E}' = \mathcal{E} \cup \mathcal{E}_{aug}$.

Lemma 7.1 (Existence of Basins) *Any weakly connected directed graph $\mathcal{G}$ that is not strongly connected must contain at least one source basin and at least one sink basin.*

Proof A weakly connected directed graph $\mathcal{G}$ can be partitioned into a set of disjoint Maximal Strongly Connected Components (MSCCs). The condensation graph $\mathcal{G}_{SCC}$ is formed by representing each MSCC as a single vertex. Since $\mathcal{G}$ is not strongly connected, $\mathcal{G}_{SCC}$ contains more than one vertex. The condensation graph is, by definition, a Directed Acyclic Graph (DAG). Any finite DAG must contain at least one source vertex (a vertex with in-degree zero) and at least one sink vertex (a vertex with out-degree zero). These source and sink vertices in $\mathcal{G}_{SCC}$ correspond directly to the source and sink basins in the original graph $\mathcal{G}$. Thus, at least one source basin and at least one sink basin must exist. □

This Theorem 7.1 establishes that a specific structural modification of a network guarantees its complete interconnectedness. The central idea is to take a network that is weakly connected in all directions and, by adding a single special cloud node, transform it into a network where every node can reach every other node via a directed path. The rules for this augmentation are precise:

- An edge is added from every existing sink node or any node in sink component to the cloud node. This means every dead-end street now has a path leading to the cloud node.
- An edge is added from the cloud node to every existing source node or any node in source component. This means the cloud node can reach every street that previously had no incoming traffic.

Theorem 7.1 (Strong Connectivity of the Augmented Network) *Let $\mathcal{G} = (\mathcal{V}, \mathcal{E})$ be a weakly connected directed graph. The augmented graph $\mathcal{G}'$, constructed according to Definition 7.5, is strongly connected.*

Proof To prove strong connectivity, we must demonstrate that for any two distinct vertices $u, w \in \mathcal{V}'$, a directed path exists from u to w and from w to u.

From the Lemma 7.1, we know that any weakly connected, non-strongly connected graph can be partitioned into Maximal Strongly Connected Components (MSCCs), which form a Directed Acyclic Graph (DAG) in the condensation graph. This ensures the existence of at least one source basin ($\mathcal{B}_S$) and one sink basin ($\mathcal{B}_T$).

Consider two arbitrary vertices $u, w \in \mathcal{V}'$.

- Case 1: Both $u, w \in \mathcal{V}$. Since $\mathcal{G}$ is weakly connected, a path exists from u to some node v_t that belongs to a sink basin, $v_t \in \mathcal{B}_T$. Similarly, a path exists from some node v_s belonging to a source basin, $v_s \in \mathcal{B}_S$, to w. In $\mathcal{G}'$, edges (v_t, z) and (z, v_s) exist by the revised definition. Therefore, a path $u \rightsquigarrow v_t \rightarrow z \rightarrow v_s \rightsquigarrow w$ establishes a path from u to w. A path from w to u is established similarly.
- Case 2: One vertex is z. Consider a path from any $u \in \mathcal{V}$ to z. Since $\mathcal{G}$ is weakly connected, there is a path from u to a node $v_t \in \mathcal{B}_T$. The edge (v_t, z) is present by construction. Therefore, a path $u \rightsquigarrow v_t \rightarrow z$ exists. A path from z to any $w \in \mathcal{V}$ is established by the edge (z, v_s) for a node $v_s \in \mathcal{B}_S$, followed by a path $v_s \rightsquigarrow w$. Thus, a path $z \rightarrow v_s \rightsquigarrow w$ exists.

Since a directed path exists between any two vertices for all cases, the augmented graph $\mathcal{G}'$ is strongly connected. □

Theorem 7.2 (Sufficiency of One Cloud [8, 9]) *For any weakly connected directed graph $\mathcal{G} = (\mathcal{V}, \mathcal{E})$ that is not strongly connected, a single external node, v_c, is sufficient but not necessary to create an augmented strongly connected graph $\mathcal{G}' = (\mathcal{V}', \mathcal{E}')$.*

Proof First, we will prove sufficiency by construction.

1. Construction of $\mathcal{G}'$: Let $\mathcal{G} = (\mathcal{V}, \mathcal{E})$ be a weakly connected, non-strongly connected directed graph. Let the set of all source basins in $mathcalG$ be $\mathcal{B}_S = \{B_{S_1}, B_{S_2}, \ldots, B_{S_m}\}$ and the set of all sink basins be $\mathcal{B}_T = \{B_{T_1}, B_{T_2}, \ldots, B_{T_n}\}$. Add a new vertex v_c to the vertex set, so that $V' = V \cup \{v_c\}$. Add a new set of edges $E_{aug} = \{(u, v_c) | \forall u \in \cup_{j=1}^{n} B_{T_j}\} \cup \{(v_c, w) | \forall w \in \cup_{i=1}^{m} B_{S_i}\}$. The augmented graph is $\mathcal{G}' = (\mathcal{V}', \mathcal{E} \cup E_{aug})$.
2. Demonstration of Strong Connectivity: We must show that for any two vertices $p, q \in V'$, a directed path exists from p to q.
   - Case 1: Both p and q are in $\mathcal{G}$. By the Lemma 7.1, there exists at least one sink basin and one source basin. Due to weak connectivity, there is a path from p to some node in a sink basin. From this node, there is an edge to v_c. From v_c, there is an edge to a source basin node. From this source basin node, there is a path to q. A path thus exists from p to q and from q to p via the cloud node v_c.
   - Case 2: One vertex is v_c. A path exists from any vertex $p \in V$ to a sink basin, and from there to v_c. A path also exists from v_c to any vertex $q \in V$ via a source basin. This completes the proof of sufficiency.

To prove non-necessity, we simply present a counterexample. Consider a directed graph with two MSCCs, S_1 and S_2, with a single edge (u, v) where $u \in S_1$ and $v \in S_2$. This graph is weakly connected but not strongly connected. We can add a single edge (x, y) where $x \in S_2$ and $y \in S_1$. This new edge creates a cycle that encompasses both components, merging them into a single, larger MSCC. The resulting graph is strongly connected. This construction was achieved by adding one edge without using a cloud node, demonstrating that a cloud node is not a necessary condition. □

7.4 Application to Urban Traffic Congestion

We now apply the transformation technique to practical scenarios involving urban traffic congestion. The goal is to create a model suitable for analysis, such as calculating congestion indices from an Ideal Flow matrix F and a capacity matrix C [7].

We provide simple numerical examples to illustrate the transformation.

Example 7.1 (A Simple Two-Node System) Consider a simple road segment connecting two intersections, A and B. Intersection A is a source that generates 100 vehicles per hour (vph). Intersection B is a sink that attracts 100 vph. The flow from A to B is 100 vph. This network is weakly connected.

We introduce an external node z. The augmented network has vertices $V' = \{A, B, z\}$. We add an edge (z, A) with flow 100, and an edge (B, z) with flow 100. The augmented network is shown in Fig. 7.3.

The flow matrix F' for the augmented network (with nodes ordered A, B, z) is:

$$\mathsf{F}' = \begin{pmatrix} 0 & 100 & 0 \\ 0 & 0 & 100 \\ 100 & 0 & 0 \end{pmatrix} \tag{7.1}$$

We can verify that this network is premagic using Eq. (8.4). The outflow vector is $\mathsf{F}'\mathbf{j} = (100, 100, 100)^T$. The inflow vector is $(\mathsf{F}'^T\mathbf{j})^T = (100, 100, 100)^T$. They are equal. The network is also strongly connected.

Example 7.2 (A Four-Node Urban Grid) Consider the network in Fig. 7.4. Node a is a source generating 150 vph. Node d is a sink attracting 150 vph. The internal flows are $f_{ab} = 100$, $f_{ac} = 50$, $f_{bd} = 100$, and $f_{cd} = 50$. Flow is conserved at the transit nodes b and c. This is a valid open network. We augment this network with node z. We add edge (z, a) with flow 150, and edge (d, z) with flow 150. Dummy links are added from any representative node in the sink component to the cloud node. Dummy links are also added from the cloud node to any representative

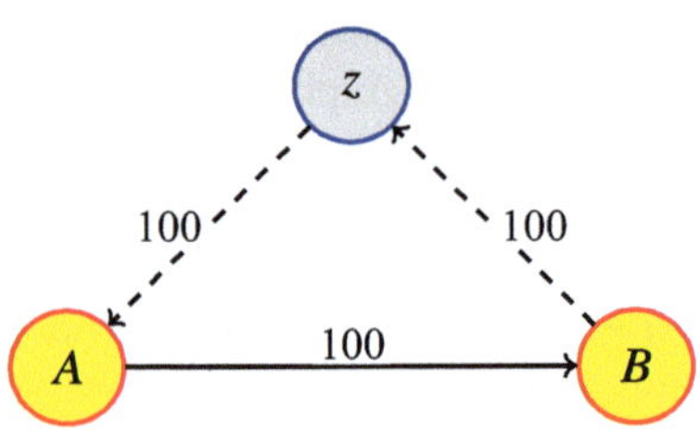

Fig. 7.3 A simple two-node network connected to an external environment. The external node z transforms the open system into a closed, strongly connected loop

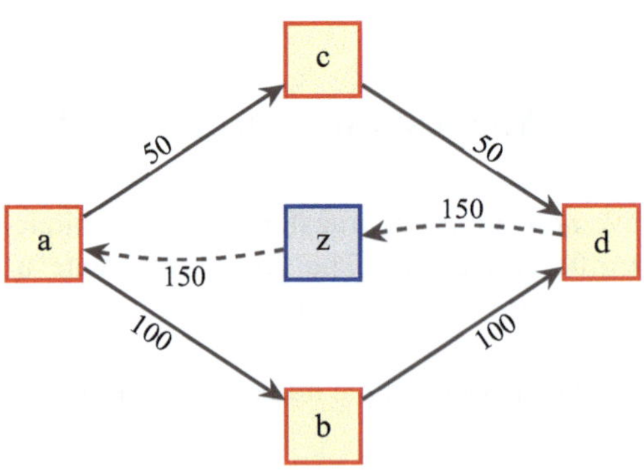

Fig. 7.4 A four-node, four-link network with an external environment. The external node z connects the source (A) and the sink (D)

node in the source component. The original network is the core network. The flow performance indices are obtained only from the core network, without dummy links or the cloud node. The augmented network is now strongly connected and premagic.

Example 7.3 (Freeway Merge Point) We now model a Y-shaped freeway merge, a common source of urban congestion. We use an IFN model based on a junction where two input links (a and b) merge into one output link (d). To create a strongly connected network, we add a conceptual return node z, which recycles flow from the exit back to the entrances, as shown in Fig. 7.5.

Example 7.4 (Lanes Capacity with Cloud Node) The typical crossing intersection network with its number of lanes is shown in the figure below. The capacity matrix consists of only eight possible values namely c_1 to c_8. The listing of the subscript numbers is from top to bottom, left to right (Fig. 7.6).

$$\mathsf{C} = \left(\begin{array}{c|ccccc} & a & b & c & d & e \\ \hline a & 0 & c_2 & 0 & 0 & 0 \\ b & c_1 & 0 & c_6 & c_7 & c_8 \\ c & 0 & c_3 & 0 & 0 & 0 \\ d & 0 & c_4 & 0 & 0 & 0 \\ e & 0 & c_5 & 0 & 0 & 0 \end{array} \right) \tag{7.2}$$

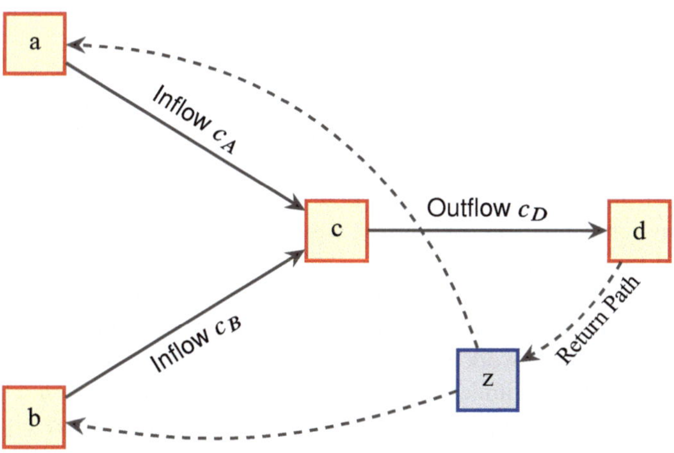

Fig. 7.5 IFN representation of a freeway merge. Nodes a and b are inputs, c is the merge point, and d is the output. The dashed links involving cloud node z close the loop

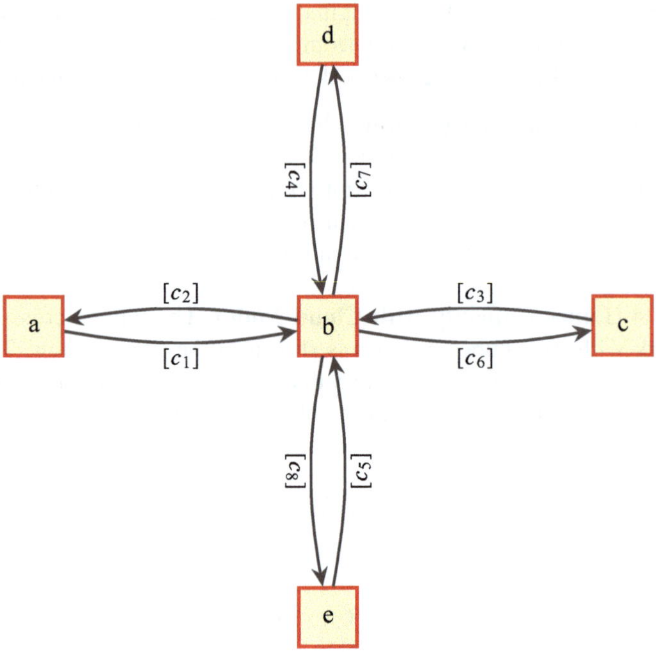

Fig. 7.6 An urban intersection represented as a directed graph

Add a cloud node and dummy links to represent the external environment. What would be the new capacity matrix?

Answer

The first step to use ideal flow model is to convert the real world network into a strongly connected network. However, our network of intersection above is already strongly connected. We don't need to add a cloud node. Having said that, we also understand from the context that the flow is originated from and going into the external environment. Thus, optionally, we can still add a cloud node to represent the external environment.

The capacity of the dummy links must follow as the extension of the actual links it represents. The capacity of the dummy link za should reflect the total flow capacity entering the network at node a. Thus, the capacity of dummy link za would be the same as link ab. Similarly, the capacity of dummy link cz would be the same as link bc. Thus, the total possible lane capacity remain the same as eight possible values (Fig. 7.7).

$$\mathsf{C} = \begin{pmatrix} & a & b & c & d & e & z \\ \hline a & 0 & c_2 & 0 & 0 & 0 & c_1 \\ b & c_1 & 0 & c_6 & c_7 & c_8 & 0 \\ c & 0 & c_3 & 0 & 0 & 0 & c_6 \\ d & 0 & c_4 & 0 & 0 & 0 & c_7 \\ e & 0 & c_5 & 0 & 0 & 0 & c_8 \\ z & c_2 & 0 & c_3 & c_4 & c_5 & 0 \end{pmatrix} \tag{7.3}$$

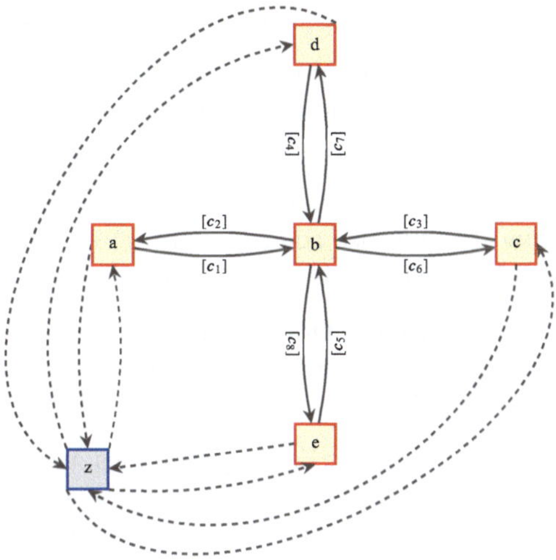

Fig. 7.7 An augmented urban intersection model for network analysis

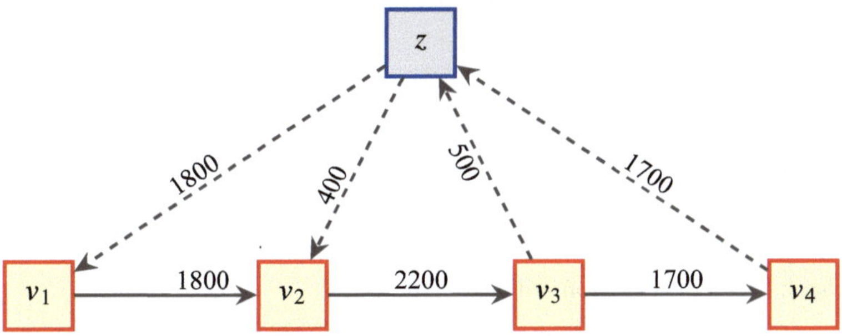

Fig. 7.8 An urban arterial road transformed into a strongly connected system

Example 7.5 (Urban Arterial Road) Consider a segment of a major urban arterial road. It has one on-ramp and one off-ramp. This is a classic open system.

- *Nodes:* v_1 (entry point), v_2 (merge point), v_3 (diverge point), v_4 (exit point).
- *Sources:* The mainline entry at v_1 and the on-ramp at v_2.
- *Sinks:* The off-ramp at v_3 and the mainline exit at v_4.

The original network is weakly connected. To analyze it, we introduce the external balancing node z. The augmented network is shown in Fig. 7.8.

Let's assign some flow values for the morning peak hour.

- Mainline entry flow at v_1: 1800 vph.
- On-ramp flow at v_2: 400 vph.
- Total Generation = $1800 + 400 = 2200$ vph.

To balance the system, total attraction must also be 2200 vph.

- Off-ramp flow at v_3: 500 vph.
- Mainline exit flow at v_4: 1700 vph.
- Total Attraction = $500 + 1700 = 2200$ vph.

In the augmented network, we add edges (z, v_1) with flow 1800 and (z, v_2) with flow 400. We also add edges (v_3, z) with flow 500 and (v_4, z) with flow 1700. The resulting network is strongly connected and premagic.

Example 7.6 (A Downtown Traffic Grid) Consider a simplified model of a downtown traffic grid, for example in Denver, USA. The grid has multiple entry and exit points.

- *Internal Nodes:* Four intersections, v_1, v_2, v_3, v_4.
- *Source:* An off-ramp from a highway enters at v_1, generating 1000 vph.
- *Sink:* A major parking garage entrance is at v_3, attracting 1000 vph.

The internal flows are balanced at the transit nodes v_2 and v_4. The augmented network is shown in Fig. 7.9.

The flows are as follows:

- *External Inflow:* $f_{z1} = 1000$.
- *External Outflow:* $f_{3z} = 1000$.
- *Internal Flows:* $f_{12} = 600, f_{14} = 400, f_{23} = 700, f_{24} = 100, f_{42} = 200, f_{43} = 300$.

Let's check flow conservation at the transit nodes:

- *Node v_2:* Inflow = $f_{12} + f_{42} = 600 + 200 = 800$. Outflow = $f_{23} + f_{24} = 700 + 100 = 800$. (Balanced)
- *Node v_4:* Inflow = $f_{14} + f_{24} = 400 + 100 = 500$. Outflow = $f_{42} + f_{43} = 200 + 300 = 500$. (Balanced)

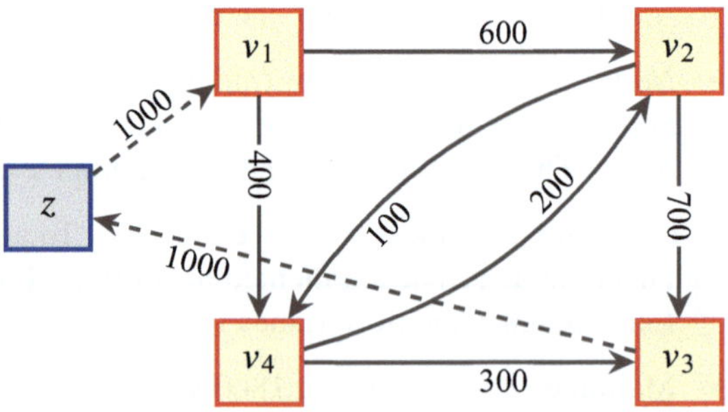

Fig. 7.9 A grid network with an external environment. The external node z models the traffic from and to the rest of the city

The model is consistent. The flow matrix F' for the augmented network (ordered v_1, v_2, v_3, v_4, z) is:

$$\mathsf{F}' = \begin{pmatrix} 0 & 600 & 0 & 400 & 0 \\ 0 & 0 & 700 & 100 & 0 \\ 0 & 0 & 0 & 0 & 1000 \\ 0 & 200 & 300 & 0 & 0 \\ 1000 & 0 & 0 & 0 & 0 \end{pmatrix}$$

This matrix is irreducible [6] and premagic. It can be used as input for further analysis, such as studying the impact of changing road capacities on overall network congestion.

7.5 Summary

We presented a formal and rigorous method for modeling open transport networks. Real-world systems, with their sources and sinks, are inherently open and thus weakly connected. This conflicts with the requirements of analytical Ideal Traffic Assignment frameworks that presuppose closed, strongly connected, and flow-balanced systems. A key contribution of this chapter is the method for analyzing open systems by augmenting them with an external balancing node, a technique that transforms them into closed, strongly connected networks amenable to IFN analysis. We demonstrated that this conflict is resolved by augmenting the network with a single external balancing node. This abstract node represents the environment beyond the system boundary. The transformation yields a new, larger network that is mathematically sound. We proved that the augmented network is guaranteed to be strongly connected. Additionally, the network is premagic, which holds provided the total flow entering the original system is equal to the total flow exiting the system. This technique is not just a mathematical convenience. It is a powerful conceptual tool. It allows us to model the interaction between a local system and its environment, preserving the internal flow dynamics while enabling the application of advanced, steady-state network analysis. This method is therefore essential for the practical and effective study of urban traffic congestion.

References

1. Gary Chartrand et al. *Graphs & Digraphs*. 7th ed. Boca Raton, FL: Chapman and Hall/CRC, 2024. ISBN: 9781032606989. https://doi.org/10.1201/9781032606989
2. Reinhard Diestel. *Graph Theory*. 5th. Berlin, Heidelberg: Springer, 2017. https://doi.org/10.1007/978-3-662-53622-3.

3. L. R. Foulds. *Graph Theory Applications*. Springer-Verlag, 1992. https://doi.org/10.1007/978-1-4612-0933-1.
4. F. Harary. *Graph Theory*. Addison-Wesley, 1969.
5. F. Harary and R. Z. Norman. “Some properties of line digraphs”. In: *Rendiconti del Circolo Matematico di Palermo* 9.2 (1960), pp. 161–168. https://doi.org/10.1007/BF02854556.
6. John G. Kemeny and J. Laurie Snell. *Finite Markov Chains*. New York: Springer, 1976. https://link.springer.com/book/9780387901923.
7. B. S. N. Reddy, H. Gonzalez, and R. J. Mondragon. “A new approach to modelling traffic flows in urban road networks”. In: *2011 3rd International Conference on Advanced and Integrated Mechatronic Systems*. 2011, pp. 1–6. https://doi.org/10.1109/AIMS.2011.6026992.
8. Kardi Teknomo. “Ideal Relative Flow Distribution on Directed Network”. In: *Journal of the Eastern Asia Society for Transportation Studies* 12 (2017), pp. 939–958. https://doi.org/10.11175/easts.12.939. https://www.jstage.jst.go.jp/article/easts/12/0/12_939/_article.
9. Kardi Teknomo. *Introduction to Ideal Flow Networks*. 2022.
10. Kardi Teknomo. “Premagic and Ideal Flow Matrices”. In: *arXiv preprint arXiv:1706.08856* (2017). https://arxiv.org/pdf/1706.08856.
11. Kardi Teknomo, R. Gardon, and C. Saloma. “Ideal Flow Traffic Analysis: A Case Study on a Campus Road Network”. In: *Philippine Journal of Science* 148.1 (2019), pp. 51–62. https://archium.ateneo.edu/discs-faculty-pubs/271/.

Chapter 8
The Core Ideal Flow Network Theory

8.1 Constructing the Stochastic Matrix from Data

The stochastic matrix is a fundamental tool for describing systems that evolve over discrete time steps, where the future state depends only on the present state, a property known as the Markov property. Recall the general definition of a stochastic matrix, presented in Definition 3.24.

Definition 8.1 (Stochastic Matrix [6]) A stochastic matrix (or a transition probability matrix) is a square matrix whose entries are all non-negative and whose rows (or columns) each sum to 1. The elements, S_{ij}, represent the probabilities of transitioning from a state i to a state j.

$$\mathsf{S}_{ij} \geq 0 \quad \text{and} \quad \sum_{j \in N} \mathsf{S}_{ij} = 1 \quad \forall i \in N \tag{8.1}$$

We can obtain a stochastic matrix from data by observing transitions between a finite number of states.

1. *Define States:* First, we must clearly define the discrete states of our system. In transportation network, the states are nodes in a road network. In an intersection, the states are left-turn, right-turn, go straights or u-turn. In a different field, states could be weather conditions (sunny, rainy, cloudy), customer loyalty levels (loyal, casual, churned), or disease stages (healthy, mild, severe).
2. *Count Transitions:* Next, we collect data by counting the number of times the system transitions from each state i to every other state j. This is often represented as a count matrix or an adjacency matrix, where the entry A_{ij} is the number of times state i was followed by state j.

K. Teknomo, *Ideal Flow Network for Transportation Research*,
Lecture Notes in Intelligent Transportation and Infrastructure,
https://doi.org/10.1007/978-981-95-4727-2_8

3. *Normalize Rows:* To convert the count matrix into a stochastic matrix, normalize each row. Divide each entry in row i by the total number of transitions originating from state i. The resulting matrix, S, will have rows that sum to 1. The entry S_{ij} then represents the empirical probability of moving from state i to state j.

This process allows the abstract mathematical concept to be applied rigorously to empirical observations in any domain with discrete states and transitions.

Example 8.1 (Stochastic from Network) In a transportation network, the states are the discrete nodes (e.g., intersections or key locations).

1. Define States: Consider a small transportation network with three nodes:
   - a: Node a
   - b: Node b
   - c: Node c
2. Count Transitions: Assume we track vehicles moving directly between these nodes over a period and count the observed transitions.
   - From Node a:
     - 40 vehicles travel to Node b (a to b)
     - 60 vehicles travel to Node c (a to c)
     - Total from a: $40 + 60 = 100$
   - From Node b:
     - 60 vehicles travel to Node a (b to a)
     - 140 vehicles travel to Node c (b to c)
     - Total from b: $60 + 140 = 200$
   - From Node c:
     - 75 vehicles travel to Node a (c to a)
     - 75 vehicles travel to Node b (c to b)
     - Total from c: $75 + 75 = 150$

   These counts form the count matrix:

$$\mathsf{C} = \left(\begin{array}{c|ccc} \nearrow & a & b & c \\ \hline a & 0 & 40 & 60 \\ b & 60 & 0 & 140 \\ c & 75 & 75 & 0 \end{array}\right)$$

   Note that entries C_{ii} are zero, as vehicles typically move between distinct nodes.

3. Normalize Rows: To obtain the Ideal Stochastic Matrix S, each entry in the count matrix C is divided by its respective row sum.
   - Row 1 (From a): Sum $= 100$
     - $\mathsf{S}_{a,a} = 0/100 = 0.0$
     - $\mathsf{S}_{a,b} = 40/100 = 0.4$
     - $\mathsf{S}_{a,c} = 60/100 = 0.6$
   - Row 2 (From b): Sum $= 200$
     - $\mathsf{S}_{b,a} = 60/200 = 0.3$
     - $\mathsf{S}_{b,b} = 0/200 = 0.0$
     - $\mathsf{S}_{b,c} = 140/200 = 0.7$
   - Row 3 (From c): Sum $= 150$
     - $\mathsf{S}_{c,a} = 75/150 = 0.5$
     - $\mathsf{S}_{c,b} = 75/150 = 0.5$
     - $\mathsf{S}_{c,c} = 0/150 = 0.0$

   The resulting Ideal Stochastic Matrix S for the transportation network is:

$$\mathsf{S} = \begin{pmatrix} 0.0 & 0.4 & 0.6 \\ 0.3 & 0.0 & 0.7 \\ 0.5 & 0.5 & 0.0 \end{pmatrix}$$

8.2 The Ideal Flow Matrix

Having established the method for constructing the Stochastic matrix S, we now address the central task of the IFN framework: deriving the Ideal Flow Matrix F. This derivation is a direct consequence of the steady-state properties of Markov chains, where the network's structure, encoded in S, dictates the equilibrium distribution of flow. The process involves two main steps: first, determining the relative flow distribution across the network, and second, scaling this distribution to an absolute magnitude using the total flow parameter, κ.

Definition 8.2 (Flow Network) A flow network is a directed graph $\mathcal{N}(\mathcal{V}, \mathcal{E})$ with a flow function $\phi : \mathcal{E} \to \mathbb{R}_{\geq 0}$, where $\mathcal{V}$ is the set of nodes and $\mathcal{E}$ is the set of directed edges called links. We associate a link flow function: for each edge $(i, j) \in \mathcal{E}$, we have a non-negative flow f_{ij}.

The following axioms are for any general IFN model.

Axiom 8.1 (Unique Identifiers) *Every node i in a network belongs to a set $\mathcal{V}$ and has a unique string identifier.*

Axiom 8.2 (Non-negative Edge Flows [2]) *The flow f_{ij} on every edge $(i, j) \in \mathcal{E}$ is non-negative. Flow changes Δf_{ij} may be negative, but $f_{ij} \geq 0$ at all times. Formally, for edge set $\mathcal{E}$, $f_{ij} \geq 0 \quad \forall (i, j) \in \mathcal{E}$.*

Axiom 8.3 (Strongly Connected Network) *The network graph $\mathcal{N}$ is* strongly connected. *This means that for every pair of nodes (i, j), there exists at least one path from i to j.*

For networks that are not naturally strongly connected, a conceptual 'cloud' node can be added to meet this requirement (see Chap. 7).

Axiom 8.4 (Flow Conservation [17]) *Flow is conserved at every node. For any node $i \in \mathcal{V}$, the total flow entering the node equals the total flow leaving it.*

$$\sum_{k \in \mathcal{V}} \mathsf{F}_{ki} = \sum_{j \in \mathcal{V}} \mathsf{F}_{ij} \quad \forall i \in \mathcal{V} \tag{8.2}$$

The total flow into any node equals the total flow out of that same node. This property, also called premagic, *ensures flow conservation at every node. Mathematically, it is expressed as:*

$$\mathsf{F}\boldsymbol{J} = \mathsf{F}^{\top}\boldsymbol{J} \tag{8.3}$$

where $\boldsymbol{J}$ is a column vector of ones. A square matrix with this property is called a premagic matrix.

Based on the axioms above, we now define the Ideal Flow Network and its matrix.

Definition 8.3 (Ideal Flow Network (IFN)) An *Ideal Flow Network (IFN)* is a strongly connected, directed graph that satisfies node-level flow conservation, with flows representing a steady-state distribution. Its Ideal Flow matrix F is nonnegative, premagic and irreducible [9].

Definition 8.4 (Ideal Flow Matrix (F)) The *Ideal Flow Matrix* F is an $n \times n$ matrix where each element F_{ij} represents the steady-state relative flow on link (i, j).

In matrix form, the flow conservation means *premagic matrix*. This would be discussed in the next sub-section.

Example 8.2 (A Simple 3-Node Network) Consider a simple 3-node network. The allowed movements (topology) define the structure. Let the outflow probabilities from each node be fixed (Fig. 8.1).

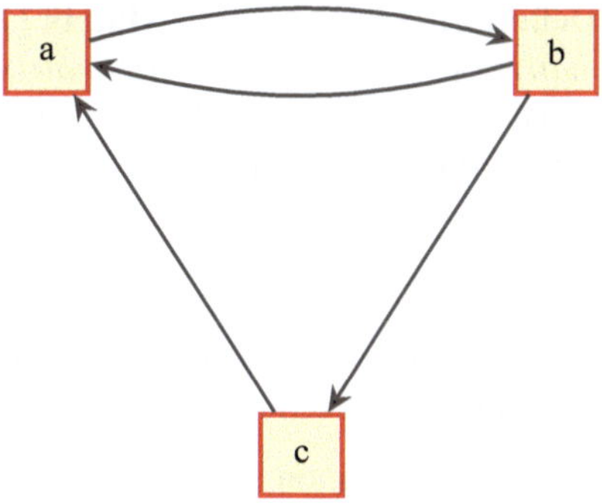

Fig. 8.1 A simple three-node, four-link strongly connected network

For example, from node 1, 50% of flow goes to node 2 and 50% to node 3. The full set of probabilities defines a row-stochastic matrix S.

$$\mathsf{S} = \left(\begin{array}{c|ccc} \nearrow & a & b & c \\ \hline a & 0 & \frac{1}{2} & \frac{1}{2} \\ b & 1 & 0 & 0 \\ c & 0 & 1 & 0 \end{array}\right)$$

To find the ideal flow matrix F, we find the smallest integer flows that satisfy these proportions and the Flow Conservation Axiom 8.4. The resulting basis flow is:

$$\mathsf{F} = \left(\begin{array}{c|ccc|c} \nearrow & a & b & c & \Sigma \\ \hline a & 0 & 1 & 1 & 2 \\ b & 2 & 0 & 0 & 2 \\ c & 0 & 1 & 0 & 1 \\ \hline \Sigma & 2 & 2 & 1 & 5 \end{array}\right)$$

We can verify the flow conservation property. For node a, the inflow is $f_{ba} = 2$. The outflow is $f_{ab} + f_{ac} = 1 + 1 = 2$. For node b, the inflow is $f_{ab} + f_{cb} = 1 + 1 = 2$. The outflow is $f_{ba} = 2$. For node c, the inflow is $f_{ac} = 1$. The outflow is $f_{cb} = 1$.

The node flow vector $\boldsymbol{\pi}$ is (2, 2, 1). This network is premagic. Total flow $\kappa = 5$.

8.2.1 The Premagic Property and Flow Conservation

In addition to connectivity, a critical property of a flow network is whether the flow is conserved at each node. This physical property, where the total flow entering a node equals the total flow leaving it, is captured algebraically by the premagic property of the network's flow matrix.

Definition 8.5 (Premagic Matrix [17]) A square matrix A is called a *premagic matrix* or *line-sum symmetric matrix* [4] if the sum of the elements in each row is equal to the sum of the elements in the corresponding column. Using the one vector $\mathbf{j}$, this condition is stated as its row sums equal the transpose of its column sums:

$$\mathsf{A}\mathbf{j} = \mathsf{A}^T\mathbf{j} \tag{8.4}$$

A non-negative premagic matrix represents node conservation of flow in a network. Total flow into each node equals total flow out of that node.

Several common matrix types are inherently premagic. Any scalar (as a 1×1 matrix), any square zero matrix, any square one matrix, any diagonal matrix, and any symmetric matrix is premagic. A *magic square*, where all row sums and column sums are equal to the same constant, is also premagic.

Example 8.3 (Testing for the Premagic Property) Show that the following matrix A is a premagic matrix.

$$\mathsf{A} = \begin{pmatrix} 1 & 0 & 5 \\ 2 & 0 & 2 \\ 3 & 4 & 1 \end{pmatrix}$$

To validate this, we compute the row sums ($\mathsf{A}\mathbf{j}$) and column sums ($\mathsf{A}^T\mathbf{j}$).

$$\mathsf{A}\mathbf{j} = \begin{pmatrix} 1 & 0 & 5 \\ 2 & 0 & 2 \\ 3 & 4 & 1 \end{pmatrix} \begin{pmatrix} 1 \\ 1 \\ 1 \end{pmatrix} = \begin{pmatrix} 6 \\ 4 \\ 8 \end{pmatrix}$$

$$\mathsf{A}^T\mathbf{j} = \begin{pmatrix} 1 & 2 & 3 \\ 0 & 0 & 4 \\ 5 & 2 & 1 \end{pmatrix} \begin{pmatrix} 1 \\ 1 \\ 1 \end{pmatrix} = \begin{pmatrix} 6 \\ 4 \\ 8 \end{pmatrix}$$

Since $\mathsf{A}\mathbf{j} = \mathsf{A}^T\mathbf{j}$, the matrix A is premagic.

Example 8.4 (Identifying Premagic Matrices) Identify which of the following matrices are premagic.

$$\mathsf{M}_1 = \begin{pmatrix} 0 & 1 & 0 & 0 \\ 0 & 0 & 7 & 0 \\ 1 & 1 & 0 & 4 \\ 0 & 4 & 0 & 0 \end{pmatrix}, \quad \mathsf{M}_2 = \begin{pmatrix} 0 & 1 & 0 & 0 \\ 0 & 0 & 6 & 0 \\ 1 & 1 & 0 & 4 \\ 0 & 4 & 0 & 0 \end{pmatrix}, \quad \mathsf{M}_3 = \begin{pmatrix} 0 & 1 & 2 & 0 \\ 1 & 0 & 5 & 0 \\ 2 & 5 & 0 & 3 \\ 0 & 0 & 3 & 0 \end{pmatrix}$$

To check, we compare the vector of row sums with the vector of column sums for each matrix.

- For M_1: Row sums are $(1, 7, 6, 4)^T$. Column sums are $(1, 6, 7, 4)^T$. These are not equal, so M_1 is not premagic.
- For M_2: Row sums are $(1, 6, 6, 4)^T$. Column sums are $(1, 6, 6, 4)^T$. These are equal, so M_2 is premagic.
- For M_3: The matrix is symmetric. All symmetric matrices are premagic. The row and column sums are both $(3, 6, 10, 3)^T$. Thus, M_3 is premagic.

8.2.2 *Irreducible Stochastic*

An irreducible stochastic matrix [15] guarantees that the Markov process it defines is ergodic, ensuring convergence to a unique stationary distribution. The stochastic matrix for the Ideal Flow Network (IFN) must be non-negative, row-stochastic, and irreducible [9].

Definition 8.6 (Irreducible Stochastic Matrix) Let the Ideal Flow Stochastic Matrix, denoted by S, be a square matrix that describes transition probabilities between nodes in a network. This matrix must satisfy three key properties:

1. *Non-negativity*: All elements of the matrix are non-negative.

$$\mathsf{S}_{ij} \geq 0 \quad \forall i, j$$

2. *Row-Stochasticity*: The sum of the elements in each row must equal one.

$$\sum_j \mathsf{S}_{ij} = 1 \quad \forall i$$

3. *Irreducibility*: The matrix is irreducible, which implies the underlying graph is strongly connected.

These properties ensure that S is not only a valid probability matrix but also one that models a network where all nodes are mutually reachable. The irreducibility is the crucial constraint that enables the use of the Perron-Frobenius theorem [5, 14]. Its irreducibility guarantees the existence of a unique stationary distribution for the system, which is a fundamental component of the IFN model. Larson and Odoni [10] applied stationary distribution analysis to urban operations research, enabling equilibrium modeling in queuing and transit systems.

The derivation of ideal flow rests on the interpretation of the transport network as a discrete-time Markov chain. In this chain, the nodes are the states, and the Behavioral Stochastic Matrix S is the one-step transition probability matrix. An element s_{ij} is the probability of transitioning from state i to state j. Because the network is assumed to be irreducible, the Perron-Frobenius theorem [5, 14] guarantees the existence of a unique, positive stationary distribution.

Theorem 8.1 (Existence of Stationary Distribution) *Given a strongly connected network, the corresponding row-stochastic matrix* S *is irreducible. According to the Perron-Frobenius theorem for stochastic matrices, there exists a unique stationary distribution vector* $\boldsymbol{\pi}$ *that is strictly positive and satisfies:*

$$\boldsymbol{\pi}^\top = \boldsymbol{\pi}^\top \mathsf{S} \tag{8.5}$$

where $\sum_i \pi_i = 1$*. The element* π_i *represents the long-run proportion of time the random [1, 11, 16] process spends at node* i*.*

Proof The proof is a standard result in the theory of Markov chains and can be found in texts such as Norris [13] or Seneta [15]. □

8.2.3 The Node Flow Vector

The node flow vector represents the stationary distribution of the irreducible row-stochastic matrix.

Definition 8.7 (Node Flow) For each node $i \in \mathcal{V}$, we define the node flow π_i as the sum of inflows into node i.

$$\pi_i = \sum_{k \in \mathcal{V} \text{ s.t. } (k,i) \in E} f_{ki} \tag{8.6}$$

Due to the Flow Conservation Axiom 8.4, the node flow π_i (defined as the sum of inflows into node i) also inherently represents the sum of outflows from node i. This is a fundamental characteristic of flow conservation within the IFN framework

The stationary distribution of all nodes in the system are represented by the *Node Flow Vector,* $\boldsymbol{\pi}$.

Definition 8.8 (Node Flow Vector) The node flow vector $\boldsymbol{\pi}^T = [\pi_1, \pi_2, \ldots, \pi_n]$ is defined as the vector of total outflows from each node.

$$\boldsymbol{\pi} = \mathsf{F}\mathbf{j} \tag{8.7}$$

From the flow conservation property, it follows that $\boldsymbol{\pi} = \mathsf{F}^T\mathbf{j}$ as well. Since F is non-negative and irreducible, every node must have at least one outgoing path, ensuring that $\boldsymbol{\pi} > \mathbf{0}$.

When the total flow is one, each element π_i represents the probability of relative flow that passes through node i. It is a probability distribution, so it must satisfy:

$$\pi_i > 0 \quad \forall i \in \mathcal{V} \quad \text{and} \quad \sum_{i=1}^{n} \pi_i = 1 \tag{8.8}$$

When the total flow κ is known, each element π_i represents the long-run proportion of total flow that passes through node i.

$$\sum_{i=1}^{n} \pi_i = \kappa, \quad \pi_i > 0 \quad \forall i \in \mathcal{V} \tag{8.9}$$

Proposition 8.1 (Ideal Flow Matrix for Total Flow) *Given a total network flow volume κ, the Ideal Flow matrix F is derived as:*

$$f_{ij} = (\kappa \times \pi_i) \times s_{ij} \tag{8.10}$$

where π_i is the i-th component of the stationary distribution vector.

8.2.4 The Eigenvector Formulation

The node vector can be computed as the stationary distribution of a Markov Chain. Kemeny and Snell [9] developed the foundational theory of finite Markov chains, including transition matrices and long-run behavior, essential for modeling stochastic flows in IFNs. To do this, find the unique eigenvector of the stochastic matrix S with unit dominant eigenvalue.

The relationship between the node flows is defined by the system of linear equations $\pi_i = \sum_k s_{ki} \times \pi_k$. In matrix form, this is the eigenvector equation:

$$\boldsymbol{\pi} = \mathsf{S}^T \boldsymbol{\pi} \tag{8.11}$$

This shows that the node flow vector $\boldsymbol{\pi}$ is the right eigenvector of the transposed stochastic matrix S^T corresponding to an eigenvalue of 1.

The *Perron-Frobenius theorem* provides the theoretical certainty for this process [12, 15]. It guarantees that for any non-negative, irreducible matrix, a condition met by the stochastic matrix of a strongly connected network, there exists a unique, strictly positive eigenvector corresponding to its largest eigenvalue (which is 1 for a stochastic matrix). This result proves that any strongly connected network has a single, inherent steady-state relative flow distribution [15].

The manual and algebraic methods in Sect. 8.3 are simply practical algorithms for finding this unique eigenvector, which in turn defines the Ideal Flow Network [12].

The vector $\boldsymbol{\pi}$ is found by solving the fundamental steady-state equation of the Markov chain as described in Theorem 8.2.

Theorem 8.2 (Steady-State Equation) *The stationary distribution vector* $\boldsymbol{\pi}$ *is the solution to the* fundamental steady-state equation, *which is a linear system of equations:*

$$\boldsymbol{\pi}^T \mathsf{S} = \boldsymbol{\pi}^T \tag{8.12}$$

This equation implies that at equilibrium, the distribution of flow across the nodes remains unchanged after one transition step. Transposing this equation gives the more common form for solving eigenvector problems:

$$\mathsf{S}^T \boldsymbol{\pi} = \boldsymbol{\pi}, \tag{8.13}$$

which can be rewritten as

$$(\mathsf{S}^T - \mathsf{I})\boldsymbol{\pi} = \mathbf{0}. \tag{8.14}$$

The solution of the linear equation for the total flow κ *is*

$$\boldsymbol{\pi} = \begin{bmatrix} \mathsf{S}^T - \mathsf{I} \\ \mathbf{j}^T \end{bmatrix} \setminus \begin{bmatrix} \mathbf{0} \\ \kappa \end{bmatrix} \tag{8.15}$$

Proof The theorem states that a stationary distribution vector $\boldsymbol{\pi}$ for an irreducible, row-stochastic matrix S is found by solving a specific system of linear equations. This is a fundamental result in Markov chain theory. The proof is based on the properties of stochastic matrices and the definition of a stationary distribution. The first part of the proof establishes the fundamental steady-state equation. A vector $\boldsymbol{\pi}$ is a stationary distribution if and only if it satisfies the equation $\boldsymbol{\pi}^T \mathsf{S} = \boldsymbol{\pi}^T$. This is the definition of a stationary distribution. It means the distribution of probability across the states remains constant after one transition. The vector $\boldsymbol{\pi}$ is also required to be a probability vector, meaning its elements are non-negative and sum to one. The proof then transforms this equation. By transposing the original equation, we get $\mathsf{S}^T \boldsymbol{\pi} = \boldsymbol{\pi}$. This is a standard eigenvector problem where $\boldsymbol{\pi}$ is the right eigenvector of S^T corresponding to the eigenvalue $\lambda = 1$. The existence and uniqueness of this solution for an irreducible row-stochastic matrix are guaranteed by the *Perron–Frobenius theorem* [5, 14]. For a transition probability matrix S, the dominant eigenvalue is always 1, and for an irreducible matrix, the corresponding eigenvector (the stationary distribution) is unique up to a scaling factor. The system $(\mathsf{S}^T - \mathsf{I})\boldsymbol{\pi} = \mathbf{0}$ is a singular system because the matrix $(\mathsf{S}^T - \mathsf{I})$ has a determinant of zero. The rows of a row-stochastic matrix sum to 1, which means the rows of $(\mathsf{S} - \mathsf{I})$ sum to 0. Therefore, the columns of $(\mathsf{S}^T - \mathsf{I})$ also sum to 0, which implies that $\mathbf{j}^T(\mathsf{S}^T - \mathsf{I}) = \mathbf{0}$. The matrix $(\mathsf{S}^T - \mathsf{I})$ is rank-deficient. Thus, the system has infinitely many solutions, all of which are scalar multiples of the unique stationary distribution vector. To obtain a unique solution, an additional constraint is needed. This is the *normalization condition*, which states that the sum of the elements of $\boldsymbol{\pi}$ must equal a constant, κ. For a probability distribution, $\kappa = 1$. In the context of Ideal Flow Networks, κ is the total flow in the network. This condition is expressed as $\mathbf{j}^T \boldsymbol{\pi} = \kappa$.

The two constraints are combined to form a single overdetermined system:

$$\begin{aligned} (\mathsf{S}^T - \mathsf{I})\boldsymbol{\pi} &= \mathbf{0} \\ \mathbf{j}^T\boldsymbol{\pi} &= \kappa \end{aligned} \tag{8.16}$$

This system can be written in a compact matrix form:

$$\mathsf{K}\boldsymbol{\pi} = \mathbf{b}$$

where

$$\mathsf{K} = \begin{bmatrix} \mathsf{S}^T - \mathsf{I} \\ \mathbf{j}^T \end{bmatrix}, \quad \mathbf{b} = \begin{bmatrix} \mathbf{0} \\ \kappa \end{bmatrix}$$

The system is overdetermined, as it consists of $n + 1$ linear equations for n unknowns. However, due to the linear dependency of the rows in $(\mathsf{S}^T - \mathsf{I})$, the system has a unique solution when the normalization constraint, $\mathbf{j}^T\boldsymbol{\pi} = \kappa$, is included. The presented method uses a full, rectangular matrix K. The system is overdetermined, and the solution is given by the left matrix division operator, which uses a generalized inverse. This approach is numerically stable. The generalized inverse, or Moore-Penrose inverse, K^{+}, provides the minimum norm least-squares solution. The solution is thus $\boldsymbol{\pi} = \mathsf{K}^{+}\mathbf{b}$. □

The backslash \ operator, which is adopted from Matlab notation, represents *left matrix division*. Specifically, for matrices K and $\mathbf{b}$, the expression $\mathsf{K} \setminus \mathbf{b}$ solves the linear system of equations $\mathsf{K} \cdot \mathbf{x} = \mathbf{b}$ for $\mathbf{x}$. This is equivalent to $\mathsf{K}^{-1}\mathbf{b}$ when K is a square, non-singular matrix. However, operator \ is generally computed based on Singular Values Decomposition (SVD) 3.23 which is more computationally efficient and numerically stable, especially for large or ill-conditioned systems. We can compute this via $\mathbf{x} = \mathsf{K}^{+}\mathbf{b}$ where K^{+} is the Generalized Left Inverse or the Moore Penrose Inverse 3.14.

$$\mathsf{K}^{+} = (\mathsf{K}^T\mathsf{K})^{-1}\mathsf{K}^T \tag{8.17}$$

Example 8.5 (Finding Node Vector from Stochastic Matrix) Given stochastic matrix of a network and total flow $\kappa = 3$,

$$\mathsf{S} = \begin{bmatrix} 0 & 1 \\ \frac{1}{2} & \frac{1}{2} \end{bmatrix}.$$

We would like to find the node vector based Eq. (8.15).

$$\boldsymbol{\pi} = \begin{bmatrix} \mathsf{S}^T - \mathsf{I} \\ \mathbf{j}^T \end{bmatrix} \setminus \begin{bmatrix} \mathbf{0} \\ \kappa \end{bmatrix}$$

First, we form matrix

$$\mathsf{S}^T - \mathsf{I} = \begin{bmatrix} 0 & \frac{1}{2} \\ 1 & \frac{1}{2} \end{bmatrix} - \begin{bmatrix} 1 & 0 \\ 0 & 1 \end{bmatrix} = \begin{bmatrix} -1 & \frac{1}{2} \\ 1 & -\frac{1}{2} \end{bmatrix}$$

Next, we form matrix

$$\mathsf{K} = \begin{bmatrix} \mathsf{S}^T - \mathsf{I} \\ \mathbf{j}^T \end{bmatrix} = \begin{bmatrix} -1 & \frac{1}{2} \\ 1 & -\frac{1}{2} \\ 1 & 1 \end{bmatrix}$$

and vector

$$\mathbf{b} = \begin{bmatrix} \mathbf{0} \\ \kappa \end{bmatrix} = \begin{bmatrix} 0 \\ 0 \\ 3 \end{bmatrix}$$

The Generalized Left Inverse 3.3.4 is

$$\mathsf{K}^+ = (\mathsf{K}^T\mathsf{K})^{-1}\mathsf{K}^T = \begin{bmatrix} -\frac{1}{3} & \frac{1}{3} & \frac{1}{3} \\ \frac{1}{3} & -\frac{1}{3} & -\frac{2}{3} \end{bmatrix}$$

The node vector is

$$\boldsymbol{\pi} = \mathsf{K}^+ \cdot \mathbf{b} = \begin{bmatrix} -\frac{1}{3} & \frac{1}{3} & \frac{1}{3} \\ \frac{1}{3} & -\frac{1}{3} & -\frac{2}{3} \end{bmatrix} \cdot \begin{bmatrix} 0 \\ 0 \\ 3 \end{bmatrix} = \begin{bmatrix} 1 \\ 2 \end{bmatrix}$$

Check that the node vector satisfies Eigenvector Eq. (8.11).

$$\boldsymbol{\pi} = \mathsf{S}^T\boldsymbol{\pi} \equiv \begin{bmatrix} 0 & \frac{1}{2} \\ 1 & \frac{1}{2} \end{bmatrix} \cdot \begin{bmatrix} 1 \\ 2 \end{bmatrix} = \begin{bmatrix} 1 \\ 2 \end{bmatrix}$$

8.2.5 Ideal Flow from Irreducible Stochastic

Theorem 8.3 (Construction of the Ideal Flow Matrix) *Given an irreducible, row-stochastic matrix* S *and its unique stationary distribution vector* $\boldsymbol{\pi}$ *such that* $\boldsymbol{\pi}^T\mathsf{S} = \boldsymbol{\pi}^T$*, the Ideal Flow matrix* F *is defined as:*

$$\mathsf{F} = diag(\boldsymbol{\pi})\mathsf{S}$$

where $diag(\boldsymbol{\pi})$ *is a diagonal matrix with the elements of* $\boldsymbol{\pi}$ *on the main diagonal.*

Proof The construction $\mathsf{F} = \text{diag}(\boldsymbol{\pi})\mathsf{S}$ can be written element-wise as:

$$f_{ij} = \pi_i s_{ij}$$

The flow from node i to node j is the product of the total flow originating from node i (π_i) and the probability of that flow moving to node j (s_{ij}). The total outflow from node i is:

$$\sum_j f_{ij} = \sum_j \pi_i s_{ij} = \pi_i \sum_j s_{ij} = \pi_i \times 1 = \pi_i$$

The total inflow into node j is:

$$\sum_i f_{ij} = \sum_i \pi_i s_{ij}$$

The steady-state equation is $\boldsymbol{\pi}^T \mathsf{S} = \boldsymbol{\pi}^T$, which can be written element-wise as $\sum_i \pi_i s_{ij} = \pi_j$. Therefore:

$$\sum_i f_{ij} = \sum_i \pi_i s_{ij} = \pi_j$$

The total outflow from node i is π_i, and the total inflow to node i is also π_i. □

From the node vector $\boldsymbol{\pi}$ and stochastic matrix S we can compute the flow matrix.

$$\mathsf{F} = \left(\boldsymbol{\pi} \cdot \mathbf{j}^T\right) \circ \mathsf{S} = diag(\boldsymbol{\pi}) \cdot \mathsf{S} \tag{8.18}$$

Corollary 8.1 (Ideal Flow Matrix Is Premagic) *The ideal flow matrix* F *is premagic because its rows and columns sum to the same vector* $\boldsymbol{\pi}$.

Proof To prove that F is premagic, we must show that the sum of each row and the sum of each column equal the same values. The sum of the rows of F is a vector where the i-th element is the sum of the i-th row. Let $\mathsf{F}\mathbf{j}$ be the vector of row sums. The i-th element is:

$$(\mathsf{F}\mathbf{j})_i = \sum_k f_{ik} = \sum_k \pi_i s_{ik} = \pi_i \sum_k s_{ik} = \pi_i \times 1 = \pi_i$$

Thus, the row sum vector is $\mathsf{F}\mathbf{j} = \boldsymbol{\pi}$. The sum of the columns of F is a vector where the j-th element is the sum of the j-th column. Let $\mathbf{j}^T\mathsf{F}$ be the vector of column sums (as a row vector). The j-th element is:

$$(\mathbf{j}^T\mathsf{F})_j = \sum_k f_{kj} = \sum_k \pi_k s_{kj}$$

From the steady-state equation $\boldsymbol{\pi}^T \mathsf{S} = \boldsymbol{\pi}^T$, we know that $\sum_k \pi_k s_{kj} = \pi_j$. Therefore, the column sum vector is $\mathbf{j}^T \mathsf{F} = \boldsymbol{\pi}^T$. Since the vector of row sums and the vector of column sums are both equal to $\boldsymbol{\pi}$ and $\boldsymbol{\pi}^T$, respectively, the matrix F is line-sum symmetric, or premagic. Thus, an *Ideal Flow Matrix* F is a nonnegative, irreducible, and premagic matrix. □

The transformation from an irreducible stochastic matrix S to an ideal flow matrix F ensures that the resulting flow distribution adheres to the principles of network equilibrium and flow conservation. Before computing the flow, we formally define the target structure.

Example 8.6 (Deriving Node Vector) Let us use the stochastic matrix S below.

$$\mathsf{S} = \begin{pmatrix} 0 & \frac{2}{3} & 0 & \frac{1}{3} \\ \frac{1}{5} & 0 & \frac{4}{5} & 0 \\ 0 & \frac{2}{5} & 0 & \frac{3}{5} \\ \frac{1}{4} & 0 & \frac{3}{4} & 0 \end{pmatrix}$$

We solve the system $(\mathsf{S}^T - \mathsf{I})\boldsymbol{\pi}^T = \mathbf{0}$ with the constraint $\sum \pi_i = 1$. Using a numerical computation tool, the computation yields the unique stationary distribution:

$$\boldsymbol{\pi}^T = \begin{pmatrix} \pi_1 \\ \pi_2 \\ \pi_3 \\ \pi_4 \end{pmatrix} \approx \begin{pmatrix} 0.113 \\ 0.230 \\ 0.387 \\ 0.270 \end{pmatrix}$$

So, the node flow vector is $\boldsymbol{\pi}^T = \frac{1}{608}[69, 140, 235, 164]$. This result means that in the long-run equilibrium, approximately 38.7% of the total flow circulating in the network passes through Node 3, making it the busiest intersection.

8.2.6 *Transition Probabilities from Ideal Flow*

The reverse process to obtain the row stochastic transition probability matrix from ideal flow is straightforward.

A row-stochastic matrix, S, can be derived from F by normalizing the rows by the node flow vector $\boldsymbol{\pi}$.

$$\mathsf{S} = (\text{diag}(\boldsymbol{\pi}))^{-1}\mathsf{F} \tag{8.19}$$

A stationary distribution of this stochastic process is a vector $\boldsymbol{\pi}$ such that $\boldsymbol{\pi}^T \mathsf{S} = \boldsymbol{\pi}^T$ which is equivalent to $\mathsf{S}^T \boldsymbol{\pi} = \boldsymbol{\pi}$.

Example 8.7 (Finding Stochastic Matrix) Suppose we have two nodes network which is an ideal flow matrix.

$$\mathbf{F} = \begin{bmatrix} 0 & 1 \\ 1 & 1 \end{bmatrix}$$

Observe that the sum of rows is the same as the sum of columns. The node vector $\pi = [1, 2]^T$. the stochastic matrix can be obtained based on Eq. (8.19).

$$\mathsf{S} = (\text{diag}(\boldsymbol{\pi}))^{-1}\mathsf{F} = \begin{bmatrix} 1 & 0 \\ 0 & \frac{1}{2} \end{bmatrix} \cdot \begin{bmatrix} 0 & 1 \\ 1 & 1 \end{bmatrix} = \begin{bmatrix} 1 & 0 \\ \frac{1}{2} & \frac{1}{2} \end{bmatrix}$$

8.2.7 *Relative Link Flow from Node Flow and Transition Probabilities*

Definition 8.9 (Relative Ideal Flow) The relative ideal flow matrix, F_{rel}, is defined by its elements:

$$f_{ij,\text{rel}} = \pi_i \times s_{ij} \tag{8.20}$$

This matrix is premagic and represents the proportional distribution of flow across the network. The total flow in this matrix, $\kappa_{\text{rel}} = \sum_{i,j} f_{ij,\text{rel}}$, is equal to the given κ.

Once the node flow vector $\boldsymbol{\pi}$ is known, calculating the flow on each link is straightforward.

Theorem 8.4 (Link Flow Calculation) *The relative flow on link (i, j), denoted F_{ij}^{rel}, is given by:*

$$f_{ij}^{rel} = \pi_i \times s_{ij} \tag{8.21}$$

In matrix form, this can be expressed as $\mathsf{F}^{rel} = diag(\boldsymbol{\pi}) \cdot \mathsf{S}$.

Example 8.8 (Deriving F) We continue the previous Example 8.6 to obtain the ideal flow matrix.

$$\mathsf{F} = \frac{1}{608}\begin{pmatrix} 69 & 0 & 0 & 0 \\ 0 & 140 & 0 & 0 \\ 0 & 0 & 235 & 0 \\ 0 & 0 & 0 & 164 \end{pmatrix} \cdot \begin{pmatrix} 0 & \frac{2}{3} & 0 & \frac{1}{3} \\ \frac{1}{5} & 0 & \frac{4}{5} & 0 \\ 0 & \frac{2}{5} & 0 & \frac{3}{5} \\ \frac{1}{4} & 0 & \frac{3}{4} & 0 \end{pmatrix} = \frac{1}{608}\left(\begin{array}{c|cccc|c} \nearrow & a & b & c & d & \Sigma \\ \hline a & 0 & 46 & 0 & 23 & 69 \\ b & 28 & 0 & 112 & 0 & 140 \\ c & 0 & 94 & 0 & 141 & 235 \\ d & 41 & 0 & 123 & 0 & 164 \\ \hline \Sigma & 69 & 140 & 235 & 164 & 608 \end{array}\right) \tag{8.22}$$

We can see that the denominator 608 is the scaling of the total flow that make integer ideal flow matrix. The matrix is premagic, the sum of rows is exactly the same as the sum of columns.

8.2.8 Scaling by Total Flow

The matrix F^{rel} represents the fundamental, structure-derived distribution of flow. To obtain absolute flow values, we must scale this relative matrix by the total flow parameter, κ.

Definition 8.10 (Absolute Ideal Flow) The absolute ideal flow matrix, F, is obtained by the scalar multiplication:

$$\mathsf{F} = \kappa \cdot \mathsf{F}^{\text{rel}} \tag{8.23}$$

Example 8.9 (Calibrating Absolute Flow) Suppose we continue our previous Example 8.8 and obtain real world flow data for the morning peak in link 2–3 to be 2352 vehicles/hour. The ideal flow in link 2–3 from Eq. (8.22) was 112. Using these information, we can compute the scaling factor to be $\varsigma = \frac{2352}{112} = 21$. This scaling factor is from the F^{rel}. Thus, the total circulating flow in the network become $\kappa = 608 \times 21 = 12{,}768$ vehicles per hour. Applying this value into Eq. (8.23), we have the absolute flow matrix.

$$\mathsf{F} = 12{,}768 \cdot \mathsf{F}^{\text{rel}} \approx \left(\begin{array}{c|cccc|c} \nearrow & a & b & c & d & \Sigma \\ \hline a & 0 & 966 & 0 & 483 & 1449 \\ b & 588 & 0 & 2352 & 0 & 2940 \\ c & 0 & 1974 & 0 & 2961 & 4935 \\ d & 861 & 0 & 2583 & 0 & 3444 \\ \hline \Sigma & 1449 & 2940 & 4935 & 3444 & 12{,}768 \end{array} \right) \text{(veh/hr)} \tag{8.24}$$

This final matrix, F, represents the model's prediction for the absolute steady-state traffic flows on each link. Observe that the flow in link 2–3 is now exactly the same as the morning peak real world flow data for the same link, which is 2352 vehicles/hour. The matrix exhibits premagic properties: its row sums are equal to its column sums.

8.2.9 Relation to the Maximum Entropy Principle

The use of the stationary distribution of a Markov chain as the basis for flow prediction has a deep justification in information theory, specifically through the *Maximum Entropy Principle*. The principle states that, given a set of constraints, the most unbiased probability distribution is the one that maximizes entropy. It can be proven that the unique stationary distribution $\boldsymbol{\pi}$ is precisely the distribution that maximizes the entropy subject to the network constraints defined by S. Therefore, by using $\boldsymbol{\pi}$ to derive the ideal flow, the Ideal Traffic Assignment framework is implicitly selecting the most probable, most random, and least presumptive flow distribution that is consistent with the underlying physical structure of the network.

8.2.10 Computational Methods for Solving Ideal Flow

For very large networks, iterative methods are more efficient than direct matrix inversion.

The most common iterative algorithm for finding the dominant eigenvector (which, for a stochastic matrix, is the stationary distribution) is the *Power Iteration method*.

Power Iteration for Stationary Distribution
Begin by initializing an arbitrary probability vector $\boldsymbol{\pi}^{(0)}$. A common choice is the uniform distribution, where $\pi_i^{(0)} = 1/n$ for all i. Then, iterate the following update for $m = 0, 1, 2, \ldots$:

$$\boldsymbol{\pi}^{(m+1)} = \mathsf{S}^T \boldsymbol{\pi}^{(m)} \tag{8.25}$$

After each iteration, compute the change in the vector using the L^1 norm:

$$\Delta = \left\| \boldsymbol{\pi}^{(m+1)} - \boldsymbol{\pi}^{(m)} \right\|_1 \tag{8.26}$$

If Δ is less than a small tolerance ϵ, terminate the process and set $\boldsymbol{\pi} = \boldsymbol{\pi}^{(m+1)}$. Otherwise, repeat the iteration step.

The Power Iteration method is guaranteed to converge to the unique stationary distribution for any irreducible, aperiodic stochastic matrix. For large networks, this iterative approach is the standard and most practical method for deriving the Ideal Flow.

Example 8.10 We will now demonstrate the Power iteration method using the provided stochastic matrix S. This process converges to the unique stationary distribution, which is the left dominant eigenvector of S. The method is an iterative numerical approach for solving the equation $\boldsymbol{\pi}^T \mathsf{S} = \boldsymbol{\pi}^T$.

The transition probability matrix for the four-node network is given as:

$$\mathsf{S} = \begin{pmatrix} 0 & \frac{2}{3} & 0 & \frac{1}{3} \\ \frac{1}{5} & 0 & \frac{4}{5} & 0 \\ 0 & \frac{2}{5} & 0 & \frac{3}{5} \\ \frac{1}{4} & 0 & \frac{3}{4} & 0 \end{pmatrix}$$

This matrix S is irreducible, ensuring that the power iteration method is guaranteed to converge to a unique stationary distribution. The matrix period is 2, thus it is not aperiodic. The power iteration begins with an arbitrary initial probability vector, $\boldsymbol{\pi}^{(0)}$. We use a uniform distribution for this example:

$$\boldsymbol{\pi}^{(0)} = \begin{bmatrix} 0.25 \\ 0.25 \\ 0.25 \\ 0.25 \end{bmatrix}$$

The iterative step is defined by the equation $\boldsymbol{\pi}^{(m+1)} = \mathsf{S}^T \boldsymbol{\pi}^{(m)}$. The algorithm proceeds by repeatedly applying this transformation until the change between successive vectors falls below a specified tolerance, $\epsilon = 1 \times 10^{-6}$.

Iteration 1: Applying the transformation yields the first state vector:

$$\boldsymbol{\pi}^{(1)} = \mathsf{S}^T \boldsymbol{\pi}^{(0)} = \begin{bmatrix} 0.11250000 \\ 0.26666667 \\ 0.38750000 \\ 0.23333333 \end{bmatrix}$$

The change in the vector is calculated using the L1-norm, $\Delta = ||\boldsymbol{\pi}^{(1)} - \boldsymbol{\pi}^{(0)}||_1 = 0.30833333$.

Iteration 2:

$$\boldsymbol{\pi}^{(2)} = \mathsf{S}^T \boldsymbol{\pi}^{(1)} = \begin{bmatrix} 0.11166667 \\ 0.23000000 \\ 0.38833333 \\ 0.27000000 \end{bmatrix}$$

The change is $\Delta = ||\boldsymbol{\pi}^{(2)} - \boldsymbol{\pi}^{(1)}||_1 = 0.07500000$. The vector is progressing towards a stable state.

Iteration 3:

$$\pi^{(3)} = \mathsf{S}^T \pi^{(2)} = \begin{bmatrix} 0.11350000 \\ 0.22977778 \\ 0.38650000 \\ 0.27022222 \end{bmatrix}$$

The change is $\Delta = ||\pi^{(3)} - \pi^{(2)}||_1 = 0.00411111$. The rate of change is decreasing.

After seven iterations, the power iteration method converges.

$$\pi^{(7)} = \mathsf{S}^T \pi^{(6)} = \begin{bmatrix} 0.11348684 \\ 0.23026307 \\ 0.38651316 \\ 0.26973693 \end{bmatrix}$$

The change is $\Delta = ||\pi^{(7)} - \pi^{(6)}||_1 = 0.00000073$. This value is less than the tolerance, so the algorithm terminates.

The final stationary distribution, π, for this Ideal Flow Network is approximately:

$$\pi \approx \begin{bmatrix} 0.11348684 \\ 0.23026307 \\ 0.38651316 \\ 0.26973693 \end{bmatrix}$$

This vector is closed to the true values which is $\pi^T = \frac{1}{608}[69, 140, 235, 164]$. This vector represents the long-term proportion of flow at each node in the network.

8.3 Practical Computation of Ideal Flow Networks

Previous sections established the rigorous theoretical foundation of the Ideal Flow Network, deriving the Ideal Flow matrix F from the steady-state properties of Markov chains [9, 13, 18]. While the eigenvector formulation provides a powerful and elegant solution, for small-scale networks, the IFN can also be computed directly through more intuitive iterative and algebraic methods. These hands-on procedures provide essential insight into the structure of the IFN solution.

Regardless of the method, the procedure yields a relative flow distribution that is unique up to a positive scaling factor, ς.

For convenience and clarity, this relative distribution is often normalized to its *cardinal IFN* form, where all link flows are minimum integers or in its *unit IFN* form where the minimum non-zero flow is exactly one.

8.3.1 The Manual Iterative Procedure

For many simple, strongly connected networks, the IFN can be determined through a step-by-step propagation of flows [16].

Manual Iterative Computation
The procedure begins with an initialization step: assign an arbitrary positive flow value (e.g., 100) to a single link in the network. This choice is arbitrary because the final result is a relative distribution.

Next, apply the following two rules iteratively until all link flows are determined:

- *Distribution Rule (Diverging Flow):* At any node i where the total inflow π_i is known, distribute this flow to its outgoing links (i, j) according to the known probabilities: $f_{ij} = s_{ij} \times \pi_i$.
- *Conservation Rule (Merging/Serial Flow):* The total inflow to any node k, given by $\pi_k = \sum_i f_{ik}$, must equal its total outflow, $\sum_j f_{kj}$. If a node has only one outgoing link, the flow on that link must equal the total inflow to the node.

After computing all link flows, verify that the flow conservation principle holds for every node: the total inflow must equal the total outflow.

Finally, normalize the flow values to obtain the unit IFN. Let $f_{\min}$ be the minimum non-zero link flow. Divide every link flow f_{ij} and node flow π_i by this value.

8.3.1.1 Illustrative Example: A Simple Urban Grid

Let's apply this procedure to the 3-node urban grid introduced previously (see Fig. 8.2). The outflow probabilities are given as:

- Node a: $s_{ab} = 1$
- Node b: $s_{ba} = \frac{1}{3}$, $s_{bc} = \frac{2}{3}$
- Node b: $s_{ca} = \frac{1}{2}$, $s_{cb} = \frac{1}{2}$

Computation Steps:

1. *Initialization:* We assign an arbitrary flow of $f_{ab} = 150$. Because this is the only inflow to node b, it implies the total flow through node b is $\pi_b = 150$.
2. *Propagation:*

   - At node b, we know $\pi_b = 150$. Using the distribution rule, $f_{ba} = s_{ba} \times \pi_b = \frac{1}{3} \times 150 = 50$. We also compute $f_{bc} = s_{bc} \times \pi_b = \frac{2}{3} \times 150 = 100$.
   - At node c, total inflow is $\pi_c = f_{bc} = 100$. Using the distribution rule, $f_{ca} = s_{ca} \times \pi_c = \frac{1}{2} \times 100 = 50$ and $f_{cb} = s_{cb} \times \pi_c = \frac{1}{2} \times 100 = 50$.
   - At node a, we use conservation rule, $\pi_a = f_{ba} + f_{ca} = 50 + 50 = 100$. Since there is only one outflow, by conservation, $f_{ab} = 100$.

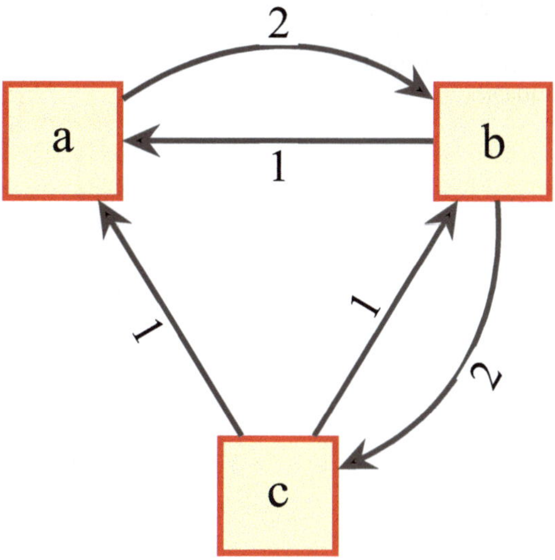

Fig. 8.2 A simple three-node, five-link directed network

3. *Verification:* Flow conservation holds at all nodes.
4. *Normalization:* The calculated flows are $\{f_{ab} = 100, f_{ba} = 50, f_{bc} = 100, f_{ca} = 50, f_{cb} = 50, \}$. The minimum non-zero flow is 50. Dividing all flows by 50 gives the unit IFN, which is also cardinal IFN (i.e. minimum integer): $\{f_{ab} = 2, f_{ba} = 1, f_{bc} = 2, f_{ca} = 1, f_{cb} = 1, \}$ The resulting flow matrix is:

$$\mathsf{F} = \begin{pmatrix} 0 & 2 & 0 \\ 1 & 0 & 2 \\ 1 & 1 & 0 \end{pmatrix}$$

8.3.1.2 The Algebraic Method for Complex Networks

If a network contains complex cycles, the iterative method may not be sufficient. In such cases, we must use an algebraic approach by setting up and solving a system of linear equations based on flow conservation.

Let's use the 4-node network example (Fig. 8.3). The outflow probabilities are uniform at nodes with two exits: $s_{ab} = s_{ad} = 0.5$ and $s_{cd} = s_{ca} = 0.5$.
Algebraic Formulation:

1. *Initialization:* Let's assign a variable to an unknown flow, for instance, let $f_{bc} = x$ and $f_{ab} = y$.
2. *Formulating Equations:* We express all other flows in terms of x and y.

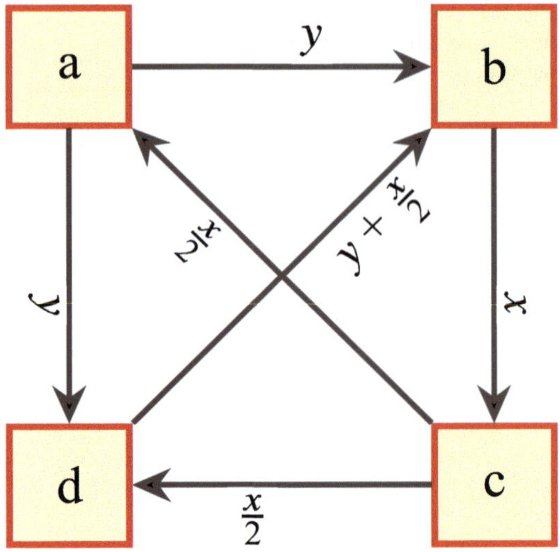

Fig. 8.3 A four-node network with a flow distribution defined by variables

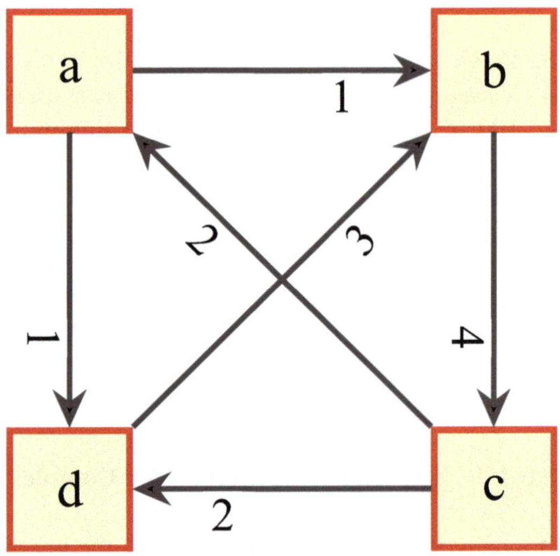

Fig. 8.4 An example of a four-node, six-link directed network with flow values

- At node c, the distribution rule implies $\pi_c = f_{cd} = f_{ca} = \frac{x}{2}$.
- At node a, the distribution rule implies $\pi_a = f_{ab} = f_{ad} = 2y = \frac{x}{2} = f_{ca}$.
- At node d, conservation requires $\pi_d = f_{ad} + f_{cd} = f_{db} = y + \frac{x}{2}$.
- At node b, conservation requires $\pi_b = f_{bc} = f_{ab} + f_{db} = y + (y + \frac{x}{2}) = x$.

3. *Solving the System:* The equation from node a simplifies to $x = 4y$. The equation from node b simplifies to $2y = \frac{x}{2}$ which is the same as $x = 4y$. The system is *underdetermined*. Both equations simplify to the same linear relationship. Both key equations are not independent; they express the same truth about the system. The single, valid linear relationship is $x - 4y = 0$. We only have one independent equation for two variables. There are infinitely many solutions. Setting $y = 1$ produces cardinal network (minimum integer IFN) as shown in Fig. 8.4.

These manual computation methods are tangible procedures for solving what is, fundamentally, an eigenvector problem [12] that was discussed in Theorem 8.2.

8.4 The Theory of Network Equivalence and Invariants

We will now build the theoretical foundation [3] of IFN equivalence. We start with the definitions. We then establish the key properties and invariants that arise from this framework.

8.4.1 Scalable Systems

The concept of scalability relates two matrices that share the same pattern but differ in magnitude by a constant factor.

Definition 8.11 (Pattern Matrix) A binary non-negative matrix $\mathsf{P} \in \{0, 1\}^{m\times n}$ is referred to as a *pattern matrix*. It specifies the structural support, i.e., the positions of potential non-zero entries, for a corresponding class of matrices.

Example 8.11 (Pattern Matrix)

$$\mathsf{R} = \begin{pmatrix} 0 & 2 & 2 & 0 \\ 3 & 0 & 3 & 0 \\ 7 & 7 & 0 & 7 \\ 0 & 0 & 9 & 0 \end{pmatrix}$$

has pattern matrix of

$$\mathsf{P} = \begin{pmatrix} 0 & 1 & 1 & 0 \\ 1 & 0 & 1 & 0 \\ 1 & 1 & 0 & 1 \\ 0 & 0 & 1 & 0 \end{pmatrix}$$

As defined in Definition 13.3, let $\mathsf{P} \in \{0, 1\}^{m\times n}$ be a pattern matrix. A matrix $\mathsf{B} \in \mathbb{R}^{m\times n}_{\geq 0}$ is called a *steady matrix* with respect to P if:

- B shares the same support as P, i.e., $\text{supp}(\mathsf{B}) = \text{supp}(\mathsf{P})$,
- All non-zero entries of B are equal to a common positive scalar $k > 0$.

Equivalently, $\mathsf{B} = k\mathsf{P}$ for some $k \in \mathbb{R}_{>0}$.

Example 8.12 Let the pattern matrix be:

$$\mathsf{P} = \begin{bmatrix} 1 & 0 & 1 \\ 0 & 1 & 0 \end{bmatrix}$$

Then a steady matrix with respect to P, for $k = 3$, is:

$$\mathsf{B} = 3\mathsf{P} = \begin{bmatrix} 3 & 0 & 3 \\ 0 & 3 & 0 \end{bmatrix}$$

This matrix B has non-zero entries only where P has ones, and all such entries are equal to the scalar 3.

Definition 8.12 (Scaled-Equivalent Matrices) Two non-negative matrices A and B with the same pattern P are called *scaled-equivalent*, denoted $\mathsf{A} \cong \mathsf{B}$, if and only if their Hadamard division results in a steady matrix. That is, there exists a scalar $k > 0$ such that:

$$\mathsf{A} \oslash \mathsf{B} = k\mathsf{P} \quad \text{and} \quad \mathsf{B} \oslash \mathsf{A} = \frac{1}{k}\mathsf{P}$$

Example 8.13 (Testing for Scaled-Equivalence) Consider the matrices A and B with pattern P.

$$\mathsf{A} = \begin{pmatrix} 0 & 2 & 3 & 0 \\ 1 & 0 & 4 & 7 \\ 0 & 6 & 0 & 0 \end{pmatrix}, \quad \mathsf{B} = \begin{pmatrix} 0 & 4 & 6 & 0 \\ 2 & 0 & 8 & 14 \\ 0 & 12 & 0 & 0 \end{pmatrix}, \quad \mathsf{P} = \begin{pmatrix} 0 & 1 & 1 & 0 \\ 1 & 0 & 1 & 1 \\ 0 & 1 & 0 & 0 \end{pmatrix}$$

We compute the Hadamard division:

$$\mathsf{A} \oslash \mathsf{B} = \begin{pmatrix} 0 & 1/2 & 1/2 & 0 \\ 1/2 & 0 & 1/2 & 1/2 \\ 0 & 1/2 & 0 & 0 \end{pmatrix} = \frac{1}{2}\mathsf{P}$$

Since the result is a steady matrix (with $k = 1/2$), we conclude that $\mathsf{A} \cong \mathsf{B}$.

8.4.2 *Fundamental Concepts of Network Equivalence*

Now we formally define equivalence.

Definition 8.13 (Equivalence IFN) Two Ideal Flow Networks, represented by their flow matrices F_1 and F_2, are *equivalent* if there exists a positive constant $\varsigma > 0$ such that:

$$\mathsf{F}_1 = \varsigma \mathsf{F}_2 \tag{8.27}$$

The constant ς is called the *global scaling factor*. We write this relationship as $\mathsf{F}_1 \equiv \mathsf{F}_2$.

This relationship is a formal equivalence relation.

Proposition 8.2 *The relationship* $\equiv$ *defined in Definition 8.13 is an equivalence relation.*

Proof We must show that the relation is reflexive, symmetric, and transitive.

1. *Reflexive:* For any IFN F_1, we have $\mathsf{F}_1 \equiv \mathsf{F}_1$ by choosing the scaling factor $\varsigma = 1$.
2. *Symmetric:* If $\mathsf{F}_1 \equiv \mathsf{F}_2$ with factor ς, then $\mathsf{F}_1 = \varsigma\mathsf{F}_2$. Since $\varsigma > 0$, we can write $\mathsf{F}_2 = (1/\varsigma)\mathsf{F}_1$. Thus, $\mathsf{F}_2 \equiv \mathsf{F}_1$ with factor $1/\varsigma$.
3. *Transitive:* If $\mathsf{F}_1 \equiv \mathsf{F}_2$ with factor ς_1 and $\mathsf{F}_2 \equiv \mathsf{F}_3$ with factor ς_2, then $\mathsf{F}_1 = \varsigma_1\mathsf{F}_2$ and $\mathsf{F}_2 = \varsigma_2\mathsf{F}_3$. Substituting gives $\mathsf{F}_1 = \varsigma_1(\varsigma_2\mathsf{F}_3) = (\varsigma_1\varsigma_2)\mathsf{F}_3$. Thus, $\mathsf{F}_1 \equiv \mathsf{F}_3$ with factor $\varsigma_1\varsigma_2$.

□

This relation partitions the set of all IFNs on a given graph into disjoint equivalence classes. All networks in a class share the same proportional flow distribution.

Proposition 8.3 (Equivalence of IFNs Implies Scaled-Equivalence) *An equivalence relationship between two Ideal Flow Networks (IFNs) is a sufficient condition for their matrices to be scaled-equivalent.*

Proof Let two IFNs be represented by their flow matrices F_1 and F_2. By Definition 8.13, if $\mathsf{F}_1 \equiv \mathsf{F}_2$, there exists a positive constant $\varsigma > 0$ such that $\mathsf{F}_1 = \varsigma\mathsf{F}_2$.

- Hadamard Division: We examine the Hadamard division of the two matrices, $\mathsf{F}_1 \oslash \mathsf{F}_2$.

$$\mathsf{F}_1 \oslash \mathsf{F}_2 = (\varsigma\mathsf{F}_2) \oslash \mathsf{F}_2$$

- Element-wise Operation: Hadamard division is an element-wise operation. For any element (i, j):

$$[\mathsf{F}_1 \oslash \mathsf{F}_2]_{ij} = \frac{[\mathsf{F}_1]_{ij}}{[\mathsf{F}_2]_{ij}} = \frac{\varsigma[\mathsf{F}_2]_{ij}}{[\mathsf{F}_2]_{ij}}$$

  This simplifies to ς for all elements where $[\mathsf{F}_2]_{ij} \neq 0$.
- Pattern Matrix: Let P be the pattern matrix of F_1 and F_2. The matrices have the same pattern, so $\mathsf{P}_{ij} = 1$ if $[\mathsf{F}_1]_{ij} > 0$ (and thus $[\mathsf{F}_2]_{ij} > 0$), and $\mathsf{P}_{ij} = 0$ otherwise.

- Steady Matrix Result: The Hadamard division results in a matrix where all nonzero elements are equal to ς. This is precisely the definition of a steady matrix. Therefore:

$$\mathsf{F}_1 \oslash \mathsf{F}_2 = \varsigma \mathsf{P}$$

- Symmetric Case: Similarly, for $\mathsf{F}_2 \oslash \mathsf{F}_1$:

$$\mathsf{F}_2 \oslash \mathsf{F}_1 = (1/\varsigma)\mathsf{P}$$

By Definition 8.12, since there exists a scalar $\varsigma > 0$ (which we denote as k in the definition) such that the Hadamard division yields a steady matrix, the matrices F_1 and F_2 are scaled-equivalent. Thus, $\mathsf{F}_1 \cong \mathsf{F}_2$. □

The equivalence of IFNs ($\mathsf{F}_1 \equiv \mathsf{F}_2$) is a specific and strong form of the more general scaled-equivalence ($\mathsf{A} \cong \mathsf{B}$), characterized by a single global scaling factor that applies uniformly to all elements of the network's flow matrix.

Corollary 8.2 (Equivalence Preserves Proportionality) *Two equivalent IFNs,* F_1 *and* F_2*, maintain the same relative flow distributions, even as their total flows may differ. The proportionality of flow on any link* (i, j) *to the total network flow is an invariant under this relationship.*

Proof The total flow of an IFN is the sum of all elements in its flow matrix. Let κ_1 and κ_2 be the total flows of F_1 and F_2, respectively.

$$\kappa_1 = \sum_{i,j} [\mathsf{F}_1]_{ij} \quad \text{and} \quad \kappa_2 = \sum_{i,j} [\mathsf{F}_2]_{ij}$$

Given $\mathsf{F}_1 = \varsigma \mathsf{F}_2$, we can relate the total flows:

$$\kappa_1 = \sum_{i,j} [\varsigma \mathsf{F}_2]_{ij} = \varsigma \sum_{i,j} [\mathsf{F}_2]_{ij} = \varsigma \kappa_2$$

The ratio of the total flows is a constant: $\kappa_1/\kappa_2 = \varsigma$. Now consider the proportional flow on any link (i, j). For F_1, the proportion is $[\mathsf{F}_1]_{ij}/\kappa_1$. For F_2, the proportion is $[\mathsf{F}_2]_{ij}/\kappa_2$. Substituting the relationships established above:

$$\frac{[\mathsf{F}_1]_{ij}}{\kappa_1} = \frac{\varsigma [\mathsf{F}_2]_{ij}}{\varsigma \kappa_2} = \frac{[\mathsf{F}_2]_{ij}}{\kappa_2}$$

□

This demonstrates that the proportionality of flow on each link remains constant across equivalent networks. This invariant property is fundamental to the concept of network equivalence, showing that the structure of the flow, rather than its absolute magnitude, is the preserved characteristic.

Corollary 8.3 (Scaled-Equivalence Is a Sufficient Condition for Equivalence IFN) *If two matrices representing Ideal Flow Networks are scaled-equivalent, they may not necessarily be equivalent. Scaled-equivalence is a broader concept, while equivalence is a more restrictive one requiring a single global scaling factor.*

Proof By Definition 8.12, two matrices A and B are scaled-equivalent if their Hadamard division results in a steady matrix, i.e., $\mathsf{A} \oslash \mathsf{B} = k\mathsf{P}$. This implies that for every non-zero element (i, j), the ratio of the corresponding entries is constant: $[\mathsf{A}]_{ij}/[\mathsf{B}]_{ij} = k$. Now, consider a different scenario where the scaling factor is not uniform across all links. For example, if we have two different scaling factors, k_1 for some links and k_2 for others. Let the matrices A and B be such that $\mathsf{A}_{ij} = k_1\mathsf{B}_{ij}$ for links in a set S_1 and $\mathsf{A}_{ij} = k_2\mathsf{B}_{ij}$ for links in a set S_2, with $k_1 \neq k_2$. The Hadamard division $\mathsf{A} \oslash \mathsf{B}$ would produce a matrix with entries k_1 and k_2, not a single constant k. This would violate the condition for scaled-equivalence, as the result is not a steady matrix. However, the converse can be true. We can define a more complex form of scaling where each row, column, or link has its own scaling factor. In such a scenario, the Hadamard division would not result in a steady matrix, and thus the matrices would not be scaled-equivalent. Yet, they might share some other structural similarity. The definition of scaled-equivalence specifically demands a single constant k. The equivalence relation ($\mathsf{F}_1 = \varsigma\mathsf{F}_2$) is a special case of this, where the single constant k is the global scaling factor ς, and it applies to every element of the matrix. □

Thus, scaled-equivalence implies that the ratio of corresponding nonzero elements is a constant. Equivalence IFN demands that the matrices themselves are scalar multiples of each other. The latter is a stricter condition.

Thus, while equivalence implies scaled-equivalence, the reverse is not generally true. Scaled-equivalence only requires that the Hadamard division of two matrices results in a steady matrix, meaning the ratio of their non-zero elements is a single constant k. However, the matrices themselves might not represent valid Ideal Flow Networks (IFNs). For instance, they might not satisfy the flow conservation property at each node, or the matrix might be reducible.

Scaled-equivalence could be a property of non-flow matrices, such as capacity or adjacency, where a single global factor may not be present, whereas equivalence is specific to the Ideal Flow Matrix and its singular scaling factor.

This finding is useful because it allowsus to leverage the properties of scaled-equivalent matrices when analyzing equivalent IFNs, knowing that the conditions for the former are met by the latter. Essentially, if two ideal flow networks are equivalent, they are guaranteed to satisfy the less stringent condition of scaled-equivalence.

8.4.3 Standard Network Representations

Within each equivalence class, we can define standard forms. These forms provide a consistent reference for analysis. The total flow in a network is $\kappa = \sum_{i,j} f_{ij}$.

The basis, cardinal, and unit networks are all standard forms of an Ideal Flow Network (IFN), each providing a unique perspective by normalizing the network's total flow, integer flows, or minimum flow, respectively.

Definition 8.14 (Standard Forms) For any equivalence class of IFNs, we define the following standard representations:

- *Basis Network:* The IFN where the total flow is one ($\kappa = 1$). It is derived by dividing any IFN in the class by its total flow κ. The basis network is $\mathsf{F}_{basis} = (1/\kappa)\mathsf{F}$.
- *Cardinal Network:* The IFN where all link flows are the smallest possible positive integers that maintain the network's flow proportionality.
- *Unit Network:* The IFN where the smallest possible link flow is one. It is derived by dividing any IFN in the class by its minimum positive flow.

The basis network is a normalized representation of an IFN by setting the total flow $\kappa = 1$. This is the most mathematically convenient representation for many analytical tasks. The flow on any given link (i, j) in this network, f_{ij}, is equal to the ratio of that link's flow to the total network flow:

$$f_{ij} = \frac{f_{ij}}{\kappa} \quad \text{since } \kappa = 1$$

This ratio, f_{ij}/κ, can be interpreted as a *link probability*. It represents the probability that a randomly chosen unit of flow from the entire network is traversing the specific link (i, j). Because the sum of all link flows equals 1, the sum of these "probabilities" across all links also equals 1, satisfying the axioms of a probability distribution.

The cardinal network represents the most fundamental, irreducible integer-based representation of an IFN's flow structure. All non-zero link flows are the smallest possible positive integers while maintaining the correct flow proportions (i.e. the same stochastic matrix). All link flows are integers. This makes the network's structure explicit in a discrete, non-decimal form, which is useful in applications where flows are quantized. The total flow of a cardinal network is the smallest possible integer total flow for the given network structure. Flows are integers, which are easy to work with and visualize. It represents the most basic, irreducible form of the network's flow distribution. The fact that the ideal flow values are rational numbers in Proposition 6.3 guarantees the existence of a cardinal network, where all link flows are represented by the smallest possible set of positive integers. However, it is not always possible to show cardinal network numerically. Especially, when the LCM of the denominator is very large.

The unit network normalizes an IFN based on its minimum link flow. The smallest positive link flow is 1. It is derived by dividing any ideal flow matrix F, by its minimum positive flow, $f_{min} = \min_{i,j;f_{ij}>0} f_{ij}$. The unit network is $\mathsf{F}_{unit} = (1/f_{min})\mathsf{F}$. If a cardinal network has a minimum flow of 1, it is also a unit network. However, a unit network is not necessarily a cardinal network, as other flows may not be integers. The unit network exists even if the flows are irrational, whereas a cardinal network is typically defined for rational flows.

Example 8.14 (Calculating Standard Forms) Consider a 3-node IFN given by the flow matrix:

$$\mathsf{F} = \begin{pmatrix} 0 & 1.5 & 4.5 \\ 3.0 & 0 & 3.0 \\ 3.0 & 4.5 & 0 \end{pmatrix} \tag{8.28}$$

This network is strongly connected and premagic. The row sums vector and column sums vector are both $[6.0, 6.0, 7.5]^T$. The total flow is $\kappa = 1.5 + 4.5 + 3.0 + 3.0 + 3.0 + 4.5 = 19.5$.

1. *Basis Network:* We divide F by $\kappa = 19.5$.

$$\mathsf{F}_{\text{basis}} = \frac{1}{19.5}\mathsf{F} = \begin{pmatrix} 0 & 1/13 & 3/13 \\ 2/13 & 0 & 2/13 \\ 2/13 & 3/13 & 0 \end{pmatrix} \tag{8.29}$$

   The total flow of $\mathsf{F}_{\text{basis}}$ is $13/13 = 1$.
2. *Cardinal Network:* We find the smallest integer representation. From $\mathsf{F}_{\text{basis}}$, the flows are proportional to the integers in the numerators.

$$\mathsf{F}_{\text{cardinal}} = \begin{pmatrix} 0 & 1 & 3 \\ 2 & 0 & 2 \\ 2 & 3 & 0 \end{pmatrix} \tag{8.30}$$

   This is equivalent to our original F, with $\varsigma = 1/1.5$.
3. *Unit Network:* We divide F by the minimum positive flow 1.5. This produces the same matrix as Cardinal Network.

$$\mathsf{F}_{\text{unit}} = \begin{pmatrix} 0 & 1 & 3 \\ 2 & 0 & 2 \\ 2 & 3 & 0 \end{pmatrix} \tag{8.31}$$

   The total flow of Unit Network is 13.

The matrices F, $\mathsf{F}_{\text{basis}}$, $\mathsf{F}_{\text{cardinal}}$ and F_{unit} all belong to the same equivalence class.

Let us give another example where the Cardinal network is not equal to Unit network.

Example 8.15 (Standard Forms from Stochastic) Consider a 3-node network given by the irreducible stochastic matrix:

$$\mathsf{S} = \begin{pmatrix} 0 & \frac{1}{2} & \frac{1}{2} \\ \frac{2}{3} & 0 & \frac{1}{3} \\ \frac{1}{4} & \frac{3}{4} & 0 \end{pmatrix} \tag{8.32}$$

1. *Basis Network:* We find the ideal flow matrix such that the total flow of $\mathsf{F}_{\text{basis}}$ is 1.

$$\mathsf{F}_{\text{basis}} = \frac{1}{55}\mathsf{F} = \begin{pmatrix} 0 & 9 & 9 \\ 14 & 0 & 7 \\ 4 & 12 & 0 \end{pmatrix} \tag{8.33}$$

   The total flow of $\mathsf{F}_{\text{basis}}$ is always 1.
2. *Cardinal Network:* We find the smallest integer representation. From $\mathsf{F}_{\text{basis}}$, the flows are proportional to the integers in the numerators.

$$\mathsf{F}_{\text{cardinal}} = \begin{pmatrix} 0 & 9 & 9 \\ 14 & 0 & 7 \\ 4 & 12 & 0 \end{pmatrix} \tag{8.34}$$

   This is equivalent to our basis network with scaling $\varsigma = 55$, which is also the total flow $\kappa = 55$.
3. *Unit Network:* We divide F by the minimum positive flow 1.5. This produces the same matrix as Cardinal Network.

$$\mathsf{F}_{\text{unit}} = \begin{pmatrix} 0 & 2.25 & 2.25 \\ 3.5 & 0 & 1.75 \\ 1 & 3 & 0 \end{pmatrix} \tag{8.35}$$

   The total flow of Unit Network is 13.75.

The matrices F, $\mathsf{F}_{\text{basis}}$, $\mathsf{F}_{\text{cardinal}}$ and F_{unit} all belong to the same equivalence class.

8.4.4 Invariants of Equivalent Networks

An important result of equivalence is that certain key properties remain constant across all networks in the same class. These are called *invariants* properties.

Theorem 8.5 (Stochastic Matrix Invariance) *All IFNs in an equivalence class have the same row-stochastic matrix* S.

Proof Let F_1 and F_2 be two equivalent IFNs, so $\mathsf{F}_1 = \varsigma \mathsf{F}_2$ for some $\varsigma > 0$. Let $\pi_i(\mathsf{F})$ be the total outflow from node i in network F. The elements of their respective stochastic matrices, S_1 and S_2, are given by $s_{ij}^{(1)} = f_{ij}^{(1)}/\pi_i(\mathsf{F}_1)$ and $s_{ij}^{(2)} = f_{ij}^{(2)}/\pi_i(\mathsf{F}_2)$. The outflow from node i in F_1 is $\pi_i(\mathsf{F}_1) = \sum_k f_{ik}^{(1)} = \sum_k \varsigma f_{ik}^{(2)} = \varsigma \sum_k f_{ik}^{(2)} = \varsigma \pi_i(\mathsf{F}_2)$. Now we write $s_{ij}^{(1)}$ in terms of F_2:

$$s_{ij}^{(1)} = \frac{\varsigma f_{ij}^{(2)}}{\varsigma \pi_i(\mathsf{F}_2)} = \frac{f_{ij}^{(2)}}{\pi_i(\mathsf{F}_2)} = s_{ij}^{(2)} \tag{8.36}$$

Since $s_{ij}^{(1)} = s_{ij}^{(2)}$ for all i, j, we have $\mathsf{S}_1 = \mathsf{S}_2$. The stochastic matrix is an invariant of the equivalence class. □

The stochastic matrix captures the distributive logic of the network. Other important invariants can be derived from it.

Theorem 8.6 (The Ratio of First-Order Statistics as a Global Invariant) *For any two equivalent Ideal Flow Networks,* F_1 *and* F_2*, the global scaling factor* ς *is equal to the ratio of any pair of corresponding first-order statistics of their non-zero link flows.*

Proof Let $\mathsf{F}_1 \equiv \mathsf{F}_2$, implying $\mathsf{F}_1 = \varsigma \mathsf{F}_2$ for some $\varsigma > 0$. The set of non-zero link flows for each network is defined as $Q_1 = \{f_{ij} | [\mathsf{F}_1]_{ij} > 0\}$ and $Q_2 = \{g_{ij} | [\mathsf{F}_2]_{ij} > 0\}$. Since the networks are equivalent, they share the same pattern, so the sets Q_1 and Q_2 have the same number of elements, m.

- *Mean Flow:* The first-order statistic of mean flow is defined as $\mu = \frac{1}{m} \sum_{i,j} f_{ij}$. As shown previously, $\mu_1 = \varsigma \mu_2$. The ratio is $\mu_1/\mu_2 = \varsigma$.
- *Median Flow:* The median is the value separating the higher half from the lower half of a flow set. For a sorted set of flows, the median is the value at the middle position. Since each flow in Q_1 is a scalar multiple of a corresponding flow in Q_2 ($f_i = \varsigma g_i$), the sorted order of the flows remains the same. Thus, the median of Q_1 will be ς times the median of Q_2. Let $Median_1$ and $Median_2$ be the medians of Q_1 and Q_2. Then $Median_1 = \varsigma \times Median_2$, and the ratio $Median_1/Median_2 = \varsigma$.
- *Minimum/Maximum Flow:* Let $f_{min}^{(1)}$ and $f_{max}^{(1)}$ be the minimum and maximum non-zero flows in F_1, and similarly for F_2. Since every flow value in F_1 is a scalar multiple of a flow in F_2, we have $f_{min}^{(1)} = \varsigma f_{min}^{(2)}$ and $f_{max}^{(1)} = \varsigma f_{max}^{(2)}$. Consequently, the ratios $f_{min}^{(1)}/f_{min}^{(2)} = \varsigma$ and $f_{max}^{(1)}/f_{max}^{(2)} = \varsigma$.

□

This logic extends to any first-order statistic that is a linear combination of the flow values or a specific order statistic (minimum, maximum, quantiles). The ratio of any such corresponding first-order statistics will be the global scaling factor ς.

Theorem 8.7 (The Ratio of Ratios of First-Order Statistics Is an Invariant) *For any two equivalent Ideal Flow Networks, the ratio of any two first-order statistics of the non-zero link flows is an invariant across the equivalence class.*

Proof Let S_A and S_B be any two first-order statistics (e.g., mean, median, max flow). For two equivalent networks, F_1 and F_2, we have from the preceding theorem:

$$S_A^{(1)} = \varsigma S_A^{(2)} \quad \text{and} \quad S_B^{(1)} = \varsigma S_B^{(2)}$$

where the superscripts denote the network. Consider the ratio of these two statistics within a single network:

$$\frac{S_A^{(1)}}{S_B^{(1)}} = \frac{\varsigma S_A^{(2)}}{\varsigma S_B^{(2)}} = \frac{S_A^{(2)}}{S_B^{(2)}}$$

The scaling factor ς cancels, demonstrating that the ratio of the first-order statistics is invariant. □

This generalizes the Coefficient of Variation (CV), which is a specific ratio of a second-order statistic (standard deviation, which scales by ς) and a first-order statistic (mean, which also scales by ς).

Examples of such invariants include the ratio of the maximum flow to the mean flow, the ratio of the interquartile range (a second-order statistic) to the median, and the ratio of the standard deviation to the mean (the CV). This confirms that it is the *relative statistical structure* of the flow, not the absolute magnitude of any single statistic, that defines an equivalence class.

Corollary 8.4 (Second-Order Statistics Invariance) *For any two equivalent Ideal Flow Networks* F_1 *and* F_2*, their respective variance and standard deviation of the set of all non-zero link flows are not invariant. However, the Coefficient of Variation (CV) is an invariant.*

Proof Let F_1 and F_2 be two equivalent IFNs, so $\mathsf{F}_1 = \varsigma\mathsf{F}_2$ for a positive constant $\varsigma > 0$. Let $f_1, \ldots, f_m$ be the non-zero elements of F_1, and $g_1, \ldots, g_m$ be the corresponding non-zero elements of F_2. By definition, $f_i = \varsigma g_i$ for all $i = 1, \ldots, m$. *Mean:* The mean of the non-zero flows in F_1 is $\mu_1 = \frac{1}{m}\sum_{i=1}^{m} f_i = \frac{1}{m}\sum_{i=1}^{m} \varsigma g_i = \varsigma\left(\frac{1}{m}\sum_{i=1}^{m} g_i\right) = \varsigma\mu_2$. The mean is not invariant. It scales with ς. *Variance:* The variance of the non-zero flows in F_1 is given by:

$$\sigma_1^2 = \frac{1}{m}\sum_{i=1}^{m}(f_i - \mu_1)^2 = \frac{1}{m}\sum_{i=1}^{m}(\varsigma g_i - \varsigma\mu_2)^2$$

$$\sigma_1^2 = \frac{1}{m}\sum_{i=1}^{m}\varsigma^2(g_i - \mu_2)^2 = \varsigma^2\left(\frac{1}{m}\sum_{i=1}^{m}(g_i - \mu_2)^2\right) = \varsigma^2\sigma_2^2$$

Since $\varsigma > 0$ and is not necessarily equal to 1, $\sigma_1^2 \neq \sigma_2^2$. The variance is not an invariant; it scales by a factor of ς^2. *Standard Deviation:* The standard deviation is the square root of the variance:

$$\sigma_1 = \sqrt{\sigma_1^2} = \sqrt{\varsigma^2 \sigma_2^2} = |\varsigma|\sigma_2 = \varsigma\sigma_2$$

The standard deviation is not an invariant; it scales by a factor of ς.

Coefficient of Variation (CV): The coefficient of variation is the ratio of the standard deviation to the mean:

$$CV_1 = \frac{\sigma_1}{\mu_1} = \frac{\varsigma\sigma_2}{\varsigma\mu_2} = \frac{\sigma_2}{\mu_2} = CV_2$$

As shown in Proposition 8.7, the CV is an invariant. This is because the scaling factor ς cancels out in the ratio.

This concludes the proof. □

The corollary establishes that while the absolute values of second-order statistics like variance and standard deviation are not invariant, their relative measure, the CV, is. This supports the broader principle that relative, not absolute, quantities are the true invariants of equivalent networks.

Corollary 8.5 (The Ratio of Second-Order Statistics Is Invariant) *For any two equivalent Ideal Flow Networks,* F_1 *and* F_2*, the ratio of their corresponding second-order statistics is the global scaling factor* ς *as the invariant.*

Proof Let $\mathsf{F}_1 \equiv \mathsf{F}_2$ with scaling factor ς. We have already shown that the variance scales as $\sigma_1^2 = \varsigma^2\sigma_2^2$ and the standard deviation scales as $\sigma_1 = \varsigma\sigma_2$. Let V_1 and V_2 be any two second-order statistics of the non-zero flows in F_1 and F_2, respectively, which are defined based on the squares of the flow values. The interquartile range (IQR) is a second-order statistic defined as the difference between the third and first quartiles: $IQR = Q_3 - Q_1$. Since $Q_3^{(1)} = \varsigma Q_3^{(2)}$ and $Q_1^{(1)} = \varsigma Q_1^{(2)}$, their difference scales as:

$$IQR_1 = Q_3^{(1)} - Q_1^{(1)} = \varsigma Q_3^{(2)} - \varsigma Q_1^{(2)} = \varsigma(Q_3^{(2)} - Q_1^{(2)}) = \varsigma IQR_2$$

The ratio of the interquartile ranges is scaled by the scaling factor: $IQR_1/IQR_2 = \varsigma$. This is a restatement of the theorem that the ratio of any two corresponding first-order statistics is the scaling factor ς. Since the IQR is defined by first-order statistics (quantiles), the ratio of the IQR values will also be the scaling factor ς, as demonstrated above. Therefore, the ratio of the second-order statistics is invariant to the global scaling factor. □

The average node entropy is a second-order statistic. It is not a first-order statistic like the mean or median because its calculation involves a non-linear operation, the logarithm, on the probabilities derived from the flows. The entropy value captures the "spread" or "uncertainty" of the flow distribution from each node

Corollary 8.6 (Entropy Is Invariant) *The Average Node Entropy (h) is an invariant for any Ideal Flow Network (IFN) equivalence class.*

Proof Let F_1 and F_2 be two equivalent IFNs, so $\mathsf{F}_1 \equiv \mathsf{F}_2$. By definition, there exists a global scaling factor $\varsigma > 0$ such that $\mathsf{F}_1 = \varsigma \mathsf{F}_2$. The Average Node Entropy is calculated directly from the stochastic matrix S.

$$h = -\frac{1}{N} \sum_{i=1}^{N} \sum_{j:s_{ij}>0} s_{ij} \log(s_{ij})$$

The stochastic matrix S is an invariant for equivalent IFNs. As established in the provided text, the elements of the stochastic matrices for F_1 and F_2 are equal:

$$s_{ij}^{(1)} = \frac{f_{ij}^{(1)}}{\pi_i(\mathsf{F}_1)} = \frac{\varsigma f_{ij}^{(2)}}{\varsigma \pi_i(\mathsf{F}_2)} = s_{ij}^{(2)}$$

Since the stochastic matrices are identical ($\mathsf{S}_1 = \mathsf{S}_2$), the value of the average node entropy, which is a function of the elements of S, must also be identical.

$$h_1 = -\frac{1}{N} \sum_{i=1}^{N} \sum_{j} s_{ij}^{(1)} \log(s_{ij}^{(1)}) = -\frac{1}{N} \sum_{i=1}^{N} \sum_{j} s_{ij}^{(2)} \log(s_{ij}^{(2)}) = h_2$$

Therefore, $h_1 = h_2$, and the average node entropy is an invariant across the equivalence class. □

Example 8.16 (Calculating Invariants) We use the networks from Example 8.14. We calculate the invariants for the equivalence class. First, we find the stochastic matrix S using $\mathsf{F}_{\text{cardinal}}$ from (8.30). The row sums are $\boldsymbol{\pi}^T = [4, 4, 5]$.

$$\mathsf{S} = \begin{pmatrix} 0/4 & 1/4 & 3/4 \\ 2/4 & 0/4 & 2/4 \\ 2/5 & 3/5 & 0/5 \end{pmatrix} = \begin{pmatrix} 0 & 0.25 & 0.75 \\ 0.5 & 0 & 0.5 \\ 0.4 & 0.6 & 0 \end{pmatrix}$$

Any network in this class, such as F from (8.28), yields the same stochastic matrix. For F, the row sums are $\boldsymbol{\pi}^T = [6, 6, 8]$. The first row of S is $[0, 1.5/6, 4.5/6] = [0, 0.25, 0.75]$, which is identical.

Now, we calculate the entropy h using natural log (nats) from (5.1) and (8.16).

$$\begin{aligned} h &= -\Big[(0.25 \ln(0.25) + 0.75 \ln(0.75)) \\ &\quad +(0.5 \ln(0.5) + 0.5 \ln(0.5)) \\ &\quad +(0.4 \ln(0.4) + 0.6 \ln(0.6))\Big] \\ &= -[-0.5623 + (-0.6931) + (-0.6730)] \\ &\approx 1.9285 \text{ nats} \end{aligned}$$

This value is constant for all networks in the class.

The core insight of this section is that while the absolute values of these statistics scale with the network, their relative measures are invariant.

8.5 Symmetric Flow and Network Reversibility

A special state of equilibrium occurs when the ideal flow matrix F is symmetric ($f_{ij} = f_{ji}$). This signifies that the volume of traffic between any two nodes is perfectly balanced in both directions. This state is deeply connected to the concept of reversibility in Markov chains [13].

Kelly [8] formalized the detailed balance condition in reversible Markov processes, essential for modeling equilibrium flows in stochastic transport systems.

Definition 8.15 (Reversible Markov Chain [1]) A Markov chain with transition probability matrix S and stationary distribution π is *reversible* if it satisfies the *detailed balance condition*:

$$\pi_i s_{ij} = \pi_j s_{ji} \quad \text{for all pairs of states } i, j \tag{8.37}$$

This condition means that in equilibrium, the steady-state probability flow from state i to state j is equal to the probability flow from j to i.

Theorem 8.8 (Equivalence of Symmetry and Reversibility) *An Ideal Flow Network has a symmetric ideal flow matrix ($\mathsf{F} = \mathsf{F}^T$) if and only if the underlying Markov chain is reversible.*

Proof First, assume the underlying Markov chain is reversible. By (8.37), $\pi_i s_{ij} = \pi_j s_{ji}$. The elements of the ideal flow matrix are defined as $f_{ij} = \pi_i s_{ij}$ and $f_{ji} = \pi_j s_{ji}$. By direct substitution, we find $f_{ij} = f_{ji}$. Thus, the matrix F is symmetric.

Second, assume the ideal flow matrix F is symmetric, so $f_{ij} = f_{ji}$. By the definition of ideal flow, this means $\pi_i s_{ij} = \pi_j s_{ji}$. This is precisely the detailed balance condition. Thus, the underlying chain must be reversible. □

Theorem 8.9 (Symmetry Stochastic Matrix implies Symmetric Ideal Flow) *A symmetric stochastic matrix always produces a symmetric ideal flow matrix.*

Proof If the stochastic matrix S is symmetric, then $s_{ij} = s_{ji}$. Symmetric stochastic matrix implies doubly stochastic because the sum of rows is now the same as the sum of columns (i.e., premagic). Based on Theorem 8.8, the ideal flow matrix is symmetric if and only if the underlying Markov chain satisfies the detailed balance condition ($\pi_i s_{ij} = \pi_j s_{ji}$) for all states i, j. Substituting these two conditions into the detailed balance equation, we get:

$$\pi_i s_{ij} = \pi_j s_{ji}$$

$$\pi_i s_{ij} = \pi_i s_{ij}$$

This equality is a logical identity, confirming that the detailed balance condition is always met. Consequently, the ideal flow matrix F is always symmetric. This relationship holds regardless of any scaling factor for the flow. □

Example 8.17 Here are examples of symmetric stochastic matrix that produce symmetric ideal flow matrix.

$$\mathbf{S} = \begin{bmatrix} 0 & \frac{1}{2} & \frac{1}{2} \\ \frac{1}{2} & 0 & \frac{1}{2} \\ \frac{1}{2} & \frac{1}{2} & 0 \end{bmatrix} \rightarrow \mathbf{F} = \begin{bmatrix} 0 & 1 & 1 \\ 1 & 0 & 1 \\ 1 & 1 & 0 \end{bmatrix}$$

$$\mathbf{S} = \begin{bmatrix} 0 & \frac{1}{2} & \frac{3}{10} & \frac{1}{5} \\ \frac{1}{2} & 0 & \frac{1}{4} & \frac{1}{4} \\ \frac{3}{10} & \frac{1}{4} & 0 & \frac{9}{20} \\ \frac{1}{5} & \frac{1}{4} & \frac{9}{20} & \frac{1}{10} \end{bmatrix} \rightarrow \mathbf{F} = \begin{bmatrix} 0 & 10 & 6 & 4 \\ 10 & 0 & 5 & 5 \\ 6 & 5 & 0 & 9 \\ 4 & 5 & 9 & 2 \end{bmatrix}$$

A symmetric flow, however, does not require a symmetric network structure. The following propositions clarify this common misconception.

Proposition 8.4 *A symmetric network structure ($\mathsf{A} = \mathsf{A}^T$) is not sufficient to guarantee a symmetric ideal flow matrix* F.

Proof If S is asymmetric, the detailed balance condition is generally not met, and thus F will not be symmetric. □

Example 8.18 Here is the example of non-symmetric stochastic matrix and non-symmetric ideal flow that both has symmetric adjacency matrix.

$$\mathsf{S} = \begin{bmatrix} 0 & \frac{1}{3} & \frac{2}{3} \\ \frac{1}{2} & 0 & \frac{1}{2} \\ \frac{1}{4} & \frac{3}{4} & 0 \end{bmatrix} \rightarrow \mathsf{F} = \begin{bmatrix} 0 & 1 & 2 \\ 2 & 0 & 2 \\ 1 & 3 & 0 \end{bmatrix}$$

The binarized pattern of both stochastic matrix and ideal flow is symmetric.

$$\mathsf{A} = \begin{bmatrix} 0 & 1 & 1 \\ 1 & 0 & 1 \\ 1 & 1 & 0 \end{bmatrix}$$

A symmetric road network (e.g., all streets are two-way) does not imply symmetric turning probabilities. Traffic regulations, signal timings, or driver preferences can make the stochastic matrix S asymmetric.

Proposition 8.5 *A symmetric ideal flow matrix* F *does not require a symmetric stochastic matrix* S.

Proof A balanced, bidirectional flow equilibrium ($f_{ij} = f_{ji}$) can exist with asymmetric turning probabilities, as long as the detailed balance condition (8.37) holds. □

Example 8.19 We demonstrate this with a numerical example.

Consider an ideal flow matrix F and its corresponding node flow vector $\boldsymbol{\pi}$:

$$\mathsf{F} = \begin{pmatrix} 0 & 20 & 30 \\ 20 & 0 & 40 \\ 30 & 40 & 0 \end{pmatrix}, \quad \boldsymbol{\pi} = \mathsf{F}\mathbf{j} = \begin{pmatrix} 50 \\ 60 \\ 70 \end{pmatrix} \tag{8.38}$$

The flow matrix F is symmetric. We derive the underlying stochastic matrix S using the relation $s_{ij} = f_{ij}/\pi_i$.

$$\mathsf{S} = \begin{pmatrix} 0 & 20/50 & 30/50 \\ 20/60 & 0 & 40/60 \\ 30/70 & 40/70 & 0 \end{pmatrix} = \begin{pmatrix} 0 & 2/5 & 3/5 \\ 1/3 & 0 & 2/3 \\ 3/7 & 4/7 & 0 \end{pmatrix} \tag{8.39}$$

The stochastic matrix S is clearly not symmetric. However, the system satisfies detailed balance by construction. For example, $\pi_1 s_{12} = 50 \times (2/5) = 20$ and $\pi_2 s_{21} = 60 \times (1/3) = 20$. Therefore, a symmetric flow F can arise from a system with an asymmetric transition probability matrix S and a non-uniform node flow vector $\boldsymbol{\pi}$. The network for this example is shown in Fig. 8.5.

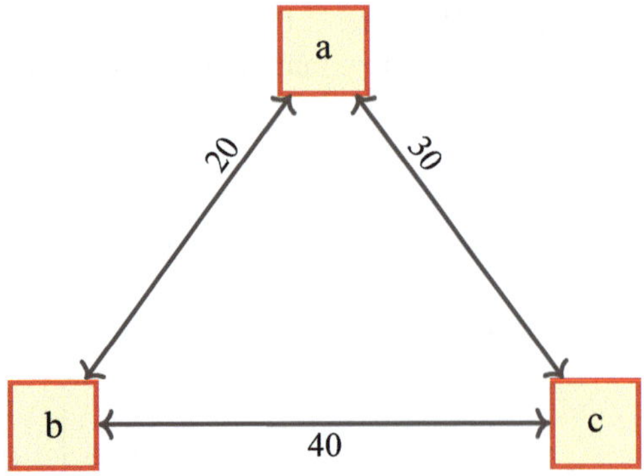

Fig. 8.5 A three-node network with symmetric flow values

8.6 Foundational Models Based on Uniform Distributions

The simplest models are based on the principle of maximum entropy [19], which suggests that, without other information, we should assume a uniform distribution of flows.

8.6.1 The Equal Outflow Model

The Equal Outflow model assumes that traffic leaving any node disperses uniformly among all available outbound links.

Definition 8.16 (Equal Outflow) The transition probability s_{ij} from node i to an adjacent node j is inversely proportional to the out-degree of node i, denoted $o_i = \sum_k a_{ik}$.

$$s_{ij} = \frac{a_{ij}}{o_i} = \frac{a_{ij}}{\sum_k a_{ik}} \tag{8.40}$$

The resulting row-stochastic matrix S is used to find the node flow vector $\boldsymbol{\pi}$, which is the stationary distribution satisfying $\boldsymbol{\pi}^T \mathsf{S} = \boldsymbol{\pi}^T$. The ideal flow matrix F is then found using $\mathsf{F} = \text{diag}(\boldsymbol{\pi})\mathsf{S}$. This model is applicable to any strongly connected network.

Example 8.20 (Urban Traffic Dispersion) Consider a network of four intersections with connectivity given by the adjacency matrix A and a total flow of $\kappa = 4000$

vehicles per hour.

$$\mathsf{A} = \begin{pmatrix} 0\,1\,1\,0 \\ 1\,0\,0\,1 \\ 1\,0\,0\,1 \\ 1\,1\,0\,0 \end{pmatrix}$$

Step 1: Construct the stochastic matrix S. The out-degrees are $o_1 = 2, o_2 = 2, o_3 = 2, o_4 = 2$. The matrix S is:

$$\mathsf{S} = \begin{pmatrix} 0 & 1/2 & 1/2 & 0 \\ 1/2 & 0 & 0 & 1/2 \\ 1/2 & 0 & 0 & 1/2 \\ 1/2 & 1/2 & 0 & 0 \end{pmatrix}$$

Step 2: Solve for the node flow vector $\boldsymbol{\pi}$. We solve $(\mathsf{S}^T - \mathsf{I})\boldsymbol{\pi} = \mathbf{0}$ subject to the constraint $\mathbf{j}^T\boldsymbol{\pi} = \kappa = 4000$. The solution is $\boldsymbol{\pi}^T \approx [1333, 1111, 667, 889]$. *Step 3: Calculate the ideal flow matrix* F. Using $\mathsf{F}_{ij} = \pi_i s_{ij}$, we distribute the node flows.

$$\mathsf{F} \approx \begin{pmatrix} 0 & 667 & 667 & 0 \\ 556 & 0 & 0 & 556 \\ 333 & 0 & 0 & 333 \\ 444 & 444 & 0 & 0 \end{pmatrix}$$

The matrix is premagic. For example, the total inflow to node 1 is $556+333+444 = 1333$, which equals its outflow, π_1. The flow distribution is shown in Fig. 8.6.

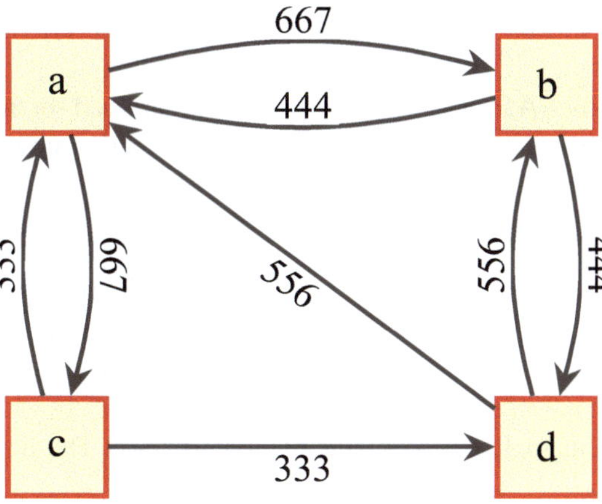

Fig. 8.6 A network example where all nodes have equal total outflow

8.6.2 The Equal Inflow Model

The Equal Inflow model assumes that traffic arriving at a node is drawn uniformly from all available inbound links.

Theorem 8.10 (Existence of Equal Inflow IFN) *A non-trivial, premagic Equal Inflow network exists if there is a non-trivial solution vector* $\mathbf{c}$ *to the homogeneous system:*

$$(\mathsf{A} - \mathsf{A}^T)\mathbf{c} = \mathbf{0} \tag{8.41}$$

The ideal flow matrix is then given by $\mathsf{F} = \mathsf{A} \circ (\mathbf{jc}^T)$, *where* $\circ$ *is the Hadamard product.*

Proof The defining property is that for any destination node j, the flow is the same on all incoming links (i, j). This means $f_{ij} = c_j$ for all i where $a_{ij} = 1$. We can write this as $\mathsf{F} = \mathsf{A} \circ (\mathbf{jc}^T)$. For F to be premagic, we require $\mathsf{F}\mathbf{j} = \mathsf{F}^T\mathbf{j}$. Substituting the expression for F leads to $(\mathsf{A} \circ (\mathbf{jc}^T))\mathbf{j} = (\mathsf{A}^T \circ (\mathbf{cj}^T))\mathbf{j}$, which simplifies to $\mathsf{A}\mathbf{c} = \mathsf{A}^T\mathbf{c}$, or $(\mathsf{A} - \mathsf{A}^T)\mathbf{c} = \mathbf{0}$. A non-trivial solution for $\mathbf{c}$ exists if the null space of the skew-symmetric matrix $(\mathsf{A} - \mathsf{A}^T)$ is non-trivial. □

Corollary 8.7 *A sufficient condition for the existence of a premagic Equal Inflow IFN is that the network structure is symmetric, i.e., its adjacency matrix* A *is symmetric* ($\mathsf{A} = \mathsf{A}^T$).

Example 8.21 (Symmetric Logistics Network) Consider a symmetric logistics network with $\kappa = 8000$ packages per day.

$$\mathsf{A} = \begin{pmatrix} 0 & 1 & 0 & 1 \\ 1 & 0 & 1 & 1 \\ 0 & 1 & 0 & 1 \\ 1 & 1 & 1 & 0 \end{pmatrix}$$

Since A is symmetric, Eq. (8.41) is satisfied for any vector $\mathbf{c}$. We choose $\mathbf{c} = \varsigma\mathbf{j}$. The flow matrix becomes $\mathsf{F} = \varsigma\mathsf{A}$. The scaling constant is $\varsigma = \kappa/\mathbf{j}^T\mathsf{A}\mathbf{j} = 8000/10 = 800$.

$$\mathsf{F} = 800\mathsf{A} = \begin{pmatrix} 0 & 800 & 0 & 800 \\ 800 & 0 & 800 & 800 \\ 0 & 800 & 0 & 800 \\ 800 & 800 & 800 & 0 \end{pmatrix}$$

This matrix is premagic by construction. The flow distribution is shown in Fig. 8.7.

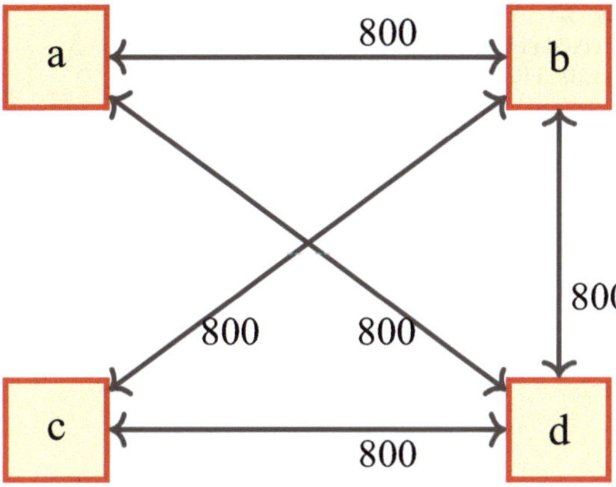

Fig. 8.7 A network example where all nodes have equal total inflow

8.7 Dual Representations: The Movement Graph

To analyze sequential movements, we introduce a dual representation of the network.

Definition 8.17 (Movement Graph) The *movement graph* (or *line digraph*) $\mathcal{L}(\mathcal{G})$ of a digraph $\mathcal{G} = (\mathcal{V}, \mathcal{E})$ is a new digraph where the vertices of $\mathcal{L}(\mathcal{G})$ are the edges of $\mathcal{G}$. A directed edge exists from vertex e_i to vertex e_j in $\mathcal{L}(\mathcal{G})$ if and only if the head of edge e_i is the same as the tail of edge e_j in $\mathcal{G}$.

Proposition 8.6 (Preservation of Connectivity) *The movement graph $\mathcal{L}(\mathcal{G})$ of a strongly connected digraph $\mathcal{G}$ is also strongly connected [7].*

The movement graph can be constructed algebraically using incidence matrices.

Definition 8.18 (Incidence Matrices) Let $\mathcal{G}$ have n vertices and m edges. The *head-incidence matrix* H is an $n \times m$ matrix where $h_{ik} = 1$ if vertex v_i is the head of edge e_k. The *tail-incidence matrix* T is an $n \times m$ matrix where $t_{ik} = 1$ if vertex v_i is the tail of edge e_k.

Proposition 8.7 (Algebraic Construction of Movement Graph) *The adjacency matrix of the movement graph, A_2, is given by $\mathsf{A}_2 = \mathsf{H}^T\mathsf{T}$.*

Theorem 8.11 (Preservation of Ideal Flow Property) *A path tracing transformation $\mathsf{F}_k \to \mathsf{F}_{k+1}$ that derives the flow on paths of length $k+1$ from the flow on paths of length k yields an ideal flow matrix F_{k+1} if and only if F_k is an ideal flow matrix.*

Proof The transformation preserves strong connectivity (and thus irreducibility) due to Proposition 8.6. The transformation is constructed to enforce flow conservation at each node of the movement graph, which ensures the resulting matrix F_{k+1} is premagic. The converse follows by contraposition. □

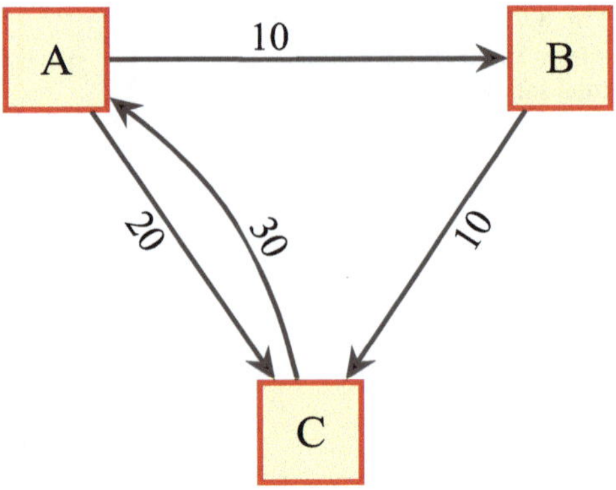

Fig. 8.8 A simple three-node directed network with specified flow values

Example 8.22 (Path Tracing) Consider the network in Fig. 8.8. The flow matrix F_1 is:

$$\mathsf{F}_1 = \begin{pmatrix} & A & B & C \\ A & 0 & 10 & 20 \\ B & 0 & 0 & 10 \\ C & 30 & 0 & 0 \end{pmatrix}$$

This matrix is an ideal flow matrix. The movements are $e_1 = $ (A,B), $e_2 = $ (A,C), $e_3 = $ (B,C), and $e_4 = $ (C,A). These form the vertices of the movement graph. The flow on paths of length two, F_2, is found by conserving flow through each node. For example, the 10 units of flow on A →B must continue to B →C. This creates a flow of 10 on the 2-path (A,B,C), which is an edge from e_1 to e_3 in the movement graph. The resulting 2-path flow matrix is:

$$\mathsf{F}_2 = \begin{pmatrix} & e_1 & e_2 & e_3 & e_4 \\ e_1 & 0 & 0 & 10 & 0 \\ e_2 & 0 & 0 & 0 & 20 \\ e_3 & 0 & 0 & 0 & 10 \\ e_4 & 10 & 20 & 0 & 0 \end{pmatrix}$$

The row and column sums are $[10, 20, 10, 30]^T$, matching the flows on the corresponding edges in F_1. F_2 is also an ideal flow matrix. For the primary network shown in Fig. 8.8, the movement graph that matches the connectivity shown in Fig. 8.9.

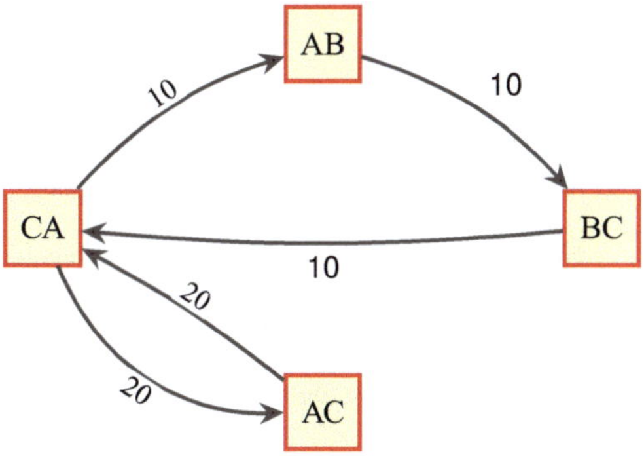

Fig. 8.9 A network representing the movement graph flow from matrix F_1

8.8 Summary

This chapter presented several advanced theoretical models for Ideal Flow Networks. We began by establishing the fundamental definitions of network theory, including directed graphs, connectivity, and the premagic property, which is central to the IFN framework. We then explored models based on uniform flow distribution, namely the Equal Outflow and Equal Inflow models, which serve as foundational tools for analysis when detailed data is unavailable.

Finally, we detailed the dual network representation using movement graphs (line digraphs), providing an algebraic method to trace and analyze multi-step paths while preserving the core properties of an ideal flow. These models provide a powerful and versatile toolkit for the theoretical analysis and practical design of efficient and balanced transport systems. We lays out the rigorous mathematical foundation of the Ideal Traffic Assignment framework. It formally defines a transportation network as a directed graph and introduces the key structural matrices for capacity, link length, and maximum speed. The chapter establishes core definitions for an Ideal Flow Network, including its representation as a strongly connected, directed graph, and defines the Ideal Flow and Congestion matrices. The crucial connection between the flow distribution and the eigenvector problem is explained, with the Perron-Frobenius theorem providing the theoretical guarantee for a unique, strictly positive eigenvector. This proves that any strongly connected network has a single, inherent steady-state relative flow distribution. The chapter concludes by showing that the manual and algebraic computation methods are practical algorithms for finding this unique eigenvector, which in turn defines the Ideal Flow pattern.

References

1. David Aldous and James Allen Fill. *Reversible Markov Chains and Random Walks on Graphs*. Available at: https://www.stat.berkeley.edu/~aldous/RWG/book.html. 2002.
2. Abraham Berman and Robert J. Plemmons. *Nonnegative Matrices in the Mathematical Sciences*. Philadelphia: Society for Industrial and Applied Mathematics, 1994. https://doi.org/10.1137/1.9781611971262. https://epubs.siam.org/doi/book/10.1137/1.9781611971262.
3. Norman Biggs. *Algebraic Graph Theory*. 2nd. Cambridge University Press, 1993. https://doi.org/10.1017/CBO9780511608704.
4. B. C. Eaves et al. "Line-sum-symmetric scalings of square nonnegative matrices". In: *Mathematical Programming Studies*. Vol. 25. Springer, 1985, pp. 124–141. https://doi.org/10.1007/BF02592021
5. G. Frobenius. "Ueber Matrizen aus nicht negativen Elementen". In: *Sitzungs-berichte der Königlich Preußischen Akademie der Wissenschaften* (1912), pp. 456–477.
6. G. R. Grimmett and D. R. Stirzaker. *Probability and Random Processes*. 3rd. Oxford University Press, 2001.
7. F. Harary and R. Z. Norman. "Some properties of line digraphs". In: *Rendiconti del Circolo Matematico di Palermo* 9.2 (1960), pp. 161–168. https://doi.org/10.1007/BF02854556
8. Frank P. Kelly. *Reversibility and Stochastic Networks*. Cambridge University Press, 1979. https://doi.org/10.1017/CBO9780511624001. http://www.statslab.cam.ac.uk/~frank/BOOKS/kelly_book.html
9. John G. Kemeny and J. Laurie Snell. *Finite Markov Chains*. New York: Springer, 1976. https://link.springer.com/book/9780387901923
10. Richard C. Larson and Amedeo R. Odoni. *Urban Operations Research*. Prentice-Hall, 1981. https://doi.org/10.1002/9781118762427
11. László Lovász. "Random walks on graphs: A survey". In: vol. 2. Budapest: János Bolyai Mathematical Society, 1993, pp. 1–46. https://www.cs.cmu.edu/~15859n/RelatedWork/random-walks-on-graphs.pdf
12. Carl D. Meyer. *Matrix Analysis and Applied Linear Algebra*. Philadelphia: Society for Industrial and Applied Mathematics, 2000. https://doi.org/10.1137/1.9780898719512
13. J. R. Norris. *Markov Chains*. Cambridge Series in Statistical and Probabilistic Mathematics. Cambridge: Cambridge University Press, 1997. https://doi.org/10.1017/CBO9780511810633
14. Oskar Perron. "Zur Theorie der Matrices". In: *Mathematische Annalen* 64.2 (1907), pp. 248–263. https://doi.org/10.1007/BF01449896
15. Eugene Seneta. *Non-negative Matrices and Markov Chains*. Rev. ed. New York: Springer, 2006. https://doi.org/10.1007/0-387-32792-4. https://link.springer.com/book/10.1007/0-387-32792-4
16. K. Teknomo. "Ideal Flow Based on Random Walk on Directed Graph". In: *Proceedings of the 9th International Collaboration Symposium on Information, Production and Systems*. Kitakyushu: Waseda University, 2015.
17. Kardi Teknomo. "Premagic and Ideal Flow Matrices". In: *arXiv preprint arXiv:1706.08856* (2017). https://arxiv.org/pdf/1706.08856.
18. Kardi Teknomo, R. Gardon, and C. Saloma. "Ideal Flow Traffic Analysis: A Case Study on a Campus Road Network". In: *Philippine Journal of Science* 148.1 (2019), pp. 51–62. https://archium.ateneo.edu/discs-faculty-pubs/271/.
19. H. J. Van Zuylen and L. G. Willumsen. "The most likely trip matrix estimated from traffic counts". In: *Transportation Research Part B: Methodological* 14.3 (1980), pp. 281–293. https://doi.org/10.1016/0191-2615(80)90008-9.

Chapter 9
Ideal Flow Network for Transport Analysis

9.1 Introduction

Urban traffic congestion is a systemic challenge with significant economic and social costs. While conventional interventions such as adding road capacity or implementing public transit are designed for long-term impact, their effectiveness is often challenged by sustained urban growth.

The preceding chapters developed a formal methodology for deriving the Ideal Flow Network (IFN) matrix, F, from the network's underlying stochastic structure. This Ideal Flow matrix predicts steady-state traffic volumes across all links. In this chapter, we investigate this question through the lens of the Ideal Traffic Assignment (ITA) framework. The ITA is based on the Ideal Flow matrix. This matrix serves as a foundational tool for analyzing how structural optimization might yield more sustainable traffic flow.

We shift focus from predicting *how much* flow exists to measuring what that flow does. Specifically, we look at its *impact* on how well the network performs. We will define these concepts mathematically and establish the functional relationships that convert the abstract notion of flow into the tangible costs experienced by network users. The framework adjusts the abstract model to reflect physical road features, like link capacity, length, and free flow (maximum) speed. The ITA framework outputs core performance metrics: congestion, speed, travel time, and delay on each link. These concepts will be defined mathematically, establishing functional relationships that translate abstract flow into the tangible costs borne by users.

These help us evaluate how efficiently a transport system operates. This lets us analyze traffic patterns and congestion more realistically. Our goal is to construct a rigorous foundation for the advanced analyses that follow in later chapters.

The ITA framework is useful for analyzing transport networks using their built-in features. It relies on structural properties like topology and link capacity. Ideal Traffic Assignment framework explores an alternative approach, focusing

K. Teknomo, *Ideal Flow Network for Transportation Research*,
Lecture Notes in Intelligent Transportation and Infrastructure,
https://doi.org/10.1007/978-981-95-4727-2_9

on understanding the network's structural dynamics as a primary determinant of flow patterns, and using this insight to evaluate policy interventions. The policy intervention would be discussed more detail in Chap. 14. In this chapter, we will first introduce the Ideal Traffic Assignment framework. Then, we will build the theory with mathematical rigor and finally apply it to practical scenarios to demonstrate its use in urban transport planning.

Upon completion of this chapter, the following competencies will be established:

- Foundational principles of the Ideal Traffic Assignment framework will be articulated and contextualized.
- The ideal flow distribution for a specified transport network will be derived through formal methods.
- Key performance indicators—including congestion, speed, travel time, and delay—will be computed from the ideal flow configuration.

9.2 Ideal Traffic Assignment Framework

The Ideal Traffic Assignment framework (ITA) constitutes a systematic procedure for leveraging the Ideal Flow Network (IFN). This process facilitates the estimation of absolute traffic volumes and various network performance metrics. The process begins with the network structure and link capacities and free flow speed then proceeds to calculate steady-state flows, congestion link speeds, travel time and delays. The overall framework is depicted in Fig. 9.2.

Figure 9.1 presents the fundamental computational process of the Ideal Traffic Assignment. It outlines the sequence of operations that convert network capacities into flow assignments and congestion metrics. The diagram consists of four main components: the capacity matrix C, the stochastic matrix S, the relative flow matrix F_{rel}, and the absolute flow matrix F_{abs}.

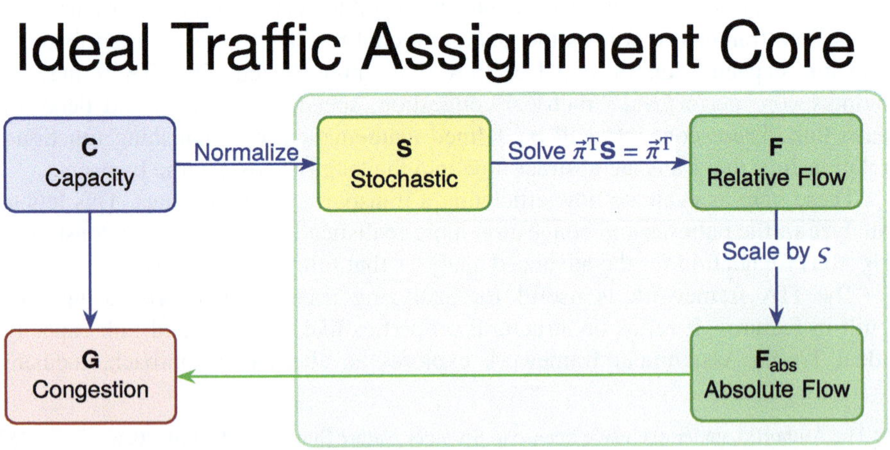

Fig. 9.1 Core of Ideal Traffic Assignment

The process begins with the capacity matrix. This matrix represents the maximum flow rate for each edge in the network. The capacity matrix is then normalized to create the stochastic matrix. The stochastic matrix is a row-stochastic matrix, where each row sums to one. It describes the probability of a traveler moving from one node to an adjacent node. The stochastic matrix is crucial for calculating the relative flow. The relative flow is determined by solving the fundamental steady-state equation of Markov Chain (8.12).

$$\pi^T \mathsf{S} = \pi^T$$

The row vector π^T represents the stationary distribution of travelers across the network's nodes. This distribution, when arranged into the relative flow matrix F, quantifies the proportion of total network flow on each edge. This relative flow is then scaled by a global scaling factor, ς, to produce the absolute flow matrix F_{abs}. The absolute flow matrix contains the actual number of vehicles per unit of time on each edge.

Finally, the absolute flow matrix and the original capacity matrix are used to compute the congestion matrix G. The congestion matrix quantifies the level of congestion on each network edge. This core computational block forms the basis for the more comprehensive frameworks.

In Fig. 9.2, now we expand the core ITA block above by integrating inputs and outputs.

The framework is partitioned into three conceptual stages. The first stage consists of input matrices: the adjacency matrix A, the length matrix L, and the free-flow speed matrix U. The adjacency matrix matrix defines network structure which is the topology and connectivity on which nodes are connected by which edges. The length matrix L matrix specifies the length distance of each edge, while the free-flow speed matrix U matrix gives the maximum speed of travel on each edge under

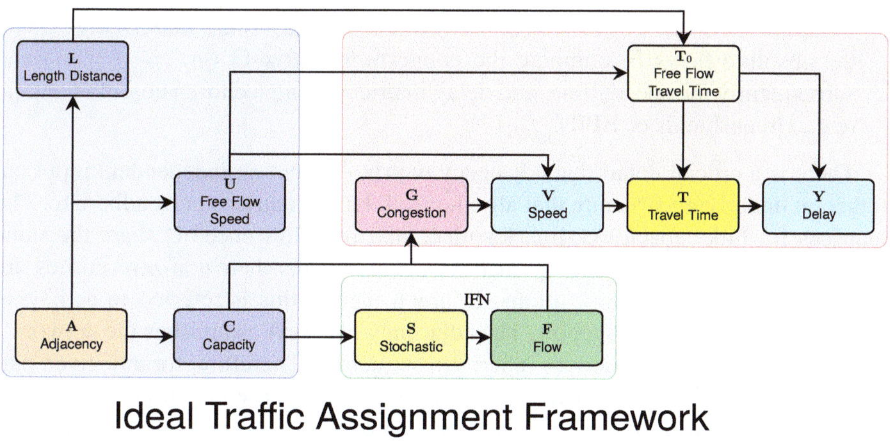

Fig. 9.2 Basic Framework for Ideal Traffic Assignment

ideal, uncongested conditions. These inputs are used to create the capacity matrix C, which then feeds into the IFN core.

The IFN block, containing the stochastic matrix S and the flow matrix F, is the second stage. This block calculates the flow distribution and congestion as described previously as the core of ITA.

The third stage involves the calculation of various network performance metrics. The congestion matrix G, derived from the IFN block, is used to determine the actual speed of travel (V) for each edge. This speed, along with the length matrix L, determines the total travel time T. Simultaneously, the length and free-flow speed matrices are used to compute the free-flow travel time T^0. The delay Y is then calculated as the difference between the actual travel time T and the free-flow travel time T^0. This framework provides a one-way calculation of traffic metrics based on a static network state.

The steps of Ideal Traffic Assignment can be summarized as follows:

1. *Define Inputs*: Start with the network's adjacency matrix A, link capacity matrix C (veh/hr), link length matrix L (km), and maximum (free-flow) speed matrix U (km/hr).
2. *Derive Stochastic Matrix*: Calculate the stochastic transition probabilities S based on proportional capacity or derive it from behavioral data such as turning distributions in each intersection.
3. *Compute Relative Ideal Flow*: Solve for the stationary distribution π of the Markov chain and compute the relative ideal flow matrix. The values in F_{rel} are proportional, not absolute flows.
4. *Scale to Absolute Flow*: The relative flows can be scaled to absolute flows F_{abs}. One way is to calibrate the relative flow based on real world traffic flow data. Another way is done by anchoring the network to an assumed a certain level of either maximum congestion, average speed or average travel time at certain designated links. More detail treament calibrating IFN would be discussed in Chap. 11.
5. *Compute Network Performance: Congestion, Speed, Time, and Delay*: Using the absolute flows F, compute the congestion matrix G ($g_{ij} = f_{ij}/c_{ij}$) and subsequently the speed, time, and delay matrices using a chosen cost-flow model (e.g., Greenshields or BPR).

There is a crucial detail that adjacency matrix A is not an independent input but rather an underlying structure that all other link-based matrices must adhere to. The matrices for link capacity C, link length L, and free-flow speed U share the same sparsity pattern as the adjacency matrix A. This means their non-zero entries are located in the exact same positions. In graph theory, this is referred to as having the same graph-theoretic support. The adjacency matrix A establishes the network's topology, and the other matrices inherit this structure. Therefore, for any given pair of nodes (i, j), if a directed link does not exist (i.e., the entry in A is zero), the corresponding entries in C, L, and U must also be zero. This condition ensures that physical properties are only defined for existing connections.

In many comparative analyses, the explicit calculation of absolute flow is not necessary. When comparing different network scenarios or design alternatives, the relative ideal flow matrix F_{rel} is sufficient. Because it preserves the proportional relationships between flows on each link, any changes in network structure or capacity will be reflected in the relative flows. By comparing the F_{rel} matrices of different scenarios, one can assess the relative impact on flow distribution without needing to calibrate for a specific absolute total network flow. This approach is valuable for design evaluation and sensitivity analysis, as it focuses on the internal dynamics of the network rather than a single, scaled value. The calibration to absolute flow is only required when a quantitative measure of total network throughput or specific link volumes is needed, such as for a direct comparison with observed traffic counts.

9.2.1 Supply-Demand Dynamics in Traffic Networks

In traditional traffic management paradigm, demand determines supply magnitude. When road is congested, we tend to view it in short term perspective that to add more supplies or restrict demand to cope with the increasing demand. Thus, congestion prompts expansion of road capacity. Policy options are either increase supply (i.e. road width) or restrict demand (such as congestion pricing). However, population and economic growth intensify demand in the long-term.

In contrast, in the long-term perspective of Ideal Traffic Assignment, the demand may control the supply as human past wisdom. Historical demand shaped existing infrastructure. It suggests that supply does not merely respond to demand; it also influences it. Adjusting supply may induce behavioral shifts in demand. Supply correction leads to network-wide equilibrium. Demand redistribution emerges from supply modification. The presence of a new or expanded road, an increase in supply, can itself generate new demand. This occurs as drivers alter their behavior, perhaps by choosing new routes, making more trips, or moving to areas previously considered too remote.

Proposition 9.1 (Dynamic Interplay Between Supply and Demand) *In a transport network, altering supply can dynamically influence and re-balance demand, leading to a new network equilibrium.*

Proof First, consider a traffic network in an initial equilibrium state. According to User Equilibrium Axiom 9.3, all drivers are on their minimum travel time paths, given the current network conditions and travel times. Now suppose a change is made to the supply. This change modifies the travel time function for that segment for the given flow. The supply change disrupts the initial equilibrium. The paths that include the altered segment now have a higher or lower travel time compared to other paths. This violates the condition of User Equilibrium Axiom 9.3, as some drivers now have an incentive to switch path. The demand would be rebalanced.

This shift in route choice constitutes a re-balancing of demand across the network, not just for the link where the policy was affected. This re-routing process continues until a new equilibrium is reached. In this new state, all drivers are once again on their minimum travel time paths, but the distribution of flow across the network, the demand pattern, is fundamentally different from the initial state. The increased flow on the altered segment will eventually cause its travel time to rise, but it will stabilize at a new value consistent with the equilibrium condition. □

For example, a new lane is added to a road segment e. This change modifies the travel time function for that segment, $t_e(f_e)$. For a given flow, the new travel time $t_e'(f_e)$ is now less than the old travel time $t_e(f_e)$. When the supply change, the path that include the altered segment e now have a lower travel time compared to other paths. The flow on the altered segment e will increase, while the flow on other segments may decrease. Drivers who were previously using more congested or longer routes will now re-route to the path with the reduced travel time. Over time, it would reach a new equilibrium.

Proposition 9.1 demonstrates that a change in network supply (e.g., adding a lane) directly affects the travel time functions of specific network segments. This change disrupts the initial equilibrium, causing drivers to alter their route choices. This alteration in driver behavior is the mechanism by which demand is re-balanced. The process continues until a new equilibrium is established, proving that altering supply can indeed control and re-distribute demand. This proposition idea that demand can be rebalanced via supply changes is a fundamental principle of *dynamic traffic assignment* [13] and is central to understanding the phenomenon of *induced demand* [7], refers to the phenomenon where increasing road capacity leads to higher traffic volumes due to behavioral and land-use responses.

When a specific edge or node becomes congested, its cost or impedance (i.e., travel time or delay) increases. This increase in cost acts as a signal. In a free-flow network, drivers would choose the path with the least cost. However, when congestion increases the cost on one path, drivers may shift to alternative routes, effectively redistributing the demand across the network.

The policy represents "correction of the supply" corresponds to a change in the network's capacity or a modification of the flow dynamics. This can be achieved not just by widening a road, but by introducing policies that modify the effective cost of travel. For example, congestion pricing (charging a fee to use a road during peak hours) or road diet (reduce the number of lanes) are changes the perceived cost for drivers. This change in cost, in turn, alters their route choices and travel times, rebalancing the demand across the entire network. This rebalancing leads to a new equilibrium state. This equilibrium is not necessarily a simple Nash equilibrium [15, 16] where each driver acts selfishly. Rather, it can be a system-wide optimization where a central authority (the policy maker) sets the network parameters (the "supply correction") to achieve a desired state, such as minimizing total network-wide travel time or maximizing average flow efficiency. This conceptual shift from a one-way causal relationship to a feedback loop is a central tenet of modern transportation science.

9.3 Mathematical Foundations

In this section, we establish the mathematical foundations of the Ideal Traffic Assignment (ITA). We use a formal structure of axioms, definitions and propositions. This approach ensures that our model is logical and consistent.

9.3.1 Transport Network Representation

We begin by defining a transport network and the core axioms that govern its behavior in our model.

Definition 9.1 (Transport Network) A transport network is is a directed graph $\mathcal{N}(\mathcal{V}, \mathcal{E})$ with a flow function $\phi : \mathcal{E} \to \mathbb{R}_{\geq 0}$, where $\mathcal{V}$ is a finite set of n nodes representing intersections or conflict points, and $\mathcal{E}$ is the set of directed edges called links representing road segments connecting pairs of nodes in V. We associate a link flow function: for each edge $(i, j) \in \mathcal{E}$, we have a non-negative flow f_{ij}.

Definition 9.2 (Transport Network Structure) The network's structure of a transport network is represented by an $n \times n$ adjacency matrix A, where $a_{ij} = 1$ if a link exists from node i to node j, and $a_{ij} = 0$ otherwise.

Definition 9.3 (Structural Input Matrices ($\mathsf{C}, \mathsf{L}, \mathsf{U}, \mathsf{S}$)) The physical structure of the network is defined by three $n \times n$ input matrices:

- The *Capacity Matrix* (C), where c_{ij} is the maximum flow rate of link (i, j).
- The *Link Length Matrix* (L), where L_{ij} is the physical length distance of link (i, j).
- The *Maximum Speed Matrix* (U), where U_{ij} is the free-flow speed of link (i, j).
- The *Stochastic Matrix* (S), where S_{ij} is the probability of a vehicle traveling from node i to node j.

If no link exists from node i to j, the corresponding matrix elements are zero. The capacity matrix C and stochastic matrix S are considered the primary structural input.

Proposition 9.2 *For an ideal traffic assignment framework, the set of structural matrices* $\{\mathsf{C}, \mathsf{L}, \mathsf{U}, \mathsf{S}\}$ *must conform to the network's adjacency matrix* A. *Specifically, the non-zero entries of* C, L, U *and* S *must be a strict subset of the non-zero entries of* A.

Proof If a link does not exist from node i to node j, then the adjacency matrix entry a_{ij} is 0. Since there is no link, there can be no physical properties associated with it, such as capacity, length, or speed. Therefore, the corresponding matrix entries c_{ij}, l_{ij}, u_{ij} and s_{ij} must also be zero. This implies that for every entry (i, j), the condition $c_{ij} \neq 0 \implies a_{ij} \neq 0$, $l_{ij} \neq 0 \implies a_{ij} \neq 0$, and $u_{ij} \neq 0 \implies a_{ij} \neq 0$ and $s_{ij} \neq 0 \implies a_{ij} \neq 0$ must hold. The set of non-zero entries for C, L, U

and S is, therefore, a subset of the non-zero entries for A. In practice, these sets are identical because the model assumes all existing links have defined capacity, length, and speed. The adjacency matrix A thus provides the essential structural constraint for all other link-specific matrices. □

The structure input matrices C, L, U and S are constrained by A. The non-zero entries of these matrices are defined on the support of A. This relationship can be expressed using the concept of element-wise product or the Hadamard product. Given a directed graph with adjacency matrix A, where $a_{ij} = 1$ if a link exists from node i to j, and $a_{ij} = 0$ otherwise, any structural matrix $\mathsf{X} \in \{\mathsf{C}, \mathsf{L}, \mathsf{U}\}$ must satisfy the following condition:

$$\mathsf{X} = \mathsf{X} \circ \mathsf{A} \tag{9.1}$$

Here, $\circ$ denotes the Hadamard product. This equation means that the entry x_{ij} can only be non-zero if the corresponding entry a_{ij} is also non-zero. If $a_{ij} = 0$, the product will force x_{ij} to be 0, effectively ensuring that the structural matrices only contain information for existing links. This functional form expresses the dependency of the structural matrices on the network's topology as defined by A.

Definition 9.4 (Policy) A policy is a deliberate action that alters the physical or operational characteristics of a transportation network. These changes are defined by modifying at least one of the network's input matrices: the Capacity Matrix (C), the Link Length Matrix (L), or the Free Flow Speed Matrix (U). By changing these matrices, policy interventions directly affect the network's supply, which, in turn, influences and re-balances travel demand.

Notice that in ITA framework, a policy is defined as *supply side alteration* rather than from demand side. A change in supply is a change in the physical or operational characteristics of a road segment. This could be adding a lane (increasing capacity, thus lowering travel time for a given flow), reducing lane ("road diet") in the light of congestion rebalancing, implementing a new traffic light signal plan [19], or introducing a toll (increasing the generalized cost of travel) and so on. Generalized cost [18] combines time, monetary cost, and other disutilities into a single metric for mode and route choice modeling.

9.3.2 Axioms for Ideal Traffic Assignment

The Ideal Traffic Assignment framework is built upon a set of foundational axioms. These axioms govern the network's structure, the behavior of its components, and the mathematical properties of the model. The following axioms are central to the Ideal Traffic Assignment framework and are presented here as an expansion of the core IFN theory that have been discussed in previous Chap. 8

The following structural axioms below define the fundamental graph-theoretic properties of the network.

Axiom 9.1 (Strong Connectivity) *A functional transport network must be strongly connected. This means that every node is reachable from every other node through a path of directed links. This property ensures that the adjacency matrix* A *is irreducible [12].*

As explained more detail in Chap. 7, for networks that are not naturally strongly connected (e.g., it has source or sink nodes), a conceptual "cloud" node can be added to establish strong connectivity. We add links from this cloud node to all source nodes and from all sink nodes to the cloud node. This transformation creates a strongly connected graph without altering the original flow dynamics.

Axiom 9.2 (Steady-State Equilibrium) *When observed over a sufficiently long period, a transport network approaches a steady-state equilibrium. In this condition, the aggregate flow patterns are stable and do not change over time. The model focuses on analyzing this equilibrium state rather than transient flow dynamics.*

The axioms below describe the behavioral assumptions related to flow and its relationship with network properties.

Axiom 9.3 (User Equilibrium Principle) *In a network with a fixed travel demand, drivers will choose routes that minimize their individual travel time. At equilibrium, no driver can unilaterally switch routes to reduce their travel time. This axiom is also known as Wardrop's First Principle [24]. In Game Theory, this is known as Nash equilibrium [15].*

A convex optimization method [4, 8] widely used in traffic assignment problems to solve user equilibrium.

Axiom 9.4 (Existence of Behavioral Stochastic Matrix) *The collective, time-invariant behavior of drivers at any node i can be described by $n \times n$ row-stochastic matrix* S, *referred to as the* behavioral stochastic matrix S. *An element s_{ij} represents the probability that a vehicle at node i will choose link (i, j) for its next movement. This matrix is row-stochastic, meaning all entries are non-negative, and the sum of probabilities for each row equals one ($\sum_j s_{ij} = 1$). This axiom provides a stable, probabilistic representation of driver behavior in equilibrium.*

This Axiom 9.4 establishes that the collective route choices of drivers can be formally modeled. It posits that driver decisions at any given node are not random but can be represented by a behavioral stochastic matrix S. The matrix element s_{ij} represents the turning movement probability that a vehicle will choose to move from intersection node i to node j. This formalizes the idea of a stable, probabilistic driver behavior within the network. The axiom further defines the mathematical properties of this matrix: all probabilities must be non-negative, and the sum of probabilities for all possible outbound links from a node must equal one. This ensures that the matrix is valid for use in a Markov chain analysis.

This Axiom 9.4, focusing specifically on the probabilistic nature of turning movements in each intersection. It states that the collective behavior of drivers at any intersection (node) can be described by a fixed time-invariant probability distribution of a vehicle moving to an adjacent node. This assumption simplifies the traffic model by treating driver choices as stable probabilities over time, rather than as constantly fluctuating, unpredictable decisions. The row-stochastic nature of the matrix is again emphasized, highlighting that from any given node, the probabilities of all possible turns sum to one, representing the certainty that a driver will choose one of the available outbound links.

Axiom 9.5 (Capacity Proportionality) *In the absence of detailed behavioral data, the transition probability s_{ij} is assumed to be directly proportional to the capacity c_{ij} of the outbound link. This is represented by the equation $s_{ij} = c_{ij}/\sum_k c_{ik}$. The traffic distributes itself in proportion to the available road space. This assumption, based on the* Maximum Entropy Principle *[11], provides the least biased and most objective method for defining the stochastic matrix when only capacity information is available.*

This Axiom 9.5 provides a method for constructing the stochastic matrix when empirical data on driver behavior is unavailable. It is a fundamental modeling choice based on the Maximum Entropy Principle. The principle states that when making an assumption with incomplete information, one should choose the least biased distribution. This axiom assumes that in the absence of other data, drivers distribute themselves proportionally to the available road space. The probability of choosing a link is equal to the link's capacity divided by the total outbound capacity from that node. This provides a default, objective method for generating the stochastic matrix from a known network structure.

These axioms below define the properties of performance metrics and the concept of network equivalence (see Sect. 8.4).

Axiom 9.6 (Well-Behaved Travel Time Functions) *The travel time on any link, t_{ij}, is a continuous, a monotonically increasing, and strictly convex function of the congestion on that link, g_{ij}. These properties are fundamental for ensuring that travel time increases smoothly with flow and that equilibrium solutions are stable and unique. As flow increases, travel time increases. This is a standard assumption in traffic flow theory.*

Axiom 9.7 (The Total Flow Parameter (κ) and Network Equivalence [2]) *The absolute magnitude of network flow is determined by a scalar input, the* Total Flow parameter *κ, where $\kappa = \sum_{(i,j)\in\mathcal{E}} f_{ij}$. Networks that share the same structure and relative flow patterns, differing only by this scalar, κ, are defined as* equivalent networks *[2, 22]. These networks share the same underlying stochastic matrix,* S.

9.3.3 *Behavioral Stochastic Matrix*

The foundational definitions, axioms and propositions from the previous sections provide all the necessary components to develop the Ideal Traffic Assignment from first principles. It is a sequential process that transforms the physical description of the network (the capacity matrix C) into a complete, steady-state flow pattern (F) up to the network performances.

The first step of the Ideal Traffic Assignment is to translate the physical structure of the network into a model of aggregate driver behavior. Axiom 9.4 posits the existence of a Behavioral Stochastic Matrix, S, that represents the equilibrium turning movement probabilities at each node.

The ITA framework proposes that, in the absence of more detailed information, these probabilities are primarily a function of the network's capacity. The rationale is that network capacity is the long-term, physical embodiment of historical demand based on Proposition 9.1. High-capacity links exist because they have historically served high volumes of traffic, and drivers will, in aggregate, tend to favor these higher-capacity routes.

The stochastic matrix S can be derived from two primary methods:

- *Behavioral data derivation*: The transition probabilities can be directly inferred from empirical observations or behavioral models. This method uses real-world data, such as turning movements at intersections, to capture the probabilistic route choices of drivers, based on Axiom 9.4.
- *Capacity-based computation*: In absence of the behavioral data of the probability of turning movements at intersections, based on Axiom 9.5, the transition probabilities can be determined by normalizing the capacity matrix C. This approach assumes that traffic flow is distributed proportionally to the capacity of the available links, reflecting an idealized, structural property of the network.

We would discuss these two methods the next Sects. 9.3.4 and 9.3.5.

9.3.4 Probability of Turning Movements at Intersections

Transition probabilities are derived from empirical data or behavioral models, leveraging observed turning movements at intersections to reflect drivers' probabilistic route choices.

Most common current technology is based on Traffic cameras coupled with Computer Vision. How to get probability of turning movements at intersections? To get the probability of turning movements at intersections using video analytics, the process involves a series of steps that move from raw visual data to a statistical measure. The raw video footage is the input, and a statistical probability is the output. This is a primary function of computer vision and deep learning in Intelligent Transportation Systems.

1. Vehicle Detection and Classification: The process begins with real-time detection of vehicles in the video feed. Computer vision algorithms, often using deep learning architectures like Convolutional Neural Networks (CNNs), are trained to identify vehicles and distinguish them from other objects. These systems can also classify vehicles by type (e.g., car, truck, bus).

2. Object Tracking: Once a vehicle is detected, it is continuously tracked across multiple video frames as it moves toward and through the intersection. The system assigns a unique identifier to each vehicle and builds a trajectory, a path of consecutive positions, for it.
3. Trajectory Analysis: The trajectory of each tracked vehicle is analyzed to determine its path. Virtual "detection zones" or "trip wires" are often defined in the software for each lane and potential exit from the intersection. For example, a vehicle entering the intersection from the north and exiting to the west would be classified as a left-turn movement. A vehicle that enters from the north and exits to the south would be classified as a through movement.
4. Data Aggregation and Probability Calculation: The classified movements are then aggregated over a specific time period (e.g., per hour, per day). The total number of vehicles entering the intersection from a specific approach is counted, along with the number of vehicles making each type of turn (left, right, through).

The probability of a specific turning movement is then calculated using the following formula:

$$P(\text{turning movement}) = \frac{\text{Number of vehicles making that turn}}{\text{Total vehicles approaching from same entry link}} \tag{9.2}$$

This calculation is performed for each entry link at the intersection to create a complete set of turning movement probabilities, which can then be used to populate the Markov transition probability matrix S.

A range of technologies listed below can also be used to gather the necessary data for developing a Markov transition probability matrix, particularly for turning movements. These lists are not exhausted. These technologies include:

- *Radar sensors*: These use microwave signals to detect the speed and presence of vehicles. By mounting them at an intersection, it's possible to track vehicles as they approach and move through the junction, thus identifying their turning movements. They are non-intrusive and can function effectively in various weather conditions.
- *Lidar (Light Detection and Ranging)*: Lidar sensors emit pulsed laser light to measure distances to objects. By creating a precise 3D map of the intersection and tracking the position of vehicles over time, Lidar can accurately determine their trajectories and turning movements.
- *Magnetometers*: These small, in-ground sensors detect changes in the Earth's magnetic field caused by the presence of a vehicle. While typically used for simple vehicle counts, an array of magnetometers placed strategically at an intersection can track a vehicle's path to infer its turning movement.
- *Acoustic sensors*: These use microphones and advanced signal processing to detect and classify vehicles based on their sound signatures. Although less common for precise turning movement counts, they can contribute to a broader understanding of traffic patterns and flow.

9.3.5 *The Generalized Proportional Capacity Model*

When behavioral data is unavailable, transition probabilities are estimated using Axiom 9.5. The capacity matrix C is normalized to derive these probabilities. This method assumes traffic distributes proportionally to link capacities. It reflects an idealized structural property of the network.

The relationship between capacity and turning probability is formalized by the *Generalized Proportional Capacity Model*. This model provides a flexible functional form to derive the Behavioral Stochastic Matrix, S, from the capacity matrix, C.

Proposition 9.3 (Generalized Proportional Capacity Model) *The probability s_{ij} of a driver at intersection node i choosing to travel to an adjacent node j is given by:*

$$s_{ij} = \frac{w_{ij}}{\sum_{k \in \mathcal{V}} w_{ik}} \tag{9.3}$$

where w_{ij} is the "attractiveness" or weight of out-link (i, j), defined as:

$$w_{ij} = c_{ij}^{\alpha} e^{\beta c_{ij}} \tag{9.4}$$

The parameters α and β are dimensionless calibration parameters that modulate the relationship between capacity and choice probability.

The two parameters allow the model to capture a range of behaviors:

- *The α parameter (Power Law):* This parameter models a proportional preference. When $\beta = 0$, a link with twice the capacity of another is considered 2^{α} times as attractive.
- *The β parameter (Exponential Law):* This parameter models a stronger, non-linear preference for high-capacity links, often reflecting a preference for highway-like facilities.

A crucial special case is the *simple proportional model*, where $\alpha = 1$ and $\beta = 0$. In this case, $w_{ij} = c_{ij}$, and the turning probability is simply the link's capacity divided by the total outgoing capacity from the node. This assumption is also often called *Markov Splitting* [17] or *Proportional Capacity*.

$$s_{ij} = \frac{c_{ij}}{\sum_{k \in \mathcal{V}} c_{ik}} \tag{9.5}$$

This represents the most direct and unbiased assumption: that flow distributes itself in direct proportion to the available capacity.

9.3.6 *Calibration of Parameters in Generalized Proportional Capacity Model*

The Generalized Capacity-Based Model provides significant flexibility, but its predictive power is only realized when the parameters (α, β) are chosen to reflect real-world driver behavior. *Calibration* is the rigorous, data-driven process of estimating the optimal values of these parameters. It is the crucial step that grounds the theoretical IFN model in empirical reality, tailoring it to the unique characteristics of a specific transportation network. The fundamental goal of calibration is to find the parameter values (α^*, β^*) that cause the model's predicted turning probabilities, $p_{ij}^{\text{model}}(\alpha, \beta)$, to most closely match the turning proportions observed in the field.

9.3.6.1 Data Requirements

The primary input for calibration is a set of *observed turning counts* from a representative sample of intersections. For each observed intersection i, we need the number of vehicles, N_{ij}, that were counted making the turn onto each outgoing link (i, j).

Example 9.1 consider 4-node network to illustrate the calibration challenge for Proportional Capacity Model. We will use observed counts from three intersections: Nodes 1, 2, and 4.

As the Table 9.1 shows, the simple proportional model ($\alpha = 1, \beta = 0$) provides a reasonable first guess but does not perfectly match the observed behavior.

Our task is to find a single pair of (α, β) that provides the best possible fit across *all* these observations simultaneously. This can be done either with Least Squares Error (LSE) Minimization or Maximum Likelihood Estimation (MLE).

Table 9.1 Observed turning counts and proportions

From Node	To Node	Capacity (c_{ij})	Observed Counts (N_{ij})	Observed Prop. ($p_{ij}^{\mathbf{obs}}$)	Simple Model ($p_{ij}^{\mathbf{model}}(\alpha = 1, \beta = 0)$)
1	2	1800	650	0.650	0.600
1	4	1200	350	0.350	0.400
Total		*3000*	*1000*	*1.000*	*1.000*
2	1	1800	800	0.400	~ 0.429
2	3	2400	1200	0.600	~ 0.571
Total		*4200*	*2000*	*1.000*	*1.000*
4	1	1200	500	0.250	~ 0.286
4	3	3000	1500	0.750	~ 0.714
Total		*4200*	*2000*	*1.000*	*1.000*

9.3.6.2 Least Squares Error (LSE) Minimization

Least Square [21] is a fundamental estimation technique minimizing the sum of squared errors. This method frames calibration as a problem of minimizing the geometric distance between the vector of observed proportions and the vector of model-predicted proportions. The objective is to find the parameters (α, β) that minimize the sum of the squared differences (residuals) between them. The *Least Squares Error (LSE)* objective function is:

$$LSE(\alpha, \beta) = \sqrt{\sum_{i \in K} \sum_{j \in \text{out}(i)} [p_{ij}^{\text{obs}} - p_{ij}^{\text{model}}(\alpha, \beta)]^2} \tag{9.6}$$

where K is the set of intersections with observed data. The optimal parameters (α^*, β^*) are those that minimize this function. This highly non-linear function must be solved using a numerical optimization algorithm.

9.3.6.3 Maximum Likelihood Estimation (MLE)

Maximum Likelihood Estimation (MLE) [6] is a powerful statistical method that seeks the parameters that make the observed counts the most probable outcome. The set of turning movements from a single intersection i can be modeled as a draw from a Multinomial distribution. The observed count of vehicles N_{ij} represent the number of vehicles making a specific turning movement from intersection i to a subsequent link j. The objective is to maximize the Log-Likelihood function:

$$LL(\alpha, \beta) = \sum_{i \in K} \sum_{j \in \text{out}(i)} N_{ij} \ln(p_{ij}^{\text{model}}(\alpha, \beta)) \tag{9.7}$$

This function is also solved using numerical optimization.

While LSE and MLE are different, for well-behaved problems they often yield very similar parameter estimates.

9.4 Performance Metrics

9.4.1 Ideal Flow Network to Congestion

Once the stochastic matrix, S, has been determined (either by assumption or calibration), the next step is to calculate the long-run, equilibrium distribution of activity across the network's nodes. This is the *stationary distribution vector*, $\boldsymbol{\pi}$, of the Markov chain defined by S.

The stationary distribution $\boldsymbol{\pi}$ is the unique positive left eigenvector of the stochastic matrix S corresponding to the eigenvalue of 1. It is the solution to the equilibrium equation is based on the fundamental steady-state equation (8.12).

$$\boldsymbol{\pi}^T \mathsf{S} = \boldsymbol{\pi}^T$$

subject to the normalization constraint $\sum_i \pi_i = 1$. The i-th component, π_i, represents the proportion of total network flow that passes through node i at steady state.

The existence and uniqueness of this vector for any strongly connected network (Axiom 9.1) is guaranteed by the Perron-Frobenius theorem [9, 20]. From the stationary distribution, we can define the relative ideal flow. The next step combines the node-level activity ($\boldsymbol{\pi}$) with the link-level choice probabilities (S) to produce the complete link flow pattern. The flow on any link (i, j) is simply the total flow passing through its origin node i, multiplied by the probability of choosing that specific link.

As we have discussed at length in Chap. 8, an Ideal Flow Network (IFN) is a model of a network in a state of perfect flow equilibrium. It is described by an $n \times n$ non-negative flow matrix F, where each element f_{ij} represents the volume of flow on the link from node i to node j. The matrix F must satisfy two key properties:

1. *Irreducibility:* The matrix F is irreducible, which is inherited from the strong connectivity of the network graph.
2. *Premagic Property [23]:* The total flow into any node equals the total flow out of that same node. This property ensures flow conservation at every node.

Bell and Iida [1] formalized flow conservation principles in transport networks, providing foundational algorithms for modeling node balance and route assignment in IFN systems.

The ideal flow matrix represents the most probable *relative flow distribution* given the network structure. Relative flow values must be converted to absolute vehicle volumes for quantitative traffic analysis. In the absence of empirical flow data, scaling can be performed using an assumed maximum congestion threshold (e.g., $\xi = 0.75$).

Definition 9.5 (Absolute Ideal Flow) Let ξ be the maximum allowable congestion (flow-to-capacity ratio) for any link in the network. The scaling factor, ς, is calculated as:

$$\varsigma = \frac{\xi}{\max_{i,j:c_{ij}>0}\left(\frac{f_{ij,\mathrm{rel}}}{c_{ij}}\right)} \tag{9.8}$$

The absolute ideal flow matrix F is then:

$$\mathsf{F}_{abs} = \varsigma \cdot \mathsf{F}_{rel} \tag{9.9}$$

The total flow in the network is $\kappa = \sum_{i,j} f_{ij}$.

The congestion on a link is defined as the ratio of the flow on that link to the link's maximum capacity.

Definition 9.6 (The Congestion Matrix (G)**)** The *Congestion Matrix* G is an $n \times n$ matrix where each element g_{ij} represents the congestion ratio (also known as the *volume-to-capacity ratio* or degree of saturation) on link (i, j). It is computed via the Hadamard (element-wise) division of the Ideal Flow Matrix F by the Capacity Matrix C:

$$g_{ij} = f_{ij}/c_{ij} \quad \text{for all links where } c_{ij} > 0 \tag{9.10}$$

In matrix notation, this is expressed as:

$$\mathsf{G} = \mathsf{F} \oslash \mathsf{C} \tag{9.11}$$

To preserve dimensional consistency, the computation of absolute flow and congestion requires the use of the absolute capacity matrix.

The congestion ratio g_{ij} is a critical performance indicator.

- *Dimensionless Measure:* As a pure ratio, it is dimensionless. This allows for direct comparison of the operational stress on different links, regardless of their absolute capacities.
- *Utilization or Stress:* The value of g_{ij} represents the degree to which a link's capacity is being utilized by the predicted demand flow. Its interpretation is standard in traffic engineering:
  - $g_{ij} < 1.0$ *(Under-saturation):* The predicted traffic demand is less than the link's maximum capacity. Flow is generally stable. For instance, $g_{ij} = 0.95$ is heavily utilized but still technically under-saturated.
  - $g_{ij} = 1.0$ *(Saturation):* The demand is exactly equal to the capacity. The link is operating at its theoretical maximum throughput.
  - $g_{ij} > 1.0$ *(Over-saturation):* The predicted demand exceeds the link's physical capacity. In the real world, this state is physically impossible to sustain. It implies that the demand arriving at the link is greater than the link's departure rate, resulting in the formation and growth of a vehicle queue. The cost-flow functions of BPR [5] and the modified Greenshields models are specifically designed to account for this state. The original Greenshields model [10] is not mathematically defined for this condition.

The Congestion Matrix, G, has several key mathematical properties.

- *Non-negativity:* The Ideal Flow f_{ij} and Capacity c_{ij} are non-negative by definition for all links. Therefore, it follows directly that $g_{ij} \geq 0$ for all $(i, j) \in \mathcal{E}$.
- *Domain:* The domain of g_{ij} is a crucial distinction between models.
  - For the *original Greenshields model* [10], the mathematical domain is $[0, 1]$, as the square root term $\sqrt{1-g}$ is undefined for $g > 1$.

  - For the *BPR and modified Greenshields models*, the domain of g_{ij} is $[0, \infty)$. These models are explicitly designed to handle values greater than 1, which represents a breakdown of stable flow conditions in physical reality.

- *Input for Cost Functions:* The most important role of the Congestion Matrix is to serve as the primary independent variable for the link cost functions that determine travel time and delay. The matrix G provides the standardized, dimensionless measure of this stress, forming the crucial link between the model's flow prediction and the real-world costs borne by travelers.

Example 9.2 (Capacity to Congestion) Consider a simple 3-node network with lane capacity shown in Fig. 9.3 illustrates the relative lane capacities assigned to each link in the 3-node network, forming the basis for computing absolute vehicular capacities.. The link capacities in number of lanes are given in the diagram. The relative capacity matrix in number of lanes is:

$$\mathsf{C}_{rel} = \begin{pmatrix} 0 & 3 & 0 \\ 1 & 0 & 2 \\ 0 & 2 & 0 \end{pmatrix}$$

Suppose per lane has capacity from Highway Capacity Manual [3] is 1800 vehicles per hour (vph). The actual absolute link capacities in vehicles per hour are as follow:

- $c_{ab} = 3 \times 1800 = 5400$ vph,
- $c_{ba} = 1 \times 1800 = 1800$ vph,
- $c_{bc} = 2 \times 1800 = 3600$ vph,
- $c_{cb} = 2 \times 1800 = 3600$ vph

Then the absolute capacity matrix in vehicle per hour is:

$$\mathsf{C}_{abs} = \begin{pmatrix} 0 & 5400 & 0 \\ 1800 & 0 & 3600 \\ 0 & 3600 & 0 \end{pmatrix}$$

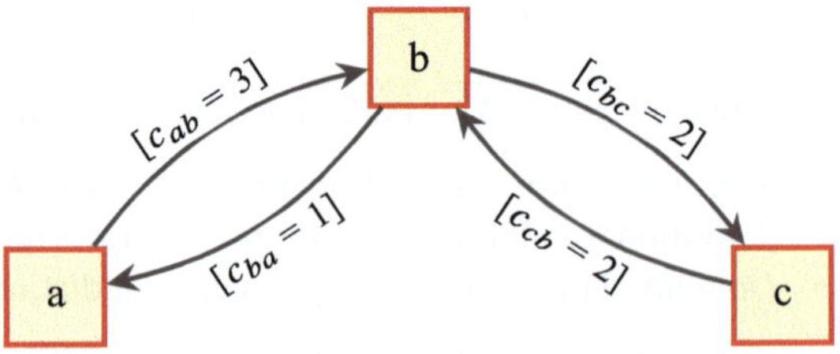

Fig. 9.3 A simple 3-node network with relative link capacities in number of lanes

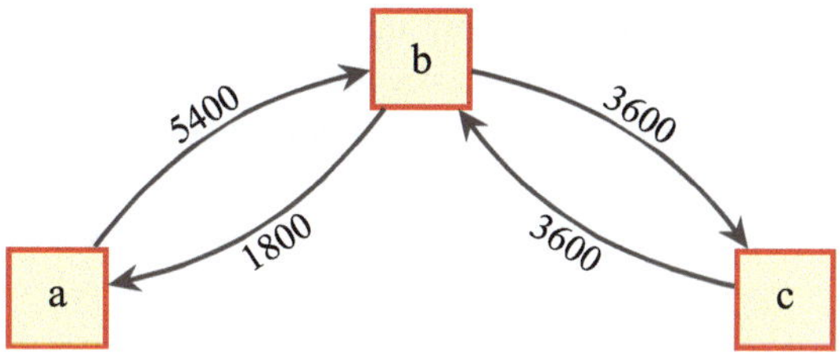

Fig. 9.4 A simple 3-node network with absolute link capacities in vehicles per hour (veh/hr)

Figure 9.4 presents the same 3-node network with link capacities expressed in vehicles per hour, calculated using a per-lane standard of 1800vph.

1. *Compute Stochastic Matrix* S*:* Since we use the same multiplier factor of capacity of 1800 vehicles per hour per lane, we can simplified to use lanes rather than the actual absolute capacity in using Eq. (9.5):
   - Node a: Total outgoing capacity = $c_{ab} = 3$ lanes. $s_{ab} = \frac{c_{ab}}{c_{ab}} = 1$.
   - Node b: Total outgoing capacity = $c_{ba} + c_{bc} = 1 + 2 = 3$ lanes. $s_{ba} = \frac{c_{ba}}{c_{ba}+c_{bc}} = \frac{1}{3}$, $s_{bc} = \frac{c_{bc}}{c_{ba}+c_{bc}} = \frac{2}{3}$.
   - Node c: Total outgoing capacity = $c_{cb} = 2$ lanes. $s_{cb} = \frac{c_{cb}}{c_{cb}} = 1$.

   The stochastic matrix is:

$$\mathsf{S} = \begin{pmatrix} 0 & 1 & 0 \\ \frac{1}{3} & 0 & \frac{2}{3} \\ 0 & 1 & 0 \end{pmatrix}$$

2. *Compute Stationary Distribution* $\boldsymbol{\pi}$*:* We solve $\boldsymbol{\pi}^\top = \boldsymbol{\pi}^\top \mathsf{S}$ with $\sum \pi_i = 1$. This gives the system: $\pi_a = \frac{1}{3}\pi_b$, $\pi_b = \pi_a + \pi_c$, $\pi_c = \frac{2}{3}\pi_b$. Solving this system yields the unique solution:

$$\boldsymbol{\pi}^\top = \left(\frac{1}{6}\ \frac{1}{2}\ \frac{1}{3}\right) = \frac{1}{6}\left(1\ 3\ 2\right)$$

3. *Compute Relative Flow* F_{rel}*:* Setting $\kappa = 6$ as our total flow, we can use Eq. (8.20), $f_{ij,\text{rel}} = \pi_i s_{ij}$ to get the integer relative flow matrix:

$$\mathsf{F}_{\text{rel}} = \begin{pmatrix} 0 & 1 & 0 \\ 1 & 0 & 2 \\ 0 & 2 & 0 \end{pmatrix}$$

   This matrix is premagic, as the row sums equal the column sums: $(1, 3, 2)$. This represents the most probable flow distribution given the network structure.

   To maintain the same unit, to get the absolute flow in veh/hr, we need to use the absolute capacity matrix in veh/hr.

The relative flow must be scaled to absolute vehicle volumes to be useful for traffic analysis. In the absence of real traffic flow data, we scale the flow based on assumed maximum acceptable level of congestion in the network (e.g., $\xi = 0.75$).

4. *Find Maximum Relative Congestion:* We compute $f_{ij,\text{rel}}/c_{ij}$ for each link:

   - Link ab: $\frac{1}{5400} = 0.000185185$
   - Link ba: $\frac{1}{1800} = 0.000555556$
   - Link bc: $\frac{2}{3600} = 0.000555556$
   - Link cb: $\frac{2}{3600} = 0.000555556$

   The maximum value is $\frac{2}{3600} = 0.000555556$ on links (ba, bc, cb).

   Let the maximum allowable congestion be $\xi = 0.75$.

5. *Compute Scaling Factor ς:* Using Eq. (9.8):

$$\varsigma = \frac{\xi}{\max\left(\frac{f_{ij,rel}}{c_{ij}}\right)} = \frac{0.75}{\frac{1}{1800}} = 1350$$

6. *Compute Absolute Flow F_{abs}:* Using Eq. (9.9), we multiply F_{rel} by ς:

$$\mathsf{F}_{abs} = 1350 \cdot \begin{pmatrix} 0 & 1 & 0 \\ 1 & 0 & 2 \\ 0 & 2 & 0 \end{pmatrix} = \begin{pmatrix} 0 & 1350 & 0 \\ 1350 & 0 & 2700 \\ 0 & 2700 & 0 \end{pmatrix} \text{ (veh/hr)}$$

   The total flow is $\kappa = 1350 + 1350 + 2700 + 2700 = 8150$ veh/hr.

7. *Compute Congestion* G*:* Using Eq. (9.12), we divide F_{abs} by C_{abs} via Hadamard Division:

   - $g_{ab} = 1350/5400 = 0.25$
   - $g_{ba} = 1350/1800 = 0.75$
   - $g_{bc} = 2700/3600 = 0.75$
   - $g_{cb} = 2700/3600 = 0.75$

   As designed by the maximum allowable congestion, the maximum congestion is indeed 0.75 on the most constrained links.

   The congestion matrix is:

$$\mathsf{G} = \begin{pmatrix} 0 & \frac{1}{4} & 0 \\ \frac{3}{4} & 0 & \frac{3}{4} \\ 0 & \frac{3}{4} & 0 \end{pmatrix}$$

9.4.2 Link Performance Metrics

Once we have the congestion and absolute flow, we can compute link performance metrics [14]. The first and most fundamental link performance metric is the congestion ratio itself. It quantifies the relationship between the predicted traffic flow and the physical capacity of the roadway. We represent this relationship using the Congestion Matrix, G.

Definition 9.7 (Performance Metrics) Given the absolute flow matrix F and capacity matrix C:

- *Congestion Matrix* G*:* The congestion on each link is its flow-to-capacity ratio (also called volume-to-capacity or v/c ratio).

$$g_{ij} = \frac{f_{ij}}{c_{ij}} \tag{9.12}$$

- *Speed Matrix* V*:* Link speed is a function of congestion. A common model is the Bureau of Public Roads (BPR) function [5]:

$$v_{ij} = \frac{u_{ij}}{1 + \gamma (g_{ij})^{\eta}} \tag{9.13}$$

  where u_{ij} is the free-flow speed, and γ and η are parameters (e.g., $\gamma = 0.15$, $\eta = 4$).
- *Travel Time Matrix* T*:* Travel time is link length l_{ij} divided by speed v_{ij}.
- *Delay Matrix* Y*:* Delay is the difference between the travel time on a congested link and its free-flow travel time. Free-flow travel time is the link length l_{ij} divided by the Free-flow (maximum) speed u_{ij}.

$$y_{ij} = t_{ij} - \frac{l_{ij}}{u_{ij}} = \frac{l_{ij}}{u_{ij}} \cdot \gamma (g_{ij})^{\eta} \tag{9.14}$$

9.4.3 Travel Time Functions: Converting Congestion to Cost

With the Congestion Matrix G formally defined, we can now model the functional relationship between network utilization and travel time. This step is critical as it translates the physical state of the network into a direct, tangible cost. The travel time on a link is not constant; it increases non-linearly as the flow on the link approaches its capacity. This section details the standard models used to capture this fundamental relationship.

Definition 9.8 (Travel Time Matrix (T)) The *Travel Time Matrix* T is an $n \times n$ matrix where t_{ij} is the actual (congested) travel time on link (i, j). It is a function of the link's free-flow speed, link distance and its congestion.

To analyze traffic performance, a mathematical model is required. This model must relate the flow f and the travel time, or cost, t. The travel time on a link is the link's cost. In this section we discuss two Cost-Flow models: The Greenshields and Bureau of Public Roads (BPR).

The relationship between flow, speed, and density forms the basis of traffic flow theory. Practical applications in traffic assignment typically rely on well-behaved, empirical functions that are computationally efficient.

9.4.3.1 Greenshields Model

The Greenshields model [10] is a fundamental linear model. It provides relationships for macroscopic variables of flow, speed, and density.

Definition 9.9 (Greenshields Model) The *Greenshields model* posits a linear relationship between traffic speed (v) and density (k). Speed is a function of density, denoted $v(k)$.

$$v(k) = u\left(1 - \frac{k}{\rho}\right) \tag{9.15}$$

In this equation, u is the *free-flow speed*, which is the speed at zero density. ρ is the *jam density*, which is the density at zero speed. The fundamental traffic flow relationship states that flow (f) is the product of speed and density:

$$f = kv \tag{9.16}$$

The flow is also a function of density, $f(k)$. By substituting Eq. (9.15) into Eq. (9.16), the flow-density relationship is established.

$$f(k) = kv(k) = u\left(k - \frac{k^2}{\rho}\right) \tag{9.17}$$

The maximum possible flow is the *capacity*, c. This capacity occurs at $k = \rho/2$ and $v = u/2$.

$$c = \frac{u\rho}{4} \tag{9.18}$$

The congestion, or flow-to-capacity ratio, is defined as $g = f/c$.

Proposition 9.4 (Speed from Flow) *For a given flow $f < c$ on a link, the corresponding speed v can be found by first solving for density k from Eq. (9.17), which is a quadratic equation in k: $k^2 - \rho k + \frac{\rho f}{u} = 0$. This yields two solutions for density:*

$$k = \frac{\rho}{2}\left(1 \pm \sqrt{1 - \frac{4f}{u\rho}}\right) = \frac{\rho}{2}\left(1 \pm \sqrt{1 - \frac{f}{c}}\right) = \frac{\rho}{2}\left(1 \pm \sqrt{1-g}\right) \tag{9.19}$$

The corresponding speeds are found using Eq. (9.15). The higher speed corresponds to the uncongested state, and the lower speed to the congested state.

For a given flow $f < c$, two distinct states are possible. One state is the "uncongested" branch, characterized by low density and high speed. The other state is the "congested" branch, defined by high density and low speed. This duality is critical for understanding network performance.

Proposition 9.5 (Greenshields Travel Time from Flow) *For a link of length l with a given flow $f < c$, the travel time t can be expressed as a function of the flow, $t(f)$. The Greenshields functional form is*

$$t(f) = \frac{2l}{u}\left(1 \pm \sqrt{1-g}\right)^{-1} \tag{9.20}$$

where the plus sign corresponds to the uncongested state and the minus sign corresponds to the congested state.

Proof By definition, travel time t is the link length l divided by the speed v.

$$t = \frac{l}{v}$$

From the Greenshields model, the speed-density relationship is $v = u(1 - k/\rho)$, which implies $v = u - uk/\rho$. The flow-density relationship is $f = kv$. From this, density can be expressed as $k = f/v$. Substituting this into the speed-density equation gives

$$v = u - \frac{u(f/v)}{\rho} \implies v = u - \frac{uf}{\rho v}$$

Multiplying by v and rearranging, a quadratic equation in v is obtained:

$$v^2 - uv + \frac{uf}{\rho} = 0$$

Solving for v using the quadratic formula yields

$$v = \frac{u \pm \sqrt{u^2 - 4(1)\left(\frac{uf}{\rho}\right)}}{2} = \frac{u}{2}\left(1 \pm \sqrt{1 - \frac{4f}{u\rho}}\right)$$

From Eq. (9.18), $c = u\rho/4$, which means $1/c = 4/(u\rho)$. The expression under the radical can be written as $1 - f/c$. Thus,

$$v = \frac{u}{2}\left(1 \pm \sqrt{1 - \frac{f}{c}}\right)$$

Finally, substituting this expression for v into the definition of travel time $t = l/v$ provides the proof:

$$t = \frac{l}{\frac{u}{2}\left(1 \pm \sqrt{1 - \frac{f}{c}}\right)} = \frac{2l}{u}\left(1 \pm \sqrt{1 - \frac{f}{c}}\right)^{-1}$$

The plus sign corresponds to the uncongested state (higher speed, lower time) and the minus sign to the congested state (lower speed, higher time). This concludes the proof. □

Example 9.3 (Cost-Flow Relationship Example) Consider a single lane link with length $l = 10$ km. The free-flow speed is $u = 60$ km/hr and the jam density is $\rho = 120$ veh/km. The link's capacity is $c = (60 \times 120)/4 = 1800$ veh/hr. Let us analyze the link's performance for an observed flow of $f = 1350$ veh/hr. The congestion ratio is $g = f/c = 1350/1800 = 0.75$.

Using Eq. (9.19), we find the two possible densities: $k = \frac{120}{2}(1 \pm \sqrt{1 - 0.75}) = 60(1 \pm 0.5)$.

- Uncongested density: $k_{unc} = 60(1 - 0.5) = 30$ veh/km.
- Congested density: $k_{con} = 60(1 + 0.5) = 90$ veh/km.

The corresponding speeds are calculated using Eq. (9.15):

- Uncongested speed: $v_{unc} = 60(1-30/120) = 45$ km/hr. (Check: $f = 30 \times 45 = 1350$).
- Congested speed: $v_{con} = 60(1 - 90/120) = 15$ km/hr. (Check: $f = 90 \times 15 = 1350$).

The same flow can thus occur under two very different operating conditions.

Using Proposition 9.5, we calculate the two possible travel times:

- Uncongested Travel Time:

$$t_{unc} = \frac{2 \times 10}{60}\left(1 + \sqrt{1 - 0.75}\right)^{-1} = \frac{1}{3}(1 + 0.5)^{-1} = \frac{1}{3} \times \frac{2}{3} = \frac{2}{9} \text{ hours}$$

This corresponds to $t_{unc} \approx 13.33$ minutes.
- Congested Travel Time:

$$t_{con} = \frac{2 \times 10}{60}\left(1 - \sqrt{1 - 0.75}\right)^{-1} = \frac{1}{3}(1 - 0.5)^{-1} = \frac{1}{3} \times 2 = \frac{2}{3} \text{ hours}$$

This corresponds to $t_{con} = 40$ minutes.

This example demonstrates that a single flow value can result in two distinct travel times, representing the uncongested and congested states of the network. This highlights the non-linear nature of cost in a traffic network.

The travel time on a link, denoted $t(g)$, is a function of the congestion ratio $g = f/c$. The Greenshields model provides two distinct expressions for travel time, corresponding to uncongested and congested states. The Greenshields cost-flow function for a link of length l with free-flow speed u and capacity c is given by $t(g) = \frac{2l}{u}\left(1 \pm \sqrt{1 - g}\right)^{-1}$ where the plus sign corresponds to the uncongested state and the minus sign to the congested state. We now prove that this function satisfies the well-behaved axiom only for the uncongested branch.

Proposition 9.6 (Greenshields Uncongested Cost-Flow Properties) *The Greenshields cost-flow function for the uncongested state, $t_{unc}(g)$, is continuous, monotonically increasing, and strictly convex with respect to the congestion ratio $g \in [0, 1)$.*

Proof The travel time function for the uncongested state is

$$t_{unc}(g) = \frac{2l}{u}\left(1 + \sqrt{1 - g}\right)^{-1}$$

1. *Continuity:* The function is a composition of elementary continuous functions: a constant, the square root function, addition, and the inverse function. The domain of the square root is restricted to $g \leq 1$, and the denominator $1 + \sqrt{1 - g}$ is non-zero for all $g \in [0, 1)$. Therefore, $t_{unc}(g)$ is a continuous function on the interval $[0, 1)$.

2. *Monotonically Increasing:* To prove that the function is monotonically increasing, we must show that its first derivative with respect to g is positive.

$$\begin{aligned}\frac{dt_{unc}}{dg} &= \frac{d}{dg}\left[\frac{2l}{u}\left(1+\sqrt{1-g}\right)^{-1}\right] \\ &= \frac{2l}{u}\cdot(-1)\left(1+\sqrt{1-g}\right)^{-2}\cdot\frac{d}{dg}\left(1+\sqrt{1-g}\right) \\ &= -\frac{2l}{u}\left(1+\sqrt{1-g}\right)^{-2}\cdot\left(\frac{1}{2\sqrt{1-g}}\cdot(-1)\right) \\ &= \frac{l}{u\sqrt{1-g}\left(1+\sqrt{1-g}\right)^2}\end{aligned}$$

For $g \in [0, 1)$, all terms in the expression are positive: $l > 0$, $u > 0$, $\sqrt{1-g} > 0$, and $(1+\sqrt{1-g})^2 > 0$. Thus, $\frac{dt_{unc}}{dg} > 0$, which proves that the function is monotonically increasing.

3. *Strict Convexity:* To prove strict convexity, we must show that the second derivative is positive.

$$\begin{aligned}\frac{d^2t_{unc}}{dg^2} &= \frac{d}{dg}\left[\frac{l}{u}\frac{1}{\sqrt{1-g}(1+\sqrt{1-g})^2}\right] \\ &= \frac{l}{u}\frac{d}{dg}\left[(1-g)^{-1/2}(1+(1-g)^{1/2})^{-2}\right]\end{aligned}$$

Let $y(g) = \sqrt{1-g}$. Then $t_{unc}(g) = \frac{2l}{u}(1+y)^{-1}$, and $\frac{dy}{dg} = -\frac{1}{2\sqrt{1-g}}$. The first derivative is $\frac{dt_{unc}}{dg} = \frac{l}{u}\frac{1}{\sqrt{1-g}(1+\sqrt{1-g})^2}$. By applying the quotient rule to the first derivative, it can be shown that

$$\frac{d^2t_{unc}}{dg^2} = \frac{l}{u}\frac{4\sqrt{1-g}-(3g-4)}{2(1-g)^{3/2}(1+\sqrt{1-g})^3}$$

For $g \in [0, 1)$, the denominator is always positive. The numerator is $4\sqrt{1-g} - (3g-4)$. For $g \in [0, 1)$, the term $(3g-4)$ is always negative, ranging from -4 to -1. The term $4\sqrt{1-g}$ is always positive, ranging from 4 to 0. The difference between a positive term and a negative term is always positive. Thus, the numerator is always positive. Therefore, $\frac{d^2t_{unc}}{dg^2} > 0$, which proves that the function is strictly convex.

The congested branch of the Greenshields model, which corresponds to the negative sign in Eq. (9.20), exhibits a non-increasing travel time function. This branch therefore does not satisfy the well-behaved axiom. □

9.4.3.2 Modified Greenshields Cost-Flow Function

A modification to the Greenshields speed formula is proposed, resulting in a piecewise function for speed v. This function is defined with respect to the congestion ratio $g = f/c$.

$$v = \begin{cases} \frac{1}{2}u(1 + \sqrt{1-g}) & \text{if } g \leq 1 \\ \frac{1}{2}u(1 - \sqrt{g-1}) & \text{if } g > 1 \end{cases} \tag{9.21}$$

From the definition of travel time, $t = l/v$, where l is the link length, the corresponding cost-flow function $t(g)$ is derived:

$$t(g) = \begin{cases} \frac{2l}{u}(1 + \sqrt{1-g})^{-1} & \text{if } g \leq 1 \\ \frac{2l}{u}(1 - \sqrt{g-1})^{-1} & \text{if } g > 1 \end{cases} \tag{9.22}$$

This example applies the modified Greenshields formula to the same traffic scenario. This modified function provides a single travel time for any given flow, unlike the original model.

Example 9.4 (Modified Greenshields Cost-Flow Relationship) Consider a single lane link with length $l = 10$ km. The free-flow speed is $u = 60$ km/hr and the jam density is $\rho = 120$ veh/km. The link's capacity is $c = (60 \times 120)/4 = 1800$ veh/hr. Let us analyze the link's performance for an observed flow of $f = 1350$ veh/hr. The congestion ratio is $g = f/c = 1350/1800 = 0.75$.

The modified Greenshields cost-flow function is given by

$$t(g) = \begin{cases} \frac{2l}{u}(1 + \sqrt{1-g})^{-1} & \text{if } g \leq 1 \\ \frac{2l}{u}(1 - \sqrt{g-1})^{-1} & \text{if } g > 1 \end{cases} \tag{9.23}$$

Since the congestion ratio is $g = 0.75$, which is less than or equal to 1, the upper branch of the formula is used. This corresponds to the uncongested state in the original Greenshields model. The travel time is calculated as follows:

$$t = \frac{2 \times 10}{60}\left(1 + \sqrt{1 - 0.75}\right)^{-1} = \frac{1}{3}(1 + 0.5)^{-1} = \frac{1}{3} \times \frac{2}{3} = \frac{2}{9} \text{ hours}$$

This corresponds to a single travel time of approximately 13.33 minutes. This example demonstrates that the modified formula produces a single, unambiguous travel time for a given flow, avoiding the dual-state issue of the original Greenshields model for flows below capacity.

This Modified Greenshields function is evaluated against the properties of the Well-Behaved Travel Time Axiom (9.6) in the following Proposition 9.7.

Proposition 9.7 (Properties of the Modified Greenshields Function) *The modified cost-flow function $t(g)$ is continuous and monotonically increasing for all $g \geq 0$. However, it is not strictly convex on this entire domain.*

Proof

1. *Continuity:* We must demonstrate that the two branches of the function connect smoothly at the transition point $g = 1$.

   - As g approaches 1 from below (uncongested branch):

   $$\lim_{g \to 1^-} t(g) = \lim_{g \to 1^-} \frac{2l}{u}(1 + \sqrt{1-g})^{-1} = \frac{2l}{u}(1 + \sqrt{1-1})^{-1} = \frac{2l}{u}$$

   - As g approaches 1 from above (congested branch):

   $$\lim_{g \to 1^+} t(g) = \lim_{g \to 1^+} \frac{2l}{u}(1 - \sqrt{g-1})^{-1} = \frac{2l}{u}(1 - \sqrt{1-1})^{-1} = \frac{2l}{u}$$

   Since the left-hand and right-hand limits are equal, the function is continuous at $g = 1$. It is also continuous on the open intervals $(0, 1)$ and $(1, \infty)$ because it is composed of elementary continuous functions. Therefore, the function is continuous for all $g \geq 0$.
2. *Monotonically Increasing:* We must show that the first derivative with respect to g is positive for all $g \geq 0$.

   - For the uncongested branch ($g < 1$), the first derivative is:

   $$\frac{dt}{dg} = \frac{l}{u\sqrt{1-g}(1 + \sqrt{1-g})^2}$$

   This expression is always positive for $g \in [0, 1)$.
   - For the congested branch ($g > 1$), the first derivative is:

   $$\frac{dt}{dg} = \frac{d}{dg}\left[\frac{2l}{u}(1 - \sqrt{g-1})^{-1}\right] = \frac{l}{u\sqrt{g-1}(1 - \sqrt{g-1})^2}$$

   This expression is also always positive for $g > 1$.

   Since the derivative is positive for all $g \geq 0$, the function is monotonically increasing.
3. *Strict Convexity:* We must demonstrate that the second derivative is positive.

   - For the uncongested branch ($g < 1$), the function is strictly convex, as previously proven.

- For the congested branch ($g > 1$), we analyze the second derivative. The behavior of the first derivative as g increases from 1 to infinity is a key indicator. At $g \to 1^+$, the derivative $\frac{dt}{dg} \to \infty$. As $g \to \infty$, the derivative $\frac{dt}{dg} \to 0$. Since the slope of the function decreases as g increases, the function is strictly concave on the interval $(1, \infty)$.

The function's strict convexity holds only on the interval $[0, 1)$. Because the function is strictly concave for $g > 1$, it fails to satisfy the strict convexity requirement for its entire domain. □

9.4.3.3 The Bureau of Public Roads (BPR) Function

The most widely used function for modeling congested travel time is the empirical formula developed by the U.S. Bureau of Public Roads (BPR) [5].

Definition 9.10 (BPR Travel Time Function) The actual travel time on a link, t, is a function of its free-flow travel time, its congestion ratio, and two calibration parameters, γ and η.

$$t = t^0 \left(1 + \gamma \left(\frac{f}{c}\right)^{\eta}\right) \tag{9.24}$$

Using congestion ratio $g = f/c$, this is expressed more cleanly as:

$$t = t^0(1 + \gamma(g)^{\eta}) \tag{9.25}$$

Parameter Interpretation:

- η *(eta):* This exponent controls the severity of the delay. Standard values in practice are often near 4.0.
- γ *(gamma):* This coefficient scales the function. A standard value is 0.15. It determines the travel time when the link is at capacity ($g = 1.0$), which will be $(1 + \gamma)$ times the free-flow travel time.

Yang and Bell [25] analyzed the sensitivity of BPR-based travel time functions in equilibrium models, providing calibration strategies for realistic network assignment.

Example 9.5 (BPR Cost-Flow Relationship) We use the same network parameters to calculate travel time with the BPR function. The BPR function [5] is defined as:

$$t = t^0(1 + \gamma(g)^{\eta}) \tag{9.26}$$

The input values are:

- Free-flow travel time, $t^0 = l/u = 10/60 = 1/6$ hours (10 minutes).
- Congestion ratio, $g = f/c = 0.75$.
- Calibration parameters, $\eta = 4$ and $\gamma = 0.15$.

The travel time is calculated as:

$$t = t^0(1 + \gamma(g)^\eta) = \frac{1}{6}\left[1 + 0.15(0.75)^4\right]$$

First, we evaluate the term for the congestion ratio: $(0.75)^4 \approx 0.3164$. Substituting this value back into the equation:

$$t = \frac{1}{6}[1 + 0.15(0.3164)] = \frac{1}{6}(1 + 0.04746) = \frac{1}{6}(1.04746)$$

This results in a travel time of $t \approx 0.17458$ hours. This corresponds to a single, unambiguous travel time of approximately 10.47 minutes.

Example 9.6 (Calculating Congested Travel Time) Let us use the standard BPR parameters, $\eta = 4$ and $\gamma = 0.15$. We can now calculate the actual travel times for our network using the congestion ratios 0.95 in link da and congestion 1.58 in link ab.

- *For the under-saturated link da:*
  - $t^0_{da} \approx 0.0096$ hours. $g_{da} = 0.95$.
  - $t_{da} = 0.0096 \times (1 + 0.15 \times (0.95)^4) \approx 0.0108$ hours (a 12.2% increaseas delay).
- *For the over-saturated link ab:*
  - $t^0_{ab} \approx 0.0086$ hours. $g_{12} \approx 1.58$.
  - $t_{ab} = 0.0086 \times (1 + 0.15 \times (1.58)^4) \approx 0.0166$ hours (a 93.6% increase as delay).

Applying this calculation to all links yields the complete Travel Time Matrix T (in hours).

Axiom 9.6 establishes the properties required for a travel time function to be considered well-behaved. It stated that for stable equilibrium modeling, the travel time function must be continuous, strictly increasing, and strictly convex with respect to congestion. The Bureau of Public Roads (BPR) function is a standard model that satisfies these conditions.

Proposition 9.8 (Properties of the BPR Function) *The BPR travel time function, given by $\tau(x) = t^0(1+\gamma x^\eta)$, is a continuous, strictly increasing, and strictly convex function of the congestion ratio $x = g_{ij}$, for the domain $x > 0$, assuming standard parameters ($t^0 > 0$, $\gamma > 0$, and $\eta > 1$).*

Proof The proof proceeds by verifying each of the three required properties.

1. *Continuity:* The function $\tau(x) = t^0 + t^0\gamma x^\eta$ is a polynomial function of the congestion ratio x. Polynomials are continuous functions over the set of all real numbers. Therefore, the BPR function is continuous on its domain of interest, which is $x > 0$.
2. *Strict Monotonicity:* To establish strict monotonicity, the first derivative with respect to the congestion ratio x must be shown to be strictly positive for $x > 0$.

$$\frac{d\tau}{dx} = \frac{d}{dx}[t^0(1 + \gamma x^\eta)] = t^0\gamma\eta x^{\eta-1}$$

By the assumption of standard parameters, $t^0 > 0$, $\gamma > 0$, and $\eta > 1$. For any $x > 0$, the term $x^{\eta-1}$ is also positive. The product of these positive terms is strictly positive. Thus, $\frac{d\tau}{dx} > 0$ for all $x > 0$. The function is strictly increasing.
3. *Strict Convexity:* To establish strict convexity, the second derivative with respect to x must be shown to be strictly positive for $x > 0$.

$$\frac{d^2\tau}{dx^2} = \frac{d}{dx}[t^0\gamma\eta x^{\eta-1}] = t^0\gamma\eta(\eta - 1)x^{\eta-2}$$

Under the assumption of standard parameters, t^0, γ, η are positive, and importantly, $\eta > 1$, which makes $(\eta - 1)$ positive. For any $x > 0$, the term $x^{\eta-2}$ is also positive. The product of these positive terms is strictly positive. Thus, $\frac{d^2\tau}{dx^2} > 0$ for all $x > 0$. The function is strictly convex.

The BPR function, when used with typical parameter values, rigorously satisfies all three mathematical properties required for a well-behaved cost function. □

9.4.3.4 Parameter Calibration for BPR Functions

While standard parameters ($\eta = 4, \gamma = 0.15$) are widely used, they are merely default values. For accurate, localized modeling, these parameters should be calibrated using empirical data for different types of roads.

Methodology: Calibration is a regression problem. The goal is to find the parameters (η, γ) that make the BPR function best fit a set of observed data points $(G_k^{\text{obs}}, T_k^{\text{obs}})$. This is achieved using a *non-linear least squares* algorithm to minimize the sum of squared errors between the observed travel times and the travel times predicted by the BPR function:

$$\min_{\eta,\gamma} \sum_k \left(t_k^{\text{obs}} - t^0(1 + \gamma \times (g_k^{\text{obs}})^\eta)\right)^2 \tag{9.27}$$

9.4.4 Delay

We begin by defining the free-flow travel time on a link, which occurs under zero-congestion conditions.

Definition 9.11 (Free-Flow Travel Time Matrix (T_0)) The *Free-Flow Travel Time Matrix* T_0 is an $n \times n$ matrix where each element t_{ij}^0 represents the theoretical minimum travel time on link (i, j). It is achieved when a vehicle can traverse the link at its maximum permissible speed, unimpeded by other traffic. It is calculated as the link's length divided by its maximum speed:

$$t_{ij}^0 = \frac{l_{ij}}{u_{ij}} \tag{9.28}$$

In matrix notation, this is computed via the Hadamard division:

$$\mathsf{T}_0 = \mathsf{L} \oslash \mathsf{U} \tag{9.29}$$

While the Travel Time Matrix T provides the total time cost for traversing a link, for performance evaluation and economic analysis, it is often more insightful to isolate the portion of time cost that is attributable solely to congestion. This additional time is termed *delay*.

Link delay is formally defined as the difference between the actual, congested travel time and the theoretical minimum (free-flow) travel time.

Definition 9.12 (The Delay Matrix (Y)) The *Delay Matrix* Y is an $n \times n$ matrix where each element y_{ij} represents the additional travel time on link (i, j) caused by congestion. It is calculated as the element-wise difference between the Travel Time Matrix T and the Free-Flow Travel Time Matrix T_0:

$$y_{ij} = t_{ij} - t_{ij}^0 \tag{9.30}$$

In matrix notation, this is a simple matrix subtraction:

$$\mathsf{Y} = \mathsf{T} - \mathsf{T}_0 \tag{9.31}$$

The Delay Matrix Y is a direct and powerful measure of network inefficiency.

- *Direct Cost of Congestion:* Delay isolates the penalty imposed on users by the presence of other vehicles. It is the most tangible measure of the user-perceived cost of congestion.
- *Performance Evaluation and Bottleneck Identification:* Transportation agencies use delay metrics to pinpoint network bottlenecks. High delay per vehicle, identifying it as a significant problem spot.

- *Economic Analysis:* Delay is the cornerstone of the economic evaluation of transport projects. By multiplying delay by an accepted *value of time (VOT)*, an economic measure of the monetary worth people place on their time, planners can quantify the total economic cost of congestion.

9.4.5 System-Wide Performance Metrics

While link-level delay is useful for diagnostics, aggregate measures are needed to assess the performance of the entire system.

Definition 9.13 (Total System Delay ($Z_{\mathbf{delay}}$)) The *Total System Delay*, often measured in *vehicle-hours of delay (VHD)*, is the sum of all individual vehicle delays across the entire network over a given period. It is calculated by multiplying the delay on each link by the number of vehicles traversing that link and summing the results.

$$Z_{\text{delay}} = \sum_{(i,j)\in E} f_{ij} \times y_{ij} \tag{9.32}$$

This is equivalent to summing all the elements of the Hadamard product of the Flow and Delay matrices:

$$Z_{\text{delay}} = \sum_{i,j} (\mathsf{F} \circ \mathsf{Y})_{ij} \tag{9.33}$$

Other useful system-wide metrics can be derived:

- *Total System Travel Time (Z_{T}):* Also known as Vehicle-Hours Traveled (VHT), this is the sum of all travel times for all vehicles in the network.

$$Z_{\mathrm{T}} = \sum_{(i,j)\in E} f_{ij} \cdot t_{ij} = \sum_{i,j} (\mathsf{F} \circ \mathsf{T})_{ij} \tag{9.34}$$

This can be decomposed into a free-flow component and a delay component:

$$Z_{\mathrm{T}} = \sum (\mathsf{F} \circ (\mathsf{T}_0 + \mathsf{Y})) = \sum (\mathsf{F} \circ \mathsf{T}_0) + \sum (\mathsf{F} \circ \mathsf{Y}) = Z_0 + Z_{\text{delay}} \tag{9.35}$$

where Z_0 is the total system travel time if all vehicles could travel at free-flow speeds.

- *Average Delay per Vehicle:* This provides a network-wide average of the delay experienced per vehicle per unit of flow. It is calculated as the total delay divided by the total flow:

$$\text{Average Delay} = \frac{Z_{\text{delay}}}{\sum_{(i,j)\in E} f_{ij}} = \frac{\sum(\mathsf{F} \circ \mathsf{Y})}{\kappa} \tag{9.36}$$

- *Vehicle-Kilometers Traveled (VKT):* This metric measures the total distance traveled by all vehicles in the network. It is the sum of the products of link flow and link length.

$$\text{VKT} = \sum_{(i,j)\in E} f_{ij} \cdot l_{ij} = \sum_{i,j} (\mathsf{F} \circ \mathsf{L})_{ij} \tag{9.37}$$

- *Average Speed:* The average speed across the network can be derived by dividing the total VKT by the total VHT (Z_{T}).

$$\text{Average Speed} = \frac{\text{VKT}}{Z_{\text{T}}} = \frac{\sum(\mathsf{F} \circ \mathsf{L})}{\sum(\mathsf{F} \circ \mathsf{T})} \tag{9.38}$$

- *Entropy of Flow Distribution:* Entropy is a measure of the dispersion or uniformity of flow across the network. A high entropy value indicates a more dispersed flow pattern, suggesting that traffic is distributed over many links, while a low entropy value suggests that flow is concentrated on a few links.

$$h(\mathsf{F}) = -\sum_{i,j} \frac{f_{ij}}{\kappa} \log\left(\frac{f_{ij}}{\kappa}\right) \tag{9.39}$$

  where κ is the total system flow. This metric provides insight into the efficiency and robustness of the network's traffic distribution.

The flow-weighted average speed, calculated as $\frac{\text{VKT}}{Z_{\text{T}}}$, is typically lower than the simple arithmetic average of link speeds. This discrepancy arises from the weighting mechanism. The flow-weighted average speed is not a simple average. It is a ratio of total distance traveled to total time spent. Links with high traffic volumes and high congestion tend to have lower speeds. These links contribute disproportionately to the denominator (Z_{T} or VHT), which represents the total time spent traveling.

Conversely, a simple average of link speeds treats all active links equally, regardless of their traffic volume. It does not account for the number of vehicles on each link. A high-speed, low-traffic link contributes as much to the average as a low-speed, high-traffic link. Since congested links with low speeds are often the ones carrying a high volume of vehicles, they heavily influence the flow-weighted average, pulling it downward. The flow-weighted average is a more accurate representation of the average experience of a vehicle in the network. It accounts for both the speed on each link and the number of vehicles experiencing that speed.

9.5 Mini Study

This section provides a comprehensive mini study of a 4-node and 9-link network, as illustrated in Fig. 9.5.

The diagram Fig. 9.5 shows the network's physical properties: link capacities in number of lanes, link lengths in kilometers, and free-flow speeds in km/hour.

Based on the diagram, the structural input matrices C_{rel}, L, and U are formed. The relative capacity matrix, in number of lanes, is:

$$\mathsf{C}_{\text{rel}} = \begin{pmatrix} 0 & 1 & 0 & 1 \\ 1 & 0 & 2 & 0 \\ 2 & 1 & 0 & 2 \\ 1 & 0 & 2 & 0 \end{pmatrix}$$

Assuming a standard capacity of 1800 vehicles/hour per lane, the absolute capacity matrix C is:

$$\mathsf{C} = 1800 \cdot \mathsf{C}_{\text{rel}} = \left(\begin{array}{c|cccc} \nearrow & a & b & c & d \\ \hline a & 0 & 1{,}800 & 0 & 1{,}800 \\ b & 1{,}800 & 0 & 3{,}600 & 0 \\ c & 3{,}600 & 1{,}800 & 0 & 3{,}600 \\ d & 1{,}800 & 0 & 3{,}600 & 0 \end{array}\right)$$

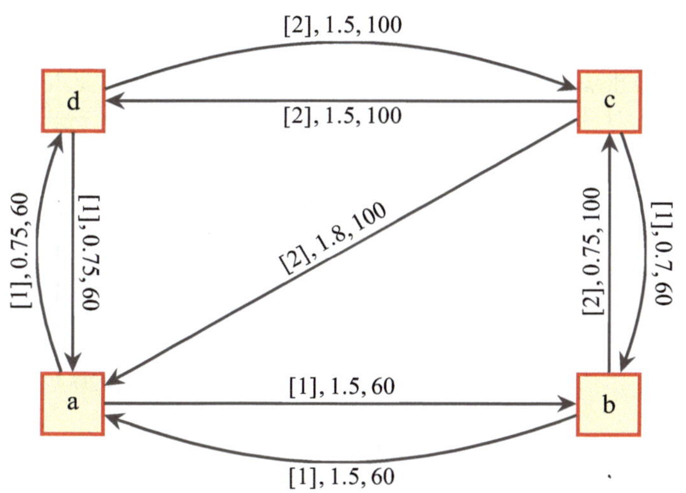

Fig. 9.5 A 4-node network with capacity (lanes), lengths (km), and free-flow speed (km/hr)

The link free-flow speed matrix U and link length matrix L are:

$$\mathsf{U} = \begin{pmatrix} 0 & 60 & 0 & 60 \\ 60 & 0 & 100 & 0 \\ 100 & 60 & 0 & 100 \\ 60 & 0 & 100 & 0 \end{pmatrix} \quad \text{and} \quad \mathsf{L} = \begin{pmatrix} 0 & 1.5 & 0 & 1.0 \\ 1.5 & 0 & 1.0 & 0 \\ 1.8 & 1.0 & 0 & 1.5 \\ 1.0 & 0 & 1.5 & 0 \end{pmatrix}$$

Based on relative capacity matrix C_{rel}, we use simple proportional capacity Eq. (9.5) to derive the stochastic transition probability matrix S. The stochastic transition probability matrix is:

$$\mathsf{S} = \begin{pmatrix} 0 & \frac{1}{2} & 0 & \frac{1}{2} \\ \frac{1}{3} & 0 & \frac{2}{3} & 0 \\ \frac{2}{5} & \frac{1}{5} & 0 & \frac{2}{5} \\ \frac{1}{3} & 0 & \frac{2}{3} & 0 \end{pmatrix}$$

From this, a cardinal (minimum integer) relative ideal flow matrix F_{rel} is obtained, with a total flow of 204 units.

$$\mathsf{F}_{rel} = \left(\begin{array}{c|cccc|c} \nearrow & a & b & c & d & \Sigma \\ \hline a & 0 & 27 & 0 & 27 & 54 \\ b & 13 & 0 & 26 & 0 & 39 \\ c & 24 & 12 & 0 & 24 & 60 \\ d & 17 & 0 & 34 & 0 & 51 \\ \hline \Sigma & 54 & 39 & 60 & 51 & 204 \end{array}\right)$$

Based on the calibration on the real flow (see Chap. 11) it was found that the optimum total flow is $\kappa = 21{,}420$ veh/hr. The ideal total flow is scaled to a realistic value of $\kappa = 21{,}420$ veh/hr, resulting in a scaling factor $\varsigma = \frac{21{,}420}{240} = 105$. The absolute Ideal Flow Matrix F is then:

$$\mathsf{F} = \varsigma \cdot \mathsf{F}_{rel} \approx \left(\begin{array}{c|cccc|c} \nearrow & a & b & c & d & \Sigma \\ \hline a & 0 & 2835 & 0 & 2835 & 5670 \\ b & 1365 & 0 & 2730 & 0 & 4095 \\ c & 2520 & 1260 & 0 & 2520 & 6300 \\ d & 1785 & 0 & 3570 & 0 & 5355 \\ \hline \Sigma & 5670 & 4095 & 6300 & 5355 & 21{,}420 \end{array}\right)$$

Note: The total flow here is $\kappa = 21{,}420$.

The Congestion Matrix G is calculated by the Hadamard division of F by C:

$$\mathsf{G} = \mathsf{F} \oslash \mathsf{C} \approx \left(\begin{array}{c|cccc} \nearrow & a & b & c & d \\ \hline a & 0 & 1.575 & 0 & 1.575 \\ b & 0.758 & 0 & 0.758 & 0 \\ c & 0.700 & 0.700 & 0 & 0.700 \\ d & 0.992 & 0 & 0.992 & 0 \end{array}\right)$$

This matrix shows that links like (a, b) and (a, d) are over-saturated ($g > 1$).

The Free-Flow Travel Time Matrix T_0 is calculated by the Hadamard division of L by U:

$$\mathsf{T}_0 = \mathsf{L} \oslash \mathsf{U} = \begin{pmatrix} 0 & 0.0250 & 0 & 0.0167 \\ 0.0250 & 0 & 0.0100 & 0 \\ 0.0180 & 0.0167 & 0 & 0.0150 \\ 0.0167 & 0 & 0.0150 & 0 \end{pmatrix} \text{(hours)}$$

Using the BPR function with $\eta = 4$ and $\gamma = 0.15$, the Travel Time Matrix T is

$$\mathsf{T} = \mathsf{T}_0 \circ (1 + 0.15 \circ \mathsf{G}^4) \approx \begin{pmatrix} 0 & 0.0481 & 0 & 0.0321 \\ 0.0262 & 0 & 0.0105 & 0 \\ 0.0186 & 0.0173 & 0 & 0.0155 \\ 0.0191 & 0 & 0.0172 & 0 \end{pmatrix} \text{(hours)}$$

The Link Speed matrix is computed based on Travel Time Matrix and Link Distance matrix

$$\mathsf{V} \approx \begin{pmatrix} 0 & 31.20 & 0 & 31.20 \\ 57.16 & 0 & 95.27 & 0 \\ 96.52 & 57.91 & 0 & 96.52 \\ 52.40 & 0 & 87.33 & 0 \end{pmatrix} \text{(km/hours)}$$

The average link speed for the network is ≈ 67.28 km/hour.

The Delay Matrix $\mathsf{Y} = \mathsf{T} - \mathsf{T}_0$, in hours, is:

$$\mathsf{Y} \approx \begin{pmatrix} 0 & 0.0231 & 0 & 0.0154 \\ 0.0012 & 0 & 0.0005 & 0 \\ 0.0006 & 0.0006 & 0 & 0.0005 \\ 0.0024 & 0 & 0.0022 & 0 \end{pmatrix} \text{(hours)}$$

For link (a, b), a vehicle experiences a delay of approximately 0.0231 hours (1.38 minutes), which is a significant increase $\approx$ (92.3%) over its free-flow time of 1.5 minutes.

We can now compute the system-wide metrics to assess the network's overall performance.

- *Total System Delay (Z_{delay}):* This is the sum of all elements of the Hadamard product $\mathsf{F} \circ \mathsf{Y}$.

$$Z_{\text{delay}} = \sum_{i,j} (\mathsf{F} \circ \mathsf{Y})_{ij} \approx 127.91 \text{ vehicle-hours of delay per hour}$$

  The network generates approximately 128 hours of "wasted" time due to congestion every hour.
- *Total System Travel Time (Z_{T}):* The Vehicle-Hours Traveled (VHT) or Total System Travel is the sum of all elements of the Hadamard product $\mathsf{F} \circ \mathsf{T}$.

$$Z_{\mathrm{T}} = \sum_{i,j} (\mathsf{F} \circ \mathsf{T})_{ij} \approx 494.92 \text{ vehicle-hours traveled per hour}$$

  Vehicles in the network collectively spend 495 hours traveling over a one-hour period.
- *Vehicle-Kilometers Traveled (VKT):* This is the sum of all elements of the Hadamard product $\mathsf{F} \circ \mathsf{L}$.

$$\text{VKT} = \sum_{i,j} (\mathsf{F} \circ \mathsf{L})_{ij} \approx 28{,}581 \text{ vehicle-kilometers traveled per hour}$$

  The total distance covered by all vehicles in the network is 28,581 km in one hour.
- *Average Speed (Flow-Weighted):* The average speed is the total VKT divided by the total VHT (Z_{T}).

$$\text{Average Speed} = \frac{\text{VKT}}{Z_{\mathrm{T}}} = \frac{28{,}581}{494.92} \approx 57.75 \text{ km/hr}$$

  The average speed across the entire network, considering both traffic volume and distance, is approximately 57.75 km/hr. This is a robust and accurate representation of the system's performance. In general, the flow-weighted average speed is smaller than simple arithmetic average of average link speed for the network which is approximately 67.28 km/hour.
- *Entropy of Flow Distribution:* The entropy measures how uniformly flow is distributed across the network.

$$h(\mathsf{F}) = -\sum_{i,j} \frac{f_{ij}}{\kappa} \log\left(\frac{f_{ij}}{\kappa}\right) \approx 3.10\ (97.81\% \text{ of max entropy})$$

The entropy value of 3.10 suggests a high degree of flow dispersion, indicating that traffic is not overly concentrated on a few "bottleneck" links, even with the presence of over-saturated links. This is a useful metric for network resilience. Alternatively, we can also use average node entropy, which will yields $\approx$ 0.775 (95.34% of max entropy)

9.6 Summary

The Ideal Traffic Assignment (ITA) framework provides a complete, self-contained method for generating a physically consistent (premagic) and behaviorally plausible traffic flow matrix from the network's structural properties alone. This is modulated by a small number of observable parameters. All derived metrics, including ideal flow, congestion, speed, travel time, and delay, are functions of the network's foundational structural inputs: the capacity matrix C, link length matrix L, and free-flow speed matrix U. These matrices are inherently constrained by the network's underlying topology, as defined by the adjacency matrix A. The framework demonstrates a deterministic sequence of operations that transforms these physical properties into comprehensive performance metrics.

References

1. Michael G.H. Bell and Yasunori Iida. *Transportation Network Analysis*. John Wiley & Sons, 1997. https://doi.org/10.1002/9781118782548
2. Norman Biggs. *Algebraic Graph Theory*. 2nd. Cambridge University Press, 1993. https://doi.org/10.1017/CBO9780511608704
3. Transportation Research Board, Engineering National Academies of Sciences, and Medicine. *Highway Capacity Manual 7th Edition: A Guide for Multimodal Mobility Analysis*. Washington, DC: The National Academies Press, 2022. ISBN: 978-0-309-08766-7. https://doi.org/10.17226/26432. https://nap.nationalacademies.org/catalog/26432/highway-capacity-manual-7th-edition-a-guide-for-multimodal-mobility
4. Stephen Boyd and Lieven Vandenberghe. *Convex Optimization*. Cambridge University Press, 2004. https://doi.org/10.1017/CBO9780511804441
5. Bureau of Public Roads. *Traffic Assignment Manual*. Tech. rep. U.S. Department of Commerce, Urban Planning Division, 1964.
6. George Casella and Roger Berger. *Statistical Inference*. 2nd ed. Chapman and Hall/CRC, 2024. https://doi.org/10.1201/9781003456285
7. Gilles Duranton and Matthew A. Turner. "The Fundamental Law of Road Congestion: Evidence from US Cities". In: *American Economic Review* 101.6 (2011), pp. 2616–2652. https://doi.org/10.1257/aer.101.6.2616
8. Marguerite Frank and Philip Wolfe. "An algorithm for quadratic programming". In: *Naval Research Logistics Quarterly* 3.1–2 (1956), pp. 95–110. https://doi.org/10.1002/nav.3800030109
9. G. Frobenius. "Ueber Matrizen aus nicht negativen Elementen". In: *Sitzungs-berichte der Königlich Preußischen Akademie der Wissenschaften* (1912), pp. 456–477.

10. B. D. Greenshields. "A study of traffic capacity". In: *Highway Research Board Proceedings* 14 (1935), pp. 448–477.
11. Edwin T. Jaynes. "Information Theory and Statistical Mechanics". In: *Physical Review* 106.4 (1957), pp. 620–630. https://doi.org/10.1103/PhysRev.106.620
12. John G. Kemeny and J. Laurie Snell. *Finite Markov Chains*. New York: Springer, 1976. https://link.springer.com/book/9780387901923
13. Hani S. Mahmassani and Srinivas Peeta. "System Optimal Dynamic Assignment for Electronic Route Guidance in a Congested Traffic Network". In: *Urban Traffic Networks*. Ed. by Nathan H. Gartner and Gennaro Improta. Berlin, Heidelberg: Springer, 1995, pp. 3–37. ISBN: 9783642796418. https://doi.org/10.1007/978-3-642-79641-8_1. https://link.springer.com/chapter/10.1007/978-3-642-79641-8_1
14. P. Van Mieghem. *Performance Analysis of Complex Networks and Systems*. Cambridge University Press, 2014. https://doi.org/10.1017/CBO9781107415874
15. J. F. Nash. "Non-Cooperative Games". In: *Annals of Mathematics* 54.2 (1951), pp. 286–295. https://doi.org/10.2307/1969529
16. John F. Nash. "Equilibrium Points in N-Person Games". In: *Proceedings of the National Academy of Sciences* 36.1 (1950), pp. 48–49. https://doi.org/10.1073/pnas.36.1.48
17. E. Nummelin. "A Splitting Technique for Harris Recurrent Markov Chains". In: *Zeitschrift für Wahrscheinlichkeitstheorie und verwandte Gebiete* 43.4 (1978), pp. 309–318. https://doi.org/10.1007/BF00534764
18. Juan de Dios Ortúzar and Luis G. Willumsen. *Modelling Transport*. 4th. Chichester: John Wiley & Sons, 2011. https://doi.org/10.1002/9781119993308
19. M. Papageorgiou et al. "Review of road traffic control strategies". In: *Proceedings of the IEEE* 91.12 (2003), pp. 2043–2067. https://doi.org/10.1109/JPROC.2003.819610
20. Oskar Perron. "Zur Theorie der Matrices". In: *Mathematische Annalen* 64.2 (1907), pp. 248–263. https://doi.org/10.1007/BF01449896
21. George A. F. Seber and Alan J. Lee. *Linear Regression Analysis*. 2nd ed. Wiley, 2003. ISBN: 9780471415404
22. K. Teknomo. "Ideal Flow Based on Random Walk on Directed Graph". In: *Proceedings of the 9th International Collaboration Symposium on Information, Production and Systems*. Kitakyushu: Waseda University, 2015.
23. Kardi Teknomo. "Premagic and Ideal Flow Matrices". In: *arXiv preprint arXiv:1706.08856* (2017). https://arxiv.org/pdf/1706.08856
24. J. G. Wardrop. "Some theoretical aspects of road traffic research". In: *Proceedings of the Institution of Civil Engineers* 1.3 (1952), pp. 325–362. https://doi.org/10.1680/iicep.1952.11362. https://www.researchgate.net/publication/227555985_Wardrop_Equilibria
25. Hai Yang and Michael G.H. Bell. "Models and algorithms for road network design: a review and some new developments". In: *Transport Reviews* 18.3 (1998), pp. 257–278. https://doi.org/10.1080/01441649808717016

Chapter 10
Functional Analysis in Network Models

10.1 The Foundations of Functional Analysis

This section introduces the functional forms and properties essential for network analysis It provides the core conceptual toolkit for understanding how variables within a network are interconnected.

In functional analysis [3, 8], always ask "what inputs does this quantity actually depend on," and then write it that way. Our purpose is to make what depends on what" explicit, using functions that map inputs to outputs along the modeling pipeline. By the end, reading the chapter would enable interpretation and construction of expressions such as $\mathsf{S}(\mathsf{C}; \alpha, \beta)$ and $\mathsf{F}(\mathsf{S}; \kappa)$, their composition into $\mathsf{F}(\mathsf{C}; \alpha, \beta, \kappa)$, and the application of Jensen's inequality to formally explain why balanced loading minimizes total travel time.

Functional notation [4] is a mathematical tool that reveals a system's dependencies. We write an *output* as a function of its inputs with parameters separated by a semicolon, for example $\mathsf{S}(\mathsf{C}; \alpha, \beta)$. When we want to stress the *function* itself, we may write $\mathsf{S} = f(\mathsf{C}, \alpha, \beta)$ or $\mathsf{S} = \mathsf{S}_{\alpha,\beta}(\mathsf{C})$. Both are correct, and we will choose the version that makes the dependency clearest in each context. Think of $\mathsf{S}_{\alpha,\beta}(\cdot)$ as the machine, and $\mathsf{S}(\mathsf{C}; \alpha, \beta)$ as what comes out of the machine for a particular input. We should be explicit about which context are *states* that change across scenarios and which are *parameters* that are fixed within a scenario. That is because, depending on the context, in functional form, we can also hide the parameters to focus on the actual input and output. Thus, we can write $\mathsf{S}(\mathsf{C}; \alpha, \beta) = \mathsf{S}(\mathsf{C})$.

Example 10.1 (Markov Splitting) Hadamard division $\mathsf{X} \oslash \mathsf{Y}$ is elementwise division with strictly positive denominators. Matrix products describe how splits propagate flow. Normalization operators make rows or columns sum to one when we build S. One of the applications of Hadamard division is Markov Splitting [6].

K. Teknomo, *Ideal Flow Network for Transportation Research*,
Lecture Notes in Intelligent Transportation and Infrastructure,
https://doi.org/10.1007/978-981-95-4727-2_10

Based on Markov Splitting or Proportional Capacity, we will have this formula in *scalar form*. When we care about any per-link capacity, we write c_{ij}.

$$s_{ij} = \frac{c_{ij}}{\sum_j c_{ij}} \tag{10.1}$$

In *matrix form*, it can be writen as Hadamard division.

$$\mathsf{S} = \mathsf{C} \oslash ((\mathbf{Cj})\mathbf{j}^T) \tag{10.2}$$

In *functional form*, we can write it as a very simple notation, abstracting away the functional details.

$$\mathsf{S}(\mathsf{C}) \tag{10.3}$$

10.2 Objects and Spaces for the Dependency Graph

Let $\mathcal{A}$ be the set of directed links, and let $|\mathcal{A}| = n$. The *capacity matrix* $\mathsf{C} \in \mathbb{R}_+^{n\times n}$ stores the feasible throughput per link in consistent units. The *stochastic transition probability matrix* is $\mathsf{S} \in [0, 1]^{n\times n}$, and it encodes how relative flow leaving one link splits to downstream links under rules tuned by (α, β). The *ideal flow matrix* is $\mathsf{F} \in \mathbb{R}_+^{n\times n}$, and it represents the link-wise flow throughputs implied by S and a scale parameter κ. The *congestion matrix* is $\mathsf{G} \in \mathbb{R}_+^{n\times n}$, and it is the elementwise flow relative to capacity. The *speed-limit matrix* is $\mathsf{U} \in \mathbb{R}_+^{n\times n}$, and the *length matrix* is $\mathsf{L} \in \mathbb{R}_+^{n}$. The *travel-time matrix* is $\mathsf{T} \in \mathbb{R}_+^{n\times n}$, and the *operating-speed matrix* is $\mathsf{V} \in \mathbb{R}_+^{n\times n}$.

10.3 Simple Properties of Functional Form

The functional forms employed in network analysis exhibit several core properties. These properties are fundamental for rigorous model validation and analysis.

10.3.1 Monotonicity

Definition 10.1 (Monotone Increase) A functional Φ is monotone increasing in an argument X if $\mathsf{X} \preceq \mathsf{X}'$ implies $\Phi(\ldots, \mathsf{X}, \ldots) \preceq \Phi(\ldots, \mathsf{X}', \ldots)$ elementwise.

If we turn a knob up, the output should not go down.

Proposition 10.1 (Monotonicity of Delay in Congestion) *If* $\mathsf{T} = \mathsf{T}(\mathsf{G}, \mathsf{U}, \mathsf{L})$ *is defined entrywise by a scalar travel time map that is increasing in its first argument, then* $\mathsf{G} \preceq \mathsf{G}'$ *implies* $\mathsf{T}(\mathsf{G}, \mathsf{U}, \mathsf{L}) \preceq \mathsf{T}(\mathsf{G}', \mathsf{U}, \mathsf{L})$.

Proof The claim holds entrywise by the scalar monotonicity and therefore holds elementwise for the matrix. □

Example 10.2 (Monotonicity of T **in** G**)**

We have functional form $\mathsf{T}(\mathsf{G}, \mathsf{U}, \mathsf{L})$.

If congestion rises everywhere from G to $\mathsf{G}' \succeq \mathsf{G}$ and U, L are fixed, then any reasonable $\mathsf{T}(\mathsf{G}, \mathsf{U}, \mathsf{L})$ will increase elementwise.

Let $\mathsf{G} = \begin{pmatrix} 0 & 0.6 \\ 0.8 & 0 \end{pmatrix}$ and $\mathsf{G}' = \begin{pmatrix} 0 & 0.7 \\ 0.9 & 0 \end{pmatrix}$ with $\mathsf{U} = \begin{pmatrix} 0 & 60 \\ 60 & 0 \end{pmatrix}$ and $\mathsf{L} = \begin{pmatrix} 0 & 1 \\ 1 & 0 \end{pmatrix}$.

Assume an entrywise BPR-lite form $\mathsf{T} = \mathsf{L} \oslash \mathsf{U} \circ \left(1 + \gamma\, \mathsf{G}^{\circ\eta}\right)$ with $\gamma, \eta > 0$ and elementwise power $\circ$.

Then $\mathsf{G} \preceq \mathsf{G}'$ gives $\mathsf{T}(\mathsf{G}, \mathsf{U}, \mathsf{L}) \preceq \mathsf{T}(\mathsf{G}', \mathsf{U}, \mathsf{L})$ entrywise. Assuming U is in km/hour and L in km and using the usual $\gamma = 0.15, \eta = 4$ we have travel time matrix $\mathsf{T}(\mathsf{G}, \mathsf{U}, \mathsf{L}) = \begin{pmatrix} 0 & 61.17 \\ 63.69 & 0 \end{pmatrix}$ and $\mathsf{T}(\mathsf{G}', \mathsf{U}, \mathsf{L}) = \begin{pmatrix} 0 & 62.16 \\ 65.90 & 0 \end{pmatrix}$ in seconds, which is larger as stated by *monotonicity property*.

Example 10.3 (Monotonicity with Heterogeneous Lengths)

Consider a link length matrix

$$\mathsf{L} = \begin{pmatrix} 0 & 2 \\ 1 & 0 \end{pmatrix}$$

which assigns different lengths in km to different links in the network.

Let the uniform free flow speed matrix be

$$\mathsf{U} = 60\mathsf{J}$$

where J is a matrix of ones (i.e., every link has the same max-speed of 60 km/hour).

Suppose a perturbed congestion matrix G' is slightly larger than a baseline G, satisfying

$$\mathsf{G} \preceq \mathsf{G}' + 0.1\mathsf{J}$$

meaning each entry in G is less than or equal to the corresponding entry in G' plus 0.1.

Now apply any travel time function that increases with flow (i.e., is monotonic entrywise like BPR or Greenshields). Using $\mathsf{G} = \begin{pmatrix} 0 & 0.6 \\ 0.8 & 0 \end{pmatrix}$, we have $\mathsf{G}' = \begin{pmatrix} 0 & 0.7 \\ 0.9 & 0 \end{pmatrix}$. Using BPR function with the usual parameters $\gamma = 0.15, \eta = 4$, we have travel time matrix $\mathsf{T}(\mathsf{G}, \mathsf{U}, \mathsf{L}) = \begin{pmatrix} 0 & 122.33 \\ 63.69 & 0 \end{pmatrix}$ and $\mathsf{T}(\mathsf{G}', \mathsf{U}, \mathsf{L}) = \begin{pmatrix} 0 & 124.32 \\ 65.90 & 0 \end{pmatrix}$ in seconds. As the result, the travel time matrix T computed under G' will be larger at every entry than under G, even though the link lengths in L scale the travel times differently across the network. Observe in this example, only one of the inputs, which is congestion matrix G changes to G', while both other inputs U, L remain the same.

10.3.2 Homogeneity and Invariance

Definition 10.2 (Positive Homogeneity) A functional Φ is *positively homogeneous of degree k* in X if $\Phi(\ldots, \lambda\mathsf{X}, \ldots) = \lambda^k \Phi(\ldots, \mathsf{X}, \ldots)$ for all $\lambda > 0$.

If we scale an input, the output scales in a predictable way.

Proposition 10.2 (Steady Ratio Under Proportional Scaling) *If* F *scales with* C *by the same positive factor on its support, then* $\mathsf{G} = \mathsf{F} \oslash \mathsf{C}$ *is invariant on that support.*

Proof For any used entry (i, j) we have $\mathsf{G}_{ij} = (\lambda\mathsf{F}_{ij})/(\lambda\mathsf{C}_{ij}) = \mathsf{F}_{ij}/\mathsf{C}_{ij}$. □

Definition 10.3 (Proportional and Inversely Proportional Forms)
Proportionality means $\mathsf{F} = \sigma\, \mathsf{C}$ for some scalar $\sigma > 0$ on the support of use.
Inverse proportionality means a quantity is of the form $\mathsf{Q} = \gamma\, \mathsf{C}^{\oslash 1}$ entrywise for some $\gamma > 0$, where $\oslash 1$ denotes the elementwise reciprocal.

Doubling both the total flow and capacity leaves the congestion unchanged.

Example 10.4 (Uniform Expansion at a Balanced Operating Point)
Let $\mathsf{C} = \begin{pmatrix} 0 & 10 \\ 10 & 0 \end{pmatrix}$ and $\mathsf{F} = \begin{pmatrix} 0 & 6 \\ 6 & 0 \end{pmatrix}$, so $\mathsf{G} = \begin{pmatrix} 0 & 0.6 \\ 0.6 & 0 \end{pmatrix}$.
Scale capacities and flows by $\lambda = 1.5$ to $\mathsf{C}' = 1.5 \times \mathsf{C}$ and $\mathsf{F}' = 1.5 \times \mathsf{F}$.
Then $\mathsf{G}' = \mathsf{F}' \oslash \mathsf{C}' = \mathsf{G}$ by Proposition 10.2.

Example 10.5 (Uniform Expansion at a Balanced Operating Point)
Let $\mathsf{C} = \begin{pmatrix} 0 & 1 & 2 \\ 2 & 0 & 4 \\ 1 & 3 & 0 \end{pmatrix}$ and $\mathsf{F} = \begin{pmatrix} 0 & 1 & 2 \\ 2 & 0 & 3 \\ 1 & 4 & 0 \end{pmatrix}$ which is irreducible premagic. So $\mathsf{G} = \begin{pmatrix} 0 & 1 & 1 \\ 1 & 0 & \frac{3}{4} \\ 1 & \frac{4}{3} & 0 \end{pmatrix}$. Expand uniformly to $\mathsf{C}' = 2 \times \mathsf{C} = \begin{pmatrix} 0 & 2 & 4 \\ 4 & 0 & 8 \\ 2 & 6 & 0 \end{pmatrix}$ and preserve

proportionality in operations to $\mathsf{F}' = 2 \times \mathsf{F} = \begin{pmatrix} 0 & 2 & 4 \\ 4 & 0 & 6 \\ 2 & 8 & 0 \end{pmatrix}$, and the congestion remains $\mathsf{G}' = \mathsf{G} = \begin{pmatrix} 0 & 1 & 1 \\ 1 & 0 & \frac{3}{4} \\ 1 & \frac{4}{3} & 0 \end{pmatrix}$.

Proportional means "a constant times," and inverse proportional means "a constant divided by." A *proportional form* describes a linear, scalar relationship where one variable is a constant multiple of another.

On the other hand, *inverse proportionality* describes a relationship where one variable is a constant divided by another.

Inverse proportionality means that as one variable increases, the other decreases in such a way that their product remains constant. Mathematically, if variable A is inversely proportional to variable B, then:

$$A = \frac{k}{B}$$

where k is a constant.

In the context of transportation, this often applies to travel time and speed. If a vehicle travels a fixed distance, then:

$$\text{Travel Time} = \frac{\text{Distance}}{\text{Speed}}$$

Here, travel time is inversely proportional to speed. As speed increases, travel time decreases, and vice versa, assuming the distance stays the same.

In the context of ideal flow network, the proportionality is a core property: the ideal flow matrix F is directly proportional to the stochastic transition probability matrix S and total flow in the network κ. This is expressed as $\mathsf{F}(\mathsf{S}; \kappa)$ where κ is a constant scalar.

10.3.3 Symmetry

Definition 10.4 (Permutation Invariance) A functional Φ is *permutation invariant* across a set of equal links if for any permutation matrix P acting on those indices, we have $\Phi(\mathsf{P}\mathsf{X}\mathsf{P}^{\top}) = \mathsf{P}\,\Phi(\mathsf{X})\,\mathsf{P}^{\top}$.

If two links are identical, the model should not invent a difference between them. This is the mathematical backbone for why grids behave predictably.

Proposition 10.3 (Equivariance Under Permutations [7]) *If* $\mathsf{S}(\mathsf{C})$ *is built by symmetric rules on identical outgoing links, then for any permutation* P *of those links,* $\mathsf{S}(\mathsf{PCP}^\top) = \mathsf{P}\,\mathsf{S}(\mathsf{C})\,\mathsf{P}^\top$.

Proof Identical links receive identical treatment in the construction of S, so permuting inputs just permutes outputs in the same way. □

Example 10.6 (Permutation Invariance of a Symmetric Stochastic Matrix)

We index the three outgoing links in the order $\{self, branch1, branch2\}$. Consider the row-stochastic transition probability matrix

$$\mathsf{S} = \begin{pmatrix} 0 & \frac{1}{2} & \frac{1}{2} \\ 1 & 0 & 0 \\ 1 & 0 & 0 \end{pmatrix},$$

where the first row splits flow from *self* equally to the two identical branches, and each branch deterministically returns to *self.*
Let the permutation matrix that swaps the two branches be

$$\mathsf{P} = \begin{pmatrix} 1 & 0 & 0 \\ 0 & 0 & 1 \\ 0 & 1 & 0 \end{pmatrix}, \quad \text{so that it leaves } \textit{self} \text{ fixed and exchanges branch 1 with branch 2.}$$

Left-multiplication by P permutes rows, and right-multiplication by $\mathsf{P}^\top$ permutes columns, so the conjugation $\mathsf{P}\,\mathsf{S}\,\mathsf{P}^\top$ applies the same swap to both indices.
A direct calculation gives

$$\mathsf{P}\,\mathsf{S}\,\mathsf{P}^\top = \begin{pmatrix} 0 & \frac{1}{2} & \frac{1}{2} \\ 1 & 0 & 0 \\ 1 & 0 & 0 \end{pmatrix} = \mathsf{S},$$

because rows 2 and 3 are identical and the first row has equal halves, so swapping columns 2 and 3 leaves it unchanged.
Therefore the stochastic structure is unchanged apart from relabeling, which is precisely permutation invariance on identical links.

In constrast, if now we use asymmetrical stochastic matrix

$$\mathsf{S}' = \begin{pmatrix} 0 & \frac{1}{4} & \frac{3}{4} \\ 1 & 0 & 0 \\ 1 & 0 & 0 \end{pmatrix},$$

then the same calculation gives

$$\mathsf{P}\,\mathsf{S}'\,\mathsf{P}^\top = \begin{pmatrix} 0 & \frac{3}{4} & \frac{1}{4} \\ 1 & 0 & 0 \\ 1 & 0 & 0 \end{pmatrix} \neq \mathsf{S}',$$

10.3.4 Continuity

Definition 10.5 (Continuity) A functional Φ is *continuous* if for every $\varepsilon > 0$ there exists $\delta > 0$ such that $\|\mathsf{X} - \mathsf{X}'\|_F < \delta$ implies $\|\Phi(\mathsf{X}) - \Phi(\mathsf{X}')\|_F < \varepsilon$, where $\|\cdot\|_F$ is the Frobenius norm.

The Frobenius norm [1] is a matrix norm used in optimization and error analysis, particularly in least squares and SVD contexts (see Sect. 3.7.1.2).

When data wiggle, outputs should not jump.

Proposition 10.4 (Entrywise Continuity Lifts to Matrix Continuity) *If* $\mathsf{T}(\mathsf{G}, \mathsf{U}, \mathsf{L})$ *is obtained by applying a scalar continuous map entrywise, then* T *is continuous in* G *(with* U, L *fixed).*

Proof Continuity holds at each entry and therefore for the matrix as a whole in any entrywise norm, including the Frobenius norm. □

Example 10.7 (Small Perturbations of Congestion Produce Small Time Changes)

Let $\mathsf{G}' = \mathsf{G} + 0.001\mathsf{J}$ and let $\mathsf{T} = \mathsf{L} \oslash \mathsf{U} \circ \left(1 + \gamma\, \mathsf{G}^{\circ\eta}\right)$.

By Proposition 10.4, $\|\mathsf{T}(\mathsf{G}', \mathsf{U}, \mathsf{L}) - \mathsf{T}(\mathsf{G}, \mathsf{U}, \mathsf{L})\|_F$ shrinks to zero with the 0.01 perturbation. To make it concrete, consider a link length matrix

$$\mathsf{L} = \begin{pmatrix} 0 & 2 \\ 1 & 0 \end{pmatrix}$$

which assigns different lengths in km to different links in the network.

Let the uniform free flow speed matrix be

$$\mathsf{U} = 60\mathsf{J}$$

where J is a matrix of ones (i.e., every link has the same max-speed of 60 km/hour).

Suppose a perturbed congestion matrix G' is slightly larger than a baseline G, satisfying

$$\mathsf{G}' = \mathsf{G} + 0.001\mathsf{J}$$

Using $\mathsf{G} = \begin{pmatrix} 0 & 0.6 \\ 0.8 & 0 \end{pmatrix}$, we have $\mathsf{G}' = \begin{pmatrix} 0 & 0.601 \\ 0.801 & 0 \end{pmatrix}$ and then using BPR function we using the usual parameters $\gamma = 0.15, \eta = 4$ we have travel time matrix $\mathsf{T}(\mathsf{G}, \mathsf{U}, \mathsf{L}) = \begin{pmatrix} 0 & 122.33 \\ 63.69 & 0 \end{pmatrix}$ and $\mathsf{T}'(\mathsf{G}', \mathsf{U}, \mathsf{L}) = \begin{pmatrix} 0 & 122.35 \\ 63.70 & 0 \end{pmatrix}$ in seconds. We can see the difference $\|\mathsf{T} - \mathsf{T}'\|_F < \delta = 0.5$ seconds, which is very small and it would even become smaller if we make the perturbation smaller.

10.3.5 Closure and Other Properties

Definition 10.6 (Closure of the Nonnegative Cone) Let $\mathcal{M}_+$ be the set of entrywise nonnegative matrices of a fixed size. We say $\mathcal{M}_+$ is *closed under an operation* if applying the operation to members of $\mathcal{M}_+$ yields a matrix still in $\mathcal{M}_+$.

Proposition 10.5 (Basic Closure) *$\mathcal{M}_+$ is closed under addition and Hadamard product, and it is closed under Hadamard division when denominators are strictly positive entrywise.*

Proof Sums and elementwise products of nonnegative entries remain nonnegative, and division by strictly positive entries preserves nonnegativity. □

Closure properties are foundational to ensuring that a series of operations remains within a defined set. In IFN analysis, for example, the matrices must remain non-negative, as negative flow or capacity is physically meaningless. The set of non-negative matrices is closed under addition.

Additionally, while other properties such as *transitivity* may apply to relationships within the model, it is important to note that properties like *commutativity* do not generally hold for matrix operations in this context.

If flows and capacities are nonnegative, then so are their sums, their elementwise products, and their load ratios as long as capacities are positive where flows are positive.

Example 10.8 (Closure Checks with IFN Quantities) If $\mathsf{F}, \mathsf{C} \in \mathcal{M}_+$ and C is strictly positive on the support of F, then $\mathsf{G} = \mathsf{F} \oslash \mathsf{C} \in \mathcal{M}_+$. If two scenarios have $\mathsf{F}^{(1)}$ and $\mathsf{F}^{(2)}$, then their sum is a feasible nonnegative matrix for a combined demand case.

$$\begin{pmatrix} a & b \\ c & d \end{pmatrix} \times \begin{pmatrix} e & f \\ g & h \end{pmatrix} \neq \begin{pmatrix} e & f \\ g & h \end{pmatrix} \times \begin{pmatrix} a & b \\ c & d \end{pmatrix} \tag{10.4}$$

Proposition 10.6 (Matrix Multiplication Is Not Commutative) *In general,* $\mathsf{AB} \neq \mathsf{BA}$ *for matrices of the same size.*

Proof Counterexample: $\mathsf{A} = \begin{pmatrix} 0 & 1 \\ 2 & 0 \end{pmatrix}$ and $\mathsf{B} = \begin{pmatrix} 1 & 0 \\ 0 & 2 \end{pmatrix}$ give $\mathsf{AB} = \begin{pmatrix} 0 & 2 \\ 2 & 0 \end{pmatrix}$ and $\mathsf{BA} = \begin{pmatrix} 0 & 1 \\ 4 & 0 \end{pmatrix}$, which are different. □

10.3.6 Convexity and Jensen's Inequality

Definition 10.7 (Convex Function and Jensen's Inequality) [3, 8]
A scalar map f is convex if $f(\lambda x + (1-\lambda)y) \le \lambda f(x) + (1-\lambda) f(y)$ for x, y and $\lambda \in [0, 1]$. If $w_a \ge 0$ with $\sum_a w_a = 1$, then $f\big(\sum_a w_a x_a\big) \le \sum_a w_a f(x_a)$. This can be seen in Fig. 10.1.

For convex travel time curves, spreading the same total load more evenly reduces the total travel time. This one line powers many of our later results about balance.

Proposition 10.7 (Entrywise Jensen for Matrices) *Let f be convex and applied entrywise to matrices. For $0 \le \lambda \le 1$ and matrices X, Y of the same size, we have $f\big(\lambda \mathsf{X} + (1-\lambda)\mathsf{Y}\big) \preceq \lambda f(\mathsf{X}) + (1-\lambda) f(\mathsf{Y})$ elementwise.*

Proof Apply scalar Jensen at each entry. □

Corollary 10.1 (Even Congestions Reduce the Sum of Convex Travel Times) *Let $\mathsf{G}^{(1)}$ and $\mathsf{G}^{(2)}$ be two congestion matrices with the same entrywise average, and let travel time function be f entrywise convex. Then the matrix $\mathsf{G}^{\text{avg}} = \frac{1}{2}(\mathsf{G}^{(1)} + \mathsf{G}^{(2)})$ yields elementwise $f(\mathsf{G}^{\text{avg}}) \preceq \frac{1}{2}\big(f(\mathsf{G}^{(1)}) + f(\mathsf{G}^{(2)})\big)$ and hence a smaller or equal sum of travel times.*

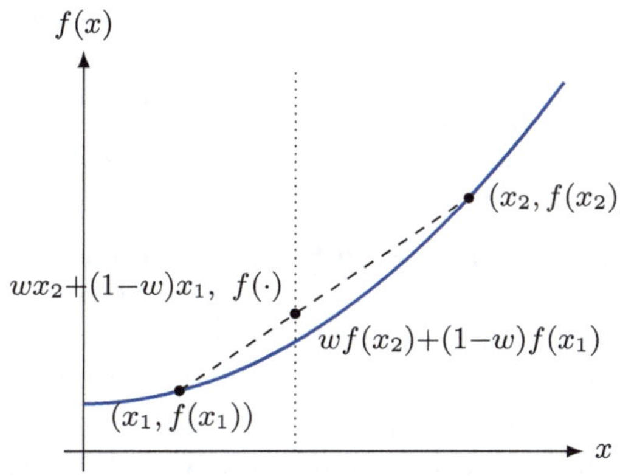

For convex f, the value at the weighted average of inputs lies *below* the weighted average of outputs.

Fig. 10.1 A geometric illustration of Jensen's inequality for a convex function

Proof Use Proposition 10.7 with $\lambda = \frac{1}{2}$ and then sum all entries. □

Example 10.9 (Two-Link Comparison at Fixed Average Load)

Let $\mathsf{G}^{(A)} = \begin{pmatrix} 0 & 0.8 \\ 0.8 & 0 \end{pmatrix}$ and $\mathsf{G}^{(B)} = \begin{pmatrix} 0 & 0.4 \\ 1.2 & 0 \end{pmatrix}$ so both have the same average over used entries. With function $f(x) = 1 + \gamma x^{\eta}$ we get $f(\mathsf{G}^{\text{avg}}) \preceq \frac{1}{2}\left(f(\mathsf{G}^{(A)}) + f(\mathsf{G}^{(B)})\right)$ entrywise and a smaller total sum for case (A).

To make this a concrete example, let us use a link length matrix

$$\mathsf{L} = \begin{pmatrix} 0 & 2 \\ 1 & 0 \end{pmatrix}$$

which assigns different lengths in km to different links in the network. Let the uniform free flow speed matrix be

$$\mathsf{U} = 60\mathsf{J}$$

where J is a matrix of ones (i.e., every link has the same max-speed of 60 km/hour).

$$\mathsf{G}^{\text{avg}} = \frac{1}{2}\left(\begin{pmatrix} 0 & 0.8 \\ 0.8 & 0 \end{pmatrix} + \begin{pmatrix} 0 & 0.4 \\ 1.2 & 0 \end{pmatrix}\right) = \begin{pmatrix} 0 & 0.6 \\ 1.0 & 0 \end{pmatrix}$$

BPR function with the usual parameters $\gamma = 0.15, \eta = 4$ we have travel time matrix $\mathsf{T}^{(A)}(\mathsf{G}^{(A)}, \mathsf{U}, \mathsf{L}) = \begin{pmatrix} 0 & 127.37 \\ 63.69 & 0 \end{pmatrix}$ and $\mathsf{T}^{(B)}(\mathsf{G}^{(B)}, \mathsf{U}, \mathsf{L}) = \begin{pmatrix} 0 & 120.46 \\ 78.66 & 0 \end{pmatrix}$ in seconds. The weighted average output is $\frac{1}{2}\left(f(\mathsf{G}^{(A)}) + f(\mathsf{G}^{(B)})\right) = \frac{1}{2}(\mathsf{T}^{(A)} + \mathsf{T}^{(B)}) = \begin{pmatrix} 0 & 123.92 \\ 71.17 & 0 \end{pmatrix}$. The function of average input $f(\mathsf{G}^{\text{avg}}) = \mathsf{T}^{\text{avg}} = \begin{pmatrix} 0 & 122.33 \\ 69.00 & 0 \end{pmatrix}$. Observe that in all entries, the function of average input is always smaller than the weighted average output $f(\mathsf{G}^{\text{avg}}) \preceq \frac{1}{2}\left(f(\mathsf{G}^{(A)}) + f(\mathsf{G}^{(B)})\right)$.

Example 10.10 (Balanced Grid Flavor Beats Skewed Corridor)

Compare $\mathsf{G}^{\text{bal}} = \begin{pmatrix} 0.8 & 0.8 & 0.8 \\ 0.8 & 0.8 & 0.8 \\ 0.8 & 0.8 & 0.8 \end{pmatrix}$ with $\mathsf{G}^{\text{skew}} = \begin{pmatrix} 1.2 & 1.0 & 0.6 \\ 0.6 & 1.0 & 0.2 \\ 0.6 & 0.2 & 1.2 \end{pmatrix}$, which have equal means over the nine entries. For convex f, Jensen gives $f(\mathsf{G}^{\text{bal}}) \preceq \frac{1}{2}\left(f(\mathsf{G}^{\text{bal}}) + f(\mathsf{G}^{\text{skew}})\right)$ entrywise and therefore a strictly smaller total travel time for the balanced case unless f is linear on the interval.

10.4 Functional Analysis of the Ideal Traffic Assignment Framework

This section establishes a functional dependency between the network's structural input matrices and the framework's core performance metrics [5]. We will demonstrate that all derived matrices, including ideal flow, congestion, speed, travel time, and delay, are functions of the initial structural matrices, which are inherently constrained by the network's underlying topology.

First, let use defining each component of the pipeline of Ideal Traffic Assignment (ITA) framework: S, F, G, T, V. We give minimalist definitions that are easy to use.

10.4.1 Stochastic Matrix

Definition 10.8 (Stochastic Functional) The stochastic matrix is $\mathsf{S} = \mathsf{S}(\mathsf{C}; \alpha, \beta)$, and it maps network structure and tuning parameters to normalized routing probabilities at nodes and diverges.

S turns geometry and rules into shares of where flow goes next.

Example 10.11 (Two Exits, Symmetric Choice) At a node with two identical outgoing links, a symmetric rule gives row $(0.5, 0.5)$ of S for that node, independent of the absolute scale of C.

10.4.2 Ideal Flow

Definition 10.9 (Flow Functional) The ideal flow is $\mathsf{F} = \mathsf{F}(\mathsf{S}; \kappa)$, and it aggregates the stochastic rules through the network under conservation, scaled by network total flow κ to match the analysis period.

F is how much ends up on each link" when the stochastic rules apply everywhere.

Lemma 10.1 (Composition Rule) $\mathsf{F}(\mathsf{C}; \alpha, \beta, \kappa) = \mathsf{F}\big(\mathsf{S}(\mathsf{C}; \alpha, \beta); \kappa\big)$.

Proof S depends only on $(\mathsf{C}; \alpha, \beta)$, and F depends only on $(\mathsf{S}; \kappa)$, so substitution gives the claim. □

We may write $\mathsf{F}(\mathsf{S}, \kappa)$ or $\mathsf{F}(\mathsf{C}, \alpha, \beta, \kappa)$, and they are the same by composition.

Example 10.12 (Y-Split With Rejoin) Figure 10.2 shows a source link splits equally to two middle links and then rejoins to a sink. Conservation implies the source and sink carry twice the flow of each middle link, and symmetry keeps the middle pair equal.

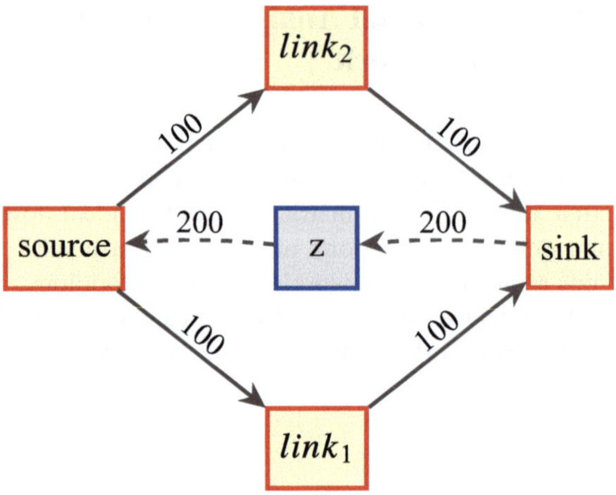

Fig. 10.2 A network with flow splitting and rejoining, connected to an external node

10.4.3 Congestion

Definition 10.10 (Congestion Functional) Congestion is $\mathsf{G} = \mathsf{G}(\mathsf{F}, \mathsf{C}) = \mathsf{F} \oslash \mathsf{C}$ on the support where $\mathsf{C} > 0$.

G is the fraction of capacity we are using, link by link.

Corollary 10.2 (Steady Congestion When F Is Proportional To C) *If* $\mathsf{F} = \lambda\mathsf{C}$ *on the actively used links, then* G *is constant on those links.*

Proof Apply Proposition 10.2 componentwise. □

10.4.4 Travel Time and Speed

Definition 10.11 (Travel Time Functional) Travel time is $\mathsf{T} = \mathsf{T}(\mathsf{G}, \mathsf{U}, \mathsf{L})$, and it increases with G, decreases with U, and is proportional in L up to units.

Definition 10.12 (Speed Functional) Operating speed is $\mathsf{V} = \mathsf{V}(\mathsf{L}, \mathsf{T})$, and with consistent units we usually compute $\mathsf{V} = \mathsf{L} \oslash \mathsf{T}$.

More crowding means more time, higher limits help, longer links take longer, and speed is just distance divided by time.

Example 10.13 (BPR-Lite Sensitivity) A classic sensitivity analysis: the rate of change of travel time with respect to flow is governed by the first derivative, and the result is scaled by the size of the perturbation (10%).

Given:

$$T_a = T_a^0 \left(1 + \gamma G_a^\eta\right)$$

To estimate how T_a changes with a small change in G_a, we compute:

$$\frac{dT_a}{dG_a} = T_a^0 \cdot \gamma \cdot \eta \cdot G_a^{\eta-1}$$

So, a 10% increase in G_a (i.e., $\Delta G_a / G_a = 0.1$) leads to an approximate change in T_a of:

$$\Delta T_a \approx \frac{dT_a}{dG_a} \cdot \Delta G_a = T_a^0 \cdot \gamma \cdot \eta \cdot G_a^{\eta-1} \cdot \Delta G_a$$

Divide both sides by T_a^0 to express the relative change:

$$\frac{\Delta T_a}{T_a^0} \approx \gamma \eta G_a^{\eta-1} \cdot \Delta G_a$$

Since $\Delta G_a / G_a = 10\%$, this becomes:

$$\text{Relative increase in } T_a \approx \gamma \eta G_a^{\eta-1} \times 10\%$$

Thus, if $T_a = T_a^0 (1 + \gamma G_a^\eta)$ with $T_a^0 = L_a / U_a$, then a 10% increase in G_a raises T_a by roughly $\gamma \eta G_a^{\eta-1} \times 10\%$ when G_a is not negligible.

10.4.5 The Dependency DAG

In the Ideal Flow Network (IFN) model, the stochastic matrix S is computed from the capacity matrix C and scalar parameters α and β. We represent this relationship as $\mathsf{S}(\mathsf{C}; \alpha, \beta)$. This notation indicates that S is a function of these specific inputs. Similarly, the ideal flow matrix F is computed from the stochastic matrix S and the total flow parameter κ, which can be written as $\mathsf{F}(\mathsf{S}; \kappa)$.

This framework allows us to trace a full chain of dependencies within the network as shown in Fig. 10.3. The congestion matrix G, for example, is the Hadamard division of the ideal flow matrix F by the capacity matrix C. This can be written as $\mathsf{G}(\mathsf{F}, \mathsf{C})$. Since F is itself a function of C, α, β, and κ, it follows that G can also be expressed as a function of these core parameters: $\mathsf{G}(\mathsf{C}; \alpha, \beta, \kappa)$. This hierarchical dependency extends to all key matrices. The travel time matrix T, which depends on G, link length L, and maximum speed U, is ultimately a function of the core inputs.

The pipeline is

$$\mathsf{S}(\mathsf{C}; \alpha, \beta) \rightarrow \mathsf{F}(\mathsf{S}; \kappa) \rightarrow \mathsf{G}(\mathsf{F}, \mathsf{C}) \rightarrow \mathsf{T}(\mathsf{G}, \mathsf{U}, \mathsf{L}) \rightarrow \mathsf{V}(\mathsf{L}, \mathsf{T}).$$

By composition we may also write $\mathsf{F}(\mathsf{C}; \alpha, \beta, \kappa) = \mathsf{F}\big(\mathsf{S}(\mathsf{C}; \alpha, \beta); \kappa\big)$.

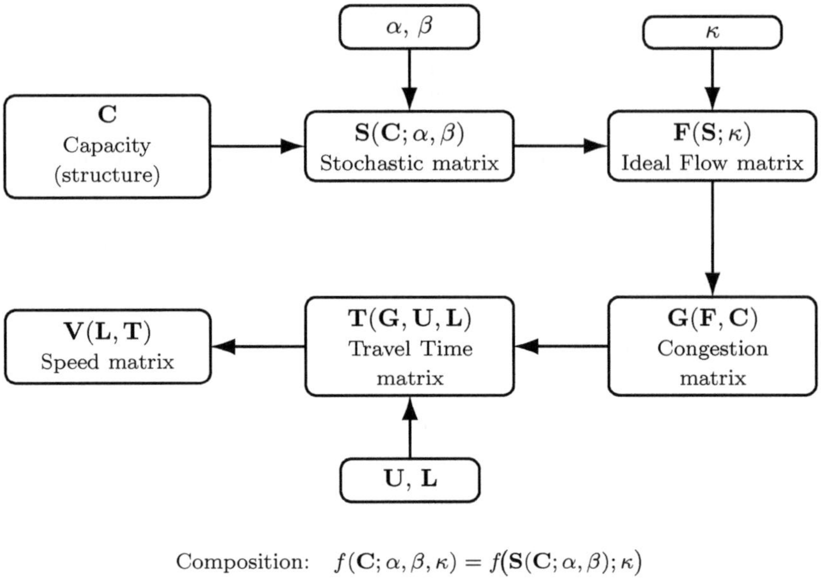

Fig. 10.3 A directed acyclic graph (DAG) illustrating the functional pipeline for network analysis

10.4.6 Functional Relationships and Hierarchical Dependencies

The functional dependencies of the Ideal Flow Network (IFN) matrices can be traced hierarchically. First, the capacity matrix C and link lengths L are determined by the network's adjacency matrix, A. Free-flow speeds U are an inherent property of the links, depending on their capacity. The stochastic matrix S is a function of capacity and tuning parameters, $\mathsf{S}(\mathsf{C}; \alpha, \beta)$. The ideal flow matrix F is then computed from S and the total flow κ, yielding $\mathsf{F}(\mathsf{S}; \kappa)$. By composition, this can be expressed as $\mathsf{F}(\mathsf{C}; \alpha, \beta, \kappa)$. This leads to the congestion matrix G, a function of both F and C, which ultimately depends on $\mathsf{C}, \alpha, \beta, \kappa$. The travel time matrix T depends on $\mathsf{G}, \mathsf{U}, \mathsf{L}$, which functionally connects T to the core inputs $\mathsf{C}, \alpha, \beta, \kappa$. This full chain culminates in the operating speed V, which is also a function of these core parameters.

Proposition 10.8 (Stochastic Matrix as a Function of Capacity) *The stochastic matrix* S *is a function of the capacity matrix* C*. This relationship is established through a row-normalization process.*

Proof The stochastic matrix S represents the transition probabilities of an Ideal Flow Network (IFN), where each entry s_{ij} is the probability of flow transitioning from node i to node j. The capacity matrix C provides the maximum flow rate c_{ij} for each link. A fundamental premise of the IFN as stated in Axiom 9.4 is that flow is distributed proportionally to a link's capacity. Therefore, the transition probability s_{ij} is defined as the ratio of the capacity of link (i, j) to the total outflow capacity from node i. This is given by the functional form $\mathsf{S} = \mathsf{S}(\mathsf{C})$. This expression demonstrates that each element of S is derived directly from the elements of C. Consequently, the entire matrix S is a function of the matrix C, which can be denoted as $\mathsf{S}(\mathsf{C})$. □

Theorem 10.1 (Ideal Flow and Performance Matrices as Functions of Structural Inputs) *The Ideal Flow Matrix* F *and all subsequent performance matrices, including congestion* G*, speed* V*, travel time* T*, and delay* Y*, are ultimately defined as functions of the network's foundational structural inputs: the capacity matrix* C*, link length matrix* L*, and free-flow speed matrix* U*.*

Proof By establishing a chain of functional dependencies that begins with the structural matrices and concludes with the performance metrics, we can prove that the entire Ideal Traffic Assignment framework is intrinsically based on the network's defined structure.

- Ideal Flow as a Function of Capacity:
  - The relative ideal flow matrix F_{rel} is derived from the stationary distribution $\boldsymbol{\pi}$ of the Markov chain defined by S.
  - From Proposition 10.8, we established $\mathsf{S} = \mathsf{S}(\mathsf{C})$.
  - Therefore, the relative flow is a function of the stochastic matrix, $\mathsf{F}_{rel} = \mathsf{F}_{rel}(\mathsf{S})$.
  - By substitution, $\mathsf{F}_{rel} = \mathsf{F}_{rel}(\mathsf{S}(\mathsf{C}))$. This shows that the relative flow is a function of the capacity matrix.
  - The absolute flow matrix F is obtained by scaling the relative flow matrix F_{rel} by a factor ς. This factor, which can be calibrated based on observed data or set by policy, does not introduce new independent structural variables into the dependency chain. Therefore, the overall flow distribution remains a function of the capacity matrix, $\mathsf{F} = \mathsf{F}(\mathsf{C})$.
- Congestion as a Function of Capacity:
  - The congestion matrix G is calculated as the Hadamard division of the absolute flow matrix F by the capacity matrix C: $\mathsf{G} = \mathsf{F} \oslash \mathsf{C}$.
  - As shown in the previous step, $\mathsf{F} = \mathsf{F}(\mathsf{C})$.
  - By substitution, $\mathsf{G} = \mathsf{F}(\mathsf{C}) \oslash \mathsf{C}$.
  - Therefore, the congestion matrix is exclusively a function of the capacity matrix, $\mathsf{G} = \mathsf{G}(\mathsf{C})$.
- Link Speed as a Function of the Structural Inputs:
  - The link speed matrix V is determined by a traffic flow model, such as Greenshields or BPR, which relates link speed to congestion. Using BPR, the travel time matrix T is a function of the length L, free flow speed U, and congestion G. The speed is simply $\mathsf{V} = \mathsf{L} \oslash \mathsf{T}$. It means $\mathsf{V} = \mathsf{V}(\mathsf{L}, \mathsf{U}, \mathsf{C})$.

  - The Greenshields model [2], for example, defines a relationship where speed is a function of the flow-to-capacity ratio, which is the definition of congestion. It is a foundational speed-density relationship in traffic flow theory, proposing a linear decline in speed as density increases. Therefore, the link speed matrix V is a function of the free-flow speed matrix U and the congestion matrix G, which can be denoted as $\mathsf{V} = \mathsf{V}(\mathsf{U}, \mathsf{G})$.
  - By substituting the dependency of G on C from the previous step, we have $\mathsf{V} = \mathsf{V}(\mathsf{U}, \mathsf{G}(\mathsf{C}))$.
  - This demonstrates that the link speed matrix is a function of the structural inputs: $\mathsf{V} = \mathsf{V}(\mathsf{U}, \mathsf{C})$.

- Travel Time as a Function of Structural Inputs:

  - From Definition 9.8, the travel time matrix T is a function of link length L, maximum speed U, and congestion G.
  - This functional relationship can be expressed as $\mathsf{T} = \mathsf{T}(\mathsf{L}, \mathsf{U}, \mathsf{G})$.
  - From the previous step, we established that $\mathsf{G} = \mathsf{G}(\mathsf{C})$.
  - By substitution, $\mathsf{T} = \mathsf{T}(\mathsf{L}, \mathsf{U}, \mathsf{G}(\mathsf{C}))$.
  - Therefore, the travel time matrix is a function of the structural inputs: $\mathsf{T} = \mathsf{T}(\mathsf{L}, \mathsf{U}, \mathsf{C})$.

- Delay as a Function of Structural Inputs:

  - The free-flow travel time matrix T^0 is defined as the Hadamard division of the link length matrix L by the maximum speed matrix U: $\mathsf{T}^0 = \mathsf{L} \oslash \mathsf{U}$. This makes $\mathsf{T}^0 = \mathsf{T}^0(\mathsf{L}, \mathsf{U})$.
  - The delay matrix Y is the difference between the actual travel time matrix T and the free-flow travel time matrix T^0: $\mathsf{Y} = \mathsf{T} - \mathsf{T}^0$.
  - By substitution of the relationships established in previous step, we obtain $\mathsf{Y} = \mathsf{T}(\mathsf{L}, \mathsf{U}, \mathsf{C}) - \mathsf{T}^0(\mathsf{L}, \mathsf{U})$.
  - Therefore, the delay matrix is also a function of the structural inputs: $\mathsf{Y} = \mathsf{Y}(\mathsf{L}, \mathsf{U}, \mathsf{C})$.

Each performance matrix of the Ideal Traffic Assignment framework is derived from a preceding matrix, forming a deterministic sequence of operations that originates from the foundational structural matrices (C, L, U). Since these structural matrices are constrained by the network's adjacency matrix A (as proven in Theorem 10.1), the entire framework is intrinsically based on the network's defined structure. □

This is formalized through a functional dependency chain, proving that the entire framework is intrinsically based on the network's defined structure.

10.5 Summary

The rigorous language of functional analysis enables formal proofs that unite the theory. As conclusion, all these variables in Ideal Traffic Assignments (ITA) are functionally depends on the network topological structure represented by the adjacency matrix. This proved our claim that Ideal Traffic Assignments framework is functionally based on network structure.

The derivation of Ideal Traffic Assignment provides a complete, self-contained method for generating a physically consistent (premagic) and behaviorally plausible traffic flow matrix from the network's structural properties alone, modulated by a small number of observable parameters.

References

1. Gene H. Golub and Charles F. Van Loan. *Matrix Computations*. 4th ed. Johns Hopkins University Press, 2013. ISBN: 9781421407944.
2. B. D. Greenshields. "A study of traffic capacity". In: *Highway Research Board Proceedings* 14 (1935), pp. 448–477.
3. Erwin Kreyszig. *Introductory Functional Analysis with Applications*. John Wiley & Sons, Inc., 1978. ISBN: 0-471-50731-8.
4. David G. Luenberger. *Optimization by Vector Space Methods*. Wiley, 1997. ISBN: 9780471181170. http://sites.science.oregonstate.edu/~show/old/142_Luenberger.pdf
5. P. Van Mieghem. *Performance Analysis of Complex Networks and Systems*. Cambridge University Press, 2014. https://doi.org/10.1017/CBO9781107415874
6. E. Nummelin. "A Splitting Technique for Harris Recurrent Markov Chains". In: *Zeitschrift für Wahrscheinlichkeitstheorie und verwandte Gebiete* 43.4 (1978), pp. 309–318. https://doi.org/10.1007/BF00534764
7. Eugene Seneta. *Non-negative Matrices and Markov Chains*. Rev. ed. New York: Springer, 2006. https://doi.org/10.1007/0-387-32792-4. https://link.springer.com/book/10.1007/0-387-32792-4
8. John J. Sopka. "Introductory Functional Analysis with Applications (Erwin Kreyszig)". In: *SIAM Review* 21.3 (1979), pp. 412–413. https://doi.org/10.1137/1021075.

Chapter 11
Calibration of Ideal Traffic Assignment

11.1 Bridging Theory to Reality: The Calibration Problem

Mathematical models are essential tools for urban transportation planning. They facilitate the understanding and prediction of traffic flow. The Ideal Traffic Assignment (ITA) framework [8], as defined in previous chapter (Chap. 9), provides a powerful theoretical foundation. It derives a unique, self-consistent pattern of relative traffic flows from a network's structural properties. The derived Ideal Flow matrix F satisfies the core principles of flow conservation and represents a theoretical steady-state distribution.

This elegant theory presents a practical challenge. The model, when based on network structure alone, determines the relative proportions of flow but not the absolute volume. For all practical applications such as congestion analysis on corridor performance evaluation or strategic network planning, absolute flow values are required. Calibration is the systematic process of anchoring this theoretical model to the physical world. It transforms the relative patterns of the Ideal Flow Network into a tool for estimating absolute traffic volumes.

This chapter presents a systematic and rigorous approach to this calibration problem. The methodology is designed to function with parsimonious data, a common condition in real-world environments. We will detail methods for finding the optimal scaling factor that converts relative flow to absolute flow. This is achieved by leveraging a limited number of empirical observations, which may include traffic counts, travel times, or average speeds.

Upon completion of this chapter, the following analytical capabilities will be established:

- Theoretical foundations of global scaling and demand invariance will be articulated.
- Calibration methods based on observed traffic counts will be formulated and applied.

K. Teknomo, *Ideal Flow Network for Transportation Research*,
Lecture Notes in Intelligent Transportation and Infrastructure,
https://doi.org/10.1007/978-981-95-4727-2_11

- Calibration methods utilizing observed travel times and average speeds will be formulated and applied.
- The coefficient of determination, R^2, will be employed as a diagnostic metric for assessing data consistency.
- K-fold cross-validation will be implemented to evaluate model performance robustness.
- Comparative scenario analysis will be conducted to assess short-term policy and infrastructure interventions.

11.2 Axioms and Definitions of Calibration

The calibration of an Ideal Flow Network (IFN) is a process of reconciling a theoretical model with empirical reality. This section establishes the fundamental properties and principles used in the calibration process.

11.2.1 IFN Equivalence Class

The Ideal Flow Network model, when derived from a network's structural properties, does not produce a unique flow matrix. Instead, it defines an entire equivalence class of flow matrices.

Definition 11.1 (IFN Equivalence Class [5, 7]) Two IFN matrices, F_1 and F_2, belong to the same equivalence class if and only if one is a scalar multiple of the other.

$$\mathsf{F}_1 \equiv \mathsf{F}_2 \iff \mathsf{F}_1 = \varsigma \mathsf{F}_2, \quad \text{for some positive scalar } \varsigma \in \mathbb{R}^+ \tag{11.1}$$

The ideal flow matrix F is thus a representation of a relative flow pattern. It provides a unique distribution of turning proportions but lacks an inherent magnitude. All members of this equivalence class are consistent with the underlying network structure and its premagic properties [6].

11.2.2 Short-Term Demand Invariance

The total volume of traffic circulating within a network, known as the total flow κ, is a function of travel demand. In the context of short-term policy evaluations, it is a reasonable assumption that the total demand does not change immediately in response to minor alterations in the physical network structure, such as adding a new lane or a traffic signal.

Axiom 11.1 (Short-Term Demand Invariance) *For comparative analysis of short-term policy changes (e.g., re-routing, lane modifications), the total flow* κ *is a system invariant. Structural modifications to the network affect the flow distribution but do not alter the total demand for travel.*

This axiom is critical for using the IFN framework in policy analysis. It permits the use of a single, empirically derived scaling factor across multiple scenarios to isolate and evaluate the purely structural impact of a proposed change.

11.2.3 Global Scaling Parameter

The global scaling parameter is the scalar value that resolves the ambiguity of the IFN equivalence class. It is the key to converting the model's relative flow patterns into absolute traffic volumes.

Definition 11.2 (Global Scaling Parameter) The global scaling parameter, denoted by ς, is the scalar that maps the relative ideal flow matrix F_{rel} to the matrix of absolute traffic flows F_{abs}. The calibrated flow matrix F_{abs} is obtained by multiplying the relative ideal flow matrix by this parameter.

$$\mathsf{F}_{abs} = \varsigma \mathsf{F}_{rel} \tag{11.2}$$

The value of ς is not determined by the network structure. It is an empirical parameter derived from observed data. The calibration process is thus defined as the search for the optimal value ς^* that best aligns the model's predictions with the available data.

11.3 Calibration with Empirical Data

This section details the most fundamental method for calibrating an Ideal Flow Network (IFN) by using a limited set of empirical traffic counts. The procedure finds a single optimal global scaling parameter ς^* that best fits the model's inherent relative flow pattern to a set of absolute observations.

11.3.1 Calibration with Observed Traffic Counts

The calibration process is fundamentally data-driven. It requires obtaining accurate measurements of traffic volumes for a subset of links in the network. The data required are absolute flow counts, typically measured in vehicles per hour (vph), for a select set of links, $\mathcal{K}$.

Practical sources for this observed link flow data are diverse and include both traditional and modern technologies including Intelligent Transport Systems [4]:

- *Fixed Inductive Loop Detectors:* These are sensors embedded in the road that detect passing vehicles, providing continuous time-series data for specific locations on major roads.
- *Video Analytics:* Traffic cameras coupled with computer vision algorithms can automatically count vehicles, classify them by type, and record turning movements at intersections.
- *Mobile GPS Traces:* Data from GPS devices and mobile phones can be aggregated to estimate traffic flows. This *floating car data* offers broad spatial coverage but requires significant processing to convert individual trajectories into accurate link volumes.

A robust calibration procedure must be able to handle this noisy and incomplete data to produce a reasonable estimate of the network's state.

11.3.1.1 Minimizing the Sum of Squared Errors

The core objective of calibration is to find the single value of the global scaling parameter ς that minimizes the total discrepancy between the model's predicted flows and the observed flows. A common and mathematically tractable measure for this discrepancy is the Sum of Squared Errors (SSE).

Definition 11.3 (Sum of Squared Errors [3]) Let F be the ideal relative flow matrix and F^{obs} the matrix of observed link flows. For a subset of links $\mathcal{K} \subseteq \mathcal{E}$ where observations are available, the *Sum of Squared Errors (SSE)* for a global scaling parameter ς is defined as:

$$\text{SSE}(\varsigma) = \sum_{k \in \mathcal{K}} (f_k^{\text{model}} - f_k^{\text{obs}})^2 = \sum_{k \in \mathcal{K}} (\varsigma f_k - f_k^{\text{obs}})^2 \tag{11.3}$$

where f_k is the relative flow from F and f_k^{obs} is the observed flow from F^{obs} on link k.

Proposition 11.1 (Minimizing the Sum of Squared Errors) *The optimal global scaling parameter ς^* is the value that minimizes SSE(ς) over the observed links $\mathcal{K}$.*

Minimizing the SSE yields the best linear fit between the model and the data in a least squares sense.

11.3.1.2 The Optimal Global Scaling Parameter

The value of ς^* that minimizes the SSE can be derived analytically.

Theorem 11.1 (Optimal Scaling Parameter) *Given a set of r observed link flows, f_k^{obs} for $k \in \mathcal{K}$, and the corresponding relative flows from the model, f_k, the optimal scaling parameter ς^* that minimizes the Sum of Squared Errors is given by:*

$$\varsigma^* = \frac{\sum_{k\in\mathcal{K}} f_k^{obs} \times f_k}{\sum_{k\in\mathcal{K}} (f_k)^2} \tag{11.4}$$

Proof To find the value of ς that minimizes the SSE, we take the first derivative of the SSE function with respect to ς and set it to zero.

$$\frac{d(\text{SSE})}{d\varsigma} = \frac{d}{d\varsigma}\sum_{k\in\mathcal{K}}(\varsigma f_k - f_k^{\text{obs}})^2 = 0 \tag{11.5}$$

Applying the chain rule, we obtain:

$$\sum_{k\in\mathcal{K}} 2(\varsigma f_k - f_k^{\text{obs}}) f_k = 0 \tag{11.6}$$

Rearranging the terms to solve for ς:

$$\sum_{k\in\mathcal{K}} (\varsigma f_k^2 - f_k^{\text{obs}} f_k) = 0 \tag{11.7}$$

$$\varsigma \sum_{k\in\mathcal{K}} f_k^2 = \sum_{k\in\mathcal{K}} f_k^{\text{obs}} f_k \tag{11.8}$$

$$\varsigma^* = \frac{\sum_{k\in\mathcal{K}} f_k^{\text{obs}} f_k}{\sum_{k\in\mathcal{K}} f_k^2} \tag{11.9}$$

The second derivative, $\frac{d^2(\text{SSE})}{d\varsigma^2} = 2\sum(f_k)^2$, is always positive, confirming that this value is a minimum. □

11.3.2 Coefficient of Determination

The quality of the calibration fit is quantitatively measured by the coefficient of determination, R^2. This metric evaluates how well the calibrated model's predictions account for the total variance in the observed data.

Definition 11.4 (Coefficient of Determination (R^2) [9]) The coefficient of determination, R^2, is a measure of the proportion of the variance in the observed data that is predictable from the model. It is defined as:

$$R^2 = 1 - \frac{\text{SSE}}{\text{SST}} = 1 - \frac{\sum_{k \in \mathcal{K}} (f_k^{\text{calibrated}} - f_k^{\text{obs}})^2}{\sum_{k \in \mathcal{K}} (f_k^{\text{obs}} - \bar{f}^{\text{obs}})^2} \tag{11.10}$$

Here, SSE is the Sum of Squared Errors from the calibrated flows, and SST is the Total Sum of Squares of the observed data around its mean, $\bar{f}^{\text{obs}}$. The value of R^2 is typically in the range [0, 1] for a well-fitting model. A value closer to 1 indicates a superior fit, while a low or negative value suggests a poor fit, where the model performs worse than simply predicting the mean of the observed data.

The optimal scaling parameter that minimizes the Sum of Squared Errors (SSE) is mathematically identical to the one that maximizes the coefficient of determination (R^2). Therefore, the formulas are the same, and the results of the numerical examples remain unchanged.

Proposition 11.2 (The Equivalence of Minimizing SSE and Maximizing R-squared) *The optimal global scaling parameter ς^* that maximizes the coefficient of determination, R^2, is the same as the parameter that minimizes the Sum of Squared Errors (SSE).*

Proof

- The coefficient of determination, R^2, is defined as the proportion of the variance in the observed data that is predictable from the model. Its formula is given by: $R^2 = 1 - \frac{\text{SSE}}{\text{SST}}$ where SSE is the Sum of Squared Errors and SST is the Total Sum of Squares.
- The Sum of Squared Errors (SSE) is defined as $\text{SSE}(\varsigma) = \sum_{k \in \mathcal{K}} (\varsigma f_k - f_k^{\text{obs}})^2$. It is a function of the scaling parameter ς. The Total Sum of Squares (SST) is defined as $\text{SST} = \sum_{k \in \mathcal{K}} (f_k^{\text{obs}} - \bar{f}^{\text{obs}})^2$. It is a constant with respect to ς because it depends only on the observed data, not the scaling parameter.
- To maximize the value of R^2, we must minimize the value of the term $\frac{\text{SSE}}{\text{SST}}$. Since SST is a positive constant and does not depend on ς, minimizing the ratio $\frac{\text{SSE}}{\text{SST}}$ is equivalent to minimizing the numerator, SSE.
- The optimization problem of maximizing R^2 is mathematically identical to the problem of minimizing SSE. The solution to both is the value of ς^* that satisfies $\frac{d(\text{SSE})}{d\varsigma} = 0$. This is the very same derivation presented in Theorem 11.1, which leads to the formula: $\varsigma^* = \frac{\sum_{k \in \mathcal{K}} f_k^{\text{obs}} \cdot f_k}{\sum_{k \in \mathcal{K}} (f_k)^2}$

This proves that the optimal scaling parameter for both objectives is the same. □

11.3.3 Comparison of SSE and Coefficient of Determination

While the optimal scaling parameter is the same for both minimization objectives, SSE and R^2 serve different purposes in model evaluation.

Sum of Squared Errors (SSE) is an absolute measure of the discrepancy between the model's predictions and the observed data. Its value is dependent on the scale of the flows and the number of observations. SSE is useful for comparing the performance of different models on the *same* dataset. Coefficient of Determination (R^2) is a normalized measure that indicates the proportion of the observed data's variance that is explained by the model. Its value, typically ranging from 0 to 1, provides a fit quality metric that is independent of the data's scale. A high R^2 value (close to 1) indicates a superior fit, while a low or negative R^2 can serve as a diagnostic tool for inconsistent data.

The best practice is to use the minimization of SSE as the objective function for the calibration process itself, as it is a direct and mathematically tractable optimization problem. Subsequently, use the R^2 value as a diagnostic and reporting metric (see Sect. 11.4) to evaluate the quality of the calibration fit, which is especially useful when comparing the results across different datasets or studies.

11.3.3.1 Numerical Examples Absolute Flow Calibration

In this section, we present several numerical examples to demonstrate the application of the global scaling methodology for calibrating an IFN model with absolute flow observations.

Example 11.1 (Calibration with a Single Observation) Consider a network where the Ideal Flow matrix gives a relative flow of $f_{CD} = 3$ on link CD. Suppose we have a single observed flow on this link, $f_{CD}^{\text{obs}} = 900$ vph. Using Theorem 11.1 with a single data point ($r = 1$), the formula simplifies to:

$$\varsigma^* = \frac{f_{CD}^{\text{obs}} \cdot f_{CD}}{(f_{CD})^2} = \frac{f_{CD}^{\text{obs}}}{f_{CD}} = \frac{900}{3} = 300$$

The optimal global scaling parameter is $\varsigma^* = 300$. The calibrated flow on link CD is $f_{CD}^{\text{cal}} = 300 \cdot 3 = 900$ vph, with a resulting error of zero.

The Sum of Squared Errors (SSE) and the Coefficient of Determination (R^2) are calculated as follows:

$$\text{SSE} = (f_{CD}^{\text{cal}} - f_{CD}^{\text{obs}})^2 = (900 - 900)^2 = 0$$

The mean of the observed flows is $\bar{f}^{\text{obs}} = 900$.

$$\text{SST} = (f_{CD}^{\text{obs}} - \bar{f}^{\text{obs}})^2 = (900 - 900)^2 = 0$$

Since both SSE and SST are zero, the fit is perfect. The concept of R^2 is not well-defined in this case, but the fit is exact.

Example 11.2 (Calibration with Multiple Observations)
Now, suppose we have two observations on the same network: a flow of $f_{BC}^{\text{obs}} =$ 500 vph on link BC and a flow of $f_{CD}^{\text{obs}} = 900$ vph on link CD. The corresponding relative flows from the Ideal Flow matrix are $f_{BC} = 2$ and $f_{CD} = 3$. Each observation suggests a different individual scaling factor:

- From link BC: $\varsigma_{BC} = \frac{f_{BC}^{\text{obs}}}{f_{BC}} = \frac{500}{2} = 250$
- From link CD: $\varsigma_{CD} = \frac{f_{CD}^{\text{obs}}}{f_{CD}} = \frac{900}{3} = 300$

To find the single, optimal global scaling parameter that minimizes the total error, we apply Theorem 11.1:

$$\varsigma^* = \frac{(f_{BC}^{\text{obs}} \cdot f_{BC}) + (f_{CD}^{\text{obs}} \cdot f_{CD})}{(f_{BC})^2 + (f_{CD})^2}$$

$$\varsigma^* = \frac{(500 \cdot 2) + (900 \cdot 3)}{2^2 + 3^2} = \frac{1000 + 2700}{4 + 9} = \frac{3700}{13} \approx 284.62$$

The optimal scaling parameter is $\varsigma^* \approx 284.62$. The calibrated flows for these links are:

- $f_{BC}^{\text{cal}} = 284.62 \cdot 2 = 569.24$ vph
- $f_{CD}^{\text{cal}} = 284.62 \cdot 3 = 853.86$ vph

This single value, ς^*, represents the best linear compromise across all available data points.

The Sum of Squared Errors and the Coefficient of Determination are calculated to evaluate the fit: The errors are $e_{BC} = 569.24 - 500 = 69.24$ and $e_{CD} = 853.86 - 900 = -46.14$.

$$\text{SSE} = (69.24)^2 + (-46.14)^2 \approx 4794.18 + 2128.90 \approx 6923.08$$

The mean of the observed flows is $\bar{f}^{\text{obs}} = \frac{500+900}{2} = 700$.

$$\begin{aligned}\text{SST} &= (500 - 700)^2 + (900 - 700)^2 = (-200)^2 + (200)^2 \\ &= 40{,}000 + 40{,}000 = 80{,}000\end{aligned}$$

The coefficient of determination is:

$$R^2 = 1 - \frac{\text{SSE}}{\text{SST}} = 1 - \frac{6923.08}{80{,}000} \approx 1 - 0.0865 = 0.9135$$

The high R^2 value indicates that the calibrated model provides a very strong fit to the observed data. While the calculation for ς^* is performed to minimize SSE. The resulting value of $\varsigma^* \approx 284.62$ simultaneously maximizes R^2 for that dataset. The final R^2 value of 0.9135 quantifies the goodness-of-fit for this optimal scaling.

Example 11.3 (Calibration with Multiple Observations) In practice, a calibration procedure will have multiple, and likely inconsistent, data points. Consider a network with a ideal flow matrix that includes a relative flow of $f_1 = 3$ on link 1, $f_2 = 2$ on link 2 and $f_3 = 3$ on link 3. Suppose we have three observations: $f_1^{\text{obs}} = 750$ vph, $f_2^{\text{obs}} = 600$ vph and $f_3^{\text{obs}} = 900$ vph. Each observation individually suggests a different scaling factor:

- From link 1: $\varsigma_1 = 750/3 = 250$
- From link 2: $\varsigma_2 = 600/2 = 300$
- From link 3: $\varsigma_3 = 900/3 = 300$

The SSE-minimizing formula finds the single best compromise value. Applying Theorem 11.1:

$$\varsigma^* = \frac{(f_1^{\text{obs}} \cdot f_1) + (f_2^{\text{obs}} \cdot f_2) + (f_3^{\text{obs}} \cdot f_3)}{(f_1)^2 + (f_2)^2 + (f_3)^2} = \frac{(750 \cdot 3) + (600 \cdot 2) + (900 \cdot 3)}{(3)^2 + (2)^2 + (3)^2} \tag{11.11}$$

$$\varsigma^* = \frac{2250 + 1200 + 2700}{9 + 4 + 9} = \frac{6150}{22} \approx 279.54 \tag{11.12}$$

This single value, $\varsigma^* \approx 279.54$, is the optimal scaling factor that provides the best linear fit to all available data. The calibrated flows would be $f_1^{\text{cal}} = 279.54 \cdot 3 \approx 838.62$, $f_2^{\text{cal}} = 279.54 \cdot 2 \approx 559.08$ and $f_3^{\text{cal}} = 279.54 \cdot 3 \approx 838.62$.

The SSE and R^2 are calculated as follows: The mean of the observed flows is $\bar{f}^{\text{obs}} = \frac{750+600+900}{3} = 750$.

$$\text{SST} = (750-750)^2+(600-750)^2+(900-750)^2 = 0+(-150)^2+(150)^2 = 45{,}000$$

The errors are $e_1 = 838.62 - 750 = 88.62$, $e_2 = 559.08 - 600 = -40.92$, and $e_3 = 838.62 - 900 = -61.38$.

$$\begin{aligned}\text{SSE} &= (88.62)^2 + (-40.92)^2 + (-61.38)^2 \approx 7853.51 \\ &+ 1674.49 + 3767.49 \approx 13{,}295.49\end{aligned}$$

The coefficient of determination is:

$$R^2 = 1 - \frac{\text{SSE}}{\text{SST}} = 1 - \frac{13{,}295.49}{45{,}000} \approx 1 - 0.29545 = 0.70455$$

The R^2 value of approximately 0.70 indicates a good fit, suggesting the model's structure is reasonably consistent with the observed data.

11.3.4 Calibration with Observed Link Performance Data

This section presents an advanced calibration methodology. It uses observed link performance data, such as travel time and speed. These observations help infer traffic flow. The inferred flow is then used to determine the global scaling factor. This approach is essential when traffic counts are unavailable, but other sensor data, such as GPS traces, provide speed or travel time measurements.

11.3.4.1 The BPR Speed and Travel Time Functions

The relationship between traffic flow and link performance is described by the Bureau of Public Roads (BPR) functions. These functions model travel time and speed as a non-linear function of congestion.

Definition 11.5 (BPR Travel Time Function [1]) The travel time t_{ij} on a link (i, j) is a function of its free-flow travel time t_{ij}^0, capacity c_{ij}, and the absolute flow f_{ij}.

$$t_{ij} = t_{ij}^0 \left(1 + \gamma \left(\frac{f_{ij}}{c_{ij}}\right)^\eta\right) \tag{11.13}$$

The free-flow travel time t_{ij}^0 is determined by the link's length l_{ij} and free-flow speed u_{ij}, where $t_{ij}^0 = \frac{l_{ij}}{u_{ij}}$. The parameters γ and η are empirical constants.

Definition 11.6 (BPR Speed Function) The average speed v_{ij} on a link (i, j) is a function of its free-flow speed u_{ij}, capacity c_{ij}, and the absolute flow f_{ij}.

$$v_{ij} = \frac{u_{ij}}{1 + \gamma \left(\frac{f_{ij}}{c_{ij}}\right)^\eta} \tag{11.14}$$

Both of these functions link the observable performance metrics (t_{ij} or v_{ij}) to the unobserved absolute flow f_{ij}, providing a pathway for calibration.

Example 11.4 (BPR Function Application for a Single Link) Consider a specific link in a network with the following characteristics:

- Link Length (l_{ij}): 5 km
- Free-Flow Speed (u_{ij}): 60 km/h
- Link Capacity (c_{ij}): 2000 vehicles/h
- BPR Parameters: $\gamma = 0.15$ and $\eta = 4$

Suppose an absolute flow of $f_{ij} = 1500$ vehicles/h is observed on this link.

1. Calculate Free-Flow Travel Time (t_{ij}^0):

$$t_{ij}^0 = \frac{l_{ij}}{u_{ij}} = \frac{5 \text{ km}}{60 \text{ km/h}} = 0.0833 \text{ h} = 5 \text{ min}$$

2. Calculate Congestion: The congestion is the ratio of the observed flow to the link's capacity.

$$\frac{f_{ij}}{c_{ij}} = \frac{1500 \text{ vph}}{2000 \text{ vph}} = 0.75$$

3. Calculate Congested Travel Time (t_{ij}): Using the BPR travel time function, the congested travel time is:

$$t_{ij} = 5 \text{ min} \left(1 + 0.15\,(0.75)^4\right)$$

$$t_{ij} = 5 \text{ min} \left(1 + 0.15 \cdot 0.3164\right) = 5.2375 \text{ min}$$

4. Calculate Congested Speed (v_{ij}): Using the BPR speed function, the average speed on the link under the given flow is:

$$v_{ij} = \frac{60 \text{ km/h}}{1 + 0.15\,(0.75)^4} = \frac{60 \text{ km/h}}{1.0475} \approx 57.28 \text{ km/h}$$

11.3.4.2 Derivation of Flow from Travel Time

Given the observed travel time on a link, we can algebraically invert the BPR travel time function to determine the absolute flow.

Lemma 11.1 (Flow from Travel Time) *The absolute flow f_{ij} on a link (i, j) can be derived from its observed travel time t_{ij}, free-flow travel time t_{ij}^0, capacity c_{ij}, and empirical BPR parameters γ and η.*

$$f_{ij} = c_{ij} \left[\frac{1}{\gamma} \left(\frac{t_{ij}}{t_{ij}^0} - 1 \right) \right]^{\frac{1}{\eta}} \tag{11.15}$$

Proof Starting with the BPR travel time function (Eq. 11.13), we solve for the flow f_{ij}.

$$\frac{t_{ij}}{t_{ij}^0} = 1 + \gamma \left(\frac{f_{ij}}{c_{ij}}\right)^\eta \tag{11.16}$$

$$\frac{1}{\gamma}\left(\frac{t_{ij}}{t_{ij}^0} - 1\right) = \left(\frac{f_{ij}}{c_{ij}}\right)^\eta \tag{11.17}$$

$$\left[\frac{1}{\gamma}\left(\frac{t_{ij}}{t_{ij}^0} - 1\right)\right]^{\frac{1}{\eta}} = \frac{f_{ij}}{c_{ij}} \tag{11.18}$$

Multiplying by c_{ij} yields the final result. □

The Ideal Flow Network (IFN) model provides the following relative flows. We have observed travel times from a real-time monitoring system, and we have the network characteristics for each link. Our objective is to determine the optimal global scaling parameter ς^* that best calibrates the model to this data.

Example 11.5 (Calibration from Travel Time on Multiple Links) Consider a network with three links, 1, 2, and 3.

- BPR Parameters: $\gamma = 0.15$ and $\eta = 4$
- Link 1: Relative flow $f_1 = 3$, observed travel time $t_1^{obs} = 5.24$ min, free-flow travel time $t_1^0 = 5.0$ min, and capacity $c_1 = 1800$ vph.
- Link 2: Relative flow $f_2 = 2$, observed travel time $t_2^{obs} = 7.1$ min, free-flow travel time $t_2^0 = 7.0$ min, and capacity $c_2 = 1500$ vph.
- Link 3: Relative flow $f_3 = 3$, observed travel time $t_3^{obs} = 4.03$ min, free-flow travel time $t_3^0 = 4.0$ min, and capacity $c_3 = 2500$ vph.

1. Infer Absolute Flows (f_k^{obs}) from Travel Times: We use the formula from Lemma 11.1 to infer the absolute flow on each link from the observed travel times.
   - Link 1: $f_1^{obs} = 1800\left[\frac{1}{0.15}\left(\frac{5.24}{5.0} - 1\right)\right]^{1/4} \approx 1353.6$ vph
   - Link 2: $f_2^{obs} = 1500\left[\frac{1}{0.15}\left(\frac{7.1}{7.0} - 1\right)\right]^{1/4} \approx 831.0$ vph
   - Link 3: $f_3^{obs} = 2500\left[\frac{1}{0.15}\left(\frac{4.03}{4.0} - 1\right)\right]^{1/4} \approx 1180.0$ vph

2. Calculate the Optimal Global Scaling Parameter (ς^*): Using the inferred absolute flows and the ideal relative flows, we apply the formula from Theorem 11.1.

$$\varsigma^* = \frac{(1353.6 \cdot 3) + (831.0 \cdot 2) + (1180.0 \cdot 3)}{3^2 + 2^2 + 3^2}$$

$$\varsigma^* = \frac{4060.8 + 1662.0 + 3540.0}{22} = \frac{9262.8}{22} \approx 421.04$$

3. Calculate Calibrated Flows (f_k^{cal}): The calibrated flows for each link are:
   - Link 1: $f_1^{\text{cal}} = 421.04 \cdot 3 \approx 1263.12$ vph
   - Link 2: $f_2^{\text{cal}} = 421.04 \cdot 2 \approx 842.08$ vph
   - Link 3: $f_3^{\text{cal}} = 421.04 \cdot 3 \approx 1263.12$ vph

4. Calculate SSE and SST: The errors are $e_1 = 1263.12 - 1353.6 = -90.48$, $e_2 = 842.08 - 831.0 = 11.08$, and $e_3 = 1263.12 - 1180.0 = 83.12$.

$$\text{SSE} = (-90.48)^2 + (11.08)^2 + (83.12)^2 \approx 15{,}218.3$$

   The mean of the observed flows is $\bar{f}^{\text{obs}} = \frac{1353.6+831.0+1180.0}{3} \approx 1121.53$ vph.

$$\text{SST} = (1353.6 - 1121.53)^2 + (831.0 - 1121.53)^2 + (1180.0 - 1121.53)^2 \approx 141{,}597.0$$

5. Calculate R^2:

$$R^2 = 1 - \frac{\text{SSE}}{\text{SST}} = 1 - \frac{15{,}218.3}{141{,}597.0} \approx 1 - 0.1075 \approx 0.8925$$

   The high R^2 value of approximately 0.89 indicates a strong fit, suggesting that the model's inherent relative flow pattern is highly consistent with the performance data observed on these links.

11.3.4.3 Derivation of Flow from Average Speed

Similarly, if we have observations of a link's average speed, we can derive the absolute flow by inverting the BPR speed function.

This Lemma 11.2 establishes the mathematical relationship for inferring the absolute traffic flow on a link from its observed average speed.

Lemma 11.2 (Flow from Average Speed) *The absolute flow f_{ij} on a link (i, j) can be derived from its observed average speed v_{ij}, free-flow speed u_{ij}, capacity c_{ij}, and empirical BPR parameters γ and η.*

$$f_{ij} = c_{ij}\left[\frac{1}{\gamma}\left(\frac{u_{ij}}{v_{ij}} - 1\right)\right]^{1/\eta} \tag{11.19}$$

Proof Starting with the BPR speed function (Eq. 11.14), we solve for the flow f_{ij}.

$$v_{ij}\left(1 + \gamma\left(\frac{f_{ij}}{c_{ij}}\right)^{\eta}\right) = u_{ij} \tag{11.20}$$

$$1 + \gamma\left(\frac{f_{ij}}{c_{ij}}\right)^{\eta} = \frac{u_{ij}}{v_{ij}} \tag{11.21}$$

$$\left(\frac{f_{ij}}{c_{ij}}\right)^{\eta} = \frac{1}{\gamma}\left(\frac{u_{ij}}{v_{ij}} - 1\right) \tag{11.22}$$

Taking the η-th root and multiplying by c_{ij} yields the desired result. □

Example 11.6 (Calibration from Average Speed)

Consider a network with three links: A, B, and C. The IFN model provides the following ideal relative flows, and we are given network data and observed average speeds for each link.

- BPR Parameters: $\gamma = 0.15$ and $\eta = 4$
- Link A: $f_A = 3$, $u_A = 60$ km/h, $c_A = 2000$ vph, $v_A^{\text{obs}} = 55$ km/h
- Link B: $f_B = 2$, $u_B = 80$ km/h, $c_B = 1500$ vph, $v_B^{\text{obs}} = 70$ km/h
- Link C: $f_C = 4$, $u_C = 50$ km/h, $c_C = 2500$ vph, $v_C^{\text{obs}} = 45$ km/h

Our goal is to find the optimal global scaling parameter ς^* that best calibrates the model to this data.

1. Infer Absolute Flows (f_k^{obs}) from Speeds: We use Lemma 11.2 to infer the absolute flow on each link from the observed speeds.

   - Link A: $f_A^{\text{obs}} = 2000\left[\frac{1}{0.15}\left(\frac{60}{55} - 1\right)\right]^{1/4} = 2000(0.606)^{1/4} \approx 2000(0.883) \approx 1766$ vph
   - Link B: $f_B^{\text{obs}} = 1500\left[\frac{1}{0.15}\left(\frac{80}{70} - 1\right)\right]^{1/4} = 1500(1.0476)^{1/4} \approx 1500(1.0116) \approx 1517$ vph
   - Link C: $f_C^{\text{obs}} = 2500\left[\frac{1}{0.15}\left(\frac{50}{45} - 1\right)\right]^{1/4} = 2500(0.7407)^{1/4} \approx 2500(0.927) \approx 2318$ vph

2. Calculate the Optimal Global Scaling Parameter (ς^*): Using the inferred absolute flows and the ideal relative flows, we apply the formula from Theorem 11.1.

$$\varsigma^* = \frac{\sum(f_k^{obs} \cdot f_k)}{\sum(f_k)^2} = \frac{(1766 \cdot 3) + (1517 \cdot 2) + (2318 \cdot 4)}{3^2 + 2^2 + 4^2}$$

$$\varsigma^* = \frac{5298 + 3034 + 9272}{9 + 4 + 16} = \frac{17{,}604}{29} \approx 607.03$$

3. Calculate Calibrated Flows (f_k^{cal}): The calibrated flows for each link are:
   - Link A: $f_A^{cal} = 607.03 \cdot 3 \approx 1821.09$ vph
   - Link B: $f_B^{cal} = 607.03 \cdot 2 \approx 1214.06$ vph
   - Link C: $f_C^{cal} = 607.03 \cdot 4 \approx 2428.12$ vph

4. Calculate SSE and SST: The errors are: $e_A = 1821.09 - 1766 = 55.09$, $e_B = 1214.06 - 1517 = -302.94$, $e_C = 2428.12 - 2318 = 110.12$.

   $$\text{SSE} = (55.09)^2 + (-302.94)^2 + (110.12)^2 \approx 3035 + 91{,}773 + 12{,}126 = 106{,}934$$

   The mean of the observed flows is $\bar{f}^{obs} = \frac{1766+1517+2318}{3} = \frac{5601}{3} = 1867$ vph.

   $$\text{SST} = (1766 - 1867)^2 + (1517 - 1867)^2 + (2318 - 1867)^2$$

   $$\text{SST} = (-101)^2 + (-350)^2 + (451)^2 \approx 10{,}201 + 122{,}500 + 203{,}401 = 336{,}102$$

5. Calculate R^2:

   $$R^2 = 1 - \frac{\text{SSE}}{\text{SST}} = 1 - \frac{106{,}934}{336{,}102} \approx 1 - 0.318 \approx 0.682$$

   The R^2 value of approximately 0.68 indicates a moderately good fit, suggesting that the model's inherent relative flow pattern is reasonably consistent with the performance data observed on these links.

11.3.4.4 The Data-Driven Calibration Algorithm

The lemmas above provide a pathway to convert observed performance data into set of absolute flows. This set of inferred flows can then be used in conjunction with Theorem 11.1 to calculate the optimal global scaling parameter ς^*.

First, we use the absolute link capacity matrix C, link distance matrix, L and link free-flow speed matrix U to obtain ideal relative flow matrix F_{rel} and relative congestion matrix G_{rel}. Then we apply the following procedure:

Inputs:

- F_{rel}: The ideal relative flow matrix.
- F^{obs} or V^{obs} or T^{obs}: A matrix of observed flows, speeds or travel times on a subset of links $\mathcal{K}$.
- γ, η: Empirical BPR parameters.

Output: The calibrated absolute flow matrix F.
Procedure:

1. *Infer Absolute Flows:* For each link (i, j) in the set of monitored links $\mathcal{K}$, use either Lemma 11.1 or Lemma 11.2 to calculate the observed absolute flow f_{ij}^{obs}.
2. *Calculate Global Scaling Factor:* With the set of inferred absolute flows, apply Theorem 11.1 to compute the optimal global scaling parameter ς^*.
3. *Compute Calibrated Flow Matrix:* Multiply the ideal relative flow matrix F by ς^* to obtain the calibrated absolute flow matrix $\mathsf{F} = \varsigma^* \mathsf{F}_{rel}$.
4. *Compute Calibrated Congestion Matrix:* $\mathsf{G} = \mathsf{F} \oslash \mathsf{C}$.
5. *Compute Calibrated Speed and Travel Time Matrices:* This step is to validate that the observed speeds or travel times is closed to the estimation.

This procedure demonstrates how the IFN framework can leverage a variety of data types, moving beyond the need for direct traffic counts.

11.4 The Diagnostic Power of Calibration

The calibration process is not merely a tool for estimating unknown flows; it is also a powerful diagnostic instrument. By evaluating the quality of the fit between the calibrated model and the observed data, we can validate the consistency of the empirical measurements and the foundational assumptions of the Ideal Flow Network (IFN) model itself.

11.4.1 Data Consistency and Coefficient of Determination

The R^2 value is a critical diagnostic indicator. A negative R^2 is of particular significance, as it reveals a fundamental conflict between the observed data and the structural properties of the network.

Corollary 11.1 (Data Consistency and R^2) *A negative value for R^2 indicates that the observed data is fundamentally inconsistent with the flow conservation property of the Ideal Flow Network. This suggests that the empirical observations violate the axioms of a premagic flow matrix, making it impossible to find a valid global scaling* *factor that can provide a reasonable fit. The errors are larger than the variance of the data itself, implying a conflict that cannot be resolved by simple scaling.*

Justification: For a valid IFN, the scaled flows must, by definition, maintain the premagic property, including the total inflow at each node equaling the total outflow. If observed data violates this fundamental conservation law, no single global scaling factor can rectify the discrepancy. The SSE of any calibrated model will be larger than the SST, yielding a negative R^2 value. This is not a failure of the calibration method but a diagnostic signal that the input data contains a logical contradiction with respect to the network's structure.

Example 11.7 (Detecting and Correcting Data Conflicts) Consider a simple 3-node network. The ideal relative flow matrix is given as:

$$\mathsf{F} = \begin{pmatrix} 0 & 2 & 1 \\ 1 & 0 & 2 \\ 2 & 1 & 0 \end{pmatrix} \tag{11.23}$$

Suppose we observe the following flows on a subset of links: $f_{12}^{\text{obs}} = 600$, $f_{23}^{\text{obs}} = 300$, and $f_{31}^{\text{obs}} = 100$.

First, we calculate the optimal global scaling parameter ς^* using Theorem 11.1.

$$\varsigma^* = \frac{(600 \cdot 2) + (300 \cdot 2) + (100 \cdot 2)}{(2)^2 + (2)^2 + (2)^2} = \frac{1200 + 600 + 200}{4 + 4 + 4} = \frac{2000}{12} \approx 166.67 \tag{11.24}$$

The calibrated flows are $f_{12}^{\text{cal}} = 166.67 \cdot 2 = 333.34$, $f_{23}^{\text{cal}} = 166.67 \cdot 2 = 333.34$, and $f_{31}^{\text{cal}} = 166.67 \cdot 2 = 333.34$. Now, we calculate the SSE and SST to find R^2. SSE $= (333.34 - 600)^2 + (333.34 - 300)^2 + (333.34 - 100)^2 = 71{,}115.56 + 1111.56 + 54{,}435.56 = 126{,}662.68$. The mean of the observed flows is $\bar{f}^{\text{obs}} = \frac{600+300+100}{3} = \frac{1000}{3} = 333.33$. SST $= (600 - 333.33)^2 + (300 - 333.33)^2 + (100 - 333.33)^2 = 71{,}115.56 + 1111.56 + 54{,}435.56 = 126{,}662.68$. In this case, $R^2 = 1 - \frac{126{,}662.68}{126{,}662.68} = 0$.

Now, suppose we observe a conflicting data set: $f_{12}^{\text{obs}} = 450$, $f_{23}^{\text{obs}} = 600$, and $f_{31}^{\text{obs}} = 900$. Let's analyze the flow conservation at the nodes, as dictated by the premagic property. The relative flows are $f_{12} = 2$, $f_{23} = 2$, $f_{31} = 2$.

- Outflow from Node 1 is $f_{12} + f_{13}$.
- Inflow to Node 1 is $f_{21} + f_{31}$.

The observed data states inflow to node 1 is 900 from link 31. The outflow from node 1 is 450 to link 12. This violates conservation. Let's calculate the R^2 for this conflicting data set. $\varsigma^* = \frac{(450 \cdot 2)+(600 \cdot 2)+(900 \cdot 2)}{(2)^2+(2)^2+(2)^2} = \frac{900+1200+1800}{12} = \frac{3900}{12} = 325$. Calibrated flows are $f_{12}^{\text{cal}} = 325 \cdot 2 = 650$, $f_{23}^{\text{cal}} = 650$, $f_{31}^{\text{cal}} = 650$. SSE $= (650 - 450)^2 + (650 - 600)^2 + (650 - 900)^2 = 200^2 + 50^2 + (-250)^2 = 40{,}000 + 2500 + 62{,}500 = 105{,}000$. The mean of the observed flows is $\bar{f}^{\text{obs}} = \frac{450+600+900}{3} = 650$.

$\text{SST} = (450-650)^2+(600-650)^2+(900-650)^2 = (-200)^2+(-50)^2+250^2 = 40{,}000+2500+62{,}500 = 105{,}000$. In this case, the R^2 is also zero. This example shows that while a negative R^2 is a clear sign of conflict, an R^2 of zero can also occur when the data is conflicting, but the deviations from the mean are identical to the deviations from the model's prediction. The key takeaway is the non-ideal behavior of the model fit. When the model consistently fails to align with the data (indicated by a low or negative R^2), it is a direct signal that the data is not consistent with the underlying structural assumptions of a premagic IFN.

Example 11.8 (Detecting Travel Time Data Conflicts) Consider a network with three links: (1,2), (2,3), and (3,1). We are given the following ideal relative flows from the IFN model and observed travel times from a real-time monitoring system.

- BPR Parameters: $\gamma = 0.15$ and $\eta = 4$
- Link (1,2): $f_{12} = 3$, $t_{12}^{\text{obs}} = 6.2$ min, $c_{12} = 1800$ vph, $t_{12}^0 = 5.0$ min
- Link (2,3): $f_{23} = 2$, $t_{23}^{\text{obs}} = 8.5$ min, $c_{23} = 1500$ vph, $t_{23}^0 = 7.0$ min
- Link (3,1): $f_{31} = 4$, $t_{31}^{\text{obs}} = 4.1$ min, $c_{31} = 2500$ vph, $t_{31}^0 = 4.0$ min

1. Infer Absolute Flows (f_k^{obs}): We use the formula from Lemma 11.1 to find the observed absolute flow on each link.

   - Link (1,2):

$$f_{12}^{\text{obs}} = 1800\left[\frac{1}{0.15}\left(\frac{6.2}{5.0}-1\right)\right]^{1/4} =$$

$$1800\left[\frac{0.24}{0.15}\right]^{1/4} = 1800(1.6)^{1/4} \approx 1800(1.1247) \approx 2024.43 \text{ vph}$$

   - Link (2,3):

$$f_{23}^{\text{obs}} = 1500\left[\frac{1}{0.15}\left(\frac{8.5}{7.0}-1\right)\right]^{1/4} =$$

$$1500\left[\frac{0.2143}{0.15}\right]^{1/4} = 1500(1.4287)^{1/4} \approx 1500(1.0933) \approx 1639.90 \text{ vph}$$

   - Link (3,1):

$$f_{31}^{\text{obs}} = 2500\left[\frac{1}{0.15}\left(\frac{4.1}{4.0}-1\right)\right]^{1/4} =$$

$$2500\left[\frac{0.025}{0.15}\right]^{1/4} = 2500(0.1667)^{1/4} \approx 2500(0.6389) \approx 1597.36 \text{ vph}$$

2. Calculate the Optimal Global Scaling Parameter (ς^*): We now use the derived observed flows and the ideal relative flows to find the scaling factor.

$$\varsigma^* = \frac{\sum(f_k^{\text{obs}} \cdot f_k)}{\sum(f_k)^2} = \frac{(2024.46 \cdot 3) + (1640.10 \cdot 2) + (1597.25 \cdot 4)}{3^2 + 2^2 + 4^2}$$

$$\varsigma^* = \frac{6073.38 + 3280.20 + 6389.00}{9 + 4 + 16} = \frac{15{,}742.58}{29} \approx 542.85$$

3. Calculate Calibrated Flows (f_k^{cal}): The calibrated flows for each link are:

- Link (1,2): $f_{12}^{\text{cal}} = 542.85 \cdot 3 = 1628.54$ vph
- Link (2,3): $f_{23}^{\text{cal}} = 542.85 \cdot 2 = 1085.69$ vph
- Link (3,1): $f_{31}^{\text{cal}} = 542.85 \cdot 4 = 2171.38$ vph

4. Calculate SSE and SST: The errors are $e_{12} = 1628.54 - 2024.43 = -395.89$, $e_{23} = 1085.69 - 1639.90 = -554.21$, and $e_{31} = 2171.38 - 1597.36 = 574.02$.

$$\text{SSE} = (-395.89)^2 + (-554.21)^2 + (574.02)^2 \approx 793{,}379.55$$

The mean of the observed flows is $\bar{f}^{\text{obs}} = \frac{2024.43+1639.90+1597.36}{3} = \frac{5261.68}{3} \approx 1753.89$ vph.

$$\text{SST} = (2024.43 - 1753.89)^2 + (1639.90 - 1753.89)^2 + (1597.36 - 1753.89)^2$$

$$\text{SST} = (270.53)^2 + (-114.00)^2 + (-156.54)^2 \approx 110{,}687.82$$

5. Calculate R^2:

$$R^2 = 1 - \frac{\text{SSE}}{\text{SST}} = 1 - \frac{793{,}379.55}{110{,}687.82} \approx 1 - 7.168 \approx -6.168$$

The negative R^2 value indicates a significant conflict in the data. The observed travel times are highly inconsistent with the structural flow conservation properties assumed by the IFN model. This suggests that the model's structure cannot adequately explain the observed data.

Example 11.9 (Detecting and Correcting Data Conflicts) Consider a network with nodes A, B, C, D, and E. The ideal relative flow matrix is premagic and satisfies flow conservation. Suppose we observe the following flows: $f_{AB}^{\text{obs}} = 450$, $f_{BC}^{\text{obs}} = 900$, and $f_{CD}^{\text{obs}} = 600$. The observed data is invalid for the network structure because it violates flow conservation. We can demonstrate this by examining the flow balance at each node.

Let the absolute flow matrix be denoted by F, where f_{ij} represents the flow from node i to node j. The flow conservation principle, which is a necessary condition for a premagic matrix, states that for every node, the total inflow must equal the total outflow.

- At node B, the total inflow is $f_{AB} = 450$. The total outflow is $f_{BC} + f_{BD} + \cdots = 900 + f_{BD} + \ldots$. Since flows must be non-negative, the outflow from node B is at least 900. This is a direct contradiction: $450 \neq 900 + f_{BD} + \ldots$.
- Similarly, at node C, the total inflow is $f_{BC} + f_{EC} + \cdots = 900 + f_{EC} + \ldots$, which is at least 900. The total outflow is $f_{CD} + f_{CA} + \cdots = 600 + f_{CA} + \ldots$, which is at most 600 (since other flows are unknown). This again demonstrates a conflict.

The observed data cannot satisfy the fundamental premagic constraints of the IFN. This conflict would manifest as a negative R^2 value when attempting to calibrate the network, signaling that the errors of the calibrated model are larger than the variance of the data itself. A negative R^2 is not a model failure, but a powerful diagnostic indicator that the data itself contains a logical contradiction with the physical network structure.

The analysis can be presented more formally using a matrix representation.

$\nearrow$	A	B	C	D	E	Σ
A	0	450	0	u	0	$450 + u$
B	0	0	900	v	0	$900 + v$
C	0	0	0	600	0	600
D	w	0	0	0	x	$w + x$
E	0	0	y	0	0	y
Σ	w	450	$900 + y$	$600 + u + v$	x	κ

From this matrix, we can see the inflows and outflows for each node. The row sums must equal the column sums. For Node B, we have $450 = 900 + v$. Since $v \geq 0$, this is impossible. This matrix representation provides a direct visual and mathematical proof of the data conflict.

11.5 Model Validation with K-fold Cross-Validation

After calibrating the model with a set of observed data, it is necessary to validate its performance to ensure that the chosen parameters and assumptions hold for unseen data. Calibration and validation are two distinct but interconnected steps in the modeling process. *Calibration* is the process of fitting a model to a given dataset. Its goal is to find the parameter values (in this case, ς^*) that provide the best possible fit to the available observations. *Validation*, on the other hand, is the process of assessing a model's performance on data it has not seen before. Its purpose is to measure the model's generalizability and to guard against overfitting. A model can be well-calibrated (meaning it fits the training data perfectly) but perform poorly in validation if it fails to predict new observations accurately. K-fold cross-validation is a resampling technique used to assess model generalizability by partitioning data into k subsets. Widely applied in transport demand modeling and machine learning. K-fold cross-validation is a robust procedure for this purpose, as it systematically assesses the model's predictive accuracy without relying on a single training-testing split.

11.5.1 The K-fold Validation Algorithm

The k-fold validation [2] procedure provides a method to robustly evaluate the model's calibration by rotating a dataset through training and testing cycles.

Inputs:

- F: The ideal relative flow matrix.
- F^{obs}: A vector of N observed link flows.
- K: The number of folds for cross-validation.

Output: A total error metric, such as the sum of squared errors from all test sets.
Procedure:

1. Divide the observed data F^{obs} into K equal-sized blocks. Let these blocks be $B_1, B_2, \ldots, B_K$.
2. Initialize a variable for the total test error: TotalError $= 0$.
3. For each fold $k = 1, \ldots, K$:

   a. *Training Set:* The training data consists of all blocks except B_k.
   b. *Test Set:* The test data is the single block B_k.
   c. *Calibrate:* Using the training set, compute the optimal global scaling parameter ς_k^* with Theorem 11.1.
   d. *Predict:* Compute the calibrated flows for the links in the test set using ς_k^*.
   e. *Test:* Calculate the sum of squared errors for the test set. $\text{TestError}_k = \sum_{j \in B_k} (f_j^{\text{calibrated}} - f_j^{\text{obs}})^2$.
   f. *Aggregate:* Add the test error to the total: $\text{TotalError} = \text{TotalError} + \text{TestError}_k$.

4. The final output is the TotalError. This value can be used to compare the performance of different models or calibration parameters. The goal is to find the parameter set that minimizes this total error, providing the most generalizable model fit.

This procedure ensures that every data point is used for both training and testing, and that the performance metric is not biased by a particular data split.

Example 11.10 (3-Fold Cross-Validation) This numerical example illustrates the K-fold cross-validation procedure using a simple 3-node network. We perform a 3-fold validation ($K = 3$) on a dataset of 3 observed link flows.

Network and Data:

- *Ideal Relative Flow Matrix* (F):

$$\mathsf{F} = \begin{pmatrix} 0 & 2 & 1 \\ 1 & 0 & 2 \\ 2 & 1 & 0 \end{pmatrix}$$

- *Observed Data* ($N = 3$):
  - $f_1^{\text{obs}} = 600$ vph (corresponding to a relative flow of 2 on link (1,2))
  - $f_2^{\text{obs}} = 300$ vph (corresponding to a relative flow of 1 on link (1,3))
  - $f_3^{\text{obs}} = 500$ vph (corresponding to a relative flow of 1 on link (2,1))
- *K-folds:* $K = 3$. The data is divided into three blocks, each containing a single data point.
  - $B_1 = \{f_1^{\text{obs}} = 600\}$
  - $B_2 = \{f_2^{\text{obs}} = 300\}$
  - $B_3 = \{f_3^{\text{obs}} = 500\}$

Procedure: Iterating Through the Folds

1. *Fold 1: Test on* B_1
   - *Training Set:* Data from B_2 and B_3. Relative flows are $f_2 = 1, f_3 = 1$. Observed flows are $f_2^{\text{obs}} = 300$, $f_3^{\text{obs}} = 500$.
   - *Calibrate* (ς_1^*):

$$\varsigma_1^* = \frac{(300 \cdot 1) + (500 \cdot 1)}{(1)^2 + (1)^2} = 400$$

   - *Predict:* Use $\varsigma_1^* = 400$ to predict the flow for the test set, B_1. $f_1^{\text{cal}} = \varsigma_1^* \cdot f_1 = 400 \cdot 2 = 800$ vph.
   - *Test and Aggregate:* $\text{TestError}_1 = (f_1^{\text{cal}} - f_1^{\text{obs}})^2 = (800 - 600)^2 = 40{,}000$. $\text{TotalError} = 0 + 40{,}000 = 40{,}000$.
2. *Fold 2: Test on* B_2
   - *Training Set:* Data from B_1 and B_3. Relative flows are $f_1 = 2, f_3 = 1$. Observed flows are $f_1^{\text{obs}} = 600$, $f_3^{\text{obs}} = 500$.
   - *Calibrate* (ς_2^*):

$$\varsigma_2^* = \frac{(600 \cdot 2) + (500 \cdot 1)}{(2)^2 + (1)^2} = 340$$

   - *Predict:* Use $\varsigma_2^* = 340$ to predict the flow for the test set, B_2. $f_2^{\text{cal}} = \varsigma_2^* \cdot f_2 = 340 \cdot 1 = 340$ vph.

- *Test and Aggregate:* $\text{TestError}_2 = (f_2^{\text{cal}} - f_2^{\text{obs}})^2 = (340 - 300)^2 = 1600$. $\text{TotalError} = 40{,}000 + 1600 = 41{,}600$.

3. *Fold 3: Test on B_3*

- *Training Set:* Data from B_1 and B_2. Relative flows are $f_1 = 2, f_2 = 1$. Observed flows are $f_1^{\text{obs}} = 600$, $f_2^{\text{obs}} = 300$.
- *Calibrate (ς_3^*):*

$$\varsigma_3^* = \frac{(600 \cdot 2) + (300 \cdot 1)}{(2)^2 + (1)^2} = 300$$

- *Predict:* Use $\varsigma_3^* = 300$ to predict the flow for the test set, B_3. $f_3^{\text{cal}} = \varsigma_3^* \cdot f_3 = 300 \cdot 1 = 300$ vph.
- *Test and Aggregate:* $\text{TestError}_3 = (f_3^{\text{cal}} - f_3^{\text{obs}})^2 = (300 - 500)^2 = 40{,}000$. $\text{TotalError} = 41{,}600 + 40{,}000 = 81{,}600$.

Result and Interpretation: The final output of the K-fold validation procedure is the total error of 81,600. This value is a robust performance metric that quantifies the model's ability to generalize to unseen data. It is the metric to be minimized when comparing different models or parameter sets.

The procedure does not yield a single optimal scaling parameter ς^* for the entire network. Instead, it produces a sequence of parameters, $\varsigma_1^* = 400$, $\varsigma_2^* = 340$, $\varsigma_3^* = 300$, each of which is optimal for a specific training subset. The purpose of validation is not to find this final parameter, but to rigorously test the model's predictive strength. Once the model's performance has been validated, a single final calibration is performed on the entire dataset of observed flows to obtain the definitive ς^* for practical application.

11.6 Calibration for Scenario Analysis

A significant advantage of the Ideal Traffic Assignment calibration methodology is its power in scenario analysis. Under the axiom of short-term demand invariance, we can use a single calibrated scaling factor from a base scenario to predict the full network state for a modified network. This allows for rapid and parsimonious policy evaluation without the need to re-estimate complex travel demand models.

11.6.1 Procedure: Comparative Scenario Analysis

The core of the ITA policy framework is the comparison of a base scenario with a proposed alternative. The absolute flow in the network is not a known value;

therefore, we must derive it. The Ideal Flow Network (IFN) provides a relative flow of network performance, but for practical policy evaluation, it is often necessary to scale the model to an absolute flow value. This section introduces the concept of scaling based on a maximum allowable congestion level, transforming the comparative analysis into a quantifiable metric for urban transportation planning.

The foundation of this technique is the assumption that a network's performance is constrained by its most congested link. We can define a maximum acceptable congestion level as a planning criterion. This criterion allows us to convert the relative flow values from the Ideal Flow Network model into an absolute, real-world flow.

Definition 11.7 (Absolute Ideal Flow) Let a network be defined by its capacity matrix C and its relative ideal flow matrix F_{rel}. Let ξ be a predetermined, maximum allowable congestion (flow-to-capacity ratio) for any link in the network. The scaling factor, ς, is calculated as:

$$\varsigma = \frac{\xi}{\max_{i,j:c_{ij}>0}\left(\frac{f_{ij,\text{rel}}}{c_{ij}}\right)} \tag{11.25}$$

The absolute ideal flow matrix, F_{abs}, is then given by:

$$\mathsf{F}_{abs} = \varsigma \cdot \mathsf{F}_{rel}$$

The total flow in the network, κ, is the sum of all elements in F_{abs}, i.e., $\kappa = \sum_{i,j} f_{ij,abs}$.

Think of the ideal flow model as a perfect blueprint of traffic distribution, but without a sense of scale. The scaling factor, ς, is like a magnifying glass that adjusts the entire blueprint until its most crowded part reaches a specific level of crowdedness that we, as planners, consider acceptable. This gives us a concrete number for total traffic, κ, which can be managed by the network without exceeding our congestion threshold.

The scaling method provides a robust way to compare different policy scenarios. The central idea is to use the total flow, κ, as an invariant between the base and alternative scenarios. This ensures that the comparison is fair, as both scenarios are evaluated under the same traffic demand.

Proposition 11.3 (Absolute Ideal Flow) *The absolute ideal flow for a given network can be derived by assuming a maximum allowable congestion level, ξ. The scaling factor ς is a function of the most congested link in the network.*

Proof By definition, the congestion on any link (i, j) is the ratio of its flow to its capacity: $g_{ij} = f_{ij}/c_{ij}$. The congestion matrix G is a function of the flow and capacity matrices. A network's performance is limited by its most congested link, $\max(\mathsf{G})$. To establish a base case, we assume a maximum allowable congestion level, ξ. The scaling factor, ς, scales the relative ideal flow to ensure that the maximum congestion in the base scenario does not exceed ξ. This is given by the formula Eq. (11.25). The absolute ideal flow matrix, F_{abs}, is then the scaled relative flow matrix, and the total flow, κ, is the sum of all elements in F_{abs}. The total flow, κ, is considered an invariant between the base and alternative scenarios for short-term policy analysis. □

Proposition 11.4 (Total Flow Invariance) *For a short-term policy analysis (e.g., within one year), the total flow in the network, κ, is considered an invariant between the base scenario and any alternative scenario.*

Proof The total flow, κ, represents the total demand in the network. This demand is influenced by external, uncontrollable factors such as population growth, economic development, and travel patterns. A policy intervention, which is a change to the network's structure ($\mathsf{A}, \mathsf{C}, \mathsf{L}, \mathsf{U}, \mathsf{S}$), does not directly alter these external factors. Therefore, over a short temporal horizon where these factors are assumed constant, the total flow can be treated as an invariant. □

This Proposition 11.4 is inline with the Definition of Short-Term 14.9.

Proposition 11.5 (Scenario-Based Performance Comparison) *An alternative policy scenario is demonstrably better than a base scenario if its maximum congestion, under the same total flow invariant, is less than the base scenario's maximum congestion.*

Proof Let $\mathcal{N}_b$ be the base network with capacity matrix C_b and relative flow $\mathsf{F}_{rel,b}$. Let ξ be the target maximum congestion, and let κ_1 be the absolute total flow calculated from the scaling factor $\varsigma_b = \xi / \max(\mathsf{G}_{b,rel})$. Now, let a policy be applied to create an alternative network $\mathcal{N}_a$ with capacity matrix C_a. We can then compute the absolute flow for this alternative network, $\mathsf{F}_{abs,a}$, such that the total flow is kept constant at κ_1. The maximum congestion for the alternative scenario, $\max(\mathsf{G}_{a,abs})$, can then be calculated. If $\max(\mathsf{G}_{a,abs}) < \max(\mathsf{G}_{b,abs}) = \xi$, the policy is successful because it achieves a more efficient flow distribution for the same amount of demand. □

We can't know the exact amount of traffic in a city, but we can say, "Let's find out the maximum traffic a network can handle before its worst bottleneck hits a certain level of congestion, say, 80%." That total traffic number, κ, is our benchmark. Now, for our proposed new road or policy, we ask, "Can this new network handle the same amount of traffic, κ, but with a better outcome, meaning with a maximum congestion of less than 80%?" If the answer is yes, we know the policy is an improvement, regardless of the actual, unknown total traffic.

The procedure for comparative scenario analysis is as follows:

Inputs:

- F_{base}: The ideal relative flow matrix for the current (base) network structure.
- $\mathbf{f}_{\text{obs}}$: A vector of observed absolute flows for a subset of links in the base network.
- C_{alt}: The capacity matrix for the proposed alternative network structure.

- L_{alt}: The link distance matrix for the alternative network.
- U_{alt}: The free-flow speed matrix for the alternative network.
- γ, η: Empirical BPR parameters.

Output: The predicted absolute flows, congestion, speeds, travel times, and delays for the alternative network.

Procedure:

1. *Calibrate the Base Scenario:* Using the observed flows $\mathbf{f}_{\text{obs}}$ on the base network, compute the optimal global scaling parameter $\varsigma^*_{\text{base}}$ with Theorem 11.1.
2. *Compute the Total System Flow (κ):* The total flow is the invariant quantity across scenarios. It is calculated by scaling the sum of the relative flows from the base scenario.

$$\kappa = \varsigma^*_{\text{base}} \sum_{i,j} \mathsf{F}_{\text{base},ij} \tag{11.26}$$

3. *Model the Alternative Scenario:* The structural change is reflected in the alternative matrices C_{alt}, L_{alt}, and U_{alt}. From these, derive the new ideal relative flow matrix, F_{alt}.
4. *Calculate the New Scaling Factor (ς^*_{alt}):* The scaling factor for the alternative network must be chosen such that the total flow remains invariant.

$$\varsigma^*_{\text{alt}} = \frac{\kappa}{\sum_{i,j} \mathsf{F}_{\text{alt},ij}} \tag{11.27}$$

5. *Predict Absolute Flows:* The predicted absolute flow matrix for the new network design is calculated as:

$$\mathsf{F}_{\text{predicted}} = \varsigma^*_{\text{alt}} \cdot \mathsf{F}_{\text{alt}} \tag{11.28}$$

6. *Predict Performance Metrics:* The calibrated flows $\mathsf{F}_{\text{predicted}}$ can now be used to predict the network's performance. For each link (i, j) in the alternative network, we can compute the congestion, travel time, and speed using the BPR functions.

This procedure allows transportation planners to rapidly estimate the system-wide impact of various infrastructure projects on absolute traffic flows, congestion, and travel times, providing a robust tool for policy evaluation.

11.7 Summary

This chapter has provided a detailed methodology for calibrating the Ideal Flow Network (IFN) model, transitioning it from a theoretical construct of relative flows to a practical tool for estimating absolute traffic volumes. The principles established in this chapter lead to several deductions regarding the characteristics of traffic flow and the utility of the ITA framework.

The ITA framework distinguishes between network structure and travel demand. The physical topology of the network determines the relative distribution of flow, which is captured in the scale-invariant ideal flow matrix, F. The total magnitude of traffic, which is a function of external travel demand, is independently captured by a single, empirically derived global scaling factor, ς^*.

The efficacy of calibrating a full network from a small subset of observed links demonstrates that traffic flow patterns possess a high degree of structural determinism. The strong connectivity and flow conservation constraints inherent to the network impose a logical structure on flow distribution. As a result, observing a few key parts provides sufficient information to infer the state of the entire network with high accuracy, which is quantifiable through metrics such as the coefficient of determination (R^2). This finding supports the validity of the data-parsimonious approach as a method rooted in the network's structural properties.

The calibration process serves a diagnostic function. A negative R^2 value is not a failure of the model's predictive capability. Rather, it is an indicator that the observed data contains a fundamental contradiction with the network's physical and conservation constraints. This outcome signals that the empirical data is not consistent with the axioms of a premagic flow matrix, and thus cannot be accurately represented by a calibrated IFN model.

The practical application of the IFN framework is contingent on abstract network theory. The concept of IFN equivalence provides the logical justification for applying a single scaling factor. The calibration process transforms the IFN from a theoretical model into a data-driven and verifiable framework. This methodology provides a consistent approach for understanding and managing urban transportation systems.

References

1. Bureau of Public Roads. *Traffic Assignment Manual*. Tech. rep. U.S. Department of Commerce, Urban Planning Division, 1964.
2. T. Hastie, R. Tibshirani, and J. Friedman. *The Elements of Statistical Learning: Data Mining, Inference, and Prediction*. 2nd. Springer, 2009. https://doi.org/10.1007/978-0-387-84858-7.
3. George A. F. Seber and Alan J. Lee. *Linear Regression Analysis*. 2nd ed. Wiley, 2003. ISBN: 9780471415404.
4. Joseph M. Sussman. *Perspectives on Intelligent Transportation Systems (ITS)*. Springer, 2005. ISBN: 978-0-387-23257-7. https://doi.org/10.1007/b101063.
5. Kardi Teknomo. *Introduction to Ideal Flow Networks*. 2022.
6. Kardi Teknomo. "Premagic and Ideal Flow Matrices". In: *arXiv preprint arXiv:1706.08856* (2017). https://arxiv.org/pdf/1706.08856.
7. Kardi Teknomo. *The Signatures of Ideal Flow Networks* 2024. https://arxiv.org/abs/2408.06344.
8. Kardi Teknomo, R. Gardon, and C. Saloma. "Ideal Flow Traffic Analysis: A Case Study on a Campus Road Network". In: *Philippine Journal of Science* 148.1 (2019), pp. 51–62. https://archium.ateneo.edu/discs-faculty-pubs/271/
9. J.M. Wooldridge. *Introductory Econometrics: A Modern Approach*. ISE-International Student Edition. South-Western Cengage Learning, 2009. ISBN: 9780324788907. https://books.google.co.id/books?id=LjeGPgAACAAJ

Chapter 12
Premagic Patterns and Network Topology

12.1 Introduction to Premagic and Eulerian Networks

This chapter provides a foundational theory for understanding the relationship between mathematical patterns and network topology. It introduces the premagic matrix [11], an abstract concept that represents a state of perfect balance and conservation in a system. This chapter develops a rigorous understanding of premagic matrices, beginning with their definition and fundamental properties, including the conditions under which symmetric and diagonal matrices exhibit premagic behavior. The analysis then situates the premagic property within the theoretical framework of Ideal Flow Networks (IFN), establishing its direct link to the principle of flow conservation. The premagic structure of a network's adjacency matrix is shown to characterize Eulerian topologies, networks that are inherently balanced. Consequences of this balanced topology are explored, with emphasis on its potential to yield uniform flow and optimal congestion under specific conditions.

12.2 The Premagic Matrix in Nature and Mathematics

While abstract, the premagic matrix [11] represents a beautiful, underlying pattern of balance found in nature. Analogous to how Fibonacci numbers describe patterns in plant growth, the premagic property describes a state of harmony in a system. In physical systems, this property represents a state where the total "weight" or "flow" entering a point is perfectly equal to the total exiting it. This natural equilibrium, often unseen, provides a foundation for understanding ideal systems.

K. Teknomo, *Ideal Flow Network for Transportation Research*,
Lecture Notes in Intelligent Transportation and Infrastructure,
https://doi.org/10.1007/978-981-95-4727-2_12

12.3 Premagic Matrices: Foundational Theory

The premagic property is foundational to the theory of Ideal Flow Networks (IFN). In this context, an ideal flow matrix is a non-negative, irreducible, and premagic matrix. The premagic property of this matrix is a direct representation of flow conservation at every node. This means that for any node, the total inflow equals the total outflow, ensuring the network can sustain a steady-state flow distribution without accumulation or loss.

A *premagic matrix* [11] is a non-negative square matrix where the sum of the elements in each row is equal to the sum of the elements in the corresponding column. This property signifies a state of mathematical balance or conservation. The condition for a premagic matrix C can be expressed using a column vector of ones $\mathbf{j}$ as:

$$\mathsf{C}\mathbf{j} = \mathsf{C}^T\mathbf{j}$$

Several types of matrices inherently possess the premagic property:

- *Symmetric matrices* are always premagic because the elements are mirrored across the diagonal, naturally making row sums equal to column sums.
- *Diagonal matrices* are premagic, as the row and column sums are each equal to the value of the single non-zero element in that row or column.
- A square matrix where all elements have the same value is a premagic matrix.
- *Magic matrices* are a subset of premagic matrices.

The set of premagic matrices is closed under addition, subtraction, and scalar multiplication. The core concept of a premagic matrix is mathematically precise and serves as an axiom for ideal flow analysis.

Definition 12.1 (Premagic Matrix) A non-negative square matrix $\mathsf{A} \in \mathbb{R}^{n \times n}$ is defined as premagic if and only if the sum of the elements in each of its rows is equal to the sum of the elements in its corresponding column. This property is stated algebraically as:

$$\sum_{j=1}^{n} a_{ij} = \sum_{j=1}^{n} a_{ji} \quad \forall i \in \{1, \ldots, n\}$$

Using matrix notation, where $\mathbf{j}$ is a column vector of ones, this is concisely expressed as:

$$\mathsf{A}\mathbf{j} = \mathsf{A}^T\mathbf{j}$$

Axiom 12.1 (Flow Conservation) *The premagic property is the algebraic representation of the principle of flow conservation. In a network, a matrix representing connections or flows is premagic if and only if the total inflow to any node equals the total outflow from that node.*

Proposition 12.1 (Symmetric Matrices) *Every symmetric matrix is premagic.*

Proof A matrix B is symmetric if and only if $\mathsf{B} = \mathsf{B}^T$. Multiplying both sides of this equality by the vector of ones $\mathbf{j}$ yields $\mathsf{B}\mathbf{j} = \mathsf{B}^T\mathbf{j}$. This equation is the formal definition of a premagic matrix. Thus, every symmetric matrix satisfies the premagic property. □

A symmetric matrix exhibits perfect balance, where elements are mirrored across the main diagonal. This inherent symmetry ensures that summing elements along a row produces the same result as summing along the corresponding column. The premagic property is a generalization of this, requiring only that the row and column sums are equal, without needing element-wise equality.

Example 12.1 (Symmetric Matrix) Consider the following symmetric matrix B.

$$\mathsf{B} = \begin{bmatrix} 1 & 2 & 3 \\ 2 & 4 & 5 \\ 3 & 5 & 6 \end{bmatrix}$$

The row sum vector is $\mathsf{B}\mathbf{j} = \begin{bmatrix} 1+2+3 \\ 2+4+5 \\ 3+5+6 \end{bmatrix} = \begin{bmatrix} 6 \\ 11 \\ 14 \end{bmatrix}$. The column sum vector is $\mathsf{B}^T\mathbf{j} = \begin{bmatrix} 1+2+3 \\ 2+4+5 \\ 3+5+6 \end{bmatrix} = \begin{bmatrix} 6 \\ 11 \\ 14 \end{bmatrix}$. Since $\mathsf{B}\mathbf{j} = \mathsf{B}^T\mathbf{j}$, the matrix is premagic.

Proposition 12.2 (Closure Properties) *The set of premagic matrices is closed under addition, subtraction, and scalar multiplication.*

Proof Let A_1 and A_2 be two premagic matrices, and let $\alpha \in \mathbb{R}$ be a scalar.

1. *Addition:* By definition, $\mathsf{A}_1\mathbf{j} = \mathsf{A}_1^T\mathbf{j}$ and $\mathsf{A}_2\mathbf{j} = \mathsf{A}_2^T\mathbf{j}$. Consider the matrix $\mathsf{A}_3 = \mathsf{A}_1 + \mathsf{A}_2$. Then, $\mathsf{A}_3\mathbf{j} = (\mathsf{A}_1 + \mathsf{A}_2)\mathbf{j} = \mathsf{A}_1\mathbf{j} + \mathsf{A}_2\mathbf{j} = \mathsf{A}_1^T\mathbf{j} + \mathsf{A}_2^T\mathbf{j} = (\mathsf{A}_1^T + \mathsf{A}_2^T)\mathbf{j} = \mathsf{A}_3^T\mathbf{j}$. Thus, A_3 is premagic.
2. *Scalar Multiplication:* Consider the matrix $\mathsf{A}_4 = \alpha\mathsf{A}_1$. Then, $\mathsf{A}_4\mathbf{j} = (\alpha\mathsf{A}_1)\mathbf{j} = \alpha(\mathsf{A}_1\mathbf{j}) = \alpha(\mathsf{A}_1^T\mathbf{j}) = (\alpha\mathsf{A}_1^T)\mathbf{j} = \mathsf{A}_4^T\mathbf{j}$. Thus, A_4 is premagic.

Since both addition and scalar multiplication preserve the property, linear combinations of premagic matrices are also premagic. □

These properties mean that if we have several systems that are individually balanced (premagic), we can combine them in various ways, by adding their matrices or scaling them, and the resulting combined system will also be balanced. This is a powerful and useful property for modeling complex networks.

Theorem 12.1 (Ideal Flow Condition) *A non-negative, irreducible matrix* F *represents an ideal flow matrix in an Ideal Flow Network (IFN) if and only if it is premagic.*

Proof This theorem is a direct consequence of Axiom 1 and the definition of an IFN. By definition, an IFN is a strongly connected directed graph where each node satisfies flow conservation (total inflow equals total outflow). The flow on the network is represented by the ideal flow matrix F.

- The total outflow from node i is given by the sum of its row: $\sum_j f_{ij}$.
- The total inflow to node i is given by the sum of its column: $\sum_j f_{ji}$.

Flow conservation at every node requires that $\sum_j f_{ij} = \sum_j f_{ji}$ for all i. This is the precise definition of a premagic matrix. Therefore, the ideal flow matrix F must be premagic. □

Example 12.2 (Simplified Network) Consider a simplified transportation network with four nodes and the following ideal flow matrix F representing traffic flow.

$$\mathsf{F} = \begin{bmatrix} 0 & 3 & 0 & 2 \\ 2 & 0 & 3 & 0 \\ 0 & 2 & 0 & 3 \\ 3 & 0 & 2 & 0 \end{bmatrix}$$

The row sums are $[5, 5, 5, 5]^T$ and the column sums are $[5, 5, 5, 5]^T$. Since the row and column sums are equal, the matrix F is premagic. This indicates that at every node, the total inflow of 5 units of traffic is equal to the total outflow of 5 units of traffic (Fig. 12.1).

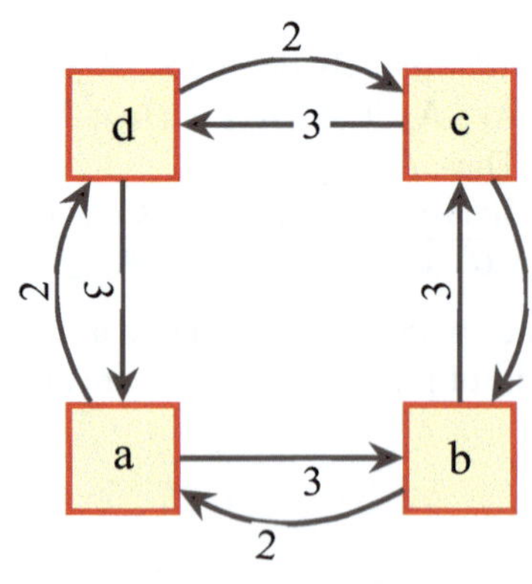

Fig. 12.1 Eulerian network example with a four-node, eight-link network with symmetric flow values

12.4 Eulerian Networks and Premagic Degrees

The *premagic degree* property defines a network in which the in-degree of every node is equal to its out-degree. This is the graphical equivalent of the adjacency matrix being premagic. A directed graph is *Eulerian* if it is strongly connected and its adjacency matrix is premagic. For example, a city with a perfect grid layout is a classic representation of an Eulerian network. Bang-Jensen and Gutin [1] provide comprehensive coverage of cycles in directed graphs, including Hamiltonian and Eulerian cycles, essential for modeling recurrent flows in transport networks. Diestel [5] discusses Eulerian circuits and conditions for Eulerian networks in graph theory, which are essential for modeling closed-loop flows in transport systems.

Definition 12.2 (Eulerian Digraph) A directed graph $D = (V, E)$ is Eulerian if it is strongly connected and for every vertex $v \in V$, the in-degree, $deg^{-}(v)$, equals the out-degree, $deg^{+}(v)$.

Definition 12.3 (Premagic Degree) A network is said to have a premagic degree if its adjacency matrix A is premagic. This property implies that the in-degree of every node is equal to its out-degree. The in-degree of node i is $\sum_j a_{ji}$ (sum of column i) and the out-degree is $\sum_j a_{ij}$ (sum of row i).

Newman [9] discussed degree distributions and flow conservation in complex networks, offering statistical tools to analyze premagic degree conditions in directed transport graphs

Theorem 12.2 (Eulerian Characterization) *A directed graph is Eulerian if and only if it is strongly connected and its adjacency matrix* A *is premagic.*

Proof This theorem is a fundamental result in graph theory. The proof has two directions.

- ($\Longrightarrow$) *If D is Eulerian, then* A *is premagic.* By Definition 2, an Eulerian digraph is strongly connected and has $deg^{+}(v) = deg^{-}(v)$ for every vertex v. In matrix terms, the out-degree of vertex i is the sum of row i of the adjacency matrix, $\sum\_j a_{ij}$. The in-degree of vertex i is the sum of column i, $\sum\_j a_{ji}$. The condition $deg^{+}(v) = deg^{-}(v)$ for all v is therefore equivalent to $\sum\_j a_{ij} = \sum\_j a_{ji}$ for all i, which is the definition of a premagic matrix.
- ($\Longleftarrow$) *If* A *is premagic and D is strongly connected, then D is Eulerian.* The premagic property implies that $\sum\_j a_{ij} = \sum\_j a_{ji}$ for all i. As shown above, this is the matrix equivalent of $deg^{+}(v) = deg^{-}(v)$. A strongly connected digraph with this property is, by definition, Eulerian.

□

Example 12.3 Figure 12.2 provides a classic example of an Eulerian network. Its structure demonstrates the key properties that make it an ideal model for closed, steady-state systems. Strongly Connected: The graph is strongly connected because there is a directed path from every node to every other node. For example, to get

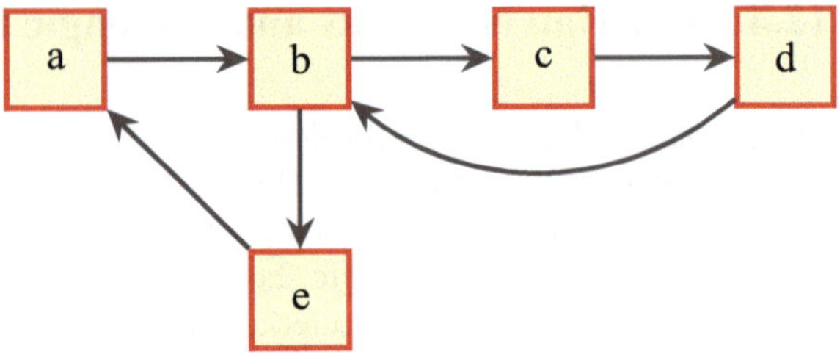

Fig. 12.2 A five-node network that has an Eulerian circuit

from node c to node a, we can follow the path $c \to d \to b \to e \to a$. Premagic Adjacency Matrix: The adjacency matrix of this network is premagic. The in-degree of every node is equal to its out-degree, which is the defining characteristic of an Eulerian network.

- Node a: In-degree = 1 (from e), Out-degree = 1 (to b).
- Node b: In-degree = 2 (from a and d), Out-degree = 2 (to c and e).
- Node c: In-degree = 1 (from b), Out-degree = 1 (to d).
- Node d: In-degree = 1 (from c), Out-degree = 1 (to b).
- Node e: In-degree = 1 (from b), Out-degree = 1 (to a).

This in-degree/out-degree balance ensures the network can sustain a continuous flow without any accumulation or loss at any node, making it a perfect example of a system in equilibrium.

This theorem formalizes the idea that a balanced network is an Eulerian network. The *inflow equals outflow* property at every intersection (node) is a necessary condition for a continuous, closed loop of traffic to exist. A strongly connected graph that satisfies this balance can support a circuit that passes through every road exactly once.

Proposition 12.3 (Uniform Flow Under Constant Capacity) *In an Eulerian network where all links have the same capacity c, the ideal flow on every link f_{ij} will be uniform and equal to $\frac{\kappa}{m}$, where κ is the total flow and m is the total number of links.*

Proof The ideal flow matrix is defined as $\mathsf{F} = \varsigma\mathsf{C}$, where C is the capacity matrix and ς is a scaling constant such that $\varsigma = \kappa/(\sum_{i,j} c_{ij})$. For an Eulerian network with uniform link capacity c, the capacity matrix is $\mathsf{C} = c\mathsf{A}$. The total capacity is the sum of all elements in C: $\sum_{i,j} c_{ij} = \sum_{i,j} ca_{ij} = c\sum_{i,j} a_{ij}$. The total number of links is $m = \sum_{i,j} a_{ij}$. Therefore, the total capacity is cm. The scaling constant is $\varsigma = \frac{\kappa}{cm}$. The flow on any link (i, j) with capacity $c_{ij} = c$ is $f_{ij} = \varsigma c_{ij} = \left(\frac{\kappa}{cm}\right)(c) = \frac{\kappa}{m}$. This value is constant for all links in the network. □

When a balanced network has equally sized roads, the ideal traffic flow distributes itself evenly. This is the most efficient state because it minimizes the total "friction" of the system. Every road is used to the same degree, preventing any single point from becoming overloaded.

Corollary 12.1 (Optimal Congestion) *In a regular Eulerian network with uniform link capacities, the resulting congestion level on all links is equal. This state represents both a* Nash Equilibrium *and a* Pareto Optimal *[6] state for the network.*

Proof The congestion level on a link is defined as $g_{ij} = f_{ij}/c_{ij}$. From Proposition 12.3, we know that in this type of network, $f_{ij} = \kappa/m$ and $c_{ij} = c$ for all links. Therefore, the congestion level is $g_{ij} = (\kappa/m)/c = \kappa/(mc)$, which is a constant for all links.

- *Nash Equilibrium [7, 8]:* Since all links have the same congestion level, the perceived travel time per unit distance is identical. A user cannot unilaterally switch to another route to decrease their travel time; all available routes offer the same conditions. This satisfies the definition of a Nash Equilibrium.
- *System Optimum:* The uniform congestion distribution is achieved when the network is perfectly balanced and capacities are uniform. In this state, the total system travel time is minimized. It is impossible to re-route flow to improve the congestion on one link without negatively impacting the congestion on another link, thereby satisfying the definition of a Pareto Optimal state.

□

Designing a network to be Eulerian leads to significant consequences for flow and congestion. One key result is that in an Eulerian network, if all links have the same capacity, the ideal flow is distributed uniformly across all links. This uniform flow results in a uniform congestion level across the entire network. This state of uniform congestion is particularly significant because it satisfies both of Wardrop's principles: User Equilibrium (UE), where no driver can unilaterally improve their travel time, and System Optimum (SO), where the total system travel time is minimized. Thus, the Eulerian network is an ideal topology that aligns individual and system-wide optimality.

12.5 Premagic Equilibrium

An extension of the above theory is the concept of a *premagic capacity matrix*. When a network's capacity matrix is premagic, it ensures that all positive entries in its congestion matrix will have the same level of congestion. This state, known as *premagic equilibrium*, is where the ideal flow distributes itself proportionally to the capacity, resulting in uniform congestion. This is an ideal state that satisfies both User Equilibrium and System Optimum simultaneously, even if the network's topology itself (its adjacency matrix) is not Eulerian. The uniform congestion of a premagic equilibrium is provably the state of minimum total travel time for the system. A network is in *premagic equilibrium* if its congestion matrix, G, is a steady matrix, meaning all active links have the same level of congestion. This operational state is a direct result of the network's structural properties. Specifically, a strongly connected network with an irreducible, premagic capacity matrix,C, will naturally achieve premagic equilibrium. The proof for this relies on the relationship $\mathsf{F} = \sigma \times \mathsf{C}$, where the ideal flow matrix is a scalar multiple of the capacity matrix. The congestion matrix is then calculated as the element-wise division of the flow by capacity: $\mathsf{G} = \mathsf{F} \oslash \mathsf{C}$. This simplifies to $\mathsf{G} = \sigma \mathsf{C} \oslash \mathsf{C}$, which results in a constant value σ for all non-zero entries. If a network has a premagic capacity matrix, the ratio of total flow (κ) to total capacity will be a constant, σ. This constant represents the uniform congestion level across the network, verifying that it is in premagic equilibrium.

Chapter 13 describes how three idealized axioms

- Premagic Equilibrium (PE)
- Steady Link Characteristics (SLC)
- Identical Path Link Counts (IPLC) [4]

can prove the convergence of User Equilibrium (UE) and System Optimum (SO) in an ideal network.

- Convergence to Nash Equilibrium: The simultaneous satisfaction of the PE, SLC, and IPLC axioms implies that the network operates as a Nash Equilibrium, or User Equilibrium. This means all active paths have equal and minimal travel times, so no driver can find a faster route by switching paths.
- Convergence to System Optimum [12]: The combination of the PE and SLC axioms proves that the network achieves a System Optimum. This state is characterized by a uniform congestion ratio, which, as proven by Jensen's Inequality, minimizes the total aggregate travel time for the entire system.
- The Ultimate Ideal Equilibrium: When all three axioms are met, the network simultaneously satisfies both the User Equilibrium and System Optimum conditions. This eliminates the conflict between what is best for an individual driver and what is best for the entire system.
- Price of Anarchy (PoA): In this ideal state, the Price of Anarchy [10], a metric that quantifies network inefficiency due to individual choices, is exactly 1. A PoA of 1 signifies that the system is perfectly efficient and that the phenomenon known as Braess' Paradox [2, 3] cannot occur.

For a more detailed explanation of the proofs and their implications, please refer to the Chap. 13 on Premagic Equilibrium and Network Optimality.

12.6 Summary

This chapter establishes the premagic matrix as a formal representation of a balanced system, defined by the condition that the vector of row sums equals the transpose of the vector of column sums. This property is shown to be algebraically equivalent to the flow conservation axiom, which ensures that total inflow equals total outflow at every node in the network.

This equivalence forms the basis of Ideal Flow Network theory, where structural balance enables reproducible and interpretable flow distributions. It is further deduced that a strongly connected network with a premagic adjacency matrix is an Eulerian network, characterized by equal in-degree and out-degree at every node.

In such networks, when link capacities are premagic, the resulting ideal flow is a scaled of the capacities across all links, yielding a constant congestion level. Crucially, this configuration satisfies both the User Equilibrium (Nash condition) and the System Optimum (Pareto condition). That is, premagic equilibrium is a sufficient condition for the simultaneous realization of individual and collective optimality.

Therefore, the premagic property serves as a design principle for constructing networks where decentralized behavior and system-wide efficiency are structurally aligned. It offers a formal guarantee that optimality is embedded in the network's algebraic foundation.

References

1. Jørgen Bang-Jensen and Gregory Z. Gutin. *Digraphs: Theory, Algorithms and Applications*. London: Springer, 2008. https://doi.org/10.1007/978-1-84800-998-1. https://link.springer.com/book/10.1007/978-1-84800-998-1.
2. D. Braess, A. Nagurney, and T. Wakolbinger. "On a paradox of traffic planning". In: *Transportation Science* 39.4 (2005), pp. 446–450.
3. Dietrich Braess. "Über ein Paradoxon aus der Verkehrsplanung". In: *Unternehmensforschung* 12.1 (1968), pp. 258–268. https://doi.org/10.1007/BF01918335.
4. Ennio Cascetta. *Transportation Systems Analysis: Models and Applications*. 2nd. New York: Springer, 2009. https://doi.org/10.1007/978-0-387-75857-2. https://link.springer.com/book/10.1007/978-0-387-75857-2.
5. Reinhard Diestel. *Graph Theory*. 5th. Berlin, Heidelberg: Springer, 2017. https://doi.org/10.1007/978-3-662-53622-3.
6. Matthias Ehrgott. *Multicriteria Optimization*. 2nd ed. Springer, 2005. ISBN: 978-3-540-21398-7. https://doi.org/10.1007/3-540-27659-9.
7. J. F. Nash. "Non-Cooperative Games". In: *Annals of Mathematics* 54.2 (1951), pp. 286–295. https://doi.org/10.2307/1969529.
8. John F. Nash. "Equilibrium Points in N-Person Games". In: *Proceedings of the National Academy of Sciences* 36.1 (1950), pp. 48–49. https://doi.org/10.1073/pnas.36.1.48.
9. Mark E. J. Newman. *Networks: An Introduction*. Oxford: Oxford University Press, 2010. https://doi.org/10.1093/acprof:oso/9780199206650.001.0001. https://academic.oup.com/book/27303.
10. T. Roughgarden and E. Tardos. "How bad is selfish routing?" In: *Proceedings 41st Annual Symposium on Foundations of Computer Science*. 2000, pp. 93–102. https://doi.org/10.1109/SFCS.2000.892069.
11. Kardi Teknomo. "Premagic and Ideal Flow Matrices". In: *arXiv preprint arXiv:1706.08856* (2017). https://arxiv.org/abs/1706.08856.
12. J. G. Wardrop. "Some theoretical aspects of road traffic research". In: *Proceedings of the Institution of Civil Engineers* 1.3 (1952), pp. 325–362. https://doi.org/10.1680/iicep.1952.11362. https://www.researchgate.net/publication/227555985_Wardrop_Equilibria.

Chapter 13
Premagic Equilibrium and Network Optimality

13.1 Introduction

A central challenge in transportation science is the inherent conflict between the uncoordinated, self-interested choices of individual travelers and the collective goal of system-wide efficiency. This divergence is captured by the concepts of User Equilibrium (UE) and System Optimum (SO). This chapter advances a unifying perspective based on *premagic capacity* [18]: when the capacity matrix is premagic, the network attains a *Premagic Equilibrium* (PE) with uniform congestion. This state represents an ideal condition for traffic distribution in transport systems. When two mild design axioms hold, *Steady Link Characteristics* (SLC) and *Identical Path Link Counts* (IPLC), the classic Wardrop equilibria align exactly: *User Equilibrium* (UE) coincides with the *System Optimum* (SO), yielding UE $=$ SO $=$ PE. These assumptions, known as the Idealized Axioms for Convergence, allow us to connect two key concepts in traffic analysis: User Equilibrium (UE) and System Optimum (SO), into a single, unified state known as Premagic Equilibrium (PE).

We will explore this theory as the critical topological condition that resolves this conflict and show its practical applications. While the ideas presented are highly theoretical and rely on some unrealistic assumptions about network uniformity, they are incredibly useful.

Think of this chapter as a theoretical exercise that explores the consequences of a highly constrained, perfect system. This isn't to say that all networks naturally achieve this state. Instead, the theory offers a design philosophy for new towns, suggesting how we can approximate these conditions to create a more efficient transportation system. The theory's focus is on the supply side of traffic, things like link properties, network topology, and the relationship between flow and capacity. It doesn't directly assume a uniform distribution of traffic demand (Origin-Destination, or OD demand). Instead, the PE assumption ($\mathsf{F} = \sigma \mathsf{C}$) implies that the most efficient flow distribution is one that is proportional to capacity. If a network's

K. Teknomo, *Ideal Flow Network for Transportation Research*,
Lecture Notes in Intelligent Transportation and Infrastructure,
https://doi.org/10.1007/978-981-95-4727-2_13

demand pattern aligns with this ideal flow, it will result in uniform congestion. This chapter will give a rigorous understanding of this powerful axioms and its ecosystem of supporting conditions, namely Premagic Equilibrium (PE), Steady Link Characteristics (SLC) and Identical Path Link Count (IPLC) and show how combining these three elements would align UE and SO.

To measure progress toward this ideal, we propose the *VarPi* index $\varpi = \bar{g}/\xi$, a congestion-balance indicator that compares capacity-weighted average congestion to peak congestion. From this, we derive the *Network Flow Identity* (NFI), $\kappa = \tilde{c}\,\varpi\,\xi$, a macro relationship that exposes the levers available to policy: expand capacity, improve operational efficiency, or adjust acceptable peak congestion.

13.2 Theoretical Foundations

We begin with the formal concepts that define a network and its equilibrium state. We build these concepts from basic axioms to more complex theorems. The framework is rooted in the Ideal Traffic Assignment (ITA) methodology, a model for predicting traffic flow based on Ideal Flow Network (IFN).

13.2.1 Core Concepts and Definitions

A transport network is a directed graph $\mathcal{G} = (\mathcal{V}, \mathcal{E})$, where $\mathcal{V}$ is a set of nodes (intersections) and $\mathcal{E}$ is a set of edges (links). The network must be strongly connected. Strongly connected means there is a directed path from every node to every other node. Therefore, the pattern of its adjacency matrix A is irreducible[8]. In cases where the network is not inherently strongly connected, this structural requirement can be met by introducing a conceptual external node that balances flows across disconnected components. This extension preserves the conditions necessary for irreducible matrix and is discussed in detail in Chap. 7.

Definition 13.1 (Capacity Matrix) The capacity of the network is given by a non-negative irreducible square matrix $\mathsf{C} = [c_{ij}]$. The element $c_{ij} > 0$ is the capacity of the directed link from node i to node j. If there is no direct link, $c_{ij} = 0$. The total capacity of the network $\tilde{c}$ is the sum of all its elements.

Definition 13.2 (Flow and Congestion Matrices) The ideal flow in the network is given by a matrix $\mathsf{F} = [f_{ij}]$. The element f_{ij} is the volume of traffic on the link from node i to node j. The congestion on each link is the ratio of flow to capacity. It is represented by the congestion matrix G, defined by the Hadamard (element-wise) division:

$$\mathsf{G} = \mathsf{F} \oslash \mathsf{C} \quad \text{where } (\mathsf{A} \oslash \mathsf{B})_{ij} = a_{ij}/b_{ij} \text{ for } b_{ij} \neq 0. \tag{13.1}$$

Definition 13.3 (Steady Matrix) A matrix B is a steady matrix if all of its non-zero entries are equal to a single constant, σ. We can write this as $\mathsf{B} = \sigma \mathsf{A}$, where A is the (0,1) adjacency matrix of the network. A steady matrix represents a state of perfect uniformity.

Definition 13.4 (Premagic Capacity Matrix) A capacity matrix C is premagic if its row sums equal its corresponding column sums. Let $\mathbf{j}$ be a column vector of ones. The condition for a premagic matrix is:

$$\mathsf{C}\mathbf{j} = \mathsf{C}^T\mathbf{j} \tag{13.2}$$

This property means that, for every node i, the total capacity of all outgoing links equals the total capacity of all incoming links.

A premagic matrix represents a state of perfect balance. Imagine a system where the total weight of what goes into a node is always equal to the total weight of what comes out. For a traffic network's capacity matrix, this means the total capacity of all roads entering an intersection equals the total capacity of all roads leaving that intersection. This property describes a structurally balanced network.

Example 13.1 (Checking the Premagic Property) Consider the following relative capacity matrix C:

$$\mathsf{C} = \begin{bmatrix} 0\,3\,2\,1 \\ 1\,0\,4\,3 \\ 5\,4\,0\,2 \\ 0\,1\,5\,0 \end{bmatrix}$$

To check if C is premagic, we calculate the row and column sum vectors.

$$\mathsf{C}\mathbf{j} = \begin{bmatrix} 0\,3\,2\,1 \\ 1\,0\,4\,3 \\ 5\,4\,0\,2 \\ 0\,1\,5\,0 \end{bmatrix} \cdot \begin{bmatrix} 1 \\ 1 \\ 1 \\ 1 \end{bmatrix} = \begin{bmatrix} 6 \\ 8 \\ 11 \\ 6 \end{bmatrix}, \quad \mathsf{C}^T\mathbf{j} = \begin{bmatrix} 0\,1\,5\,0 \\ 3\,0\,4\,1 \\ 2\,4\,0\,5 \\ 1\,3\,2\,0 \end{bmatrix} \cdot \begin{bmatrix} 1 \\ 1 \\ 1 \\ 1 \end{bmatrix} = \begin{bmatrix} 6 \\ 8 \\ 11 \\ 6 \end{bmatrix}$$

Since $\mathsf{C}\mathbf{j} = \mathsf{C}^T\mathbf{j}$, the matrix C is premagic. The network capacity is balanced.

We need to distinguish the premagic capacity and premagic adjacency matrix (i.e. Eulerian Network).

Definition 13.5 (Eulerian Network) [6] A directed graph is called Eulerian network if and only if it is strongly connected and its adjacency matrix A is premagic. This property implies that the in-degree of every node is equal to its out-degree. The in-degree of node i is $\sum_j a_{ji}$ (sum of column i) and the out-degree is $\sum_j a_{ij}$ (sum of row i).

Example 13.2 (Non-premagic Capacity Eulerian Network) Here is an example of an Eulerian network which is not premagic capacity.

$$
\mathsf{C} = \left[\begin{array}{c|ccccc|c}
\nearrow & a & b & c & d & e & \Sigma \\
\hline
a & 0 & 1 & 0 & 2 & 1 & 4 \\
b & 2 & 0 & 1 & 0 & 0 & 3 \\
c & 0 & 2 & 0 & 1 & 2 & 5 \\
d & 3 & 0 & 1 & 0 & 0 & 4 \\
e & 1 & 0 & 2 & 0 & 0 & 3 \\
\hline
\Sigma & 6 & 3 & 4 & 3 & 3 & 19
\end{array}\right]
$$

The adjacency pattern of this capacity matrix is premagic, therefore it is an Eulerian network.

$$
\mathsf{A} = \left[\begin{array}{c|ccccc|c}
\nearrow & a & b & c & d & e & \Sigma \\
\hline
a & 0 & 1 & 0 & 1 & 1 & 3 \\
b & 1 & 0 & 1 & 0 & 0 & 2 \\
c & 0 & 1 & 0 & 1 & 1 & 3 \\
d & 1 & 0 & 1 & 0 & 0 & 2 \\
e & 1 & 0 & 1 & 0 & 0 & 2 \\
\hline
\Sigma & 3 & 2 & 3 & 2 & 2 & 12
\end{array}\right]
$$

Example 13.3 (Premagic Capacity and Eulerian Network) We give an example of a capacity matrix, which is Eulerian and also premagic.

$$
\mathsf{C} = \left[\begin{array}{c|ccccc|c}
\nearrow & a & b & c & d & e & \Sigma \\
\hline
a & 0 & 1 & 0 & 1 & 2 & 4 \\
b & 2 & 0 & 1 & 0 & 0 & 3 \\
c & 0 & 2 & 0 & 1 & 1 & 4 \\
d & 1 & 0 & 1 & 0 & 0 & 2 \\
e & 1 & 0 & 2 & 0 & 0 & 3 \\
\hline
\Sigma & 4 & 3 & 4 & 2 & 3 & 16
\end{array}\right]
$$

The adjacency pattern of this capacity matrix is the exactly same as in Example 13.2, which is premagic. Thus, it is an Eulerian network.

$$
\mathsf{A} = \left[\begin{array}{c|ccccc|c}
\nearrow & a & b & c & d & e & \Sigma \\
\hline
a & 0 & 1 & 0 & 1 & 1 & 3 \\
b & 1 & 0 & 1 & 0 & 0 & 2 \\
c & 0 & 1 & 0 & 1 & 1 & 3 \\
d & 1 & 0 & 1 & 0 & 0 & 2 \\
e & 1 & 0 & 1 & 0 & 0 & 2 \\
\hline
\Sigma & 3 & 2 & 3 & 2 & 2 & 12
\end{array}\right]
$$

13.2.2 *The Premagic Equilibrium and Its Conditions*

The central idea of this chapter connects the structural property of the capacity matrix to the operational state of the network. Now we can formally define the state of Premagic Equilibrium.

Definition 13.6 (Premagic Equilibrium) A network is in a state of Premagic Equilibrium if its congestion matrix G is a steady matrix. This is an operational state where every link in the network experiences the same level of congestion.

Theorem 13.1 (Condition for Premagic Equilibrium) *A strongly connected network with an irreducible, premagic capacity matrix* C *achieves a state of Premagic Equilibrium.*

Proof For a network with an irreducible and premagic capacity matrix C, the ideal flow matrix F is a scalar multiple of the capacity matrix. We can write this relationship as:

$$\mathsf{F} = \sigma \times \mathsf{C} \tag{13.3}$$

where σ is a positive scaling factor. The factor σ depends on the total flow volume in the network. Now, we use the definition of the congestion matrix G from Eq. (13.1). We substitute the relationship from Eq. (13.3):

$$\mathsf{G} = \mathsf{F} \oslash \mathsf{C} = (\sigma \times \mathsf{C}) \oslash \mathsf{C}$$

For every element (i, j) where $c_{ij} > 0$, the corresponding element in G is $g_{ij} = (\sigma c_{ij})/c_{ij} = \sigma$. Therefore, all non-zero entries of the congestion matrix G are equal to the constant g. By Definition 13.3, G is a steady matrix. By Definition 13.6, the network is in a state of Premagic Equilibrium. □

This theorem reveals that if a network's capacity is perfectly balanced (a premagic matrix), then the natural, ideal flow will be distributed in perfect proportion to that capacity. This results in uniform congestion across all links. In simple terms, a balanced network design inherently leads to a balanced operational state.

Banavar et al. [1] explored optimal transport networks with flow conservation and symmetry constraints, which mirror the structural balance required in premagic networks.

Example 13.4 (Achieving Premagic Equilibrium) Consider a 4-node network with a total traffic volume $\kappa = 41{,}850$ vehicles per hour. The absolute ideal flow matrix in vehicles per hour is given as

$$\mathsf{F} = \begin{bmatrix} 0 & 4050 & 2700 & 1350 \\ 1350 & 0 & 5400 & 4050 \\ 6750 & 5400 & 0 & 2700 \\ 0 & 1350 & 6750 & 0 \end{bmatrix}$$

Let the relative capacity matrix C_{rel} in number of lanes is premagic:

$$\mathsf{C}_{rel} = \begin{bmatrix} 0 & 3 & 2 & 1 \\ 1 & 0 & 4 & 3 \\ 5 & 4 & 0 & 2 \\ 0 & 1 & 5 & 0 \end{bmatrix}$$

Suppose the standard capacity per lane is 1800 vehicles per hour per lane. The absolute capacity matrix in vehicles per hour becomes

$$\mathsf{C} = 1800 \times \mathsf{C}_{rel} = \begin{bmatrix} 0 & 5400 & 3600 & 1800 \\ 1800 & 0 & 7200 & 5400 \\ 9000 & 7200 & 0 & 3600 \\ 0 & 1800 & 9000 & 0 \end{bmatrix}$$

The row and column sum vectors are both $[10{,}800, 14{,}400, 19{,}800, 10{,}800]^T$. The total absolute capacity is 55,800 vehicles/hour. The ratio of total flow to total capacity is dimensionless value, $\sigma = 41{,}850/55{,}800 = 0.75$. The ideal flow is $\mathsf{F} = \sigma \times \mathsf{C}$. The congestion matrix is:

$$\mathsf{G} = \mathsf{F} \oslash \mathsf{C} = \begin{bmatrix} - & 0.75 & 0.75 & 0.75 \\ 0.75 & - & 0.75 & 0.75 \\ 0.75 & 0.75 & - & 0.75 \\ - & 0.75 & 0.75 & - \end{bmatrix} = \sigma \mathsf{A}$$

The congestion matrix is a steady matrix with $\sigma = 0.75$. The network is in Premagic Equilibrium.

13.3 The Convergence of Optimalities

The state of Premagic Equilibrium has a profound connection to Wardrop's principles, but this alignment requires additional conditions. In most cases, a Premagic Equilibrium alone does not guarantee that User Equilibrium and System Optimum will coincide. We now establish the axioms that, when combined with a Premagic Equilibrium, achieve this convergence.

13.3.1 Foundational Axioms for Ideal Traffic Assignment

These axioms outline the core assumptions that govern the Ideal Traffic Assignment (ITA) framework, which we've previously discussed in detail in Chap. 9. They are the foundational principles that allow us to model and predict traffic flow in a consistent and mathematically sound way. They formalize how drivers make decisions, how flow moves through the network, and how we measure the impact of congestion.

Axiom 13.1 (Markov Splitting) *From a capacity matrix* C *with non-negative entries, we can derive a row-stochastic matrix* S *such that each entry* s_{ij} *is proportional to the corresponding entry* c_{ij}*. This is defined by the element-wise division:*

$$\mathsf{S} = \mathsf{C} \oslash ((\mathsf{C}\mathbf{j})\mathbf{j}^T)$$

where $\mathbf{j}$ *is a column vector of ones and* $\oslash$ *denotes element-wise division.*

Axiom 13.2 (Behavioral Stochastic Matrix (BSM)) *The stochastic matrix* $\mathsf{S} = [s_{ij}]$ *represents the equilibrium microscopic turning movement probabilities at each node. Each* s_{ij} *reflects the proportion of flow that chooses to exit via link* (i, j)*, based on drivers' route choices under prevailing network conditions.*

The stochastic matrix, S, models how drivers choose their paths at an intersection. Each element, s_{ij}, represents the probability that a driver arriving at node i will choose to exit via link (i, j). It reflects the collective route choices of all drivers under a given set of network conditions. In essence, it's a way to capture and predict the microscopic turning movements that, in aggregate, define the overall traffic pattern.

Axiom 13.3 (Flow Consistency (Conservation and Irreducibility)) *For every node* i *in the network, the total inflow equals the total outflow, ensuring local flow conservation. The network's flow topology, as represented by* F*, forms an irreducible Markov chain, implying that flow can eventually reach any node from any other node.*

This Axiom 13.3 ensures that the network is a closed, balanced system. First, it requires that traffic is conserved at every intersection: the total number of vehicles entering an intersection must equal the total number of vehicles leaving it. Second, it states that the network's flow topology, as represented by F, forms an irreducible Markov chain. This simply means that it's always possible for flow to get from any point in the network to any other point. This reflects a self-contained, steady-state transportation system.

Axiom 13.4 (Well-Behaved Travel Time Functions) *The travel time t_{ij} for any link (i, j) is a continuous, strictly increasing, and strictly convex function of its congestion ratio g_{ij}. It also depends on the link's intrinsic characteristics, such as length and maximum speed. Mathematically, $t_{ij} = cost(g_{ij}, \mathsf{U}_{ij}, \mathsf{L}_{ij})$ where the first and second derivatives with respect to g_{ij} are positive.*

This Axiom 13.4 is about how we quantify the cost of travel. It asserts that the travel time, t_{ij}, on any link is a predictable function of its congestion, g_{ij}, as well as its inherent characteristics like length and speed limit. The term "well-behaved" means that this function is smooth (continuous), that travel time always increases as congestion increases (strictly increasing), and that this increase accelerates as conditions worsen (strictly convex). This ensures that our model of congestion is realistic and that there's a unique and predictable cost associated with network flow. BPR function Proposition 9.8 that we have discussed in Chap. 9 satisfied this axiom.

Axiom 13.5 (Total Flow Parameter (κ)) *The total magnitude of flow in the network, $\kappa = \sum_{(i,j)\in E'} f_{ij}$, is treated as a given input parameter for determining the absolute flow magnitudes within the IFN model. The IFN model predicts the distribution of this total flow.*

The total flow parameter, κ, represents the overall volume of traffic in the entire network at a given time. The IFN model uses this parameter as an input to determine the absolute volume of traffic on each individual link, but it's the model itself that determines how that total flow is distributed across the network. As we have discussed in Chap. 11 the total flow parameter, κ, can be calibrated based on real world flow data.

13.3.2 The Idealized Axioms for Convergence

To demonstrate the convergence of User Equilibrium (UE) and System Optimum (SO), we introduce three additional, idealized network axioms.

Definition 13.7 (Active Links) A link $(i, j) \in \mathcal{E}$ is considered an *active link* if its capacity $c_{ij} > 0$ and it carries a positive ideal flow $f_{ij} > 0$. The set of all active links is denoted E'.

Definition 13.8 (Path) A path p from an origin node $o \in \mathcal{V}$ to a destination node $d \in \mathcal{V}$ is a sequence of links $(v_0, v_1), (v_1, v_2), \ldots, (v_{k-1}, v_k)$, where $v_0 = o$ and $v_k = d$. The number of links in the path, k, is defined as the path length or hop count, denoted by $|p|$.

Not all possible paths in a network are utilized by travelers. We are concerned only with paths that carry a non-zero amount of flow. An *active path* is a path composed entirely of active links.

Definition 13.9 (Active Path) Let f_p be the flow on a path p. A path p is considered an *active path* if and only if $f_p > 0$. The set of all active paths between an origin o and a destination d is denoted by $\mathcal{P}_{od}^{A}$.

An active link is defined as a link with both a positive capacity ($c_{ij} > 0$) and a positive ideal flow ($f_{ij} > 0$). The term active is serves to distinguish between links that are part of the network's topology and those that are actually utilized in the flow distribution. While an irreducible flow matrix guarantees that no node is completely isolated, it does not mean that every single potential link must carry a non-zero flow. The path-finding mechanisms of the Ideal Traffic Assignment framework may direct all flow for a given origin-destination pair through an alternative route, leaving a valid link with zero flow. The network remains strongly connected and the flow matrix remains irreducible, as other paths still ensure connectivity. With these preliminary definitions, we can formally state the central axioms. The number of links in a path, its hop count, is a fundamental topological metric. Its importance becomes paramount under specific equilibrium conditions. To understand why, we must presuppose two ideal conditions. To see why simply counting the number of road segments on a trip becomes so important, we first need to imagine a highly idealized city. We will assume two simplifying rules are in effect for our network. These rules, though unrealistic on their own, allow us to uncover a deep truth about the network's structure.

Axiom 13.6 (Premagic Equilibrium (PE)) *A network is in Premagic Equilibrium if its congestion matrix,* G*, contains a constant value for all active links. That is, for all* $(i, j) \in E'$*, the condition* $g_{ij} = \sigma$ *is satisfied, where* σ *is a positive scalar constant.*

The Premagic Equilibrium Axiom 13.6 defines a state of perfectly balanced network utilization. This axiom is critical for simplifying network behavior and enabling optimality.

Axiom 13.7 (Steady Link Characteristics (SLC)) *For all active links* $(i, j) \in E'$*, the link length,* l_{ij}*, is a constant value* $l > 0$*, and the link maximum speed,* u_{ij}*, is a constant value* $u > 0$*. This axiom creates homogeneous per-link free-flow time. There exists a constant* $\tau_0 > 0$ *such that*

$$\frac{l_{ij}}{u_{ij}} \equiv \tau_0 \quad \textit{for every active link.} \tag{13.4}$$

The Steady Link Characteristics Axiom 13.7 ensures that all active links have identical physical and operational properties, thus creating a homogeneous cost landscape. This defines a strict condition where all active links possess uniform physical properties, ensuring homogeneous cost functions for all links.

Axiom 13.8 (Identical Path Link Counts (IPLC)) *For every origin-destination (O-D) pair (o, d), all active paths between o and d consist of exactly the same number of links, denoted by k_{od}, which is a constant for all active paths of a specific O-D pair.*

The Identical Path Link Counts Axiom 13.8 imposes a topological constraint, ensuring that alternative routes are structurally equivalent. This concept [4] arises in network observability and OD matrix estimation, where multiple paths share identical link counts, complicating uniqueness in flow inference. The IPLC axiom imposes a strict rule of geometric uniformity on the routes that are actually in use. Think of it as a city regulation stating that all official taxi routes from the airport to the central station must involve exactly ten turns. The axiom doesn't concern itself with the traffic on those routes or why drivers chose them; it simply states that any route that is actively being used must conform to this single, uniform link count. It is a condition on the network's topology as experienced by its users, not a condition on the flow itself.

Definition 13.10 (Geodesic Subgraph (Shortest-Path Subgraph)) For each OD pair (s, t), define the geodesic subgraph $\mathcal{G}_{st} = \{(i, j) : \exists\ P \in \mathcal{P}_{st} \text{ with } (i, j) \in P \text{ and } |P| = d(s, t)\}$.

For each origin-destination (OD) pair (s, t), we focus only on shortest paths. The geodesic subgraph $\mathcal{G}_{st}$ includes all edges (i, j) that appear in at least one shortest path from s to t. Formally, an edge (i, j) is in $\mathcal{G}_{st}$ if there exists a path P from s to t such that (i, j) is part of P, and the length of P equals the shortest possible distance $d(s, t)$. Imagine we want to go from point s to point t using the shortest route. There might be several shortest paths, not just one. The geodesic subgraph is the collection of all road segments (edges) that appear in *any* shortest path from s to t. If a segment is part of a longer route, we ignored. It must be part of a shortest one.

Axiom 13.9 (Geodesic Gating at Splits) *Routing is restricted to geodesic edges for each OD: only arcs in $\mathcal{G}_{st}$ may carry (s, t) flow. Within $\mathcal{G}_{st}$, outflows split according to the Markov split Axiom 13.1 S (capacity-proportional split on the available geodesic arcs).*

Flow from s to t is allowed only on edges in $\mathcal{G}_{st}$. No detours or longer paths are permitted, routing is strictly shortest-path. At each node where flow splits, the split follows a Markov split Axiom 13.1 that outflow is divided proportionally to the capacity of outgoing geodesic edges. This ensures a memoryless, capacity-aware distribution of flow. Geodesic Gating means we only allow traffic from s to t to use those shortest-path segments. No wandering off the shortest route because only the geodesic edges are valid. When flow reaches a junction (a split), it divides based on how much capacity each outgoing edge has. Bigger roads get more flow, smaller ones get less, like water choosing wider pipes. This split follows a simple rule: it doesn't look ahead or remember past choices. Drivers just use local info instead of global information.

13.3.3 Propositions on Network State Under Ideal Conditions

Proposition 13.1 (Constant Link Travel Time) *The simultaneous satisfaction of conditions stated in Axioms 13.6 (PE) and 13.7 (SLC) implies that the travel time, t_{ij}, for every link in the network is a constant scalar value.*

Proof Let (i, j) be an link in the network. The travel time, t_{ij}, is a function of the congestion ratio, link length, and maximum speed. We can express this relationship as $t_{ij} = \text{cost}(g_{ij}, l_{ij}, u_{ij})$. From Axiom 13.6 (PE), we know that the congestion ratio is constant for all links: $g_{ij} = \sigma$. From Axiom 13.7 (SLC), we know that the link length and maximum speed are also constant for all links: $l_{ij} = l$ and $u_{ij} = u$. Substituting these constant values into the travel time function, we obtain:

$$t_{ij} = \text{cost}(\sigma, l, u)$$

Since σ, l, and u are all constants, the output of the function $\text{cost}(\cdot)$ is also a constant value. Let this constant link travel time be denoted as t_{link}. Therefore, for every link (i, j), $t_{ij} = t_{\text{link}}$. □

This Proposition 13.1 states that if a network satisfies both Premagic Equilibrium (PE) Axiom 13.6 and Steady Link Characteristics (SLC) Axiom 13.7, then every active link will have the exact same travel time. This occurs because PE ensures uniform congestion across all active links, while SLC guarantees that all active links are physically identical in terms of length and maximum speed. Together, these two conditions eliminate all variables in the travel time function, leaving a single, constant value.

Lemma 13.1 (Constant Path Travel Time) *The simultaneous satisfaction of conditions stated in Axioms 13.6 (PE), 13.7 (SLC), and 13.8 (IPLC) implies that the total travel time for every path between a given O-D pair is a constant scalar value.*

Proof Let (o, d) be an Origin-Destination pair. Let P be any path between o and d. The total travel time for path P, denoted t_P, is the sum of the travel times of its constituent links:

$$t_P = \sum_{(i,j)\in P} t_{ij}$$

From Proposition 13.1, under Axioms 13.6 (PE) and 13.7 (SLC), every link (i, j) has a constant travel time, $t_{ij} = t_{\text{link}}$. From Axiom 13.8 (IPLC), every path between o and d consists of exactly k_{od} links. Therefore, the total travel time for any path P between o and d is:

$$t_P = k_{od} \times t_{\text{link}}$$

Since k_{od} is a constant for a given O-D pair and t_{link} is a global constant, t_P is a constant value for all paths between (o, d). □

Building on the previous proposition, this Lemma 13.1 demonstrates that if we also add the Identical Path Link Counts (IPLC) Axiom 13.8, then every active path between a given origin-destination pair will have the same total travel time. Since all active links have a uniform travel time, and all active paths have the same number of links, the total travel time for any path is simply the constant link time multiplied by the constant number of links. This sets the stage for proving that the network is in a User Equilibrium. The proposition above provides the crucial link: if our first ideal rule (Premagic Equilibrium) holds, then travel time is decided by a single factor: the number of segments in the journey. A route with 8 segments will take exactly twice as long as a route with 4 segments. Suddenly, a simple topological feature becomes central. That feature is the hop count. It becomes the only factor a rational traveler needs to consider. Therefore, enforcing a uniform hop count (via the IPLC axiom) is equivalent to enforcing a uniform travel time, which is a cornerstone of traffic equilibrium.

Proposition 13.2 (Path Travel Time Under PE) *If a network is in a state of Premagic Equilibrium, the total travel time t_p of any active path p is directly proportional to its hop count $|p|$. Specifically, $t_p = |p| \times \bar{t}$.*

Proof Let p be an active path. By definition, every link $e \in p$ must carry non-zero flow. Under the condition of PE, the travel time t_e for every link $e \in p$ is a constant value $\bar{t}$. The total travel time t_p is the sum of the travel times of its constituent links:

$$t_p = \sum_{e \in p} t_e$$

Since there are $|p|$ links in the path, and each has a travel time of $\bar{t}$, the sum becomes:

$$t_p = \sum_{i=1}^{|p|} \bar{t} = |p| \times \bar{t}$$

Thus, the total path travel time is a direct product of its hop count and the uniform link travel time. □

The proposition above provides the crucial link: if our first ideal rule (Premagic Equilibrium) holds, then travel time is decided by a single factor: the number of segments in the journey. A route with 8 segments will take exactly twice as long as a route with 4 segments. At that moment, a single topological feature takes precedence. Hop count alone guides the rational traveler's decision. Other factors fade into the background. Therefore, enforcing a uniform hop count (via the IPLC axiom) is equivalent to enforcing a uniform travel time, which is a cornerstone of traffic equilibrium.

Proposition 13.3 (Identical Path Link Counts in a Premagic UE) *Under the conditions of a User Equilibrium (UE) and a Premagic Equilibrium (PE), it must hold that for any given origin-destination (O-D) pair* (o, d)*, all active paths* $p_1, p_2 \in \mathcal{P}^A_{od}$ *have the same hop count:* $|p_1| = |p_2|$.

Proof

1. Let the network be in a state of User Equilibrium. By Definition 13.11, the travel times on all active paths between an O-D pair must be equal. For any two active paths p_1 and p_2, we have $T_1 = T_2$.
2. Let the network also be in a state of Premagic Equilibrium. By Proposition 13.2, the travel time of any active path is directly proportional to its hop count: $t_p = |p| \times \bar{t}$.
3. Combining these two conditions, we can state that $|p_1| \times \bar{t} = |p_2| \times \bar{t}$.
4. Since the uniform link travel time $\bar{t}$ is a positive constant ($\bar{t} > 0$), we can divide both sides of the equation by $\bar{t}$.
5. This yields the necessary conclusion that $|p_1| = |p_2|$.

□

13.3.4 Theorems and Proofs of Convergence

Next, we relate PE to SO and UE. This section presents the core theoretical findings of the chapter. It demonstrates how the idealized axioms lead to the simultaneous satisfaction of both individual and system-wide optimality. The axioms serves as the critical link between local, link-level conditions and global, network-level optimality. To formalize this, we must first define the classical states of network equilibrium. To understand why the axioms are the keys, we must first define the two competing states of a traffic network. We can think of these as the network's two potential *personalities*: the selfish, individualistic personality that emerges from uncoordinated choices, and the cooperative, omniscient personality that represents perfect planning.

Definition 13.11 (User Equilibrium (UE)) A flow pattern defines a User Equilibrium if, for each O-D pair, the travel times on all active paths are equal. Furthermore, these travel times must be less than or equal to the travel time on any inactive path. This is Wardrop's First Principle, where no single user can improve their travel time by unilaterally changing their route.

Definition 13.12 (System Optimum (SO)) A flow pattern is a System Optimum if it minimizes the total system travel time. The total system travel time is the sum of the travel times on all links, weighted by the flow on each link: $Z = \sum_{e \in \mathcal{E}} f_e \times t_e(f_e)$. This represents the state of maximum operational efficiency for the network as a whole.

The divergence between UE and SO gives rise to the *Price of Anarchy* [15]. The axioms, when combined IPLC with PE and SLC, provides the precise conditions under which this divergence vanishes. It ensures that the selfish choices of individual users (leading to UE) align perfectly with the state of maximal collective good (SO). First, we will derive two theorems stating that User Equilibrium (UE) holds under certain idealized conditions. They arrive at that conclusion through slightly different assumptions and framing. First, we use shortest paths and shows longer ones are worse. Second, we use equal-length paths and shows all are equally good. Let $\mathcal{P}_{st}$ denote the set of $s \to t$ paths, and let $d(s,t)$ be the minimum number of links (edges) on any path in $\mathcal{P}_{st}$. Under uniform link time t_{link}, the cost of a path $P \in \mathcal{P}_{st}$ is $\text{cost}(P) = |P|\, t_{\text{link}}$.

Theorem 13.2 (PE Is UE Under Homogeneous Links and Geodesic Gating) *Under PE ($g \equiv \sigma$), Steady Link Characteristics (SLC) Axiom 13.7 where there exists a constant τ_0 for uniform per-link free-flow time, and geodesic gating Axiom 13.9, the Wardrop User Equilibrium holds. Indeed, for every OD (s,t) all used paths have equal minimal cost $d(s,t)\, t_{link}$, and no unused path has lower cost.*

Proof Geodesic gating ensures that any used (s,t) path has $|P| = d(s,t)$. Uniform link cost gives $\text{cost}(P) = d(s,t)\, t_{\text{link}}$ for all used P. Any unused non-geodesic path P' has $|P'| > d(s,t)$ and hence $\text{cost}(P') > |P|$, so it cannot attract flow at equilibrium. This relies on the assumption that any unused path, being non-geodesic, has a strictly higher cost, which is true under the model's assumption of uniform link times. □

Theorem 13.2 uses Uniform link cost (SLC), Geodesic gating (only shortest paths are used) and PE structure (flow conservation and balance) as the main assumptions. Imagine a network where every road takes the same time to traverse. Now suppose drivers always choose the shortest possible path (in terms of number of links). If the system is in Premagic Equilibrium (PE) and follows geodesic gating, then every driver ends up on a shortest path. In here, we show that unused paths are suboptimal. Since all shortest paths have the same cost, and longer paths cost more, no one has an incentive to switch. This matches the definition of User Equilibrium, everyone is doing the best they can given the circumstances.

Theorem 13.3 (Convergence to Nash Equilibrium [13, 14]) *The simultaneous satisfaction of conditions of being Premagic Equilibrium Axiom 13.6 (PE), Steady Link Characteristics Axiom 13.7 (SLC), and Identical Path Link Counts Axiom 13.8 (IPLC) implies that the traffic flow distribution is a Nash Equilibrium (User Equilibrium).*

Proof A traffic flow distribution is a Nash Equilibrium if two conditions are met for every Origin-Destination (O-D) pair:

1. Travel times on all utilized paths are equal and minimal.
2. No traveler can reduce their travel time by unilaterally changing their route.

Let us verify these two conditions under the three stated axioms.

1. *Equal Travel Times on Used Paths:* Consider any two (utilized) paths, P_1 and P_2, between an O-D pair (o, d). From Lemma 13.1, under Axioms 13.6, 13.7, and 13.8, the travel time for any path between (o, d) is a constant, $t = k_{od} \times t_{\text{link}}$. Therefore, the travel times on P_1 and P_2 are equal: $t_{P_1} = t_{P_2} = t$. The first condition is satisfied.
2. *No Advantage from Unused Paths:* Consider any path P' between (o, d) that is not being utilized. From Axiom 13.8 (IPLC), every path between (o, d) must consist of exactly k_{od} links. If P' were to be used, its links would be considered, and from Proposition 13.1, they would each have a travel time of t_{link}. The total travel time for P' would therefore be $t_{P'} = k_{od} \times t_{\text{link}}$. Since $t_{P'} = t$, a traveler switching to an unused path would not reduce their travel time. The second condition is also satisfied.

Because both conditions are met, the traffic flow distribution is a Nash Equilibrium.

□

Theorem 13.3 reaches the same conclusion using slightly different key assumptions. It assumes Uniform link cost (SLC), Identical path lengths (IPLC) and PE structure. Suppose every route between two points has the same number of links (IPLC). And every link takes the same time (SLC). Then every route takes the same total time. If the system is in PE, then flow is distributed in a way that respects conservation and symmetry. So, whether a driver sticks to their current route or switches to another, the travel time is the same. All paths equal, so no incentive to switch. No one can improve their situation by switching. That is a Nash Equilibrium, which in this context is equivalent to User Equilibrium. A Nash Equilibrium is a state where no driver feels they can get to their destination faster by taking a different route. This Theorem 13.3 proves that our idealized network achieves exactly that. Since all roads have the same travel time, and all possible routes between two points have the same number of roads, every route takes the exact same amount of time. There is no "faster" path to be found, so no driver has any incentive to switch. Individual self-interest leads directly to a perfectly stable outcome.

Example 13.5 (Numerical Verification of Nash Equilibrium) Let us construct an example to verify the theorem. We will use a 2x3 grid network and analyze the O-D pair from node (1) to node (6). We assume this network satisfies all three idealized axioms:

1. *PE:* Uniform congestion $\sigma = 0.9$.
2. *SLC:* All links have length $l = 1$ km and max speed $u = 60$ km/h.
3. *IPLC:* For the O-D pair (1,6), we consider two active paths, both of which have 3 links ($k_{1,6} = 3$).
   - Path A: (1) $\rightarrow$ (2) $\rightarrow$ (4) $\rightarrow$ (6)
   - Path B: (1) $\rightarrow$ (3) $\rightarrow$ (5) $\rightarrow$ (6)

First, we calculate the constant link travel time (t_{link}). Free-flow time $\mathsf{T}_{0,ij}$ = 1 km/60 km/h = 1 minute. Using the BPR function:

$$t_{\text{link}} = 1 \times (1 + 0.15 \times (0.9)^4) = 1 + 0.15 \times 0.6561 \approx 1.098 \text{ minutes} \tag{13.5}$$

Now, we numerically verify the two conditions for UE.

- *Condition 1: Equal Travel Times on Used Paths*
  - $t_A = |P_A| \times t_{\text{link}} = 3 \times 1.098 = 3.294$ minutes.
  - $t_B = |P_B| \times t_{\text{link}} = 3 \times 1.098 = 3.294$ minutes.
  - Since $t_A = t_B$, the first condition is satisfied.
- *Condition 2: No Advantage from Unused Paths*
  - Consider an alternative, unused path, Path C: (1) $\rightarrow$ (2) $\rightarrow$ (3) $\rightarrow$ (4) $\rightarrow$ (5) $\rightarrow$ (6). This path has 5 links.
  - $t_C = |P_C| \times t_{\text{link}} = 5 \times 1.098 = 5.490$ minutes.
  - We compare: $t_{\text{active}}(3.294 \text{ min}) < t_C(5.490 \text{ min})$.
  - Since the travel time on the unused path is greater, the second condition is satisfied.

Since both conditions are numerically verified, the flow distribution represents a stable Nash Equilibrium.

The next two theorems assert that Premagic Equilibrium (PE) leads to System Optimum (SO) under certain conditions. But they differ in how they arrive at that conclusion and what assumptions they rely on. The first theorem uses direct path cost comparison, basically says that if everyone takes the shortest path and roads are equal, we can't do better. the second theorem uses Jensen's Inequality on convex functions to tell us that if congestion is spread evenly and travel time grows convexly, we can't do better.

Theorem 13.4 (PE Is SO Under Homogeneous Links and Geodesic Gating) *Under the same conditions, the PE-induced assignment minimizes total system travel time (SO).*

Proof Total system travel time is

$$\text{TTT} = \sum_{(i,j)} f_{ij} t_{ij} = t_{\text{link}} \sum_{(i,j)} f_{ij} = t_{\text{link}} \sum_{(s,t)} \sum_{P \in \mathcal{P}_{st}} x_P \, |P|,$$

where x_P is the path flow and f_{ij} is the induced link flow. For each OD, minimizing $\sum_P x_P \, |P|$ subject to meeting the OD demand is achieved by placing all flow on paths with $|P| = d(s,t)$ (any convex combination thereof). Thus TTT is minimized when all OD flows are routed on geodesics; with geodesic gating and Markov splitting on those geodesics, the realized assignment is SO. □

Theorem 13.4 gives us an insight that if all drivers take the shortest possible route and all roads are equal, then the total time spent by everyone is as low as it can possibly be. Every road takes the same time to traverse (homogeneous links). Drivers are only allowed to use the shortest paths (geodesic gating). The total system travel time is minimized when all flow is routed along these shortest paths. Since PE already routes flow along geodesics (via Markov splitting), it automatically achieves the System Optimum.

Theorem 13.5 (Convergence to System Optimum) *The simultaneous satisfaction of three conditions of Premagic Equilibrium Axiom 13.6 (PE) and Steady Link Characteristics Axiom 13.7 (SLC) implies that the network is a System Optimum.*

Proof We will show that uniform congestion minimizes the total aggregate travel time. The System Optimum is the flow distribution that minimizes the total aggregate travel time for the entire system, Z. The objective function for minimization is:

$$Z = \sum_{(i,j)\in E'} f_{ij} \times t_{ij}$$

From Axiom 13.6 (PE), we have $f_{ij} = \sigma c_{ij}$. From Axiom 13.7 (SLC), we can assume a universal, strictly increasing, and strictly convex travel time function $\tau(\cdot)$ such that $t_{ij} = \tau(g_{ij}) = \tau(f_{ij}/c_{ij})$. We can redefine the objective function in terms of congestion:

$$Z = \sum_{(i,j)\in E'} c_{ij} \times \left(\frac{f_{ij}}{c_{ij}} \times t_{ij}\right) = \sum_{(i,j)\in E'} c_{ij} \times (g_{ij} \times \tau(g_{ij}))$$

Let $f(x) = x \times \tau(x)$. Since $\tau(x)$ is strictly increasing and strictly convex, its second derivative $f''(x) = 2\tau'(x) + x\tau''(x)$ is positive for $x > 0$. This confirms that $f(x)$ is a strictly convex function. This is a crucial property for proving the existence of a unique minimum. We seek to minimize Z subject to a fixed total flow,

$$\sum_{(i,j)\in E'} f_{ij} = \sum_{(i,j)\in E'} g_{ij} c_{ij} = \kappa$$

where κ is a constant. By applying Jensen's Inequality for the strictly convex function $f(x)$ with weights $w_{ij} = c_{ij}/C_{\text{total}}$, where $C_{\text{total}} = \sum c_{ij}$, we have:

$$f\left(\frac{\sum c_{ij} g_{ij}}{\sum c_{ij}}\right) \leq \frac{\sum c_{ij} f(g_{ij})}{\sum c_{ij}}$$

The equality holds if and only if all g_{ij} are equal. The inequality simplifies to $f(\kappa/C_{\text{total}}) \leq Z/C_{\text{total}}$, which gives a lower bound for Z. The minimum value of Z is achieved when the congestion ratio g_{ij} is uniform across all links, which is precisely the definition of a Premagic Equilibrium. □

Using Jensen's Inequality, we show that the total system travel time is minimized if and only if all congestion ratios, g_{ij}, are equal. This condition is precisely the definition of a Premagic Equilibrium. Therefore, a System Optimum is achieved if and only if the network is in a Premagic Equilibrium, given SLC. Theorem 13.5 stated that if traffic is spread evenly across roads, and congestion grows faster than linearly, then the total cost is minimized. PE does exactly that.

Example 13.6 (Numerical Verification of System Optimum) Consider a simple network with two parallel links.

- Total Flow: $\kappa = 2000$ veh/h.
- Link 1: Capacity $c_1 = 1500$ veh/h, Free-flow time $t_1^0 = 10$ min.
- Link 2: Capacity $c_2 = 1500$ veh/h, Free-flow time $t_2^0 = 10$ min.
- Travel Time Function: $t_i = t_i^0(1 + 0.15 \times (f_i/c_i)^4)$.

- *Case 1: Premagic Equilibrium (Uniform Congestion)* The optimal congestion ratio is $\sigma = 2000/3000 = 2/3$. This gives flows $f_1 = 1000$ and $f_2 = 1000$. The travel time on both links is $\approx$10.296 min.

$$Z_{\text{SO}} = (1000 \times 10.296) + (1000 \times 10.296) = 20{,}592 \text{ vehicle-minutes.} \tag{13.6}$$

- *Case 2: Non-Uniform Congestion* Let's shift flow to create an imbalance: $f_1 = 1200$ and $f_2 = 800$.
  - $t_1 \approx 10.614$ min.
  - $t_2 \approx 10.121$ min.

$$Z_{\text{Non-SO}} = (1200 \times 10.614) + (800 \times 10.121) = 20{,}833.6 \text{ vehicle-minutes.} \tag{13.7}$$

Verification: $Z_{\text{Non-SO}}(20{,}833.6) > Z_{\text{SO}}(20{,}592)$. This numerically confirms the theorem. The uniform congestion state (PE) is indeed the System Optimum.

Each road has a travel time that increases disproportionately with congestion (convex cost function). PE ensures that congestion is uniform across all roads. Jensen's Inequality tells us that spreading traffic evenly minimizes total travel time. By equalizing congestion, PE achieves the System Optimum. A System Optimum is about minimizing the total amount of time wasted by all drivers combined. This theorem demonstrates that in our idealized world, the most efficient way to run the network is to have every road operating at the same level of congestion.

The mathematical proof uses a concept called Jensen's Inequality, which, in this context, shows that because travel time increases disproportionately as congestion gets worse (a property of convex functions), the lowest total cost is always achieved by spreading the traffic out evenly. This prevents any single road from becoming so congested that it drastically increases the total travel time for the entire system. If we implement geodesic gating operationally (e.g., using signal control/route guidance to exclude non-shortest detours for each OD), then the capacity-proportional split S on the remaining geodesic arcs realizes both UE and SO while preserving the premagic balance on the active subgraph. If, in addition, every node's geodesic outgoing set contains multiple arcs, Markov splitting allocates those in exact capacity proportions without breaking Wardrop indifference, since all used arcs add the same t_{link}. If we guide traffic so that only shortest paths are used, and we split flow based on how much each road can handle, we get both individual fairness and system-wide efficiency. If intersections offer multiple shortest-path options, we can use a simple rule to divide traffic fairly without making anyone worse off.

Imagine we only allow travelers to take the shortest possible routes between their origin and destination (OD), and we block off any detours that aren't part of those shortest paths, using tools like traffic signals or route guidance. Now, if we divide traffic across those allowed paths in proportion to how much capacity each road segment has (i.e., wider roads get more flow), then we achieve both User Equilibrium (everyone chooses the best route for themselves) and System Optimum (overall traffic flow is efficient). We still maintain a nice premagic condition that the total flow into and out of each node (intersection) stays balanced. If every intersection has several shortest-path options to choose from, then we can use a simple probabilistic rule (Markov splitting) to divide traffic exactly according to road capacities, and this doesn't violate the principle that drivers are indifferent between all used routes because all the chosen paths take the same amount of time.

13.3.5 The Price of Anarchy and Braess' Paradox

The tension between individual and system-level optimality is well known in network theory. A classic illustration of this conflict is the Braess' Paradox [2, 3]. A key measure of a network's robustness is its immunity to paradoxes where intuitive improvements lead to systemic degradation.

Definition 13.13 (Braess' Paradox [3]) Braess' Paradox describes the phenomenon where adding a new link to a network can, counterintuitively, lead to an increase in the total system travel time under User Equilibrium.

Braess' Paradox states that adding a new road to a transportation network can, under certain conditions, increase travel time for every user. The paradox arises from the uncoordinated, selfish decisions of individual drivers. When a new, seemingly fast link is added, drivers, in pursuit of their own minimal travel time, shift to this new route. This individual action increases the congestion on the new link and its associated paths, which in turn can disrupt the flow on the original, longer paths. The network then settles into a new User Equilibrium (UE) that is less efficient than the previous state. This phenomenon provides a concrete example of how the "anarchy" of individual choices leads to a total system travel time that is sub-optimal

The paradox arises from a misalignment between individual incentives and collective optima. A network satisfying the conditions of our core theorem is structurally immune to this paradox.

Theorem 13.6 (Immunity to Braess' Paradox) *A network state that satisfies Premagic Equilibrium (PE) and the Identical Path Link Counts (IPLC) axiom is immune to Braess' Paradox.*

Proof

1. Let an initial network state satisfy PE,SLC and IPLC. For a given O-D pair, all active paths have a minimal hop count k_{min} and a corresponding travel time $T_{UE} = k_{min} \times \bar{t}$.
2. Braess' Paradox occurs when a newly added link creates an alternative path, p_{new}, that is initially perceived by users as faster, thereby drawing traffic.
3. Under PE, for p_{new} to be perceived as faster, its hop count must be strictly less than the existing active paths: $|p_{new}| < k_{min}$.
4. However, the initial equilibrium state is predicated on users already having selected the paths with the absolute minimum possible hop count, k_{min}.
5. By definition, it is topologically impossible for the addition of a new link between two existing nodes to create a new path between an O-D pair (o, d) that is shorter in hop count than the pre-existing shortest path. A new link can, at best, create a new path of equal length or reduce the minimal hop count itself.
6. If the new link creates a new path of length $|p_{new}| < k_{min}$, rational users will abandon the old paths and adopt the new shortest paths. The system will settle to a new UE with a travel time $T'_{UE} = |p_{new}| \times \bar{t}'$. Since $|p_{new}| < k_{min}$, the new equilibrium travel time will be lower than the old one. The system performance improves, and no paradox occurs.
7. The paradox occurs when the new path is selfishly attractive but globally inefficient. In a PE system, attractiveness (lower travel time) is directly and transparently tied to a topological property (lower hop count). There is no deception. A path with fewer hops is genuinely better.
8. Therefore, the paradox cannot arise. The conditions that trigger it are absent. Specifically, there is no divergence between perceived individual benefit and actual system cost.

□

Braess' Paradox is caused by a deceptive shortcut. Adding a new road tempts drivers onto a route that seems faster, but their collective choice overloads the network and makes everyone's trip longer. An network is immune to this deception because it operates under a simple, honest rule: fewer links mean a faster trip. The paradox relies on hidden costs, but in a PE system, the cost (hop count) is perfectly transparent. A new road cannot create a shortcut that is secretly worse for the system; if it's shorter, it's genuinely better. The Price of Anarchy (PoA) is a metric that formally quantifies this loss of system efficiency. PoA measures how inefficient selfish behavior (User Equilibrium, UE) is compared to the best possible outcome (System Optimum, SO). It is the ratio of the total system travel time at User

Equilibrium to the total system travel time at System Optimum. The PoA measures how much worse the system performs due to the absence of central coordination.

Definition 13.14 (Price of Anarchy (PoA)) [15] The Price of Anarchy is a scalar metric, PoA, t hat quantifies the efficiency of a network's User Equilibrium relative to its System Optimum. It is defined as the ratio of the total system travel time at User Equilibrium (Z_{UE}) to the total system travel time at System Optimum (Z_{SO}):

$$\text{PoA} = \frac{Z_{\text{UE}}}{Z_{\text{SO}}} \tag{13.8}$$

For real-world networks where the selfish choices of individuals lead to inefficiencies, the PoA is always greater than 1 ($PoA > 1$). Note that instead of using total system travel time (TSTT), sometimes, Price of Anarchy might also be computed based on Total System Delay.

13.3.6 The Unit Price of Anarchy

Perfect PE/SLC/IPLC is a theoretical ideal. Most real networks violate it. This leads to a quantifiable measure of inefficiency. To measure just how much inefficiency is caused by selfish drivers, scientists use a metric called the Price of Anarchy [15]. It's a simple ratio: how bad is the selfish outcome (UE) compared to the perfect, centrally-planned outcome (SO)? A PoA of 1.5 means the selfish traffic jam is 50% worse than it needs to be. A PoA of 1.0 is a perfect score. Unit Price of Anarchy (PoA equal 1) means UE is just as good as SO. There is no inefficiency. The Identical Path Link Counts (IPLC) Axiom 13.8 requires that for every origin-destination (O-D) pair, all *active paths* have the same number of links. The axiom's fulfillment is fundamental to proving that User Equilibrium (UE) and System Optimum (SO) coincide. We must distinguish between a network's potential topology and the actual flow distribution that satisfies this axiom.

1. *Trivial Satisfaction*: This occurs when only one path is used for an O-D pair. The condition that "all active paths have the same number of links" is trivially met because there is only one path to consider. For example, in a complete graph with uniform link travel times, drivers will always choose the direct, one-link path. All alternative, longer paths will remain inactive.
2. *Non-Trivial Satisfaction*: This is the ideal scenario where multiple paths are used, and all of them have an identical number of links. This requires a highly symmetrical network where multiple minimal-hop paths exist. Drivers distribute flow among these equally attractive paths, making all of them active simultaneously. For example, in a perfectly symmetrical grid network, the shortest paths between two non-adjacent corners are not unique in terms of links. Used in urban modeling and simulation, grid networks [5] simplify spatial

structure and support analytical tractability. This forces flow to distribute among multiple paths of the same minimal link count, satisfying the axiom in a non-trivial and powerful way.

Corollary 13.1 (The Ultimate Ideal Equilibrium) *The simultaneous satisfaction of three conditions, Premagic Equilibrium Axiom 13.6 (PE), Steady Link Characteristics Axiom 13.7 (SLC), and Identical Path Link Counts Axiom 13.8 (IPLC) implies that the traffic flow distribution simultaneously satisfies both the Nash Equilibrium (User Equilibrium) condition [13, 14] and the System Optimum (and thus Pareto Optimal [7]) condition.*

Proof The proof follows directly from the preceding theorems. From Theorem 13.3, we have proven that the three axioms collectively imply Nash Equilibrium. From Theorem 13.5, we have proven that Axioms 13.6 and 13.7 collectively imply System Optimum. Since Axiom 13.8 (IPLC) does not contradict or negate the conclusion of Theorem 13.5, the state achieved under all three axioms is both a Nash Equilibrium and a System Optimum. A System Optimum is inherently a Pareto Optimal state. Therefore, the network achieves both simultaneously. □

This Corollary 13.1 is the most profound deduction from Theorems 13.5 and 13.3. It means that in this theoretical, perfect world, what is best for the individual driver is also what is best for the entire system. There is no conflict between the goals of personal efficiency and social efficiency. By following their own self-interest, drivers naturally create a state of perfect system-wide optimization, and vice versa. In our idealized scenario, the inherent conflict between individual and system optimality is completely eliminated. The Braess' Paradox cannot occur because a perfect alignment between UE and SO is achieved.

Corollary 13.2 (Price of Anarchy Equals One) *Under PE, SLC axiom that produces τ_0, and geodesic gating, we have UE $=$ SO, and their realized link flows coincide with the Premagic Equilibrium (PE). Therefore, the Price of Anarchy is exactly one, indicating no inefficiency. The simultaneous satisfaction of conditions stated in Axioms 13.6 (PE), 13.7 (SLC), and 13.8 (IPLC) ensures this optimality.*

Proof From Corollary 13.1, we have proven that the three axioms collectively imply that the traffic flow distribution is simultaneously a Nash Equilibrium (User Equilibrium) and a System Optimum. This means that the total system travel time in the User Equilibrium state is identical to the total system travel time in the System Optimum state.

$$Z_{\text{UE}} = Z_{\text{SO}}$$

Substituting this equality into the definition of the Price of Anarchy, we obtain:

$$\text{PoA} = \frac{Z_{\text{UE}}}{Z_{\text{SO}}} = \frac{Z_{\text{SO}}}{Z_{\text{SO}}} = 1 \tag{13.9}$$

Therefore, the Price of Anarchy is exactly 1. □

In our idealized network of Premagic Equilibrium, we have proven that $Z_{UE} = Z_{SO}$, because the two states are identical. Therefore, under these conditions, the price of anarchy is precisely equal to one. The gap between individual and collective rationality is completely eliminated. The chaos and inefficiency demonstrated by the Braess' Paradox simply cannot happen under idealized network of Premagic Equilibrium. The network's structure is so perfectly balanced and its links are so uniform that any new link, or any change in routing, would not lead to a worse outcome for everyone. The self-optimizing behavior of individual drivers naturally results in a distribution of traffic that is also the most efficient for the system as a whole. The "anarchy" of selfish choices is eliminated, as all individual decisions align with the collective good. A PoA of 1 is the theoretical signature of a perfectly efficient network where individual and collective interests are in perfect harmony. This represents a state of perfect efficiency.

Example 13.7 (NE and SO Coincidence) Let us revisit the diamond grid network from Example 13.5, which satisfies all three axioms for the O-D pair (1,4).

- *Properties:* PE ($\sigma = 0.8$), SLC ($t_0 = 0.6$ min), IPLC ($k_{1,4} = 2$).
- *Derived values:* $t_{\text{link}} \approx 0.637$ min, Path travel time = 1.274 min.
- *Flow Distribution:* Let $\kappa = 1000$ veh/h. Let capacities $c_{12} = c_{13} = c_{24} = c_{34} = 625$ veh/h. The PE flow is $f_{ij} = 0.8 \times 625 = 500$ veh/h on each link.

1. *Is it a Nash Equilibrium?* As shown in Example 13.5, both paths have the same travel time (1.274 min). Therefore, it is a Nash Equilibrium.
2. *Is it a System Optimum?* The TSTT for this PE flow is $Z_{\text{PE}} = (4 \times 500) \times 0.637 = 1274$ vehicle-minutes. Consider a non-PE flow: 600 veh/h on the upper path and 400 veh/h on the lower path.
   - Path A (upper): $g = 0.96$, $t_A \approx 1.352$ min.
   - Path B (lower): $g = 0.64$, $t_B \approx 1.230$ min.

   The TSTT for this non-PE flow is $Z_{\text{Non-PE}} = (600 \times 1.352) + (400 \times 1.230) = 1303.2$ vehicle-minutes. Since $Z_{\text{Non-PE}} > Z_{\text{PE}}$, the PE flow distribution is the System Optimum.

The PE flow distribution is simultaneously the NE and the SO.

What happens when IPLC is violated? If a network provides two active paths of unequal length, say $|p_1| < |p_2|$, then achieving SO requires a violation of PE. To prevent all users from selfishly choosing the shorter path p_1, the system must induce a higher travel time on its links, such that the total path times are balanced. Imagine two groups of hikers must arrive at a campsite at the same time. One group has a short, 1-mile path; the other has a long, 2-mile path. To ensure a synchronized arrival (the System Optimum), the group on the short path must be forced to walk much slower than the group on the long path. This is what a network must do to achieve SO when IPLC is violated: it must artificially *slow down* the shorter-hop paths with congestion to make them just as unattractive as the longer-hop paths. Achieving SO in a non-IPLC world requires carefully engineered, non-uniform congestion, which is difficult to maintain. The final conclusion is a practical one. Perfect IPLC is unattainable. However, networks that approximate it by favoring symmetric, grid-like structures offer a robust strategy. These designs help minimize the Price of Anarchy. They embed optimality directly into the topology. As a result, they reduce the need for complex interventions to align individual and collective interests.

Example 13.8 (The Emergence of Anarchy When an Axiom Is Relaxed) Let's relax the *IPLC* axiom. Consider a network with two paths from O to D.

- *Properties:* $\kappa = 2000$. Both paths satisfy SLC ($t_0 = 10$ min, $C = 1500$).
- *Path 1:* A single link (highway). $|P_1| = 1$.
- *Path 2:* Two sequential links (local roads). $|P_2| = 2$.

1. *User Equilibrium State:* Drivers will ignore Path 2 because its free-flow time (20 min) is much higher than Path 1's. All 2000 vehicles will take Path 1.
   - $f_1 = 2000$, $f_2 = 0$.
   - $t_1 = 10 \times (1 + 0.15 \times (2000/1500)^4) \approx 14.74$ min. $t_2 = 20$ min.
   - The UE condition holds (used path time $<$ unused path time).
   - $Z_{\text{UE}} = (2000 \times 14.74) + (0 \times 20) = 29{,}480$ vehicle-minutes.
2. *A Better System Distribution (closer to SO):* What if a planner forced 500 vehicles onto Path 2?
   - $f_1 = 1500$, $f_2 = 500$.
   - $t_1 = 11.5$ min. $t_2 \approx 20.06$ min.
   - $Z_{\text{better}} = (1500 \times 11.5) + (500 \times 20.06) = 27{,}280$ vehicle-minutes.

   This distribution is better for the system ($Z_{\text{better}} < Z_{\text{UE}}$), but it is not a UE because drivers on Path 2 would immediately switch to the faster Path 1. The Price of Anarchy is PoA = $Z_{\text{UE}}/Z_{\text{SO}} > 29{,}480/27{,}280 \approx 1.08$. Relaxing just one axiom breaks the NE-SO convergence and introduces a Price of Anarchy.

13.4 VarPi Index: Measuring the Distance from Premagic Equilibrium

The theory of Premagic Equilibrium provides a practical goal for urban traffic management. Planners can design interventions that move a network closer to this ideal state, and we require metrics to measure this progress.

Since achieving a perfect premagic state is often not feasible, we need indicators to measure how close a network's current state is to the ideal. This is a goal-oriented approach where the ideal condition serves as a benchmark. The traditional use of average or median congestion as a performance index is often misleading [12], as it can encourage policies that solve localized problems at the expense of overall network efficiency. A correct performance index must encourage the efficient use of the entire network by measuring the gap between its current state and the ideal. There are several ways to measure the distance between a network's current congestion distribution and the uniform distribution of Premagic Equilibrium. Some well-known methods include relative entropy (Kullback-Leibler distance [9, 10]) (see Sect. 5.2.2) or Mahalanobis distance [11]. Among these metrics, the VarPi index [17] is more desirable for transportation field because can be applied to either the capacity or congestion matrices and yield exactly the same result.

Definition 13.15 (VarPi Index [17]) The VarPi (ϖ) index is a performance indicator that measures how close a congestion matrix is to the Premagic Equilibrium state. It is calculated from the positive entries of the congestion matrix G:

$$\varpi = 1 - \frac{\max(\mathsf{G}) - \text{average}(\mathsf{G})}{\max(\mathsf{G})} \tag{13.10}$$

The index ranges from 0 to 1 (or 100%). A value of 1 indicates a perfect Premagic Equilibrium, where the maximum congestion equals the (capacity-weighted) average congestion. A higher ϖ value indicates a more balanced network.

This index provides a simple, single-scalar value that summarizes the uniformity of a network's congestion. The VarPi index is very useful because to get it, we don't really need to compute congestion matrix. the following Proposition 13.4 shows that we can get the VarPi index ditectly from Capacity matrix, following Markov Splitting Axiom 13.1.

Proposition 13.4 (VarPi from Capacity Matrix) *The VarPi index computed from the congestion matrix* G *equals the VarPi index computed directly from the capacity matrix* C*:*

$$\varpi_{\mathsf{G}} = \varpi_{\mathsf{C}}$$

Proof Let ϕ be the standard capacity in vehicles/hour per lane (such as $\phi = 1800$). From the construction process of Ideal Traffic Assignment in Chap. 9:

$$\mathsf{S} = \mathsf{C} \oslash ((\mathbf{Cj})\mathbf{j}^T) \tag{13.11}$$

$$\mathsf{F} = \text{diag}(\boldsymbol{\pi})\mathsf{S} \tag{13.12}$$

$$\mathsf{G} = \mathsf{F}_{abs} \oslash \mathsf{C}_{abs} = \frac{\kappa \mathsf{F}}{\phi \mathsf{C}} = \frac{\kappa}{\phi}(\mathsf{F} \oslash \mathsf{C}) \tag{13.13}$$

Let $k = \frac{\kappa}{\phi}$ be a positive scalar constant. The VarPi index is defined as:

$$\varpi = 1 - \frac{\max(\mathsf{G}) - \text{average}(\mathsf{G})}{\max(\mathsf{G})}$$

This can be rewritten as:

$$\varpi = \frac{\text{average}(\mathsf{G})}{\max(\mathsf{G})}$$

Since $\mathsf{G} = k \times (\mathsf{F} \oslash \mathsf{C}_{rel})$ where $k > 0$:
$\max(\mathsf{G}) = k \times \max(\mathsf{F} \oslash \mathsf{C}_{rel})$
$\text{average}(\mathsf{G}) = k \times \text{average}(\mathsf{F} \oslash \mathsf{C}_{rel})$

$$\varpi_G = \frac{k \times \text{average}(\mathsf{F} \oslash \mathsf{C}_{rel})}{k \times \max(\mathsf{F} \oslash \mathsf{C}_{rel})} = \frac{\text{average}(\mathsf{F} \oslash \mathsf{C}_{rel})}{\max(\mathsf{F} \oslash \mathsf{C}_{rel})}$$

The scalar k cancels out, showing that ϖ is invariant under positive scalar multiplication. Since VarPi is scale-invariant:

$$\varpi_{\mathsf{G}} = \varpi_{\mathsf{F} \oslash \mathsf{C}}$$

Since $\mathsf{F} = \text{diag}(\boldsymbol{\pi})\mathsf{S}$ and based on Markov Splitting Axiom 13.1 S is the row-normalized version of C, we have:

$$\mathsf{F} \oslash \mathsf{C} = \frac{\text{diag}(\boldsymbol{\pi})\mathsf{S}}{\mathsf{C}} = \frac{\text{diag}(\boldsymbol{\pi})\mathsf{C} \oslash ((\mathbf{Cj})\mathbf{j}^T)}{\mathsf{C}}$$

This simplifies to:

$$\mathsf{F} \oslash \mathsf{C} = \text{diag}(\boldsymbol{\pi}) \oslash ((\mathbf{Cj})\mathbf{j}^T)$$

Since $(\mathbf{Cj})\mathbf{j}^T$ has identical values in each row, and $\text{diag}(\boldsymbol{\pi})$ contains the equilibrium distribution, the congestion pattern $\mathsf{F} \oslash \mathsf{C}$ reflects the equilibrium-weighted of the row sums of C. At Markov equilibrium, this distribution pattern is equivalent to the relative variation pattern in C itself, since both measure the same underlying network capacity heterogeneity. Therefore:

$$\varpi_{\mathsf{G}} = \varpi_{\mathsf{C}}$$

□

VarPi is a balance metric, which is also indirectly an efficiency and effectiveness target metric.

We must consider performance metrics within the context of transportation networks.

Definition 13.16 (Maximizing Throughput Efficiency) Given a fixed total flow κ and a fixed capacity distribution C, maximizing throughput efficiency corresponds to achieving the minimum possible total aggregate travel time (Z) for that flow. The state of minimum total aggregate travel time for a given total flow is the System Optimum. Our proofs that lead to Corollary 13.1 demonstrate that this minimum is achieved precisely when the congestion ratio g_{ij} is uniform across all active links. Therefore, a uniform congestion ratio is synonymous with maximal throughput efficiency under these conditions.

Definition 13.17 (Effectiveness) Effectiveness measures the degree to which a system achieves its stated goals or objectives. It is a multi-faceted concept in transportation:

- *User-Centric Effectiveness*: A network is effective if it facilitates users (i.e. driver) in making optimal choices that align with their individual utility maximization. The Nash Equilibrium is a state of high user-centric effectiveness, as no user can improve their outcome by unilaterally changing their route .
- *System-Centric Effectiveness*: A network is effective if its overall operation optimizes a defined system-wide objective function, such as minimizing total aggregate travel time. The System Optimum (which implies Pareto Optimality) is a state of high system-centric effectiveness .

In our idealized scenario where a Premagic Equilibrium holds, the network is exceptionally effective because it simultaneously achieves both user-centric and system-centric effectiveness. This makes it the ultimate theoretical ideal in terms of both efficiency and effectiveness.

13.5 The Network Flow Identity

Based on the desire to achieve Premagic Equilibrium (PE) using VarPi index as our performance metric, in this section, we derive a general *Network Flow Identity*, which expresses the relationship between *total flow*, *total capacity*, and *congestion* in road networks. This formula will help to understand how network congestion affects traffic flow in terms of its capacity and performance.

Let us start by defining again the key variables and parameters used throughout this derivation:

The *network capacity matrix* $\tilde{c} = [c_{ij}]$ represents the *capacity* of each road segment (i, j), measured in vehicles per hour. The *congestion matrix* $\mathsf{G} = [g_{ij}]$ represents the level of congestion on each road segment (i, j), where g_{ij} is the ratio of the flow f_{ij} to the capacity c_{ij}:

$$g_{ij} = \frac{f_{ij}}{c_{ij}}$$

The *total flow* κ represents the overall traffic demand across the network, which is the sum of the flows through all road segments:

$$\kappa = \sum_{i,j} f_{ij}$$

Definition 13.18 (Efficiency Indicator) The VarPi index ϖ as an *efficiency indicator* is a measure of how close the network is to *Premagic Equilibrium (PE)*, where congestion is uniform across all segments. It is defined as:

$$\varpi = 1 - \frac{\max(\mathsf{G}) - \text{average}(\mathsf{G})}{\max(\mathsf{G})} = \frac{\bar{g}}{\xi} \tag{13.14}$$

where $\xi = \max(g_{ij})$ is the maximum observed in the network and $\bar{g}$ is the capacity weighted average congestion.

$$\bar{g} = \frac{\sum_{ij} c_{ij} g_{ij}}{\sum_{ij} c_{ij}} = \frac{\sum_{ij} f_{ij}}{\sum_{ij} c_{ij}} \tag{13.15}$$

Definition 13.19 (Premagic Equilibrium (PE) State) In the *Premagic Equilibrium (PE)* state, the *congestion is uniform across all road segments*. That is, the congestion level g_{ij} is the same for all link ij, and we have:

$$g_{ij} = g \quad \forall ij \tag{13.16}$$

In this state, the *traffic flow* is proportional to the *total capacity* of the network.

Definition 13.20 (Non-premagic Equilibrium (Non-PE) State) In the *Non-Premagic Equilibrium (Non-PE)* state, the *congestion is non-uniform across road segments*. This means that *different road segments* may experience different levels of congestion. The *total flow* κ is determined by the *maximum congestion* ξ, the *efficiency indicator* ϖ, and the *total capacity* $\tilde{c}$.

We now begin with the propositional arguments that lead to the derivation of the *Network Flow Identity*.

Proposition 13.5 (Total Flow in PE State) *In the* Premagic Equilibrium (PE) *state, the congestion is uniform across all road segments, i.e.,* $g_{ij} = g$ *for all* ij*. The total flow* κ *is therefore given by the product of the* total capacity $\tilde{c}$ *and the uniform* congestion g*:*

$$\kappa = \tilde{c} \times g \tag{13.17}$$

Proof Since $g_{ij} = g$ for all links ij, the flow through each road segment is $f_{ij} = c_{ij} \times g$. Summing over all segments gives:

$$\kappa = \sum_{i,j} f_{ij} = \sum_{i,j} c_{ij} \times g = \tilde{c} \times g$$

□

In the PE state, where all roads have the same congestion, the total flow is simply the total capacity multiplied by the common congestion value.

Proposition 13.6 (Total Flow in Non-PE State) *In the* Non-Premagic Equilibrium (Non-PE) *state, the congestion is* non-uniform *across road segments. The total flow* κ *is related to the* total capacity $\tilde{c}$, maximum congestion ξ, *and the* efficiency indicator ϖ *by the following equation:*

$$\kappa = \tilde{c} \times \varpi \times \xi \tag{13.18}$$

Proof The total flow in the network is given by:

$$\kappa = \sum_{i,j} f_{ij} = \sum_{i,j} c_{ij} \times g_{ij}$$

The capacity-weighted average congestion $\bar{g}$ is related to ξ and ϖ by:

$$\bar{g} = \varpi \times \xi$$

Thus, the total flow becomes:

$$\kappa = \tilde{c} \times \bar{g} = \tilde{c} \times \varpi \times \xi$$

□

In the Non-PE state, congestion varies across the network. The total flow depends on the maximum congestion and a performance indicator ϖ, which reflects how close the network is to equilibrium.

Theorem 13.7 (The Network Flow Identity) *The* Network Flow Identity *relates* total flow κ, total capacity $\tilde{c}$, maximum congestion ξ, *and the* efficiency indicator *VarPi index* ϖ *in both PE and Non-PE states. The formula is given by:*

$$\kappa = \tilde{c} \times \varpi \times \xi \tag{13.19}$$

Proof The Network Flow Identity combines the results of the *Premagic Equilibrium (PE)* and *Non-Premagic Equilibrium (Non-PE)* propositions. Under PE, we have:

$$\kappa = \tilde{c} \times g$$

In Non-PE, the total flow is:

$$\kappa = \tilde{c} \times \varpi \times \xi$$

This proves the identity for both states. □

The Network Flow Identity (NFI) is valid in general situation (both Premagic Equilibrium and Non-Premagic Equilibrium) in steady state traffic model where constant and flow is conserved. The efficiency must be measured in VarPi index and maximum congestion indicators are measurable (e.g. through empirical calibration or simulation). It is very useful for *scenario comparisons*, where we're evaluating interventions under controlled assumptions. However, the NFI might break down if the total demand is not constant (e.g., elastic demand responding to tolls or delays), or in non-equilibrium system where flow fluctuates rapidly, or if the efficency index is ill-defined (not VarPi index) or varies across subnetworks. Think of NFI as a conservation law for network performance. If we change one part of the system (e.g., reduce congestion), we must adjust another (e.g., increase capacity or improve efficiency) to maintain the same total flow. It's especially powerful for *policy design*, because it quantifies trade-offs. For example, if we want to reduce congestion, then we need more capacity or better efficiency. If we want to cut costs then we should tolerate more congestion or reduce demand.

Example 13.9 Suppose we have a relative capacity matrix

$$\mathsf{C} = \begin{bmatrix} 0\,3\,2\,0 \\ 2\,0\,1\,2 \\ 1\,0\,3\,1 \\ 2\,0\,1\,0 \end{bmatrix}$$

The total capacity is a scalar represent the sum of all positive elements in the capacity matrix, $\tilde{c} = 18$ lanes.

Using Markov Split axioms and computing the stationary distribution of Markov chain, the cardinal Ideal Flow matrix would be

$$\mathsf{F} = \begin{bmatrix} 0 & 15 & 10 & 0 \\ 6 & 0 & 3 & 6 \\ 9 & 0 & 27 & 9 \\ 10 & 0 & 5 & 0 \end{bmatrix}$$

The total flow is the sum of all positive elements in the flow matrix, $\kappa = 100$.

The relative congestion matrix is

$$\mathsf{G} = \begin{bmatrix} 0 & 5 & 5 & 0 \\ 3 & 0 & 3 & 3 \\ 9 & 0 & 9 & 9 \\ 5 & 0 & 5 & 0 \end{bmatrix}$$

The maximum congestion is $\xi = 9$. The capacity weighted average congestion is the ratio of the sum of Hadamard product $\sum\sum(\mathsf{C} \circ \mathsf{G}) = 100$ and the sum of the capacity matrix $\tilde{c} = 18$. Thus, the capacity weighted average congestion is $\bar{g} = \frac{\sum\sum(\mathsf{C}\circ\mathsf{G})}{\tilde{c}} = 100/18 = 5.555556$. The VarPi index is $\varpi = 1 - \frac{\xi-\bar{g}}{\xi} = \frac{\bar{g}}{\xi} = \frac{\frac{100}{18}}{9} = \frac{50}{81} = 0.617284$ Now we can confirm the Network Flow Identity $\kappa = \tilde{c} \times \varpi \times \xi$ because $\tilde{c} \times \varpi \times \xi = 18 \times \frac{50}{81} \times 9 = 100 = \kappa$

In constrast, if we use arithmetic average of all positive elements in the congestion matrix, we get 5.6, which is closed to the capacity weighted average congestion of 5.555556. However, if we use this arithmetic average value to get the wrong VarPi index, the right hand side of the identity would produce $\tilde{c} \times \varpi_{wrong} \times \xi = 100.6$ instead of 100, which is not identical to the total flow in the network.

The *Network Flow Identity* equation (13.19) connects the total flow with the capacity, congestion, and efficiency of the network. It provides a *universal formula* that works for both ideal (PE) and real-world (Non-PE) traffic conditions.

Corollary 13.3 (Impact of Increasing Maximum Congestion) *If the maximum congestion ξ increases while keeping the* total capacity $\tilde{c}$ *and* efficiency *varPi index ϖ constant, the* total flow κ *will increase:*

$$\frac{\partial \kappa}{\partial \xi} > 0 \tag{13.20}$$

Proof From the Network Flow Identity equation (13.19):

$$\kappa = \tilde{c} \times \varpi \times \xi$$

Since $\tilde{c}$ and ϖ are constants, the derivative of κ with respect to ξ is positive:

$$\frac{\partial \kappa}{\partial \xi} = \tilde{c} \times \varpi > 0$$

Thus, an increase in ξ leads to an increase in κ. □

If the maximum congestion in a network increases, the total flow will also increase, assuming the network capacity and efficiency remain constant.

In this section, we have rigorously derived the *Network Flow Identity*, which relates *total flow κ*, *total capacity $\tilde{c}$*, *maximum congestion ξ*, and *efficiency indicator*

ϖ. We demonstrated how these factors influence traffic flow in both ideal and real-world conditions. The identity provides a fundamental tool for understanding and predicting traffic behavior, aiding in urban transportation planning, network design, and congestion management.

We demonstrate this with theoretical examples.

13.5.1 Theoretical Bound on Congestion Rebalancing

Assuming total flow κ remains constant (e.g., fixed demand), we equate the flow expressions before and after intervention based on Network Flow Identity.

Proposition 13.7 (Congestion Rebalancing) *Given a fixed total flow κ, if a network transitions from state 1 to state 2 with efficiency indicators ϖ_1 and ϖ_2, and maximum congestions ξ_1 and ξ_2, then:*

$$\tilde{c}_2 = \tilde{c}_1 \times \frac{\varpi_1}{\varpi_2} \times \frac{\xi_1}{\xi_2}$$

If congestion rebalancing is achieved ($\xi_2 < \xi_1$), then:

$$\tilde{c}_2 > \tilde{c}_1 \times \frac{\varpi_1}{\varpi_2} \tag{13.21}$$

If in addition, perfect efficiency is achieved ($\varpi_2 = 1 \equiv$ in PE state), then:

$$\tilde{c}_2 = \tilde{c}_1 \times \varpi_1 \times \frac{\xi_1}{\xi_2} > \tilde{c}_1 \times \varpi_1 \tag{13.22}$$

Proof Based on Network Flow Identity equation (13.19), for state 1: $\kappa = \tilde{c}_1 \times \varpi_1 \times \xi_1$, and for state 2: $\kappa = \tilde{c}_2 \times \varpi_2 \times \xi_2$. Equating gives $\tilde{c}_2 = \tilde{c}_1 \times (\varpi_1/\varpi_2) \times (\xi_1/\xi_2)$. If $\xi_2 < \xi_1$, then $\xi_1/\xi_2 > 1$, so $\tilde{c}_2 > \tilde{c}_1 \times (\varpi_1/\varpi_2)$. If $\varpi_2 = 1$, then $\tilde{c}_2 = \tilde{c}_1 \times \varpi_1 \times (\xi_1/\xi_2) > \tilde{c}_1 \times \varpi_1$. □

Proposition 13.8 (Theoretical Limit for Congestion Rebalancing) *For a network with initial efficiency ϖ_1 and fixed demand κ, the minimum capacity required to achieve a lower maximum congestion $\xi_2 < \xi_1$ is approximately:*

$$\tilde{c}_2 \approx \tilde{c}_1 \times \varpi_1 \tag{13.23}$$

assuming perfect efficiency in the new state ($\varpi_2 = 1$).

Proof From Proposition 13.7 with $\varpi_2 = 1$, $\tilde{c}_2 = \tilde{c}_1 \times \varpi_1 \times (\xi_1/\xi_2)$. For $\xi_2 < \xi_1$, $\xi_1/\xi_2 > 1$, so $\tilde{c}_2 > \tilde{c}_1 \times \varpi_1$. The minimum capacity occurs when ξ_2 is just less than ξ_1, so $\xi_1/\xi_2 \approx 1$, thus $\tilde{c}_2 \approx \tilde{c}_1 \times \varpi_1$. □

Lane reduction policy or road diet ($\Delta\mathsf{C}^{-}$) would reduce the total capacity in the network. Is it possible to reduce the total network and speed up traffic? We aim to reduce total capacity ($\tilde{c}$ decreases) while maintaining or improving the worst-case congestion ξ. That is, we want:

$$\xi_2 \leq \xi_1$$

In order to make the Assuming ideal redistribution (PE state) after intervention ($\varpi_2 = 1$), we obtain the theoretical lower bound on capacity:

$$\tilde{c}_2 > \tilde{c}_1 \times \varpi_1 \tag{13.24}$$

This inequality implies that the minimum allowable capacity after redistribution is proportional to the initial efficiency. If the initial network is inefficient ($\varpi_1 \ll 1$), substantial capacity reduction is possible without worsening congestion.

New Capacity $>$ Old Capacity $\times$ Old Efficiency

If we want to reduce total capacity but also improve the average travel time then the formula in Eq. (13.24) tell us that this is only possible under one specific condition: *our initial network must be highly inefficient*. Road network that have a low VarPi index indicates the traffic must be poorly distributed, with severe bottlenecks on a few roads while other roads are underused. The theoretical limit tell us that we cannot reduce the total capacity below our original capacity multiplied by our riginal inefficiency.

13.6 Traffic Principles and Laws Based on Network Flow Identity

We study networks whose aggregate behavior is summarized by the Network Flow Identity (NFI):

$$\kappa = \tilde{c} \times \varpi \times \xi,$$

where $\kappa > 0$ denotes total throughput, $\tilde{c} > 0$ scalar total capacity, $\varpi > 0$ a system efficiency/balance factor, and $\xi > 0$ a worst-bottleneck congestion index.

Because NFI has four variables, we can systematically derive principles by holding some variables constant and comparing two states (State 1 vs. State 2). We also define a *feasible operating region* as the set of states allowed by a policy constraint (e.g., a congestion cap).

Definition 13.21 (Feasible Operating Region) The feasible operating region is the set $\mathcal{F} \subset \{(\kappa, \tilde{c}, \varpi, \xi) \in \mathbb{R}^4_{>0}\}$ of all positive quadruples that satisfy NFI and any external policy/device constraints (e.g., a congestion cap or a fixed-capacity budget).

Lemma 13.2 (Two-State Ratio Lemma) *For any two feasible states,*

$$\frac{\kappa_2}{\kappa_1} = \frac{\tilde{c}_2}{\tilde{c}_1} \times \frac{\varpi_2}{\varpi_1} \times \frac{\xi_2}{\xi_1}.$$

Proof Divide the identities $\kappa_i = \tilde{c}_i \varpi_i \xi_i$ for $i \in \{1, 2\}$ to obtain the stated product of ratios. □

13.6.1 Combinatorial Equalities

Here we hold certain variables constant across the two states and read the exact equality constraints implied by NFI. They serve as the base identities from which all inequality (comparative-statics) results follow.

13.6.1.1 One Constant Variable

Proposition 13.9 (Constant Throughput) *Given a fixed total flow ($\kappa_1 = \kappa_2$), the change in total capacity ($\tilde{c}$) is inversely proportional to the changes in efficiency (ϖ) and maximum congestion (ξ). If $\kappa_1 = \kappa_2$, then*

$$\tilde{c}_2 = \tilde{c}_1 \times \frac{\varpi_1}{\varpi_2} \times \frac{\xi_1}{\xi_2}.$$

Proof Set $\kappa_2/\kappa_1 = 1$ in Lemma 13.2 and solve for $\tilde{c}_2$. □

This is the fundamental law of system rebalancing. To handle the same amount of traffic, any reduction in congestion or efficiency must be paid for with an increase in total capacity, and vice versa. For fixed throughput, any reduction in ϖ or ξ must be offset by more capacity, and vice versa. With the same total traffic, we can only cut congestion or reduce inefficiency if we add capacity (or trade among the three).

Proposition 13.10 (Invariant Capacity) *Given a fixed total capacity ($\tilde{c}_1 = \tilde{c}_2$), the relative change in total flow is equal to the product of the relative changes in efficiency and maximum congestion. If $\tilde{c}_1 = \tilde{c}_2$, then $\frac{\kappa_2}{\kappa_1} = \frac{\varpi_2}{\varpi_1} \times \frac{\xi_2}{\xi_1}$.*

Proof Set $\tilde{c}_2/\tilde{c}_1 = 1$ in Lemma 13.2. □

On a fixed road network, throughput can only be increased by improving operational efficiency, accepting higher peak congestion, or both.

Proposition 13.11 (Invariant Efficiency) *Given a fixed operational efficiency ($\varpi_1 = \varpi_2$), the relative change in total flow is equal to the product of the relative changes in total capacity and maximum congestion. If $\varpi_1 = \varpi_2$, then $\frac{\kappa_2}{\kappa_1} = \frac{\tilde{c}_2}{\tilde{c}_1} \times \frac{\xi_2}{\xi_1}$.*

Proof Set $\varpi_2/\varpi_1 = 1$ in Lemma 13.2. □

For a system with a given level of operational maturity, traffic growth must be absorbed by infrastructure expansion or result in higher congestion.

Proposition 13.12 (Bounded Congestion) *Given a fixed maximum congestion level ($\xi_1 = \xi_2$), the relative change in total flow is equal to the product of the relative changes in total capacity and efficiency. If $\xi_1 = \xi_2$, then $\frac{\kappa_2}{\kappa_1} = \frac{\tilde{c}_2}{\tilde{c}_1} \times \frac{\varpi_2}{\varpi_1}$.*

Proof Set $\xi_2/\xi_1 = 1$ in Lemma 13.2. □

To maintain a consistent level of service (i.e., peak congestion), any growth in traffic must be met with proportional investment in physical capacity and/or operational efficiency. If we keep the worst jam level fixed, higher flow needs either more road or smarter ops.

13.6.1.2 Two Constant Variables

Proposition 13.13 (Efficiency-Congestion Inverse Law) *For a network with constant flow and capacity ($\kappa, \tilde{c}$), operational efficiency and maximum congestion are perfectly inversely proportional. If $\kappa, \tilde{c}$ are constant, then $\varpi_2\xi_2 = \varpi_1\xi_1$.*

Proof From Lemma 13.2, $1 = (\varpi_2/\varpi_1)(\xi_2/\xi_1)$. □

This is the core principle for a traffic engineer. On a given road network with a given amount of traffic, any action that improves system-wide balance (increases ϖ) must reduce the worst bottleneck (decrease ξ). With fixed traffic and network size, smarter balance means lower peak jams, and vice versa.

Proposition 13.14 (Capacity-Congestion Inverse Law) *For a network with constant flow and efficiency (κ, ϖ), total capacity and maximum congestion are perfectly inversely proportional. If κ, ϖ are constant, then $\tilde{c}_2\xi_2 = \tilde{c}_1\xi_1$.*

Proof From Lemma 13.2, $1 = (\tilde{c}_2/\tilde{c}_1)(\xi_2/\xi_1)$. □

If the system's operational character is fixed, the only way to reduce the worst traffic jam is to add more capacity. With fixed traffic and ops, more capacity lowers the worst jam in lockstep.

Proposition 13.15 (Capacity-Efficiency Inverse Law) *For a network with constant flow and congestion (κ, ξ), total capacity and operational efficiency are perfectly inversely proportional. If κ, ξ are constant, then $\tilde{c}_2\varpi_2 = \tilde{c}_1\varpi_1$.*

Proof From Lemma 13.2, $1 = (\tilde{c}_2/\tilde{c}_1)(\varpi_2/\varpi_1)$. □

If the goal is to handle a specific amount of traffic without letting congestion get any worse, any reduction in physical capacity (e.g., a "road diet") must be compensated by a proportional increase in operational efficiency. Holding jam level and flow fixed, we can swap *concrete* for *smarts* one-for-one.

Proposition 13.16 (Throughput-Efficiency Proportionality) *For a network with constant capacity and congestion ($\tilde{c}, \xi$), total flow is directly proportional to operational efficiency. If $\tilde{c}, \xi$ are constant, then $\kappa_2/\kappa_1 = \varpi_2/\varpi_1$.*

Proof Fix $\tilde{c}_2/\tilde{c}_1 = \xi_2/\xi_1 = 1$ in Lemma 13.2. □

If we are not adding roads and are not willing to tolerate more congestion, the only way to increase throughput is by making the system smarter. No new roads and same jam: more *smarts* gives more flow.

Proposition 13.17 (Throughput-Congestion Proportionality) *For a network with constant capacity and efficiency ($\tilde{c}, \varpi$), total flow is directly proportional to maximum congestion. If $\tilde{c}, \varpi$ are constant, then $\kappa_2/\kappa_1 = \xi_2/\xi_1$.*

Proof Fix $(\tilde{c}_2/\tilde{c}_1)(\varpi_2/\varpi_1) = 1$ in Lemma 13.2. □

On a given network with a fixed operational strategy, accommodating more traffic will inevitably lead to proportionally higher peak congestion. Same roads and ops: more flow means proportionally worse peak jam.

Proposition 13.18 (Throughput-Capacity Proportionality) *For a network with constant efficiency and congestion (ϖ, ξ), total flow is directly proportional to total capacity. If ϖ, ξ are constant, then $\kappa_2/\kappa_1 = \tilde{c}_2/\tilde{c}_1$.*

Proof Fix $(\varpi_2/\varpi_1)(\xi_2/\xi_1) = 1$ in Lemma 13.2. □

If the operational rules and quality of service are held constant, the amount of traffic a network can handle is determined solely by its size. Same ops and jam: bigger network carries proportionally more traffic.

13.6.1.3 Three or More Constant Variables

There are $\binom{4}{3} = 4$ principles that can be derived by holding three variables constant. These describe the rigidity of the system.

Proposition 13.19 (Congestion Determinism) *If a network's flow, capacity, and efficiency are all constant, its maximum congestion must also be constant. If $\kappa, \tilde{c}, \varpi$ are constant, then $\xi_1 = \xi_2$.*

Proof All three ratios equal 1 in Lemma 13.2, so $\xi_2/\xi_1 = 1$. □

An unchanged system will produce an unchanged peak stress level. If nothing else changes, the worst jam cannot change.

Proposition 13.20 (Efficiency Determinism) *If a network's flow, capacity, and congestion are all constant, its efficiency must also be constant. If* $\kappa, \tilde{c}, \xi$ *are constant, then* $\varpi_1 = \varpi_2$.

Proof Analogous to Proposition 13.19. □

An unchanged system operating at the same stress level implies its operational efficiency has not changed. Same flow, roads, and jam implies same operational efficiency.

Proposition 13.21 (Capacity Determinism) *If a network's flow, efficiency, and congestion are all constant, its total capacity must also be constant. If* κ, ϖ, ξ *are constant, then* $\tilde{c}_1 = \tilde{c}_2$.

Proof Analogous to Proposition 13.19. □

If a system's demand and performance are unchanged, its underlying physical capacity must not have changed. Same demand and performance implies unchanged physical capacity.

Proposition 13.22 (Flow Determinism) *If a network's capacity, efficiency, and congestion are all constant, its total flow must also be constant. If* $\tilde{c}, \varpi, \xi$ *are constant, then* $\kappa_1 = \kappa_2$.

Proof Analogous to Proposition 13.19. □

An unchanged system operating at an unchanged performance level can handle an unchanged amount of total flow. Unchanged system $\Rightarrow$ unchanged throughput.

There is $\binom{4}{4} = 1$ thus, the following Proposition 13.23 is trivial case for completeness.

Proposition 13.23 (System Invariance) *If any three variables are identical across states, the fourth must be identical as well.*

Proof Direct from Lemma 13.2. □

A system that has not changed in any way has not changed. This confirms the logical consistency of the NFI formula.

13.6.2 Comparative Statics I: One Constant and One Inequality

We now fix one variable and perturb a second one (up or down). The resulting *strict inequalities* follow by comparing products in Lemma 13.2.

13.6.2.1 Group 1.1: Constant Total Flow (κ)

Proposition 13.24 (Induced Inefficiency) *Given constant total flow (κ), if total capacity is increased ($\tilde{c}_2 > \tilde{c}_1$), the product of efficiency and congestion in the new state must be lower than in the original state. If κ is constant and $\tilde{c}_2 > \tilde{c}_1$, then $\varpi_2\xi_2 < \varpi_1\xi_1$.*

Proof With $\kappa_2/\kappa_1 = 1$, Lemma 13.2 gives $(\varpi_2\xi_2) = (\varpi_1\xi_1)\,(\tilde{c}_1/\tilde{c}_2) < \varpi_1\xi_1$. □

This is a formal statement of induced demand's companion. If we build more roads but the total traffic volume remains the same, the network's overall performance (a combination of its efficiency and congestion) must degrade. This often manifests as traffic spreading out inefficiently to fill the new capacity. More lanes with *no* extra traffic tends to spread flow and lower the efficiency × congestion composite.

Proposition 13.25 (Efficiency Dividends) *Given constant total flow (κ), if efficiency is improved ($\varpi_2 > \varpi_1$), the product of capacity and congestion in the new state must be lower than in the original state. If κ constant and $\varpi_2 > \varpi_1$, then $\tilde{c}_2\xi_2 < \tilde{c}_1\xi_1$.*

Proof From the NFI, $\tilde{c}_1\xi_1 = \kappa/\varpi_1$ and $\tilde{c}_2\xi_2 = \kappa/\varpi_2$. Since $\varpi_2 > \varpi_1$, it follows that $1/\varpi_2 < 1/\varpi_1$, proving the inequality. □

Improving a network's operational intelligence pays a "dividend." It allows us to handle the same amount of traffic with less infrastructure and/or lower peak congestion. Smarter ops let we pay less in roads or peak jam to carry the same flow. It formalizes the dual benefit of smarter operations: less infrastructure and lower peak stress. This principle is a cornerstone for modeling algorithmic interventions around intelligent transport systems.

Proposition 13.26 (Congestion Costs) *Given constant total flow (κ), if maximum congestion is allowed to increase ($\xi_2 > \xi_1$), the product of capacity and efficiency in the new state must be lower than in the original state. If κ constant and $\xi_2 > \xi_1$, then $\tilde{c}_2\varpi_2 < \tilde{c}_1\varpi_1$.*

Proof Rearrange $\tilde{c}\varpi = \kappa/\xi$. □

Tolerating higher congestion is a policy choice that allows a network to function with less physical capacity and/or poorer operational efficiency. It quantifies the "cost" of worsening traffic jams in terms of infrastructure and intelligence savings. Tolerating worse jams lets us get by with less concrete or less smartness.

13.6.2.2 Group 1.2: Constant Total Capacity

Proposition 13.27 (Latent Demand) *Given constant total capacity ($\tilde{c}$), if total flow increases ($\kappa_2 > \kappa_1$), the product of efficiency and congestion in the new state must be greater than in the original state. If $\tilde{c}$ constant and $\kappa_2 > \kappa_1$, then $\varpi_2\xi_2 > \varpi_1\xi_1$.*

Proof From the NFI, $\varpi_1\xi_1 = \kappa_1/\tilde{c}$ and $\varpi_2\xi_2 = \kappa_2/\tilde{c}$. Since $\kappa_2 > \kappa_1$, the inequality holds. □

This is the formal law of traffic. If more cars use the same road network, the combined level of system performance (efficiency and congestion) must increase to absorb them. More cars on fixed roads increase the combined stress measure. It warns that increasing demand without adjusting capacity or efficiency will strain the system, critical for forecasting and resilience planning. It may help planners anticipate bottlenecks and justify proactive investment or operational upgrades.

Proposition 13.28 (Congestion Relief) *Given constant total capacity ($\tilde{c}$), if efficiency improves ($\varpi_2 > \varpi_1$), the ratio of flow to congestion must increase. If $\tilde{c}$ constant and $\varpi_2 > \varpi_1$, then $\kappa_2/\xi_2 > \kappa_1/\xi_1$.*

Proof From $\kappa/\xi = \tilde{c}\varpi$. □

This is the core justification for Intelligent Transport Systems (ITS) [16]. Improving a network's "smartness" guarantees that it can either handle more traffic for the same level of congestion, or the same traffic with less congestion. With the same roads, smarter control yields more flow per unit of peak jam.

Proposition 13.29 (Inefficient Throughput) *Given constant total capacity ($\tilde{c}$), if maximum congestion increases ($\xi_2 > \xi_1$), the ratio of flow to efficiency must increase. If $\tilde{c}$ constant and $\xi_2 > \xi_1$, then $\kappa_2/\varpi_2 > \kappa_1/\varpi_1$.*

Proof From $\kappa/\varpi = \tilde{c}\xi$. □

If we allow traffic jams to get worse on our network, we will be able to push more cars through it, but potentially at the cost of making the overall system less balanced and efficient. Allowing more jam can push more vehicles through, but at a balance cost.

13.6.2.3 Group 1.3: Constant Operational Efficiency (ϖ)

Proposition 13.30 (Infrastructural Stress) *Given constant operational efficiency (ϖ), if total flow increases ($\kappa_2 > \kappa_1$), the combined measure of capacity and congestion in the new state must be greater than in the original state. If ϖ constant and $\kappa_2 > \kappa_1$, then $\tilde{c}_2\xi_2 > \tilde{c}_1\xi_1$.*

Proof From $\tilde{c}\xi = \kappa/\varpi$ with ϖ fixed. From the NFI, $\tilde{c}_1\xi_1 = \kappa_1/\varpi$ and $\tilde{c}_2\xi_2 = \kappa_2/\varpi$. Since $\kappa_2 > \kappa_1$, the inequality holds. □

If a network's "smartness" is fixed, accommodating more traffic must increase the overall physical stress on the system. This stress is absorbed by a combination of new infrastructure ($\tilde{c}$) and/or higher traffic density (ξ). More traffic with same smarts raises physical stress (more capacity and/or more crowding).

Proposition 13.31 (Capacity Dividend) *Given constant operational efficiency (ϖ), if total capacity is increased ($\tilde{c}_2 > \tilde{c}_1$), the ratio of flow to congestion must increase. If ϖ constant and $\tilde{c}_2 > \tilde{c}_1$, then $\kappa_2/\xi_2 > \kappa_1/\xi_1$.*

Proof From $\kappa/\xi = \tilde{c}\varpi$ with ϖ fixed. From the NFI, $\kappa_1/\xi_1 = \tilde{c}_1\varpi$ and $\kappa_2/\xi_2 = \tilde{c}_2\varpi$. Since $\tilde{c}_2 > \tilde{c}_1$, the inequality holds. □

With a fixed operational strategy, adding capacity pays a clear "dividend" by enabling the network to handle a greater volume of traffic per unit of peak congestion. Extra lanes give more flow for the same peak jam.

Proposition 13.32 (Congestion-Fueled Density) *Given constant operational efficiency (ϖ), if maximum congestion increases ($\xi_2 > \xi_1$), the ratio of flow to capacity (the network's throughput density) must increase. If ϖ constant and $\xi_2 > \xi_1$, then $\kappa_2/\tilde{c}_2 > \kappa_1/\tilde{c}_1$.*

Proof From $\kappa/\tilde{c} = \varpi\xi$ with ϖ fixed. From the NFI, $\kappa_1/\tilde{c}_1 = \varpi\xi_1$ and $\kappa_2/\tilde{c}_2 = \varpi\xi_2$. Since $\xi_2 > \xi_1$, the inequality holds. □

This principle shows that allowing congestion to worsen is a method for increasing a network's utilization, pushing more total flow through each unit of capacity. Worse jams increase throughput *per unit of road*, i.e., denser use.

13.6.2.4 Group 1.4: Constant Maximum Congestion (ξ)

Proposition 13.33 (Required Investment) *Given a constant maximum congestion target (ξ), if total flow increases ($\kappa_2 > \kappa_1$), the combined investment in capacity and efficiency must be greater in the new state. If ξ constant and $\kappa_2 > \kappa_1$, then $\tilde{c}_2\varpi_2 > \tilde{c}_1\varpi_1$.*

Proof From $\tilde{c}\varpi = \kappa/\xi$ with ξ fixed. From the NFI, $\tilde{c}_1\varpi_1 = \kappa_1/\xi$ and $\tilde{c}_2\varpi_2 = \kappa_2/\xi$. Since $\kappa_2 > \kappa_1$, the inequality holds. □

This provides a direct measure of the investment needed to support growth while maintaining a certain quality of life. To handle more cars without worsening the worst traffic jams, a city must invest in the product of its infrastructure and its intelligence. To grow without worse peaks, we must invest in roads and/or smarter ops.

Proposition 13.34 (Capacity-Led Growth) *Given a constant maximum congestion target (ξ), if total capacity is increased ($\tilde{c}_2 > \tilde{c}_1$), the ratio of flow to efficiency must increase. If ξ constant and $\tilde{c}_2 > \tilde{c}_1$, then $\kappa_2/\varpi_2 > \kappa_1/\varpi_1$.*

Proof From $\kappa/\varpi = \tilde{c}\xi$ with ξ fixed. From the NFI, $\kappa_1/\varpi_1 = \tilde{c}_1\xi$ and $\kappa_2/\varpi_2 = \tilde{c}_2\xi$. Since $\tilde{c}_2 > \tilde{c}_1$, the inequality holds. □

When congestion is capped, adding capacity is a direct enabler of growth, allowing the network to support more flow for a given level of operational efficiency. With a jam cap, building roads directly enables more flow per unit of efficiency.

Proposition 13.35 (Efficiency-Led Growth) *Given a constant maximum congestion target (ξ), if operational efficiency is improved ($\varpi_2 > \varpi_1$), the ratio of flow to capacity must increase. If ξ constant and $\varpi_2 > \varpi_1$, then $\kappa_2/\tilde{c}_2 > \kappa_1/\tilde{c}_1$.*

Proof From $\kappa/\tilde{c} = \varpi\xi$ with ξ fixed. From the NFI, $\kappa_1/\tilde{c}_1 = \varpi_1\xi$ and $\kappa_2/\tilde{c}_2 = \varpi_2\xi$. Since $\varpi_2 > \varpi_1$, the inequality holds. □

When congestion is capped, making the system smarter is a direct enabler of growth, allowing the existing infrastructure to handle more total traffic. With a jam cap, smarter control lets the same road carry more.

13.6.3 Comparative Statics II: Two Constants and One Inequality

We now fix *two* variables and move a third. These results give sharp trade-offs and guarantees that practitioners can test in data. This subsection explores the 12 principles derived by holding two variables constant and introducing an inequality to a third. These produce more direct and powerful conclusions.

13.6.3.1 Group 2.1: Constant Flow (κ) and Capacity ($\tilde{c}$)

Proposition 13.36 (Certainty of Congestion Relief) *Given constant flow and capacity, any improvement in efficiency ($\varpi_2 > \varpi_1$)* must *result in a decrease in maximum congestion. If $\kappa, \tilde{c}$ constant and $\varpi_2 > \varpi_1$, then $\xi_2 < \xi_1$.*

Proof From Proposition 13.13, $\varpi\xi$ is constant, so raising ϖ lowers ξ. □

This provides a theoretical guarantee to traffic engineers. On a specific road network with a specific amount of traffic, a successful signal re-timing project (improving ϖ) is mathematically guaranteed to reduce the worst traffic jam. On the same roads with same demand, smarter ops *must* cut the worst jam.

Proposition 13.37 (Certainty of Inefficient Congestion) *Given constant flow and capacity, any increase in maximum congestion ($\xi_2 > \xi_1$) implies that the system's efficiency has decreased. If $\kappa, \tilde{c}$ constant and $\xi_2 > \xi_1$, then $\varpi_2 < \varpi_1$.*

Proof Inverse of Proposition 13.36. □

If our commute on the same roads with the same number of cars gets worse, it means the system is being managed less effectively than before. If peak jam got worse without more cars or fewer lanes, we ran it less efficiently.

13.6.3.2 Group 2.2: Constant Flow and Efficiency

Proposition 13.38 (Capacity Relief Guarantee) *Given constant flow and efficiency, any increase in total capacity ($\tilde{c}_2 > \tilde{c}_1$)* must *result in a decrease in maximum congestion. If κ, ϖ constant and $\tilde{c}_2 > \tilde{c}_1$, then $\xi_2 < \xi_1$.*

Proof From Proposition 13.14, $\tilde{c}\xi$ constant, so $\tilde{c} \uparrow \Rightarrow \xi \downarrow$. From the NFI with κ and ϖ fixed:

$$\tilde{c}_1 \times \xi_1 = \tilde{c}_2 \times \xi_2 \Rightarrow \frac{\tilde{c}_2}{\tilde{c}_1} = \frac{\xi_1}{\xi_2} \Rightarrow \xi_2 = \xi_1 \times \frac{\tilde{c}_1}{\tilde{c}_2}$$

Since $\tilde{c}_2 > \tilde{c}_1$, it follows that $\xi_2 < \xi_1$. □

This proposition confirms that capacity expansion, such as adding lanes or increasing throughput, can reduce peak congestion levels, assuming demand and operational efficiency remain unchanged. It provides a formal justification for infrastructure investments aimed at congestion relief. If the operational strategy is unchanged, adding new roads is guaranteed to alleviate the worst traffic jams, assuming total traffic volume does not change. With same demand and ops, building lanes *must* lower the peak jam.

Proposition 13.39 (Congestion Deficit) *For a network with constant flow (κ) and efficiency (ϖ), any increase in maximum congestion ($\xi_2 > \xi_1$) implies a decrease in total capacity (the system has lost physical capacity): If κ, ϖ constant and $\xi_2 > \xi_1$, then $\tilde{c}_2 < \tilde{c}_1$.*

Proof Inverse of Proposition 13.38. From the same network flow identity:

$$\tilde{c}_2 = \tilde{c}_1 \times \frac{\xi_1}{\xi_2}$$

If $\xi_2 > \xi_1$, then $\tilde{c}_2 < \tilde{c}_1$. □

This Proposition 13.39 is Congestion-Capacity Trade-off. If our commute gets worse under normal operations, and it's not due to more cars, it means the physical network's throughput has been reduced (e.g., a lane closure, a new bottleneck). If jam worsened with same cars and ops, some effective capacity was lost. This proposition confirms that worsening congestion under stable demand and operational conditions signals a reduction in physical capacity, whether due to lane closures, reduced throughput, or infrastructure degradation. It supports diagnostic use of congestion metrics to infer latent capacity loss.

13.6.3.3 Group 2.3: Constant Capacity ($\tilde{c}$) and Efficiency (ϖ)

Proposition 13.40 (Law of Induced Congestion) *On a network with constant physical capacity and operational efficiency, any increase in total flow* ($\kappa_2 > \kappa_1$) must *result in an increase in maximum congestion. If* $\tilde{c}, \varpi$ *constant and* $\kappa_2 > \kappa_1$, *then* $\xi_2 > \xi_1$.

Proof From Proposition 13.17, $\kappa \propto \xi$. □

This is the most fundamental law of traffic congestion. If more cars use an unchanged road system, it will get more congested. There is no escape. More cars on the same roads with same ops *must* produce more congestion.

Proposition 13.41 (Flow-Congestion Equivalence) *On a network with constant physical capacity and operational efficiency, any increase in maximum congestion* ($\xi_2 > \xi_1$) *implies that the total flow has increased. If* $\tilde{c}, \varpi$ *constant and* $\xi_2 > \xi_1$, *then* $\kappa_2 > \kappa_1$.

Proof Inverse of Proposition 13.40. □

If we observe that traffic jams are getting worse on a stable, unchanged network, it is a direct and certain indicator that the total volume of cars has increased. If peak jam rose on an unchanged system, traffic volume rose.

13.6.3.4 Group 2.4: Constant Capacity ($\tilde{c}$) and Congestion (ξ)

Proposition 13.42 (Mandate for Efficiency) *On a network with constant capacity and a fixed congestion target, any increase in total flow* ($\kappa_2 > \kappa_1$) must *be the result of an improvement in operational efficiency. If* $\tilde{c}, \xi$ *constant and* $\kappa_2 > \kappa_1$, *then* $\varpi_2 > \varpi_1$.

Proof From Proposition 13.16, $\kappa \propto \varpi$. □

This is the core business case for ITS and traffic management centers. If we cannot build more roads and refuse to let congestion worsen, the only way to accommodate growth is to become more efficient. If we won't build and won't accept more jam, only smarter ops can raise flow.

Proposition 13.43 (Efficiency Throughput Guarantee) *On a network with constant capacity and a fixed congestion target, any improvement in efficiency* ($\varpi_2 > \varpi_1$) must *result in an increase in total flow. If* $\tilde{c}, \xi$ *constant and* $\varpi_2 > \varpi_1$, *then* $\kappa_2 > \kappa_1$.

Proof Inverse of Proposition 13.42. □

This provides a guaranteed outcome for successful efficiency projects. Under controlled conditions, a smarter network is guaranteed to carry more traffic. Smarter control guarantees more throughput when capacity and jam are fixed.

13.6.3.5 Group 2.5: Constant Efficiency (ϖ) and Congestion (ξ)

Proposition 13.44 (Mandate for Capacity) *On a network with fixed operational efficiency and a fixed congestion target, any increase in total flow ($\kappa_2 > \kappa_1$)* must *be the result of an increase in total capacity. If ϖ, ξ constant and $\kappa_2 > \kappa_1$, then $\tilde{c}_2 > \tilde{c}_1$.*

Proof From Proposition 13.18, $\kappa \propto \tilde{c}$. From the NFI, $\tilde{c}_1 = \kappa_1/(\varpi\xi)$ and $\tilde{c}_2 = \kappa_2/(\varpi\xi)$. Since $\kappa_2 > \kappa_1$, it follows that $\tilde{c}_2 > \tilde{c}_1$. □

This is the core business case for civil engineering projects. If a city's operational strategy is mature and congestion is at its limit, the only way to support growth is to build more infrastructure. If ops and jam are fixed, only more road can raise flow. These are pragmatic principle. If we can't improve efficiency or reduce congestion, we must expand capacity to meet demand. They underpin infrastructure expansion decisions and help communicate trade-offs to stakeholders.

Proposition 13.45 (Capacity Throughput Guarantee) *On a network with fixed operational efficiency and a fixed congestion target, any increase in total capacity ($\tilde{c}_2 > \tilde{c}_1$)* must *result in an increase in total flow. If ϖ, ξ constant and $\tilde{c}_2 > \tilde{c}_1$, then $\kappa_2 > \kappa_1$.*

Proof Inverse of Proposition 13.44. □

Under controlled conditions, a bigger network is guaranteed to carry more traffic. Bigger network $\Rightarrow$ more throughput when ops and jam targets are fixed.

13.6.3.6 Group 2.6: Constant Flow and Congestion

These two propositions below elegantly formalize a reciprocal relationship between physical infrastructure (capacity) and operational intelligence (efficiency), under fixed throughput and congestion conditions.

Proposition 13.46 (Capacity-Efficiency Trade-Off) *If total throughput (κ) and worst-case congestion (ξ) are held constant, then increasing total capacity ($\tilde{c}_2 > \tilde{c}_1$) implies a decrease in the required system efficiency ($\varpi_2 < \varpi_1$). If κ, ξ constant and $\tilde{c}_2 > \tilde{c}_1$, then $\varpi_2 < \varpi_1$.*

Proof From Lemma 13.2, with κ, ξ fixed we have $(\tilde{c}_2/\tilde{c}_1)(\varpi_2/\varpi_1) = 1$. □

More capacity allows for lower efficiency while maintaining the same throughput and congestion. If traffic jams and flow targets are fixed, then adding lanes (increasing $\tilde{c}$) gives us more slack, we can afford less precise coordination, less optimal routing, or more uneven distribution of flow. This proposition can our understanding of how physical expansion can compensate for operational inefficiencies.

Proposition 13.47 (Efficiency-Capacity Trade-Off) *If total throughput (κ) and worst-case congestion (ξ) are held constant, then increasing system efficiency ($\varpi_2 > \varpi_1$) implies a decrease in required capacity ($\tilde{c}_2 < \tilde{c}_1$). If κ, ξ constant and $\varpi_2 > \varpi_1$, then $\tilde{c}_2 < \tilde{c}_1$.*

Proof Inverse of Proposition 13.46. □

Smarter coordination, better routing, or more balanced flow allows us to achieve the same performance with fewer lanes. If jam levels and flow targets are fixed, then improving system intelligence, through adaptive signals, dynamic pricing, or optimized routing, we can reduce the need for physical expansion and allows us to operate with fewer lanes. This proposition supports modules on algorithmic policy design and equity auditing. We can explore how smarter systems reduce reliance on costly infrastructure, and how efficiency gains can be distributed across modes or regions.

13.6.4 Selected Derived Laws

Table 13.1 presents a structured synthesis of derived principles from the Network Flow Identity (NFI). Each row captures a conditional perturbation (e.g., increasing $\tilde{c}$ or ϖ) under fixed parameters, and deduces the resulting consequence. The final column distills the core insight, offering intuitive guidance for policy design, infrastructure planning, and operational optimization.

Among these 24 principles in Table 13.1, a few stand out for their profound practical significance in transportation research, analysis, and synthesis. They represent the core challenges and opportunities in managing any transportation network.

1. *Principle 2.3.1: The Law of Induced Congestion*: More cars on an unchanged network must create more congestion. This is the most fundamental and inescapable law of traffic. It serves as the bedrock of all realistic transportation analysis. For researchers, it provides the baseline assumption for any traffic model. For policymakers, it is a crucial, non-negotiable truth that refutes any notion that traffic growth can be absorbed by an existing, unimproved system without consequence. It is the reason congestion exists and the starting point for any effort to manage it. Its simplicity and certainty make it the most important principle to understand.
2. *Principle 2.1.1: The Certainty of Congestion Relief*: Better efficiency must cut congestion on a stable network. This principle is the theoretical guarantee that justifies the entire field of Intelligent Transportation Systems (ITS), traffic signal optimization, and network management. It provides a causal link between improving a network's operational intelligence (ϖ) and achieving the tangible outcome of reduced peak congestion (ξ). For analysis, it allows researchers to isolate the effect of operational improvements. For synthesis, it gives planners a

Table 13.1 Synthesis table of NFI principles

ID	Conditions	Consequence	Core insight
1.1.1	κ constant, $\tilde{c}\uparrow$	$\varpi\xi\downarrow$	Added capacity without demand reduces $\varpi\xi$ at fixed κ
1.1.2	κ constant, $\varpi\uparrow$	$\tilde{c}\xi\downarrow$	Efficiency gains reduce required $\tilde{c}$ and/or ξ
1.1.3	κ constant, $\xi\uparrow$	$\tilde{c}\varpi\downarrow$	Tolerating more ξ allows less $\tilde{c}$ or ϖ
1.2.1	$\tilde{c}$ constant, $\kappa\uparrow$	$\varpi\xi\uparrow$	More flow raises combined performance stress
1.2.2	$\tilde{c}$ constant, $\varpi\uparrow$	$\kappa/\xi\uparrow$	Smarter ops increase flow per unit congestion
1.2.3	$\tilde{c}$ constant, $\xi\uparrow$	$\kappa/\varpi\uparrow$	Higher ξ can push more flow per unit efficiency
1.3.1	ϖ constant, $\kappa\uparrow$	$\tilde{c}\xi\uparrow$	More flow stresses capacity and/or ξ
1.3.2	ϖ constant, $\tilde{c}\uparrow$	$\kappa/\xi\uparrow$	Capacity pays a dividend vs. peak ξ
1.3.3	ϖ constant, $\xi\uparrow$	$\kappa/\tilde{c}\uparrow$	Higher ξ increases flow density
1.4.1	ξ constant, $\kappa\uparrow$	$\tilde{c}\varpi\uparrow$	Growth at fixed ξ requires investment
1.4.2	ξ constant, $\tilde{c}\uparrow$	$\kappa/\varpi\uparrow$	Capacity-led growth at capped ξ
1.4.3	ξ constant, $\varpi\uparrow$	$\kappa/\tilde{c}\uparrow$	Efficiency-led growth at capped ξ
2.1.1	$\kappa, \tilde{c}$ constant, $\varpi\uparrow$	$\xi\downarrow$	Smarter ops must cut peak ξ
2.1.2	$\kappa, \tilde{c}$ constant, $\xi\uparrow$	$\varpi\downarrow$	Worsening ξ implies lower ϖ
2.2.1	κ, ϖ constant, $\tilde{c}\uparrow$	$\xi\downarrow$	Added capacity must cut ξ
2.2.2	κ, ϖ constant, $\xi\uparrow$	$\tilde{c}\downarrow$	Higher ξ implies capacity loss
2.3.1	$\tilde{c}, \varpi$ constant, $\kappa\uparrow$	$\xi\uparrow$	More cars $\Rightarrow$ more ξ
2.3.2	$\tilde{c}, \varpi$ constant, $\xi\uparrow$	$\kappa\uparrow$	Higher ξ signals more flow
2.4.1	$\tilde{c}, \xi$ constant, $\kappa\uparrow$	$\varpi\uparrow$	Growth at capped ξ mandates efficiency
2.4.2	$\tilde{c}, \xi$ constant, $\varpi\uparrow$	$\kappa\uparrow$	Efficiency guarantees higher throughput
2.5.1	ϖ, ξ constant, $\kappa\uparrow$	$\tilde{c}\uparrow$	Growth at fixed ops/jam mandates capacity
2.5.2	ϖ, ξ constant, $\tilde{c}\uparrow$	$\kappa\uparrow$	Capacity guarantees higher throughput
2.6.1	κ, ξ constant, $\tilde{c}\uparrow$	$\varpi\downarrow$	Capacity-efficiency trade-off at fixed κ, ξ
2.6.2	κ, ξ constant, $\varpi\uparrow$	$\tilde{c}\downarrow$	Efficiency-capacity trade-off at fixed κ, ξ

powerful, non-infrastructure tool to combat traffic, providing a clear alternative or complement to simply building more roads.

3. *Principle 1.4.1: The Principle of Required Investment*: Growth without worsening congestion requires investment in capacity and/or efficiency. This is the ultimate principle for strategic planning and policymaking. It elegantly frames the central political and economic choice facing any growing city. It states that the dual goals of accommodating more traffic (κ) while maintaining quality of life (constant ξ) are not free. There is a required investment, and the NFI quantifies it as an increase in the product of capacity and efficiency ($\tilde{c}\varpi$). This principle is perfect for synthesis, as it defines the exact menu of options available to a planner: build more (increase $\tilde{c}$), get smarter (increase ϖ), or do a combination of both. It turns vague political goals into a clear, measurable engineering challenge.

4. *Principle 1.2.1: Latent Demand*: If capacity stays fixed and flow rises, then efficiency times congestion must also increase. When capacity is fixed, any increase in traffic flow inevitably raises system stress through higher inefficiency and congestion. Urban planners use this to justify demand management and transit investment, Transport engineers rely on it to diagnose performance decline, and the public experiences it as worsening traffic despite unchanged infrastructure. It's the clearest explanation for why more cars means more jam.
5. *Principle 2.5.1: Mandate for Capacity*: If efficiency and congestion are fixed, more flow can only come from more capacity. This principle forces policymakers to align growth promises with infrastructure budgets, gives engineers a solid case for expansion projects, and helps the public understand why disruptive construction is a necessary cost of economic growth.
6. *Principle 2.5.2: Capacity Throughput Guarantee*: If efficiency and congestion are fixed, more capacity must yield more flow. Planners use this to assess whether added vehicle trips align with sustainability goals, engineers quantify throughput gains post-construction, and developers frame infrastructure as a direct enabler of economic activity, not just concrete, but movement.

13.7 Applications in Urban Traffic Management

Example 13.10 (Limit of Road Diet)

- *Case 1:* $\varpi_1 = 0.5$ Then: $\tilde{c}_2 > 0.5 \times \tilde{c}_1$ Up to 50% of the original capacity may be removed, provided traffic is perfectly redistributed ($\varpi_2 = 1$).
- *Case 2:* $\varpi_1 = 1$ Then: $\tilde{c}_2 > \tilde{c}_1$ No capacity reduction is possible without increasing congestion. The only viable strategy is capacity expansion.

In the following example, we use the VarPi index to compare different scenarios and demonstrate how policy decisions can be evaluated based on their effect on network balance.

Example 13.11 (Scenario Analysis with the VarPi Index) This example uses the VarPi index to compare the effects of different capacity reduction policies on a network as shown in Fig. 13.1. The relative link capacity are shown in the figure. Link *bc* has a variation of lane capacity from $\mathsf{C}_{0,\mathsf{bc}} = 5$ lanes to $\mathsf{C}_{4,\mathsf{bc}} = 1$ lane. The goal is to evaluate reduction of capacity (*RedCap*) policy. The capacity per lane is $\phi = 1800$ vehicles/hour, and the total flow is $\kappa = 32{,}760$ vehicles/hour for all scenarios. Based on Steady Link Characteristics (SLC) Axiom 13.7 we set the free flow speed to be constant $u = 60$ km/hour and the link distance is also constant $l = 1$ km.

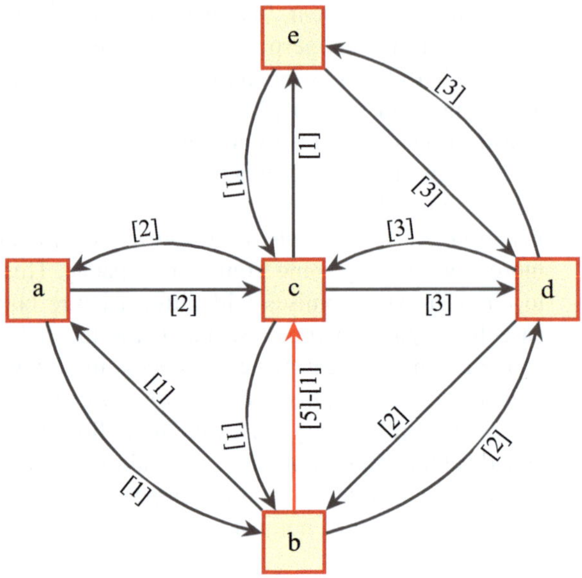

Fig. 13.1 Network for simulation of VarPi Index example: a five-node strongly connected network with a flow distribution

- Analysis of the Base Case: The initial network, represented by capacity matrix $\mathsf{C_0}$:

$$\mathsf{C_0} = \begin{bmatrix} 0 & 1 & 2 & 0 & 0 \\ 1 & 0 & 5 & 2 & 0 \\ 2 & 1 & 0 & 3 & 1 \\ 0 & 2 & 3 & 0 & 3 \\ 0 & 0 & 1 & 3 & 0 \end{bmatrix}$$

$\mathsf{C_0}$: Initial network with $\mathsf{C_{0,bc}} = 5$ lanes No policy applied:

$$\mathsf{C_0} = \text{Initial configuration}$$

The total relative capacity is 30 lanes, which is equivalent to the total absolute capacity $\tilde{c} = 1800 \times 30 = 54{,}000$ vehicle/hour. The absolute ideal flow matrix has total network flow of $\kappa = 32{,}760$ vehicle/hour.

$$\mathsf{F_0} = \begin{bmatrix} 0 & 1170 & 2340 & 0 & 0 \\ 617 & 0 & 3083 & 1233 & 0 \\ 2894 & 1447 & 0 & 4341 & 1447 \\ 0 & 2316 & 3475 & 0 & 3475 \\ 0 & 0 & 1230 & 3691 & 0 \end{bmatrix}$$

The resulting congestion matrix, computed from the ideal flow matrix, shows a non-uniform distribution of congestion.

$$\mathsf{G}_0 = \begin{bmatrix} 0 & 65.01\% & 65.01\% & 0 & 0 \\ 34.26\% & 0 & 34.26\% & 34.26\% & 0 \\ 80.39\% & 80.39\% & 0 & 80.39\% & 80.39 \\ 0 & 64.34\% & 64.34\% & 0 & 64.34\% \\ 0 & 0 & 68.35\% & 68.35\% & 0 \end{bmatrix}$$

The capacity-weighted average congestion is 60.667% and the maximum congestion is 80.39%. The VarPi index for this initial scenario is:$\varpi_0 = 1 - \frac{0.8039-0.60667}{0.8039} \approx 75.468\%$.

This VarPi index indicates a moderately unbalanced network, with a high degree of non-uniformity in its congestion distribution. We can test the correctness of this computation using Network Flow Identity: $\kappa = 32{,}760 = \tilde{c} \times \varpi \times \xi = 54{,}000 \times 0.75468 \times 0.8039$.

- Evaluating Policy Scenarios (C1–C3): Using the same total flow $\kappa = 32{,}760$ vehicles/hour, A policy of reducing lanes in a single link (*bc* in this case) is tested across three scenarios. As the capacity of this link is reduced from 5 lanes to 2 lanes (respectively from $\mathsf{C}_{0,bc}$ to $\mathsf{C}_{3,bc}$), the VarPi index consistently increases, reaching 90.59% in scenario C_3. Concurrently, the maximum congestion in the network decreases, and the total system delay is reduced. This suggests that the policy improves the overall network balance and efficiency.

  To be precise, the molecular policies are written as followed: One lane removed from link *bc*:

$$\mathsf{C}_1 = \mathsf{C}_0 + \mathcal{P}^{(-1)}_{\text{RedCap}} \quad \text{where} \quad \mathsf{C}_{1,bc} = 4$$

  Two lanes removed from link *bc*:

$$\mathsf{C}_2 = \mathsf{C}_1 + \mathcal{P}^{(-1)}_{\text{RedCap}} = \mathsf{C}_0 + \mathcal{P}^{(-2)}_{\text{RedCap}} \quad \text{where} \quad \mathsf{C}_{2,bc} = 3$$

  Three lanes removed from link *bc*:

$$\mathsf{C}_3 = \mathsf{C}_2 + \mathcal{P}^{(-1)}_{\text{RedCap}} = \mathsf{C}_0 + \mathcal{P}^{(-3)}_{\text{RedCap}} \quad \text{where} \quad \mathsf{C}_{3,bc} = 2$$

- The Premagic Equilibrium State (C4): In the final scenario, C_4, the capacity of the link is reduced to one lane, resulting in a premagic capacity matrix. Four lanes removed from link *bc*:

$$\mathsf{C}_4 = \mathsf{C}_3 + \mathcal{P}^{(-1)}_{\text{RedCap}} = \mathsf{C}_0 + \mathcal{P}^{(-4)}_{\text{RedCap}} \quad \text{where} \quad \mathsf{C}_{4,bc} = 1$$

This state achieves a perfect VarPi index of 100.00%, as the maximum congestion of 70.00% is identical to the capacity-weighted average congestion. The network is now in a state of perfect balance, where flow is distributed proportionally to capacity across all active links. Compared this to base scenario capacity matrix C_0 we have lane reduction in link *bc* from 5 lanes to just one lane. This results in a new capacity matrix C_4 which is a premagic matrix.

$$\mathsf{C}_4 = \begin{bmatrix} 0 & 1 & 2 & 0 & 0 \\ 1 & 0 & 1 & 2 & 0 \\ 2 & 1 & 0 & 3 & 1 \\ 0 & 2 & 3 & 0 & 3 \\ 0 & 0 & 1 & 3 & 0 \end{bmatrix}$$

The total lane capacity in the network is $\tilde{c} = \sum_{i,j} c_{ij} = 26$. Using Markov Split (proportional capacity) Axiom 13.1 we get the stochastic matrix to be

$$\mathsf{S}_4 = \begin{bmatrix} 0 & \frac{1}{3} & \frac{2}{3} & 0 & 0 \\ \frac{1}{4} & 0 & \frac{1}{4} & \frac{1}{2} & 0 \\ \frac{2}{7} & \frac{1}{7} & 0 & \frac{3}{7} & \frac{1}{7} \\ 0 & \frac{1}{4} & \frac{3}{8} & 0 & \frac{3}{8} \\ 0 & 0 & \frac{1}{4} & \frac{3}{4} & 0 \end{bmatrix}$$

The relative flow matrix is exactly the same as the capacity matrix.

$$\mathsf{F}_{\mathsf{rel},4} = \begin{bmatrix} 0 & 1 & 2 & 0 & 0 \\ 1 & 0 & 1 & 2 & 0 \\ 2 & 1 & 0 & 3 & 1 \\ 0 & 2 & 3 & 0 & 3 \\ 0 & 0 & 1 & 3 & 0 \end{bmatrix}$$

The resulting congestion matrix is a steady matrix with uniform congestion level, with every active link having a congestion of exactly $\sigma = 70.00\%$.

$$\mathsf{G}_4 = \begin{bmatrix} 0 & 70.00\% & 70.00\% & 0 & 0 \\ 70.00\% & 0 & 70.00\% & 70.00\% & 0 \\ 70.00\% & 70.00\% & 0 & 70.00\% & 70.00\% \\ 0 & 70.00\% & 70.00\% & 0 & 70.00\% \\ 0 & 0 & 70.00\% & 70.00\% & 0 \end{bmatrix}$$

The constant congestion value can also be obtained by $\sigma = \frac{\kappa}{\phi \times \tilde{c}} = \frac{32{,}760}{1800 \times 26} = 0.7$.

This demonstrates that when a network is in a Premagic Equilibrium, its VarPi index is always

$$\varpi_1 = 1 - \frac{0.7000 - 0.7000}{0.7000} = 100\%$$

. The capacity reduction on a single link successfully redistributed congestion and reduced the previous maximum congestion from 80.39% to a uniform 70.00%. The travel time matrix in hour is a steady matrix with uniform link travel time $t_{\text{link}} = 0.0173$ hours ≈ 62 seconds.

$$\mathsf{T}_4 = 0.0173 \times \begin{bmatrix} 0 & 1 & 1 & 0 & 0 \\ 1 & 0 & 1 & 1 & 0 \\ 1 & 1 & 0 & 1 & 1 \\ 0 & 1 & 1 & 0 & 1 \\ 0 & 0 & 1 & 1 & 0 \end{bmatrix} = t_{\text{link}} \cdot \mathsf{A}$$

Implications for Policy Decisions: The scenario analysis shows that while a perfect VarPi index of 100% represents a state of ideal balance, the *minimum total system delay* occurred in scenario C_2, not in the premagic state of C_4. It demonstrates that the state with the minimum total travel time may not be the state with perfect balance. This highlights that a policy decision based solely on maximizing network balance may not always lead to a minimum aggregate travel time. Planners must evaluate the trade-offs between achieving a highly balanced and equitable system and minimizing total system delay. The performance metrics for each scenario are summarized in the Table 13.2. The capacity is in the number of lanes (need to multiply by 1800 to get absolute total capacity in the network). The delay column is the total system delay in vehicle.hours of delay per hour. The travel time column is the Total System Travel Time (VHT) in vehicle.hours traveled per hour. The speed column is the Flow-Weighted Average Speed (VKT/VHT) in km/hour.

The simulation result validates the central idea that a strategic reduction in capacity on a single link can increase network balance, as measured by the VarPi index. However, a closer look at the results reveals a critical trade-off. While the VarPi index increases steadily toward 100% and maximum congestion decreases,

Table 13.2 Performance metrics for capacity reduction scenarios

Scenario	Capacity	VarPi	Max congestion	Delay	Travel time	Speed
C0 (Base)	30	75.468%	80.388%	18.97	564.97	57.99
C1	29	79.511%	78.931%	18.55	564.55	58.03
C2	28	84.444%	76.974%	18.22	564.22	58.06
C3	27	90.835%	74.208%	18.24	564.24	58.06
C4	26	100.00%	70.00%	19.66	565.66	57.91

the Total System Delay is not minimized in the perfectly balanced (premagic) state. The lowest delays are observed in scenarios C2, which are highly balanced but not perfectly so. Scenario C2 happened to have the lowest travel time *only among the specific scenarios tested in that particular "road diet" experiment* (C0–C4), where total capacity was changing. It was an interesting outcome of that simulation but is not a universally guaranteed optimal state. It was simply the "sweet spot" found in that sequence. In this example, we are comparing different networks, not analyzing flow on a single, fixed network.

The theory of Premagic Equilibrium posits that for a *given, fixed total capacity*, a premagic structure (C_4) represents the most efficient configuration. It aligns the User Equilibrium with the System Optimum, minimizing delay for that specific amount of infrastructure. The simulation does not analyze one fixed network. Instead, it compares a series of five distinct networks (C_0, C_1, C_2, C_3, C_4). Crucially, each step in this series involves *changing the total capacity* of the system. The reason why the simulation fail to show that Premagic capacity is the best is because IPLC ensures UE=SO alignment *for a fixed capacity*, but we forcibly reshaping the capacity C to be premagic *reduces total capacity* (C4: 26 lanes $\rightarrow$ 46,800 veh/hr; C2: 28 lanes $\rightarrow$ 50,400 veh/hr). This is not comparing an apple to a banana; it's comparing five different fruits to see which is best. The "road diet" policy from C_0 to C_4 creates two competing effects, and C_2 represents the optimal trade-off between them:

1. Positive Effect (Improved Balance): The strategic removal of lanes from link *bc* forces a more equitable flow distribution across the network. This improves the VarPi index and reduces the severity of the worst bottlenecks.
2. Negative Effect (Reduced Total Capacity): The policy also reduces the total number of lanes in the entire system, from 30 in C_0 down to 26 in C_4. This lowers the network's overall physical throughput.

This outcome of Example 13.12 is not a flaw in the simulation nor the theory. The simulation results do not contradict the theory. The theory is about flow distribution on a *fixed network*. The example is a perfect tool for illustrating this exact point. The simulation compares different networks (with different capacities). Thus, design rule must maximize total capacity first, then pursue premagic balance.

In the following Example 13.12, we do simulation of swapping lanes but the total lane capacity in the network is hold constant similar to the premagic capacity. Here we can see that indeed the premagic capacity hold the minimum travel time.

Example 13.12 (Network Reconfiguration with Constant Total Capacity) This example investigates the impact of network topology on performance while holding the total lane capacity constant. The experiment begins with the perfectly balanced, premagic network from the Example 13.11 (C_4) as the base case. Subsequent scenarios (C_5, C_6, C_7) are generated by reallocating lanes between links, keeping the total lane capacity fixed at $\tilde{c} = 26$ lanes. This approach isolates the effect of network structure on efficiency and balance distribution. The total flow remains constant at $\kappa = 32{,}760$ vehicles/hour for all scenarios.

The objective is to demonstrate how reconfiguring link capacities, even without changing the total system capacity, can significantly alter network performance metrics such as the VarPi index, maximum congestion, and total system delay.

- *Analysis of the Base Case* (C_4): The initial network is in a state of Premagic Equilibrium, represented by the symmetric, Eulerian capacity matrix C_4.

$$\mathsf{C}_4 = \begin{bmatrix} 0\,1\,2\,0\,0 \\ 1\,0\,1\,2\,0 \\ 2\,1\,0\,3\,1 \\ 0\,2\,3\,0\,3 \\ 0\,0\,1\,3\,0 \end{bmatrix}$$

  This configuration is structurally ideal because the capacity is premagic matrix. As established previously, it yields a perfect VarPi index of $\varpi_4 = 100\%$, with all active links sharing a uniform congestion level of 70.00%. The Total System Delay is 19.66 vehicle-hours. This scenario serves as the optimal benchmark for a perfectly balanced network.
- *Evaluating Reconfiguration Scenarios* ($\mathsf{C}_5 - \mathsf{C}_7$): The following scenarios are perturbations of the ideal state, created by swapping lane capacities between links.

  - *Scenario* C_5: This matrix is formed by swapping the capacities of link *cb* (from 1 to 3) and link *cd* (from 3 to 1).

$$\mathsf{C}_5 = \begin{bmatrix} 0\,1\,2\,0\,0 \\ 1\,0\,1\,2\,0 \\ 2\,3\,0\,1\,1 \\ 0\,2\,3\,0\,3 \\ 0\,0\,1\,3\,0 \end{bmatrix}$$

    Swaps 2 lanes from link *cd* to link *cb*:

$$\mathsf{C}_5 = \mathsf{C}_4 \cup \mathcal{P}^{(cd \to cb, -2)}_{\text{Realloc}}$$

    This seemingly minor reallocation breaks the premagic symmetry. The consequences are significant: the VarPi index plummets to 73.12%, and the maximum congestion escalates to 100.54%, indicating a severe new bottleneck. The Total System Delay increases by over 60% to 31.57 vehicle-hours, though the average speed decreases only slightly (2.1%).
  - *Scenario* C_6, C_7 *and* C_8: Further reallocations continue to degrade performance. In C_6, adding capacity to a non-existent link (ad) and removing it elsewhere (link *cd*) disrupts the network's Eulerian structure. This results in a VarPi index of 77.983% and a maximum congestion of 89.763%.

$$\mathsf{C}_6 = \begin{bmatrix} 0 & 1 & 2 & 2 & 0 \\ 1 & 0 & 1 & 2 & 0 \\ 2 & 1 & 0 & 1 & 1 \\ 0 & 2 & 3 & 0 & 3 \\ 0 & 0 & 1 & 3 & 0 \end{bmatrix}$$

Adds 2 lanes to link *ad* and removes 2 lanes from link *cd*:

$$\mathsf{C}_6 = \mathsf{C}_4 \cup \mathcal{P}_{\text{Realloc}}^{(cd \to ad, -2)}$$

Scenario C_6 produces slightly lower average speed (1.14%) but it yields 33% higher delay than baseline scenario C_4. Scenario C_7 represents inefficient configuration, with the lowest VarPi index (63.550%) and the highest maximum congestion (110.15%) and Total System Delay (45.03 vehicle-hours).

$$\mathsf{C}_7 = \begin{bmatrix} 0 & 1 & 2 & 2 & 0 \\ 1 & 0 & 1 & 2 & 0 \\ 2 & 1 & 0 & 1 & 1 \\ 0 & 2 & 3 & 0 & 3 \\ 0 & 0 & 3 & 1 & 0 \end{bmatrix}$$

Adds 2 lanes to link *ad* and 2 lanes to link *ec*, removing 2 lanes from link *cd* and 2 lanes from link *ce*:

$$\mathsf{C}_7 = \mathsf{C}_6 \cup \mathcal{P}_{\text{Realloc}}^{(ce \to ec, -2)}$$

Alternatively, from base:

$$\mathsf{C}_7 = \mathsf{C}_4 \cup \left(\mathcal{P}_{\text{Realloc}}^{(cd \to ad, -2)} \cup \mathcal{P}_{\text{Realloc}}^{(ce \to ec, -2)} \right)$$

Scenario C_8 represents the most inefficient configuration, with the lowest VarPi index (44.946%) and the highest maximum congestion (155.743%) and Total System Delay (101.29 vehicle-hours).

$$\mathsf{C}_8 = \begin{bmatrix} 0 & 1 & 2 & 0 & 0 \\ 1 & 0 & 5 & 2 & 0 \\ 2 & 1 & 0 & 3 & 1 \\ 0 & 2 & 1 & 0 & 3 \\ 0 & 0 & 1 & 1 & 0 \end{bmatrix}$$

Table 13.3 Performance metrics for network reconfiguration scenarios

Scenario	Capacity	VarPi	Max congestion	Delay	Travel time	Speed
C4 (Base)	26	100.00%	70.00%	19.66	565.66	57.91
C5	26	69.622%	100.543%	31.57	577.57	56.72
C6	26	77.983%	89.763%	26.16	572.16	57.26
C7	26	63.550%	110.149%	45.03	591.03	55.43
C8	26	44.946%	155.743%	101.29	647.29	50.61

Adds 4 lanes to link *bc*, removes 2 lanes from link *cd*, and removes 2 lanes from link *ce*:

$$\mathsf{C}_8 = \mathsf{C}_4 \cup \left(\mathcal{P}_{\text{Realloc}}^{(cd\to bc,-2)} \cup \mathcal{P}_{\text{Realloc}}^{(ce\to bc,-2)}\right)$$

C_7 andC_8 suffer from significantly higher delays and congestion.

Implications for Policy Decisions The performance metrics for each reconfiguration scenario are summarized in Table 13.3. The analysis reveals that network structure is as critical as total capacity. While all five scenarios possess an identical total capacity of 26 lanes, their performance varies dramatically. The *premagic* configuration of capacity matrix C_4 is unequivocally the most efficient and balanced with minimum travel time and delay and the highest average speed. Any deviation from this structure induces bottlenecks, increases delay, and reduces overall efficiency.

This experiment also validates the principle of the *Network Flow Identity* ($\kappa = \tilde{c} \times \varpi \times \xi$). Since total flow κ and total capacity $\tilde{c}$ are held constant, the identity simplifies to $\varpi \times \xi \approx$ constant. This inverse relationship is observed in the results: as the network balance performance index ϖ decreases, the maximum congestion ξ inevitably increases. Therefore, for a system with fixed capacity resources, maximizing structural balance is the key to minimizing the worst-case bottlenecks.

For a *fixed capacity*, the theory of Premagic Equilibrium guarantees that the perfectly balanced, premagic capacity configuration yields the System Optimum. It is mathematically guaranteed to provide the minimum possible travel time for that specific amount of infrastructure. With that, we can compute the Price of Anarchy (PoA).

Example 13.13 (Price of Anarchy for Network Reconfiguration with Constant Capacity) This example calculates the Price of Anarchy (PoA) for a set of network configurations where the total system capacity is held constant. The analysis is performed on scenarios C_4 through C_8, all of which operate with a fixed total capacity of 26 lanes. This approach ensures a valid "apples-to-apples" comparison, measuring the inefficiency that arises purely from suboptimal network structure, not from variations in total resources.

The PoA quantifies the cost of inefficient design relative to an optimal benchmark. The cost metric is the *Total System Delay*.

Within this group of 26-lane networks, the *System Optimum (SO)* is the premagic, perfectly balanced configuration of C_4. It achieves the lowest possible delay for this fixed amount of capacity. The PoA for any suboptimal scenario C_k in this set is therefore:

$$\text{PoA}(\mathsf{C}_k) = \frac{\text{Total System Delay of } \mathsf{C}_k}{\text{Total System Delay of } \mathsf{C}_4}$$

- *Defining the System Optimum Benchmark* (C_4)
  For a fixed system capacity of 26 lanes, scenario C_4 is the most efficient state.
  Total System Delay of C_4 (C_{SO}): 19.66 vehicle-hours.
  Its PoA is the baseline value of 1.000.

$$\text{PoA}(\mathsf{C}_4) = \frac{19.66}{19.66} = 1.000$$

- *Computation of PoA for Suboptimal Configurations* ($\mathsf{C}_5 - \mathsf{C}_8$)
  The PoA for the remaining scenarios measures the performance degradation caused by deviating from the optimal premagic structure, while keeping the number of lanes constant.
  - *Scenario* C_5: A minor structural change leads to a delay of 31.57 vehicle-hours.

$$\text{PoA}(\mathsf{C}_5) = \frac{31.57}{19.66} \approx 1.606$$

    This configuration is 60.6% less efficient.
  - *Scenario* C_6: A different suboptimal structure results in a delay of 26.16 vehicle-hours.

$$\text{PoA}(\mathsf{C}_6) = \frac{26.16}{19.66} \approx 1.331$$

    This configuration is 33.1% less efficient.
  - *Scenario* C_7: A highly inefficient structure produces a delay of 45.03 vehicle-hours.

$$\text{PoA}(\mathsf{C}_7) = \frac{45.03}{19.66} \approx 2.290$$

    The inefficiency more than doubles.

Table 13.4 Price of anarchy for reconfigurations with constant capacity

Scenario	VarPi	Max congestion	Delay	Travel time	Speed	PoA
C4 (SO)	**100.00%**	**70.00%**	**19.66**	**565.66**	**57.91**	**1.000**
C5	69.62%	100.54%	31.57	577.57	56.72	**1.606**
C6	77.98%	89.76%	26.16	572.16	57.26	**1.331**
C7	63.55%	110.15%	45.03	591.03	55.43	**2.290**
C8	44.95%	155.74%	101.29	647.29	50.61	**5.152**

- *Scenario* C_8: The most poorly designed network yields a catastrophic delay of 101.29 vehicle-hours.

$$\mathrm{PoA}(C_8) = \frac{101.29}{19.66} \approx 5.152$$

This structure is over five times less efficient than the optimal design, demonstrating a massive cost of poor resource allocation.

Implications for Policy Decisions This analysis powerfully demonstrates the economic consequences of network design choices, even when the budget (total number of lanes) is fixed. The PoA values reveal that how capacity is structured is critically important. A poorly designed network (C_8) can create over five times more delay and congestion than an optimally designed one (C_4) using the exact same amount of resources. The PoA provides a clear, quantitative justification for investing in intelligent network planning and design, rather than just raw capacity. As shown in Table 13.4, the Price of Anarchy increases significantly with inefficient network configurations, underscoring the value of intelligent design.

13.8 Summary

The presented theory establishes that a network's equilibrium state can be structurally and operationally optimal if certain idealized conditions are met. We have established the principle of Premagic Equilibrium as a tool for *flow optimization for a fixed network*, not a tool for global network design optimization. The central deduction is that the congruence of User Equilibrium (UE) and System Optimum (SO) is a direct consequence of a network's capacity structure and topology.

The first key conclusion is that a Premagic Equilibrium (PE), which results in uniform link congestion, is a sufficient condition for a system to achieve the System Optimum. The second conclusion demonstrates that this same uniform congestion, when combined with Steady Link Characteristics (SLC) and Identical Path Link Counts (IPLC), also satisfies the User Equilibrium condition. Therefore, the simultaneous satisfaction of these three conditions leads to the ultimate ideal state where UE $=$ SO $=$ PE.

This alignment is not merely theoretical, it is mathematically rigorous. Using Jensen's Inequality, the theory proves that for a fixed network and total flow, uniform congestion minimizes aggregate travel time. Thus, PE represents the most efficient flow distribution for a given network, but it does not imply that premagic networks are globally optimal in terms of total travel time or capacity. A premagic network may have lower total capacity and hence higher congestion if not properly scaled. Therefore, while PE is a tool for flow optimization, it is not a universal solution for network design optimization.

Importantly, networks satisfying these conditions are immune to classical paradoxes such as Braess' Paradox, and yield a Price of Anarchy (PoA) of exactly one, indicating perfect alignment between individual choices and collective efficiency. This provides a benchmark for evaluating network performance via the VarPi index, and a set of design principles encapsulated in the Network Flow Identity (NFI).

The final design implication is clear: while PE optimizes flow within a fixed structure, the network's topology and capacity matrix must be chosen to support this equilibrium.

References

1. Jayanth R. Banavar et al. "Structure of optimal transport networks". In: *Physical Review Letters* 84.20 (2010), p. 4745. https://doi.org/10.1103/PhysRevLett.84.4745. https://journals.aps.org/prl/abstract/10.1103/PhysRevLett.84.4745
2. D. Braess, A. Nagurney, and T. Wakolbinger. "On a paradox of traffic planning". In: *Transportation Science* 39.4 (2005), pp. 446–450.
3. Dietrich Braess. "Über ein Paradoxon aus der Verkehrsplanung". In: *Unternehmensforschung* 12.1 (1968), pp. 258–268. https://doi.org/10.1007/BF01918335.
4. Ennio Cascetta. *Transportation Systems Analysis: Models and Applications*. 2nd. New York: Springer, 2009. DOI: https://doi.org/10.1007/978-0-387-75857-2. https://link.springer.com/book/10.1007/978-0-387-75857-2.
5. Carlos F. Daganzo. "Urban Gridlock: Macroscopic Modeling and Mitigation Approaches". In: *Transportation Research Part B: Methodological* 41.1 (2007), pp. 49–62. https://doi.org/10.1016/j.trb.2006.03.001.
6. Reinhard Diestel. *Graph Theory*. 5th. Berlin, Heidelberg: Springer, 2017. https://doi.org/10.1007/978-3-662-53622-3.
7. Matthias Ehrgott. *Multicriteria Optimization*. 2nd ed. Springer, 2005. ISBN: 978-3-540-21398-7. https://doi.org/10.1007/3-540-27659-9.
8. John G. Kemeny and J. Laurie Snell. *Finite Markov Chains*. New York: Springer, 1976. https://link.springer.com/book/9780387901923.
9. S. Kullback. *Information Theory and Statistics*. Reprint of the 1959 edition. Dover Publications, 1997.
10. S. Kullback and R. A. Leibler. "On Information and Sufficiency". In: *The Annals of Mathematical Statistics* 22.1 (1951), pp. 79–86. https://doi.org/10.1214/aoms/1177729694.
11. P. C. Mahalanobis. "On the generalized distance in statistics". In: *Proceedings of the National Institute of Sciences of India* 2.1 (1936), pp. 49–55.

12. P. Van Mieghem. *Performance Analysis of Complex Networks and Systems*. Cambridge University Press, 2014. https://doi.org/10.1017/CBO9781107415874.
13. J. F. Nash. "Non-Cooperative Games". In: *Annals of Mathematics* 54.2 (1951), pp. 286–295. https://doi.org/10.2307/1969529.
14. John F. Nash. "Equilibrium Points in N-Person Games". In: *Proceedings of the National Academy of Sciences* 36.1 (1950), pp. 48–49. https://doi.org/10.1073/pnas.36.1.48.
15. T. Roughgarden and E. Tardos. "How bad is selfish routing?" In: *Proceedings 41st Annual Symposium on Foundations of Computer Science*. 2000, pp. 93–102. https://doi.org/10.1109/SFCS.2000.892069.
16. Joseph M. Sussman. *Perspectives on Intelligent Transportation Systems (ITS)*. Springer, 2005. ISBN: 978-0-387-23257-7. https://doi.org/10.1007/b101063.
17. Kardi Teknomo. *Introduction to Ideal Flow Networks*. 2022.
18. Kardi Teknomo. "Premagic and Ideal Flow Matrices". In: *arXiv preprint arXiv:1706.08856* (2017). arXiv: 1706.08856 [cs.DM]. https://arxiv.org/pdf/1706.08856.

Chapter 14
The Chemistry of Policy Analysis and Synthesis in the Ideal Traffic Assignment Framework

14.1 Introduction

The chapter establishes the Ideal Traffic Assignment (ITA) framework as a scientifically rigorous tool for transportation planning. It redefines policy formulation as a quantitative and optimizable procedure, moving it from a qualitative basis. The core of this methodology lies in defining policies as mathematical transformations on an Ideal Flow Network (IFN). The framework distinguishes between controllable variables (e.g., capacity, speed) and uncontrollable outputs (e.g., flow, congestion), which separates the active role of a planner from the passive role of a forecaster. The chapter introduces a formal language for policy interventions, categorizing them into primitive "atomic" policies. These policies are organized in a "periodic table" based on their function and permanence, providing a complete and exhaustive set of actions a planner can take. Complex, real-world interventions are then synthesized as "molecular policies," which are finite combinations of atomic policies. The framework's operators for union ($\cup$), negation ($\neg$), and sequencing ($\rightarrow$) enable the precise modeling of these interventions. Policy feasibility is rigorously defined by constraints that ensure the network remains a valid IFN, specifically, that it remains strongly connected and premagic after an intervention. The framework also provides a formal "policy conflict theorem" to identify and discard logically contradictory policy combinations. This theoretical foundation transforms policy design into a combinatorial discrete optimization problem, where an automated algorithm can generate and evaluate all feasible policies against a user-defined goal function. Finally, the chapter presents a practical method for scaling the relative IFN model to absolute, real-world flow values using a maximum allowable congestion level as a benchmark. This approach allows for meaningful scenario comparisons without requiring exhaustive real-world data, enabling planners to predict the systemic effects of their decisions and transition from reactive to proactive, data-driven planning. Furthermore, the chapter illustrated the power of ITA for long-term

K. Teknomo, *Ideal Flow Network for Transportation Research*,
Lecture Notes in Intelligent Transportation and Infrastructure,
https://doi.org/10.1007/978-981-95-4727-2_14

forecasting. By integrating the model with growth projections and proactive/reactive policy assumptions, planners can anticipate future bottlenecks and understand the systemic, often non-local, effects of their decisions.

14.2 Foundational Concepts: From Intuition to a Rigorous Framework

The Ideal Traffic Assignment (ITA) is a novel approach to traffic analysis and policy evaluation that uses the framework of an Ideal Flow Network (IFN). This model provides an alternative to traditional methods like User Equilibrium (UE) and System Optimum (SO) assignments. ITA is particularly useful for policy analysis because it provides a benchmark for the most efficient distribution of traffic flow on a network, based on its physical structure. In this chapter, we demonstrate how the ITA framework can be applied to real-world transportation policy and planning. The Ideal Traffic Assignment framework provides an approach to transportation planning to evaluate policies that modify the network's structure, such as converting a two-way road to a one-way road or adding new links, expanding or reducing road capacity.

In urban transportation planning, we need to make a crucial distinction between *planning* and *forecasting*. Forecasting involves predicting future states of the network based on existing trends and uncontrollable factors, like population growth and economic development, which in turn would influence the total demand in the network κ. We are the observers in this process, estimating what will happen if no changes are made. Planning, on the other hand, is an active process. It involves making deliberate changes to the controllable network's structural inputs to achieve specific performance goals.

It is vital to understand that the ITA is not a crystal ball. Its primary purpose is not just to predict a single, definitive future. Rather, its power lies in comparing alternative futures. This is the difference between *forecasting* and *planning*.

- *Forecasting* involves predicting outcomes under a no-action scenario. This is often called the "Base Case" or "Do Nothing" scenario. For example, a forecast might estimate that by 2050, traffic volume on the Main Street bridge will exceed its capacity by 50%. A planner can use this projection to identify a future issue: the risk of severe daily gridlock.
- *Planning* is the process of using that forecast to design and test solutions. A planner would take the forecast of gridlock and ask, "What if we build a new bypass bridge?" They would then add this new bridge to the model's network and run the four-step process again. By comparing the total system travel time in the Base Case against this "Alternative Scenario," we can quantitatively measure whether our proposed bypass is a good investment. This comparison is the heart of transportation planning.

Table 14.1 Classification of Variables in the ITA Framework

Variable	Type	Controllability	Primary context	Role in the framework
L, U, C, S, A	Input	Controllable	Planning	Define the network's structural and physical properties, which can be directly manipulated by policy. The adjacency matrix, A, is the base on which structural changes are made
κ	Input	Uncontrollable	Forecasting	Represents the total network demand; a factor to be predicted, not directly controlled by a structural policy
F, G, T^0, T, V, Y	Output	Uncontrollable	Planning	Reflect the network's performance and are dependent variables that cannot be directly manipulated

As shown in Table 14.1, the ITA framework distinguishes between controllable structural inputs, uncontrollable demand forecasts, and performance outputs that emerge from system dynamics.

The ITA model distinguishes between two fundamental types of variables. *Controllable variables* are the network's structural properties that a planner can manipulate, such as the capacity matrix C, the link distance matrix L, the free flow speed matrix U and stochastic matrix S. These are the inputs that a planner can directly and deliberately change. *Uncontrollable variables* are external factors that cannot be directly changed by a policy, such as the total flow or demand, κ. These are the outputs that are a consequence of the network's internal dynamics and the chosen inputs. This distinction is crucial for separating the active role of a planner, who designs the network, from the passive role of a forecaster, who predicts demand.

The ITA model also distinguishes between *independent variables* (the inputs) and dependent variables (the outputs). The network's structural inputs are the link length matrix L, free-flow speed matrix U and capacity matrix C, and stochastic matrix S represent the structural and physical properties of the road network. The stochastic matrix is directly related to the signal timing in intersection. These matrices are inherently constrained by the network's underlying topology, as defined by the adjacency matrix A. These are the variables that policymakers can directly control and modify. The input κ (total flow), represents the total demand in the network, which is considered an uncontrollable factor but can be calibrated using real world data.

The outputs of the ITA model are the *dependent variables* that reflect the network's performance such as flow matrix F, congestion matrix G, min travel time matrix T^0, travel time matrix T, link speed matrix V and delay matrix Y. These matrices are something that we cannot directly control. The ITA model's strength lies in its ability to simulate and predict how changes to the independent variables (i.e., policy changes) will affect these output metrics.

In ITA, planning can be framed as an optimization problem. We would like to find the combination of policy $\mathcal{P}$ changes that will lead to the desired outcomes for the goal function of dependent variables (e.g., minimizing total travel time or, more importantly, minimizing the variation in maximum congestion). A goal is not a variable to be controlled, but a criterion for evaluating the values of the uncontrollable output variables. The goal of a policy, such as "minimizing the maximum congestion," is a criterion applied to an uncontrollable variable, not a controllable variable itself. Ideal Traffic Assignment (ITA) is a framework, not just mere a model. The goal function itself is subjective and should not be fixed within a general ITA framework. It must remain a function of criteria that can be adapted by the researchers to suit a specific planning objective. We explicitly declare the goal function as an open-ended, where the researchers can define their own component of the optimization problem.

Think of it as a thermostat. We (the planner) have direct control over the temperature setting (the goal criterion). However, we do not have direct control over the room's actual temperature. The temperature (the uncontrollable output) is a result of the thermostat's settings (the controllable input) and external factors like heat loss or gain.

In the same way, the planner's goal is to find a set of controllable inputs that produces the most desirable state of the uncontrollable outputs. The goal function, $\omega(\mathcal{P})$, is a mathematical expression of this desire. It is a function of the output variables, and the optimization process seeks to find the set of controllable inputs that minimizes this function. Therefore, the planning process is controllable because the planner has authority over the inputs, but the output variables remain uncontrollable, and the goal is a target for those uncontrollable outputs.

This systematic approach transforms policy ideas from subjective opinions into testable hypotheses. Instead of relying on intuition or "common sense," policymakers can use the ITA model to simulate the effects of their ideas before implementation. This allows for a more efficient allocation of public funds and reduces the risk of ineffective or counterproductive changes. This chapter establishes a formal and rigorous framework for applying Ideal Traffic Assignment (ITA) to transportation policy and planning. The objective is to transform policy formulation from a qualitative process based on intuition to a quantitative, objective, and optimizable procedure. This is accomplished by defining policies as mathematical transformations on an Ideal Flow Network (IFN). The framework provides a methodology for evaluating interventions by comparing a base network state to an alternative scenario, allowing for policy analysis without requiring full-scale real-world data calibration. The core strength of this approach lies in its ability to simulate how changes to network inputs will affect performance outputs, providing a predictive tool for planners.

By reading this chapter, a reader will gain a foundational understanding of how to transform qualitative policy ideas into a rigorous, mathematical, and computational framework. They will learn to:

1. Systematically categorize all possible policy interventions using the "Atomic Policies."
2. Translate real-world policy problems into formal "molecular policies" composed of these atomic building blocks.
3. Formally define and identify policy conflicts using a rigorous, operator-based language.
4. Perform objective scenario analysis by using a methodology that does not require absolute real-world data, but instead scales the network based on a maximum allowable congestion level.
5. Understand the computational power of the framework to automatically generate and optimize policy combinations, moving beyond human intuition to a systematic search for solutions.

14.2.1 The Network State and Controllable Variables

The state of an Ideal Flow Network at a given time is defined by a set of inputs and is characterized by a set of outputs. We distinguish between inputs that are within the planner's control and those that are not.

Definition 14.1 (Network State) The state of an Ideal Flow Network (IFN) is defined by the tuple of matrices and scalars $(\mathsf{C}, \mathsf{L}, \mathsf{U}, \mathsf{S}, \kappa, t)$, where t is time. The network's performance is then described by a set of outputs, including the Ideal Flow matrix F, the Congestion matrix G, the Travel Time matrix T the Link Speed matrix V and the Delay matrix Y.

In modeling an Ideal Flow Network (IFN), scenario analysis serves as a structured approach for evaluating the impact of different input configurations. These scenarios are constructed by varying the network structural input, which are classified into two categories based on their amenability to policy intervention.

As previously defined in Definition 9.4, a *policy* is a deliberate action that alters the physical or operational characteristics of a transportation network. These changes are defined by modifying at least one of the network structural input: the adjacency matrix (A) the Capacity Matrix (C), the Link Length Matrix (L), the Free Flow Speed Matrix (U), the total flow (κ) and the Stochastic Matrix (S). By changing these matrices, policy interventions directly affect the network's supply, which, in turn, influences and re-balances travel demand.

A policy is symbolized by $\mathcal{P} = \{\Delta\mathsf{A}, \Delta\mathsf{L}, \Delta\mathsf{U}, \Delta\mathsf{C}, \Delta\kappa, \Delta\mathsf{S}\}$ to represents the changes network structural input of the first scenario $\mathcal{N}_1 = \{\mathsf{A}_1, \mathsf{L}_1, \mathsf{U}_1, \mathsf{C}_1, \kappa_1, \mathsf{S}_1\}$ into the second scenario $\mathcal{N}_2 = \{\mathsf{A}_1, \mathsf{L}_2, \mathsf{U}_2, \mathsf{C}_2, \kappa_2, \mathsf{S}_2\}$.

In short, we write it as

$$\mathcal{N}_1 \cup \mathcal{P} \to \mathcal{N}_2 \quad (14.1)$$

Definition 14.2 (Controllable Inputs (Policy Levers)) Controllable inputs are variables that can be directly influenced by planners or decision-makers. They correspond to the structural and operational characteristics of the network and include:

- *Capacity Matrix* C: Reflects infrastructure modifications such as road expansions or lane additions.
- *Link Distance Matrix* L: Encodes spatial layout and connectivity.
- *Maximum Speed Matrix* U: Represents regulatory constraints like speed limits.
- *Stochastic Matrix* S: Captures probabilistic routing behavior and operational adjustments.

These variables form the basis of proposed scenarios, where the objective is to evaluate the effectiveness of specific policy actions in improving network performance metrics such as congestion, travel time, and delay. While L and U are inputs, a planner's control over them is often limited. Policy changes often affect C and S more directly than L or U for existing links.

Definition 14.3 (Uncontrollable Inputs (External Factors)) Uncontrollable inputs are variables that influence the network but lie outside the scope of direct policy manipulation. The primary example is:

- *Total Flow* κ: Represents aggregate demand, which may fluctuate due to external events or long-term trends.

Additional examples include demand shocks (e.g., large public events), socio-economic shifts (e.g., population growth or employment changes), and environmental disruptions (e.g., extreme weather). These inputs are central to forecasting scenarios, which assess the network's robustness under plausible but ungovernable conditions.

This distinction is crucial for separating a planner's job from a forecaster's job. A planner controls the design of the roads (the controllable variables), while a forecaster predicts what the traffic demand will be (the uncontrollable variable). Our framework is for the planner. It says, "Given that we expect a certain amount of traffic, how should we change our roads to make the system work best?" This moves the conversation from simply reacting to traffic to proactively designing for it.

14.2.2 Scenario Analysis

Scenario comparison is a foundational tool in transportation and systems modeling, enabling the evaluation of alternative futures against a reference (base) scenario. This section formalizes the comparison of scenarios generated by potentially distinct models, possibly operating over different time horizons, input structures, and parameterizations.

Scenario analysis in Ideal Traffic Assignment framework requires a clear reference point to evaluate the impact of policy interventions or external changes. This reference is established through the baseline scenario.

Definition 14.4 (Scenario) A *scenario* $\mathcal{N}$ is a tuple $(M, \mathbf{x}, \boldsymbol{\alpha}, t)$, where: M is the model used to generate the scenario, $\mathbf{x}$ is the input vector (e.g., demand, network configuration), $\boldsymbol{\alpha}$ is the parameter vector (e.g., behavioral coefficients), t is the time horizon over which the scenario is evaluated.

Definition 14.5 (Scenario Output) The output of a scenario $\mathcal{N} = (M, \mathbf{x}, \boldsymbol{\alpha}, t)$ is denoted:

$$\mathbf{y}_{\mathcal{N}} = g(\mathbf{x}, \boldsymbol{\alpha}, t)$$

where g is the model function associated with M.

Axiom 14.1 (Comparability of Scenarios) *Two scenarios $\mathcal{N}_1$ and $\mathcal{N}_2$ are* comparable *if their outputs $\mathbf{y}_{\mathcal{N}_1}$ and $\mathbf{y}_{\mathcal{N}_2}$ lie in a common output space $\mathcal{Y}$ and are evaluated over compatible metrics.*

Even if two scenarios use different models or time horizons, we can compare them as long as their outputs are expressed in the same units or indicators.

Definition 14.6 (Baseline Scenario) The *baseline scenario* (or *base case*) represents the current or default state of the network under existing conditions, without any policy intervention or external disruption. It serves as the "do-nothing" benchmark against which all other scenarios are compared. Formally, it is defined by the observed or estimated values of the input tuple $(\mathsf{A}, \mathsf{C}, \mathsf{L}, \mathsf{U}, \mathsf{S}, \kappa, t)$, with no modifications applied. A well-defined baseline is essential for:

- Quantifying the effectiveness of proposed policies.
- Assessing the severity of uncontrollable external factors.
- Establishing a consistent reference for performance metrics such as F, G, T, V, and Y.

Definition 14.7 (Alternative Scenario) An *alternative scenario* is a modified network state derived from the *baseline scenario* through the application of a *policy* (Definition 9.4). Formally, it is constructed by introducing changes

$\Delta\mathsf{A}$, $\Delta\mathsf{C}$, $\Delta\mathsf{L}$, $\Delta\mathsf{U}$, $\Delta\mathsf{S}$ to the baseline input matrices, resulting in a new configuration:

$$(\mathsf{A} \cup \Delta\mathsf{A}, \mathsf{C} \cup \Delta\mathsf{C},\ \mathsf{L} \cup \Delta\mathsf{L},\ \mathsf{U} \cup \Delta\mathsf{U},\ \mathsf{S} \cup \Delta\mathsf{S},\ \kappa,\ t)$$

The total demand κ and time t remain unchanged in short-term planning unless the scenario also incorporates external factors.

Alternative scenarios are evaluated against a predefined *goal function*, which expresses desired outcomes in terms of the network's output variables (e.g., minimizing total travel time $\sum \mathsf{T}$ or reducing peak congestion $\max \mathsf{G}$). The goal is not a controllable variable but a criterion used to assess the effectiveness of the policy-induced changes.

An *alternative scenario* represents a modified configuration of the network that deviates from the *baseline scenario* through changes in either controllable or uncontrollable inputs. These scenarios are constructed to evaluate the impact of:

- *Proposed interventions*, such as infrastructure upgrades or regulatory changes. It is policy-driven changes to the network's structure or operations, implemented by modifying A, C, L, U, or S.
- *Forecasted conditions*, external factors such as demand surges or environmental disruptions, modeled by altering κ or introducing temporal shifts in t.

Alternative scenarios enable comparative analysis to determine whether a given change improves, degrades, or maintains network performance relative to the baseline. They are assessed using goal functions that reflect desired outcomes, such as reduced congestion or improved travel times.

14.2.3 Invariant Quantities in Scenario Analysis

In predictive modeling of transportation networks, the reliability of scenario analysis depends critically on identifying quantities that remain stable , referred to as *invariants*, across different temporal and structural configurations. The selection of appropriate invariants is contingent upon the time horizon of the analysis and the nature of the scenario being evaluated.

Definition 14.8 (Invariant Quantity) An *invariant* is a model parameter or derived quantity that is assumed to remain constant across a defined set of scenarios. Invariants serve as anchors for comparative analysis and enable consistent evaluation of policy outcomes or system responses.

Definition 14.9 (Short-Term Horizon (Days to Months)) For short-term disruptions such as road closures, public events, or temporary regulatory changes, it is reasonable to assume that:

- The total network demand κ remains constant.
- The physical structure of the network, encoded by $(\mathsf{A}, \mathsf{C}, \mathsf{L}, \mathsf{U})$, is mostly unchanged.

The model's objective in this context is to redistribute the fixed flow κ across the available paths, subject to stochastic routing behavior S and performance constraints.

Definition 14.10 (Intermediate-Term Horizon (Months to Years)) In scenarios involving infrastructure modifications (e.g., bridge construction or lane additions), the network structure itself becomes a variable. A useful invariant in this context is a *policy goal*, such as maintaining a maximum allowable congestion level:

$$\xi = \frac{f_{ij}}{c_{ij}} \leq \xi_{\max}$$

where f_{ij} is the flow on a given link and $\xi_{\max}$ is a predefined threshold (e.g., 0.85). The model evaluates whether the proposed structural changes satisfy this invariant under projected demand.

Definition 14.11 (Long-Term Horizon (Years to Decades)) For strategic planning, both demand κ and network structure $(\mathsf{A}, \mathsf{C}, \mathsf{L}, \mathsf{U})$ and routing behavior S may evolve over time. Invariants in this context are often expressed as conditional policies, such as:

Expand road capacity when congestion exceeds a critical threshold.

Long-term models incorporate demographic projections to forecast future demand, κ, simulate network performance using the ITA framework, and identify the temporal and spatial conditions under which policy triggers (e.g., capacity expansion) are activated. This enables dynamic scenario generation for infrastructure investment planning.

14.2.4 Scenario Comparison Methods

In principle, scenarios must be evaluated with a common set of metrics and a consistent time horizon. That is foundational to reliable comparative analysis.

14.2.4.1 Method 1: Direct Scenario Comparison

Direct scenario comparison involves evaluating two or more scenarios at the *same time horizon* and using the *same model*. This method is applicable when evaluating the immediate impact of a policy intervention or comparing two different policy options for a single year.

Proposition 14.1 (Direct Scenario Comparison) *Let* $\mathcal{N}_b = (M_b, \mathbf{x}_b, \boldsymbol{\alpha}_b, t_b)$ *be the base scenario and* $\mathcal{N}_a = (M_a, \mathbf{x}_a, \boldsymbol{\alpha}_a, t_a)$ *be an alternative scenario. If* $t_b = t_a$ *and both models produce outputs in* $\mathcal{Y}$*, then the scenarios are directly comparable via:*

$$\Delta \mathbf{y} = \mathbf{y}_{\mathcal{N}_a} - \mathbf{y}_{\mathcal{N}_b}$$

Proof Given $\mathbf{y}_{\mathcal{N}_b} = g_b(\mathbf{x}_b, \boldsymbol{\alpha}_b, t_b)$ and $\mathbf{y}_{\mathcal{N}_a} = g_a(\mathbf{x}_a, \boldsymbol{\alpha}_a, t_a)$, and assuming $t_b = t_a$, the difference $\Delta \mathbf{y}$ is well-defined in $\mathcal{Y}$. □

If both scenarios are evaluated over the same time frame and produce comparable outputs, we can simply subtract one from the other to see the impact of the alternative.

Example 14.1 A transportation authority is considering two alternative policies to mitigate a recurring traffic bottleneck in a city's downtown core during the evening rush hour. The first policy, $\mathcal{P}_1$, is to implement a strategic road diet ($\Delta \mathsf{C}^-$) by converting a lane to a bike path. The second policy, $\mathcal{P}_2$, is to apply an immediate signal optimization ($\delta \mathsf{S}$) at all surrounding intersections.

- Baseline Scenario ($\mathcal{N}_b$): The current network state at 5:00 PM on a weekday. The output is the maximum congestion, $\max \mathsf{G}_b$.
- Alternative Scenario 1 ($\mathcal{N}_{a1}$): The network after applying policy $\mathcal{P}_1$. The output is the new maximum congestion, $\max \mathsf{G}_{a1}$.
- Alternative Scenario 2 ($\mathcal{N}_{a2}$): The network after applying policy $\mathcal{P}_2$. The output is the new maximum congestion, $\max \mathsf{G}_{a2}$.

All three scenarios are evaluated for the same year and time, making them directly comparable. The analysis involves calculating $\Delta \mathbf{y}_1 = \max \mathsf{G}_{a1} - \max \mathsf{G}_b$ and $\Delta \mathbf{y}_2 = \max \mathsf{G}_{a2} - \max \mathsf{G}_b$ to determine which policy has the greater impact on reducing congestion.

14.2.4.2 Method 2: Time-Aligned Calibration

This method is used when scenarios are evaluated over *different time horizons*. To compare them, their outputs must be normalized or extrapolated to a *common time horizon* (t^*). This is crucial for long-term planning where different interventions have staggered implementation timelines.

Lemma 14.1 (Temporal Alignment) *If $t_b \neq t_a$, then outputs must be normalized or interpolated to a common time horizon t^* before comparison:*

$$\mathbf{y}^*_{\mathcal{N}_b} = \mathcal{T}_b(\mathbf{y}_{\mathcal{N}_b}, t_b \to t^*), \quad \mathbf{y}^*_{\mathcal{N}_a} = \mathcal{T}_a(\mathbf{y}_{\mathcal{N}_a}, t_a \to t^*)$$

Proof Let $\mathcal{T}_b$ and $\mathcal{T}_a$ be temporal transformation operators (e.g., interpolation, extrapolation, discounting). These operators map outputs to a unified time horizon t^*, enabling valid comparison. □

Theorem 14.1 (Scenario Comparison with Temporal Calibration) *Given transformed outputs $\mathbf{y}^*_{\mathcal{N}_b}$ and $\mathbf{y}^*_{\mathcal{N}_a}$, the calibrated scenario difference is:*

$$\Delta\mathbf{y}^* = \mathbf{y}^*_{\mathcal{N}_a} - \mathbf{y}^*_{\mathcal{N}_b}$$

Proof Follows directly from Lemma 14.1. The transformation ensures both outputs are temporally aligned, preserving comparability. □

If the base and alternative scenarios look at different years or time spans, we need to adjust them to the same time frame before comparing. This might involve forecasting or aggregating results.

Definition 14.12 (Residual Error) The residual error between base and alternative scenarios is defined as:

$$\mathcal{R} = \|\mathbf{y}^*_{\mathcal{N}_a} - \mathbf{y}^*_{\mathcal{N}_b}\|$$

Corollary 14.1 (Scenario Impact Assessment) *A large residual $\mathcal{R}$ indicates significant deviation of the alternative scenario from the base, suggesting high impact or structural change.*

The bigger the difference between the two scenarios, the more the alternative changes the system. This helps planners decide whether a policy or intervention is worth pursuing.

Example 14.2 A city is considering a long-term infrastructure project: building a new bridge that will be operational in 2035. They want to compare its impact against a short-term policy: implementing a series of temporary lane restrictions and signal optimizations for a major sports event in 2026. The common time horizon for comparison is set to 2035.

- Baseline Scenario ($\mathcal{N}_b$): The current network is projected to the year 2035, with demand growth modeled by κ_{2035}. The output is the forecasted congestion, $\max \mathsf{G}^*_b$.
- Alternative Scenario ($\mathcal{N}_a$): The temporary event policy is applied in 2026. The network is then modeled over time, accounting for the event's effects and subsequent demand growth, to forecast its performance in 2035. The output is $\max \mathsf{G}^*_a$.

The two scenarios are not directly comparable in terms of implementation, but their outcomes can be evaluated at the unified time horizon of 2035 by calculating the difference, $\Delta \mathbf{y}^* = \max \mathsf{G}_a^* - \max \mathsf{G}_b^*$. The result shows whether a short-term, temporary policy has a lasting effect that can compete with a long-term infrastructure project.

14.3 The Formal Language of Policy

14.3.1 Atomic Policies as Network Interventions

Contemporary transport policy is often articulated through narrative prose, administrative directives, or loosely structured plans that lack formal precision. While such approaches may suffice for broad communication, they present significant limitations in terms of reproducibility, transparency, and systemic analysis. Policies expressed in natural language are prone to ambiguity, interpretation bias, and implementation inconsistency, especially when scaled across jurisdictions, timeframes, or stakeholder groups. This lack of formal structure impedes rigorous evaluation, comparative analysis, and the development of simulation tools that could otherwise support evidence-based decision-making.

To address these limitations, we propose a paradigm shift: the formalization of transport policy as a compositional grammar, built from atomic units and assembled into molecular expressions. Atomic policies represent the smallest actionable interventions, such as adding capacity, removing access, or shifting demand, each defined with explicit targets, modes, and temporal or spatial constraints. Papadimitriou and Yannakakis [8] introduced atomic routing games where each agent controls an indivisible unit of flow, laying the inspirational foundation for atomic policy modeling in transport networks. Molecular policies are structured compositions of these atoms, enabling the representation of complex, multi-modal, and temporally layered interventions. This grammar-based approach transforms policy from a descriptive artifact into a scientific object: one that can be parsed, validated, simulated, and audited. Planners trained in this framework learn not only to analyze policy outcomes but to design interventions with formal rigor. They engage in deductive synthesis, constructing policies from first principles , and inductive interpretation, translating real-world scenarios into formal syntax. This dual capability fosters systems thinking, encourages anticipatory design, and equips learners and policy makers alike to reason about trade-offs, equity impacts, and emergent behavior within transport networks.

Moreover, the societal implications are equally compelling. Formal grammar enables participatory modeling, where communities and stakeholders can co-design policies using transparent, reproducible tools. It supports equity auditing by allowing targeted interventions for specific populations, zones, or timeframes. And it lays the foundation for interoperable policy engines that can simulate, compare, and optimize interventions across scales and contexts.

In short, the adoption of atomic and molecular grammar is not merely a technical refinement, it is a foundational reimagining of how transport policy is conceived, communicated, and enacted. It empowers planners to become designers of mobility futures, not just analysts of past failures. It invites a new generation of planners, engineers, and citizens to engage with policy as a scientific, ethical, and collaborative endeavor.

In the context of the Ideal Traffic Assignment, a policy is not a mere statement of intent. Policy is a precisely defined set of transformations or changes on the network's mathematical structure $\Delta\mathsf{L}$, $\Delta\mathsf{U}$, $\Delta\mathsf{C}$ or $\Delta\mathsf{S}$. These transformations are the fundamental tools for urban planners and transport specialists.

An atomic policy $\mathcal{A}$ is a primitive, irreducible [5] transformation applied to a single component of the network's mathematical structure. These are the most basic building blocks available to a transportation planner. The set of all possible atomic policies can be organized into a systematic framework, analogous to the periodic table in chemistry (see Sect. 14.5.3), which classifies them by their function and permanence.

Atomic Policy $\mathcal{A}$ is a single, direct action a planner can take, such as adding capacity to a single link or reversing its direction. Table 14.2 shows primitive atomic policies to encode network interventions for simulation and policy design. The list is *complete and exhaustive*. The policies cover all fundamental types of changes a planner can make to an IFN's controllable structural variables $\mathsf{L}, \mathsf{U}, \mathsf{C}, \mathsf{S}$ and network topology A, as well as to total network demand (κ). The list systematically addresses additions, reductions, and modifications for both permanent and temporary interventions. The list in the Table 14.2 is also *mutually exclusive*. Each policy represents an irreducible, single-idea transformation that cannot be further subdivided. For example, *AddCap* is distinct from *RedCap*, and neither can be broken into smaller, meaningful policy atoms.

Definition 14.13 (Atomic Policy) An *atomic policy* is a primitive, irreducible transformation applied to a single component of the network's structural matrices: the adjacency matrix A, the capacity matrix C, the link distance matrix L, the maximum speed matrix U or stochastic matrix S. A set of primitive atomic policies are defined in the following Table 14.2.

Each atomic policy represents a primitive transformation on the network's structural matrices. These transformations serve as *policy levers* in scenario design, enabling targeted interventions across capacity, connectivity, speed, and demand

Example 14.3 (Atomic Policies)

- Atomic policy *AddCap* symbolized by $\Delta\mathsf{C}^{+}$ is equivalent to *Add Lane*: $\Delta\mathsf{C}_{ij} > 0$.
- Atomic Policy *RedCap* denoted by $\Delta\mathsf{C}^{-}$ is equivalent to *Reduce Lane*: $\Delta\mathsf{C}_{ij} < 0$ for an existing link $(i, j) \in E$, with the condition that the resulting capacity remains positive.

Table 14.2 List of atomic network modification actions

Symbol	Unique name	Action	Description
$\Delta\mathsf{C}^+$	*AddCap*	Add capacity	$\Delta\mathsf{C}_{ij} > 0$
$\Delta\mathsf{C}^-$	*RedCap*	Reduce capacity	$\Delta\mathsf{C}_{ij} < 0$
$\Delta\mathsf{A}^+$	*AddLink*	Add link	Add a new link to the network
$\Delta\mathsf{A}^-$	*RemLink*	Remove link	Eliminate an existing link
$\Delta\mathsf{A}^\leftrightarrow$	*RevDir*	Reverse direction	Change a two-way street to a one-way street
$\Delta\mathcal{V}^+$	*AddNode*	Add node	Add a new node to the network
$\Delta\mathcal{V}^-$	*RemNode*	Remove node	Eliminate an existing node
$\Delta\mathsf{U}^+$	*IncSpeed*	Increase speed	$\Delta\mathsf{U}_{ij} > 0$
$\Delta\mathsf{U}^-$	*DecSpeed*	Decrease speed	$\Delta\mathsf{U}_{ij} < 0$
$\Delta\mathsf{S}$	*OptSig*	Optimize signals	Adjust outflow probabilities ($\Delta\mathsf{S}$) at a node
$\Delta\kappa^+$	*AddDem*	Add demand	$\Delta\kappa > 0$
$\Delta\kappa^-$	*RedDem*	Reduce demand	$\Delta\kappa < 0$
$\delta\mathsf{C}^+$	*TempAddCap*	Temporary add capacity	Short-term $\Delta\mathsf{C}_{ij} > 0$
$\delta\mathsf{C}^-$	*TempRedCap*	Temporary reduce capacity	Short-term $\Delta\mathsf{C}_{ij} < 0$
$\delta\mathsf{A}^+$	*TempAddLink*	Temporary add link	Short-term add link
$\delta\mathsf{A}^-$	*TempRemLink*	Temporary remove link	Short-term remove link
$\delta\mathsf{U}^+$	*TempIncSpeed*	Temporary increase speed	Short-term $\Delta\mathsf{U}_{ij} > 0$
$\delta\mathsf{U}^-$	*TempDecSpeed*	Temporary decrease speed	Short-term $\Delta\mathsf{U}_{ij} < 0$
$\delta\mathsf{S}$	*TempOptSig*	Temporary optimize signals	Short-term adjust outflow probabilities $\Delta\mathsf{S}$

- Atomic Policy *AddLink* ($\Delta\mathsf{A}^+$) introduces a new link $(i, j) \notin E$, which corresponds to setting $\mathsf{A}_{ij} = 1$ and defining new positive values for C_{ij}, L_{ij}, and U_{ij}.
- Atomic Policy *RemLink* or *Remove Link* with shorthand notation $\Delta\mathsf{A}^-$ eliminates an existing link $(i, j) \in E$ by setting $\mathsf{C}_{ij} = 0$, which corresponds to $\mathsf{A}_{ij} = 0$. This is only feasible if the network remains irreducible.
- Atomic Policy *RevDir* or *Reverse Link Direction* with shorthand notation $\Delta\mathsf{A}^\leftrightarrow$ would do the following: For a link $(i, j) \in E$, this policy sets $\mathsf{C}_{ij} = 0$ and adds a new link $(j, i) \notin E$.
- Atomic Policy *AddNode* or *Add New Node* symbolized by $\Delta\mathcal{V}^+$ would add a new node to the network, increasing the dimensions of the matrices. This policy is always followed by at least one link addition.
- Atomic Policy *RemNode* or *Delete Node*denoted by $\Delta\mathcal{V}^-$ would do the following: An existing node and all its incident links are removed from the network, reducing the dimensions of the matrices. This is only feasible if the network remains irreducible.
- Atomic Policy *Adjust Speed* includes *IncSpeed* ($\delta\mathsf{U}^+$) or *DecSpeed* ($\delta\mathsf{U}^-$) would modify the maximum speed on a link, $\Delta\mathsf{U}_{ij} \neq 0$. This can represent a traffic calming measure ($\Delta\mathsf{U}_{ij} < 0$) or a road design improvement ($\Delta\mathsf{U}_{ij} > 0$).

Definition 14.14 (Well-Formed Atomic Policy (WFAP)) Well-Formed Atomic Policy (WFAP) is based on a strict syntax that combines a base action, an optional coefficient, directionality, and spatiotemporal scope. This structure ensures that every policy statement is unambiguous and interpretable by a computational model.

A policy expression $\mathcal{A}$ is a Well-Formed Atomic Policy if and only if it conforms to the following structure:

$$\begin{aligned}\mathcal{A} := \texttt{sign}, \texttt{coefficient}, \texttt{action} - \texttt{target}^{\texttt{direction}}(\texttt{location}) \\ \times (\texttt{time} - \texttt{interval})\end{aligned} \tag{14.2}$$

This format can also be expressed as:

$$\mathcal{A} = \texttt{sign}, \texttt{coefficient}, \texttt{atomic} - \texttt{name}(\texttt{location})[@(\texttt{time} - \texttt{interval})] \tag{14.3}$$

- *Sign and Directionality*: A sign (e.g., $+$ or $-$) and directionality superscript (e.g., $+$ or $-$) are used to denote the increase or decrease of a value. The format uses a single sign or a directional superscript to avoid redundancy.
- *Coefficient*: A numeric coefficient n is used to specify the magnitude of the change. It can be an integer or a real number.
- *Action Name*: The action is the unique name of the atomic policy (e.g., *AddCap*, *RedCap*, *IncSpeed*).
- *Target Domain*: The target of the action is implicitly defined by the *action name* itself (e.g., *AddCap* targets capacity matrix C and *AddLink* targets the adjacency matrix A).
- *Location*: The location is an optional identifier that specifies the link, node, or zone to which the policy applies.
- *Temporal Scope*: The time interval, defined by two time points $[t_0, t_1]$, specifies the duration or time of application for the policy.

14.3.2 Molecular Policies as Network Interventions

A molecular policy is a combination of one or more atomic policies.

Definition 14.15 (Molecular Policy) A *molecular policy* $\mathcal{P}$ is a finite sequence or combination of one or more atomic policies, $\mathcal{P} = \{\mathcal{A}_1, \mathcal{A}_2, \ldots, \mathcal{A}_k\}$.

The Ideal Traffic Assignment (ITA) framework's formal language of operators allows for the rigorous synthesis, analysis, and evaluation of policy interventions. This approach transforms the art of policy design into a systematic, rule-based process. To express these combinations of atomic policies, we introduce a formal language with operators. The most fundamental operators are:

Definition 14.16 (Policy Operators)

- *Union Operator ($\cup$ or $+$):* Represents the combination of atomic or molecular policies. For example, a "road diet" policy, $\mathcal{P}_{RD}$, is a union of a capacity reduction and a new bike link: $\mathcal{P}_{RD} = \text{RedCap} \cup \text{AddLink}$. This is a simultaneous or sequential application of multiple atomic policies.
- *Negation Operator ($\neg$ or $-$)*: Represents the negation of a policy. For example, a policy of *preventing* a link removal is $\neg$(RemLink), which is equivalent to a policy of preservation.
- *Sequencing Operator ($\rightarrow$)*: Represents a temporal sequence of policies, where one action follows another. For example, a "new town" policy is a permanent node addition followed by permanent link additions to connect it: $\mathcal{P}_{NT} = (\text{AddNode}) \rightarrow (\text{AddLink}) \rightarrow \ldots$.

Molecular Policy represents a complex policy intervention, such as a "road diet" or a "new town" development, which cannot be described by a single elementary operation. For instance, a "split link" policy, which divides a two-way street's lanes between two opposing links, is a molecular policy comprising an "add lane" and a "reduce lane" atomic policy.

Think of atomic policies as the most basic building blocks available to a transportation planner. They are the single, most fundamental actions we can take: we can add a single lane, remove a single road, or change a speed limit. A molecular policy is simply a combination of these building blocks to create a more sophisticated project. A large project like building a new bridge is a molecular policy because it involves adding a new link, a new road, and defining its capacity and speed. This systematic approach allows us to break down any complex policy into its constituent parts, making it analyzable and optimizable.

Definition 14.17 (Formal Grammar for the Chemistry of Transport Policy) The structure of both atomic and molecular policies is rigorously defined by the Extended Backus-Naur Form (EBNF). This grammar provides the rules for parsing and validating any policy expression within the framework.

- *Atomic Policy Structure*: An atomic policy can be expressed using two primary formats:
  - *Format 1*: Use Eq. (14.2). This format is explicit and pedagogical.
  - *Format 2*: Use Eq. (14.3). This is a concise, shorthand notation.
- *Molecular Policy Structure*: Molecular policies are composed of atomic policies using operators. The EBNF defines this as a sequence of atomic policies combined by operators.

The formal specification in *Extended Backus-Naur Form (EBNF)* is shown in Fig. 14.1.

In computer science, *formal grammar* provides a rigorously defined system for specifying the syntax of languages, ranging from programming constructs to data formats. The foundational classification of such grammars was introduced by

```
policy = molecular_policy ;
molecular_policy = atomic_policy , { operator , atomic_policy } ;
atomic_policy = [ sign ] , coefficient , action , "_" , target ,
                [ "_" , location ] , direction ,
                [ "(" , time_interval , ")" ]
                | action , "(" , location , ")" , [ "@" , time_interval ] ;
action = atomic_name | atomic_symbol ;
atomic_name = "AddCap" | "RedCap" | "AddLink" | "RemLink" | "RevDir"
              | "AddNode" | "RemNode" | "IncSpeed" | "DecSpeed" | "OptSig"
              | "AddDem" | "RedDem"
              | "TempAddCap" | "TempRedCap" | "TempAddLink" | "TempRemLink"
              | "TempIncSpeed" | "TempDecSpeed" | "TempOptSig" ;
atomic_symbol = ΔC⁺ | ΔC⁻ | ΔA⁺ | ΔA⁻ | ΔA ↔
                | Δ𝒱⁺ | Δ𝒱⁻ | ΔU⁺ | ΔU⁻ | ΔS
                | Δκ⁺ | Δκ⁻
                | δC⁺ | δC⁻ | δA⁺ | δA⁻
                | δU⁺ | δU⁻ | δS ;
sign = "+" | "-" ;
coefficient = digit , { digit } ;
target = "Cap" | "Link" | "Node" | "Speed" | "Demand" | "Signal" | "Mode" |
"Group" ;
location = identifier ;
direction = "⇡" | "≙" ;
time_interval = time , "-" , time ;
operator = union_op | sequence_op | negation_op ;
union_op = "+" | ∪ ;
sequence_op = → ;
negation_op = ¬ ;
time = digit , digit , ":" , digit , digit ;
identifier = letter , { letter | digit | "_" } ;
digit = "0" | "1" | "2" | "3" | "4" | "5" | "6" | "7" | "8" | "9" ;
letter = "A" | "B" | ... | "Z" | "a" | "b" | ... | "z" ;
```

Fig. 14.1 A formal grammar defining the syntax for a transport policy using a chemistry-based analogy

Chomsky [2], whose hierarchy includes *context-free grammars*, a class pivotal for modeling syntactic structures in both natural and programming languages. One widely adopted notation for expressing context-free grammars is the *Extended Backus-Naur Form (EBNF)*, an enhancement of the original *Backus-Naur Form (BNF)*. BNF was first formalized in the ALGOL 60 report edited by Peter Naur [7], where it served to describe the syntax of ALGOL 60 and laid the groundwork for compiler design and formal language theory. EBNF refines BNF by introducing syntactic sugar and expressive constructs, and its formal specification is codified in the *ISO/IEC 14977:1996* standard [4]. This international standard ensures a precise, machine-readable, and unambiguous method for defining language rules, making it indispensable for grammar-based modeling. By adopting EBNF as defined in ISO/IEC 14977, the framework for *Ideal Traffic Assignment (ITA)* policies is anchored not in ad-hoc conventions but in the rigor of formal language theory, ensuring reproducibility, clarity, and semantic precision.

The formal grammar for the chemistry of transport policy in the form of EBNF is shown in Definition 14.17. It forms the rigorous grammar for constructing well-formed atomic and molecular policies. The structure ensures that every policy is a valid statement within the framework's logical system. The existence of both verbose names (such as *AddCap*) and compact symbols ($\Delta \mathsf{C}^+$) provides flexibility for both human-readable and machine-executable policy expressions. The use of operators enables the logical composition of these primitive policies into complex, multi-faceted interventions.

Definition 14.18 (Composition Rules for Molecular Policy Formation) The composition of molecular policies is defined by a formal grammar as defined in Definition 14.17, which uses operators to combine or sequence atomic policies. This grammar, expressed in Extended Backus-Naur Form (EBNF), provides the syntax for creating complex interventions from primitive units.

1. *Atomic to Molecular*: Any single atomic policy is a trivial molecular policy. For example,a policy to add a link, $\Delta \mathsf{A}^+$, is an atomic policy and can also be considered a molecular policy of a single element.
2. *Union of Policies*: The union operator ($\cup$ or $+$) combines policies to represent simultaneous or parallel actions. This rule is a logical AND. For the combined policy to be feasible, both $\mathcal{P}_1$ and $\mathcal{P}_2$ must be feasible, and their simultaneous application must not create a conflict on the same network element.

$$\mathcal{P} = \mathcal{P}_1 \cup \mathcal{P}_2$$

   For example, the *Bike Lane Installation* policy is a union of *RedCap* and *AddLink*. It reduces vehicle capacity and adds a new link for bicycles, both actions occurring as part of a single intervention.
3. *Sequencing of Policies*: The sequencing operator ($\rightarrow$) orders policies to represent a temporal or causal progression of actions. This rule is a logical THEN. The policy $\mathcal{P}_2$ is applied to the network state that results from the application of $\mathcal{P}_1$. This is crucial for modeling multi-stage projects, where one action depends on the completion of another.

$$\mathcal{P} = \mathcal{P}_1 \rightarrow \mathcal{P}_2$$

   For example, the *New Town* policy sequences a new node addition followed by the addition of links to connect it to the network. This temporal order is physically necessary.
4. *Negation of Policies*: The negation operator ($\neg$ or $-$) represents the reversal or prevention of a policy. A negation applied to an atomic policy creates a policy that is the inverse of the original. However, the simultaneous union of a policy and its inverse ($\mathcal{A} \cup \neg\mathcal{A}$) results in a logical contradiction and is deemed a *conflicting policy*. This is distinct from a sequential application, which results in the identity policy ($\mathcal{A} \rightarrow \mathcal{A}^{-1} = \mathcal{I}$), representing a net zero change.

$$\mathcal{P} = \neg\mathcal{A}$$

5. *Recursion and Hierarchical Composition*: Molecular policies can be composed of other molecular policies. This rule allows for the construction of highly complex policies, such as a multi-phase urban revitalization plan. The framework can model a series of phased interventions, each of which is a molecular policy itself, applied in a specific temporal order.

$$\mathcal{P} = \mathcal{P}_{molecular1} \cup (\mathcal{P}_{molecular2} \rightarrow \mathcal{P}_{molecular3})$$

A logical OR operator is deliberately excluded from the molecular policy grammar because it is not needed to model the types of policies defined within this framework. The operators already defined, union ($\cup$), negation ($\neg$), and sequencing ($\rightarrow$), are sufficient to express all of the proposed policy interventions. The framework's primary purpose is to define a set of actions that a planner *commits* to taking. A logical OR would imply a choice or a conditional action, which is better captured by scenario constructs.

A policy is a commitment. The molecular policies are fundamentally prescriptive. A policy like "build a new bridge OR widen the existing road" is not a single, executable plan. It is a decision point. The framework is designed to model a single, deterministic plan to be evaluated against a baseline. Each scenario is a concrete set of actions. The decision to choose one of two alternatives would occur *outside* the model, during the scenario selection process.

While a simple OR operator isn't present, the concept of choice is handled at the scenario analysis level. For example, a planner would create two separate scenarios:

- Scenario A: Add a bridge ($\Delta\mathsf{A}^+$).
- Scenario B: Widen the road ($\Delta\mathsf{C}^+$).

The performance of both scenarios would then be compared against the baseline, and the planner would choose the best option based on the results. This moves the logical choice from the policy definition to the analytical stage. The Power of the Union Operator ($\cup$): The union operator already functions as a form of logical AND by combining multiple policies into a single intervention. For example, the *Bike Lane Installation* policy ($\Delta\mathsf{C}^- \cup \Delta\mathsf{A}^+$) means "reduce capacity AND add a new link". All components of the policy must be executed for the policy to be considered complete.

Instead of using a logical OR within a single policy expression, the framework separates the generation and comparison steps.

1. *Policy Generation*: The automated algorithm explores the entire policy space, which is the power set of all feasible atomic policies. This process effectively generates every possible combination, including mutually exclusive options like adding a bridge and widening a road.

2. *Scenario Comparison*: Once a finite set of feasible scenarios is generated, they are all evaluated against a user-defined goal function. The best policy is selected from this set by finding the one that optimizes the goal function. This comparison process inherently fulfills the purpose of a logical OR, it selects one option from a set of alternatives.

This two-step approach allows the framework to remain mathematically precise while still providing a mechanism for complex decision-making. The logical OR is not a formal operator because it describes a choice to be made by the user, not an action to be performed by the network.

Example 14.4 (Molecular Policy Synthesis: Urban Transportation Planning Scenario)

A city plans to convert an existing two-way street, Link A-B, into a one-way street during morning rush hour to ease congestion. This street will be permanently narrowed from 4 lanes to 3 lanes to add a dedicated bike lane. A new intersection will also be added to connect a newly constructed office park to the main network. Show the Molecular Policy in shorthand notation.

Answer:

- The permanent narrowing from 4 to 3 lanes on Link A-B is a permanent reduction in capacity: $\Delta \mathsf{C}^{-}_{A,B}$.
- The addition of a dedicated bike lane is a permanent new link: $\Delta \mathsf{A}^{+}_{A,B}$.
- The conversion of a two-way street to one-way during morning rush hour is a temporary change in direction: $\delta \mathsf{A}^{\leftrightarrow}_{A,B}$.
- The construction of a new office park intersection is a permanent new node: $\Delta \mathcal{V}^{+}_{C}$.
- The connection of the office park to the network requires a new permanent link: $\Delta \mathsf{A}^{+}_{C,D}$.

Final Molecular Policy in shorthand notation:

$$\mathcal{P} = (\Delta \mathsf{C}^{-}_{A,B} \cup \Delta \mathsf{A}^{+}_{A,B} \cup \delta \mathsf{A}^{\leftrightarrow}_{A,B}) \cup (\Delta \mathcal{V}^{+}_{C} \rightarrow \Delta \mathsf{A}^{+}_{C,D})$$

14.4 Policy Analysis and Synthesis

14.4.1 Analysis: Real-World Policies as Molecular Policies

This section translates common transportation policies into the formal language of molecular policies, demonstrating how a planner's intent can be mathematically modeled. The ITA framework supports both a deductive synthesis (top-down) and inductive analysis (bottom-up) approach to policy analysis allowing for a precise evaluation of its systemic effects. Inductive Analysis works from Practice to Theory. It involves taking a complex, real-world policy and *decomposing* it

Table 14.3 Molecular policies in an inductive analysis

Real-world policy	Molecular policy	Description
Road diet	$\Delta\mathsf{C}^{-}(ij)$	Reduces link capacity to rebalance congestion
New bridge construction	$\Delta\mathsf{A}^{+}(ij)$	Adds a new physical connection to the network
Congestion pricing	$\Delta\kappa^{-}(zone_A)$	Reduces demand in a zone to improve performance
Bike lane installation	$\Delta\mathsf{C}^{-}(ij) \cup \Delta\mathsf{A}^{+}(ij_{bike})$	Reduces vehicle capacity and adds a bike link
On-street parking removal	$\Delta\mathsf{C}^{+}(ij)$	Increases effective capacity by reclaiming space
Traffic signal synchronization	$\Delta\mathsf{S}(node_1) \rightarrow \Delta\mathsf{S}(node_2) \rightarrow \ldots$	Sequential signal optimization to create a green wave
School zone speed limit	$\delta\mathsf{U}^{-}(ij)@07:00-09:00$	Temporarily reduces speed for safety during school hours
Bus Rapid Transit (BRT)	$\Delta\mathsf{C}^{-}(ij) \cup \Delta\mathsf{A}^{+}(ij_{BRT})$	Reduces general lane capacity to add a transit link
HOV lane	$\Delta\mathsf{C}^{-}(ij) \cup \Delta\mathsf{A}^{+}(ij_{HOV})$	Reallocates capacity to high-occupancy vehicles
Emergency rerouting	$\delta\mathsf{A}^{-}(ij) \cup \delta\mathsf{S}(node_{alt})$	Temporarily removes a link and adjusts signals
Night market/festival	$\delta\mathsf{A}^{-}(ij)@18:00-23:00$	Temporarily removes a street for public events

into its fundamental, irreducible atomic components. This process breaks down a practical intervention into its underlying theoretical building blocks, allowing for a rigorous examination of its constituent parts. This approach takes a real-world policy and synthesizes its component atomic parts, modeling it within the framework. It is crucial for analyzing the systemic effects of existing policies. As shown in Table 14.3, real-world interventions can be systematically represented as molecular policies, enabling structured analysis and scenario generation.

14.4.2 Synthesis: Molecular Policies as Real-World Actions

Deductive Synthesis works from Theory to Practice. This approach begins with a theoretical molecular policy and deduces its real-world application. It involves starting with a set of fundamental atomic policies and *combining them* to construct a new, complex molecular policy. This process builds a theoretical policy from the ground up, moving from simple elements to a comprehensive intervention. It is useful for designing novel policies and exploring the systemic effects of planned changes.

As shown in Table 14.4, deductive synthesis enables the translation of symbolic network operations into concrete policy interventions for planning and analysis. This Deductive Synthesis Table 14.4 works in reverse than Inductive Analysis

Table 14.4 Molecular policies in a deductive synthesis

Molecular policy	Real-world policy	Description
$\Delta \mathsf{C}^{+}(ij) \cup \Delta \mathsf{A}^{+}(ij_{flyover})$	Road widening with flyover	Increases capacity and adds a bypass link
$\Delta \mathsf{C}^{-}(ij) \cup \Delta \mathsf{A}^{+}(ij_{BRT})$	Bus Rapid Transit (BRT) Lane	Reduces general traffic capacity to add a transit link
$\Delta \mathsf{A}^{-}(ij)$	Pedestrian-only zone	Removes a vehicular link to create pedestrian space
$\Delta \mathsf{A}^{\leftrightarrow}(ij)$	One-way street conversion	Reverses directionality of a two-way link
$\Delta \mathcal{V}^{+}(nodeX) \cup \Delta \mathsf{A}^{+}(ij_{access})$	Shopping mall development	Adds a new node and connects it to the network
$\Delta \kappa^{+}(zoneX)$	New stadium construction	Introduces new demand in a zone
$\Delta \kappa^{-}(CBD)$	Congestion pricing	Reduces demand in central business district
$\Delta \mathsf{S}(nodeX)$	Traffic signal optimization	Adjusts outflow probabilities at a node
$\delta \mathsf{A}^{+}(ij_{temp})$	Temporary bridge	Adds a short-term link to the network
$\delta \mathsf{A}^{-}(ij) \cup \delta \mathsf{C}^{-}(altRoute)$	Major event rerouting	Temporarily removes a link and reduces capacity elsewhere
$\Delta \mathsf{C}^{-}(ij_1) \rightarrow \Delta \mathsf{C}^{-}(ij_2) \rightarrow \Delta \mathsf{C}^{-}(ij_3)$	"Three-Lane" Road Diet	Sequentially reduces capacity across three links

Table 14.3, demonstrating how a combination of atomic policies corresponds to a tangible planning intervention.

Note that we simplify "Traffic Signal Optimization" being a single *OptSig* policy, which is overly generalized. In reality, coordinating signals across a corridor is a molecular policy involving a sequence of *OptSig* policies applied to multiple nodes. A single *OptSig* policy refers to an atomic, irreducible transformation on the outflow probabilities of a single node. In contrast, a real-world policy like coordinating signals across a corridor is a molecular policy that consists of a sequence of *OptSig* policies applied to multiple, adjacent nodes. The redundancy in the tables is acceptable as it aids in clarity by showing both the atomic and molecular views.

14.5 Policy Validity and Feasibility

14.5.1 Feasibility and Constraints of Policies

While the atomic and molecular policy grammar introduced above is general and compatible with traditional transport modeling approaches, including four-step models and heuristic scenario analysis, the subsequent formalization of feasibility constraints, policy validity rules, and the periodic table of interventions is specific to the Ideal Traffic Assignment (ITA) framework. ITA represents a computationally rigorous paradigm in which transport policies are not merely conceptual or descriptive, but are encoded as algorithmically verifiable operations within a constrained optimization environment. This shift from general-purpose modeling to ITA-based synthesis introduces a set of restrictive techniques that, rather than limiting creativity, enhance precision and reproducibility. By embedding policies within a formally defined solution space, bounded by network topology, demand elasticity, and modal constraints, ITA enables automated validation of policy feasibility and supports the generation of alternative interventions through algorithmic reasoning. In this context, the grammar serves not only as a language of composition but as a scaffold for machine- assisted policy design, where the exploration of systemic effects is driven by structured logic rather than subjective intuition.

A policy is not valid if it violates the fundamental principles of the Ideal Traffic Assignment (ITA) framework or the underlying physical reality of a road network. These constraints below ensure that the resulting network flow state remains an Ideal Flow Network (IFN).

Bazaraa et al. [1] provide a rigorous definition of feasible sets in nonlinear programming [6], foundational for defining admissible flow configurations in optimization. Daganzo and Sheffi [3, 12] formalized flow conservation rules in network equilibrium models, ensuring that inflow equals outflow at each node in ideal flow systems. Patriksson [9, 10] stated that a fundamental principle in network modeling, stating that inflow equals outflow at each node (except sources/sinks).

Definition 14.19 (Feasible Policy) A policy $\mathcal{P}$ is feasible if the resulting network flow state, defined by the matrices $(\mathsf{C}', \mathsf{L}', \mathsf{U}', \mathsf{S}')$, remains a valid Ideal Flow Network. This requires the capacity matrix C' four conditions:

1. C' must be *irreducible*, ensuring the network remains strongly connected.
2. C' must be *non-negative*, as link capacity cannot be negative.
3. The network flow state F' resulting from C' must be *premagic*[13], a condition where the flow conservation property holds at every node.
4. The policy must adhere to the dynamic axioms of edge removal Axiom 14.2 and node removal Axiom 14.3.

A molecular policy $\mathcal{P}$ is *feasible* if and only if its application to a valid Ideal Flow Network F in $\mathcal{N}$ results in a new network flow state F' in $\mathcal{N}'$ that remains a valid IFN. A valid IFN must be strongly connected (i.e., its adjacency matrix A must be irreducible).

Axiom 14.2 (Edge Removal Condition) *An edge (i, j) is removed from the network if its flow, f_{ij}, reduces to zero. The updated edge set becomes $E' = E \setminus \{(i, j) \mid f_{ij} = 0\}$.*

Axiom 14.3 (Node Removal Condition) *A node i is removed from the network if its node flow, π_i, is zero. The removal of node i also removes all incident links. The updated sets are $N' = N \setminus \{i \mid \pi_i = 0\}$ and $E' = E \setminus \{(k, i) \mid k \in N\} \setminus \{(i, k) \mid k \in N\}$.*

These axioms introduce a dynamic element to the network. Unlike a simple graph where an unused connection still exists, in our model, an unused road segment effectively disappears. If a policy successfully diverts all traffic from a certain street, that street is considered inactive and is removed from the network's active flow computations. This makes the model more responsive to how real traffic patterns emerge and disappear. Policies must be designed to avoid a situation where a critical link is removed, making the rest of the network disconnected. For example, a policy that removes a bridge would only be feasible if other connections still exist to keep the network strongly connected. This check is crucial for a policy's real-world viability.

Proposition 14.2 (Constraint on Link Addition) *A policy of adding a link is constrained by the network's topology. If the network graph is already complete, a policy of* AddLink *is infeasible.*

Proof By definition, a complete graph is one in which every pair of distinct vertices is connected by a unique edge. A policy of *AddLink* for a link (i, j) implies that the edge (i, j) does not already exist. If the network is complete, all possible links are already present. Therefore, there are no new links to add, and the policy is infeasible. □

Proposition 14.3 (Constraint on Link Removal) *A policy of* RemLink *is feasible if and only if its application does not render the network reducible.*

Proof By definition, an IFN must be a strongly connected graph. The removal of a single link from a network can, in some cases, disconnect the network, making it reducible. A policy of *RemLink* is therefore only feasible if the resulting network state remains strongly connected. □

Proposition 14.4 (Inter-Policy Constraint) *A policy of* RedCap *on a link (i, j) is constrained by the link's role in maintaining network connectivity. If the capacity on a link (i, j) is such that a reduction would cause its flow to drop to zero, and that link is part of a minimum irreducible set of links, the policy of* RedCap *becomes infeasible.*

Proof According to the axiom of link removal, if the flow on a link reduces to zero, that link is removed from the network. If this link is a critical bridge that ensures the network's irreducibility, its removal would violate the condition that a valid IFN must be strongly connected. Therefore, a *RedCap* policy that leads to this outcome is infeasible. □

Not all combinations of atomic policies result in a feasible molecular policy. A policy is only feasible if it can be implemented without violating the fundamental principles of the Ideal Flow Network, which is irreducible premagic.

Proposition 14.5 (Feasibility of Molecular Policies) *A molecular policy $\mathcal{P} = \{\mathcal{A}_1, \ldots, \mathcal{A}_k\}$ is called feasible. This holds if and only if the following conditions are met:*

1. *Each atomic policy within the sequence $\mathcal{A}_i$ is individually feasible.*
2. *The application of the policies, in their specified sequence, results in a network flow state that is a irreducible and premagic.*

Proof By Definition 14.19, a policy is feasible if the network state remains valid. The application of any single atomic policy must result in a valid state. The sequential application of policies, as represented by the sequencing operator, maintains this validity only if the network remains connected and satisfies the premagic property after each step. Therefore, for a molecular policy to be feasible, the final state must be a valid IFN, which necessitates that all constituent policies are feasible. □

14.5.2 Conflicting Policies

Not all policy ideas are good ideas. The framework provides a logical filter to discard policies that are mathematically impossible. A planner cannot, for instance, simultaneously widen and narrow a single road. The framework's ability to identify and reject these contradictory combinations provides a fundamental layer of rigor, ensuring that only plausible and logically consistent policies are considered for analysis.

Definition 14.20 (Inverse and Identical Policy) The inverse of a policy, $\mathcal{A}$, is a policy $\mathcal{A}^{-1}$ such that applying them in sequence results in the identity policy, $\mathcal{A} \rightarrow \mathcal{A}^{-1} = I$, where I signifies no net change.

- The inverse of a capacity addition, $\Delta\mathsf{C}^+$, is a capacity reduction of the same magnitude, $(\Delta\mathsf{C}^+)^{-1} = \Delta\mathsf{C}^-$.
- The inverse of adding a link, $\Delta\mathsf{A}^+$, is removing that same link, $(\Delta\mathsf{A}^+)^{-1} = \Delta\mathsf{A}^-$.
- The inverse of a speed increase, $\Delta\mathsf{U}^+$, is a speed decrease of the same magnitude, $(\Delta\mathsf{U}^+)^{-1} = \Delta\mathsf{U}^-$.
- The inverse of adding demand, $\Delta\kappa^+$, is reducing that demand, $(\Delta\kappa^+)^{-1} = \Delta\kappa^-$.
- The inverse of optimizing a signal, $\Delta\mathsf{S}$, is adjusting the probabilities back to their original state.
- The inverse of *RevDir* is the operation itself. Applying a *RevDir* twice returns the network to its original state. Mathematically, $(\Delta\mathsf{A}^{\leftrightarrow})^2 = I$. This makes *RevDir* a self-inverse, or an involution.

Definition 14.21 (Identity Policy) An *identity policy I* is a molecular policy that, when applied, results in no net change to the network's state. The identity is achieved through a *sequential application* of a policy and its inverse, represented by the sequencing operator ($\rightarrow$).

Definition 14.22 (Conflicting Policy) A *conflicting policy* is a molecular policy that results in a logically contradictory or undefined state. This occurs when an atomic policy and its inverse are combined using the *union operator* ($\cup$), which implies a simultaneous application.

Corollary 14.2 (Conflict of Atomic Policies) *Two atomic policies, $\mathcal{A}_1$ and $\mathcal{A}_2$, are in conflict and cannot be applied simultaneously to the same network element if they are contradictory by definition.*

Proof Policies are in conflict if their application to the same element results in an undefined state. For example, applying AddCap ($\Delta C_{ij} > 0$) and RedCap ($\Delta C_{ij} < 0$) to the same link (i, j) simultaneously is a logical contradiction, as the capacity cannot both increase and decrease at the same time. The set of all policies applied to a single element must result in a single, well-defined state for that element. □

Proposition 14.6 (Identity of a Policy and Its Inverse) *Applying a molecular policy and its inverse in sequence results in the identity policy.*

Proof Let $\mathcal{A}$ be a feasible atomic policy and $\mathcal{A}^{-1}$ be its inverse. The sequential application $\mathcal{A} \rightarrow \mathcal{A}^{-1}$ returns the network to its original state. This operation is well-defined and does not lead to a logical contradiction. Thus, $\mathcal{A} \rightarrow \mathcal{A}^{-1} = I$, where I is the identity. □

Theorem 14.2 (The Policy Conflict Theorem) *A molecular policy is a conflict if and only if it applies a union of an atomic policy and its inverse applied to the same network element.*

Proof Let $\mathcal{P} = \mathcal{A} \cup \mathcal{A}^{-1}$ be a molecular policy where $\mathcal{A}$ and its inverse $\mathcal{A}^{-1}$ are applied to the same element. By the definition of the union operator, these actions are simultaneous. A simultaneous application of an action and its reversal (e.g., a capacity increase and a capacity decrease on the same link at the same time) results in a logically contradictory state. Such a state is not a valid Ideal Flow Network and is, by definition, a conflict.

Conversely, a conflict by definition is a logically contradictory operation. The only way to represent a simultaneous contradiction within this framework is through the union of a policy and its inverse. Therefore, a policy is a conflict if and only if it contains such a union. □

Example 14.5 The following are common conflicting atomic policy pairs that, when applied to the same element at the same time, render a molecular policy unfeasible.

- *AddCap* and *RedCap* on the same link.
- *IncSpeed* and *DecSpeed* on the same link.
- *AddLink* and *RemLink* on the same link (unless it is a sequential policy).
- *AddNode* and *RemNode* on the same node.
- *TempAddCap* and TempRedCap on the same link.
- *TempIncSpeed* and *TempDecSpeed* on the same link.
- *RevDir* and a two-way flow model on the same link.

14.5.3 Periodic Table of Atomic Policies

In Chemistry [11], Dmitri Mendeleev created the periodic table by arranging elements based on increasing atomic mass, while also considering their chemical properties. He noticed recurring patterns in properties and used this to predict the existence of undiscovered elements, leaving gaps in his table where they would fit. This groundbreaking work laid the foundation for the modern periodic table, which is now arranged by atomic number. Using the same analogy, we can also organizes policies based on their spatial and functional dimensions (Groups) and their temporal and action dimensions (Periods). The result is shown in Table 14.5.

- *The Groups*: These groups classify policies by the network component they affect and their core function.
  - Group I: Link Attributes (L, U): Policies that alter the physical or operational characteristics of a link.
  - Group II: Link Capacity (C): Policies that directly change the flow capacity of a link.
  - Group III: Link Topology (A): Policies that add, remove, or reverse the existence and direction of a link.
  - Group IV: Node Properties (A, S): Policies that affect a node's existence or its outflow probabilities.
  - Group V: System-Wide Variables (κ): Policies that alter the total demand on the network.
- *The Periods*: These periods categorize policies by the permanence of the action.
  - Period A: Permanent Actions: Long-term, enduring changes to the network.
  - Period B: Temporary Actions: Short-term, event-based changes.

Table 14.5 presents a structured taxonomy of atomic policies, classified by permanence and network component, enabling deductive encoding of transport interventions.

Here is the same information presented in shorthand notation in Table 14.6. It provides the same atomic policy structure in symbolic shorthand, facilitating compact representation and computational modeling.

While Tables 14.5 and 14.6 encode the same atomic policy logic, they serve complementary roles. The first table introduces named actions for conceptual clarity, while the second provides symbolic shorthand for formal modeling and algorithmic use.

The periodic table of atomic policies is not a mere organizational tool; it provides a foundational structure for the ITA framework. Its usefulness in creating molecular policies is multifaceted.

Table 14.5 Periodic table of atomic policies

Period	Group I (L, U)	Group II (C)	Group III (A)	Group IV (A, S)	Group V (κ)
A.1 (Add)	*IncSpeed* $\Delta \mathsf{U}_{ij} > 0$	*AddCap* $\Delta \mathsf{C}_{ij} > 0$	*AddLink* Add (i, j)	*AddNode* Add node n	*AddDem* $\Delta \kappa > 0$
A.2 (Reduce)	*DecSpeed* $\Delta \mathsf{U}_{ij} < 0$	*RedCap* $\Delta \mathsf{C}_{ij} < 0$	*RemLink* Remove (i, j)	·	*RedDem* $\Delta \kappa < 0$
A.3 (Modify)	*Change Length* $\Delta \mathsf{L}_{ij} \neq 0$	·	*RevDir* Reverse (i, j)	*OptSig* Adjust S row	·
B.1 (Temp Add)	*TempIncSpeed* Temp $\Delta \mathsf{U}_{ij} > 0$	*TempAddCap* Temp $\Delta \mathsf{C}_{ij} > 0$	*TempAddLink* Temp Add (i, j)	·	·
B.2 (Temp Reduce)	*TempDecSpeed* Temp $\Delta \mathsf{U}_{ij} < 0$	*TempRedCap* Temp $\Delta \mathsf{C}_{ij} < 0$	*TempRemLink* Temp Rem (i, j)	·	·
B.3 (Temp Modify)	·	·	*TempRevDir* Temp Rev (i, j)	*TempOptSig* Temp Adjust S row	·

Table 14.6 Periodic table of atomic policies in shorthand notations

Period	Group I (L, U)	Group II (C)	Group III (A)	Group IV (A, S)	Group V (κ)
A.1 (Add)	ΔU^+	ΔC^+	ΔA^+	$\Delta \mathcal{V}^+$	$\Delta\kappa^+$
A.2 (Reduce)	ΔU^-	ΔC^-	ΔA^-	·	$\Delta\kappa^-$
A.3 (Modify)	$\Delta L^\pm$	·	$\Delta A^\leftrightarrow$	ΔS	·
B.1 (Temp Add)	δU^+	δC^+	δA^+	·	·
B.2 (Temp Reduce)	δU^-	δC^-	δA^-	·	·
B.3 (Temp Modify)	·	·	$\delta A^\leftrightarrow$	δS	·

- *Completeness and Exhaustiveness*: The table ensures that all fundamental, irreducible policy actions are explicitly defined. By exhaustively listing all possible atomic policies for controllable network variables (L, U, C, A, S) and system-wide variables (κ), it provides the complete set of building blocks from which all policies can be constructed. This prevents the omission of potential interventions.
- *Systematic Policy Synthesis*: The table's structure guides the deliberate and systematic creation of molecular policies. A planner can select actions from different groups and periods to synthesize a policy that addresses multiple aspects of a problem. For example, a policy that requires both permanent capacity changes and temporary signal adjustments can be constructed by combining elements from Period A, Group II (*AddCap*), and Period B, Group IV (*TempOptSig*).
- *Pedagogical Clarity and Conceptual Rigor*: The periodic table provides a simple, memorable analogy that grounds the complex process of policy design. It formalizes the distinction between different types of interventions (e.g., permanent vs. temporary, structural vs. operational) and clarifies their underlying mathematical basis. This rigor is essential for transitioning policy design from a qualitative art to a quantitative science.
- *Identification of Infeasible Policies*: The deliberate gaps (·) in the table, which are not random but a result of the framework's axioms, highlight logically or physically impossible policies. For instance, a temporary policy for adding a node (*TempAddNode*) is conceptually infeasible, as a physical node is a permanent piece of infrastructure. This feature helps planners avoid contradictory or unworkable interventions from the outset, thus streamlining the policy exploration process.

The "·" in Table 14.5 or in Table 14.6 has a dual meaning: it denotes that a certain type of atomic policy is either *logically infeasible* based on the framework's axioms or *not applicable* due to the nature of the variable. It also functions, by its absence, to highlight areas for future research and conceptual discovery.

The table's is in its precise classification. The "·" entries are not random gaps; they are a result of applying the framework's own rules.

- *Group IV, Period A.2 (Remove Node)*: This is not an atomic policy because a node's removal is a consequence, not a direct action. According to the framework's Axioms 14.3, a node is removed automatically if its total flow becomes zero. A planner's action would be to implement a policy (e.g., a *RemLink* policy) that leads to this zero-flow condition. The removal itself is a secondary, axiomatic outcome.
- *Groups I, II, III, IV, V, Period A.3 (Modify)*: "Modify" implies a change that is neither a simple addition nor reduction. For variables in these groups, a change is always a change in value, which can be expressed as either an addition or a reduction. For example, changing the capacity of a link, $\Delta \mathsf{C}_{ij} \neq 0$, is inherently either an increase or a decrease, already covered by *AddCap* or *RedCap*. The one exception is *RevDir*, which is a special type of structural change that fundamentally alters a link's flow direction, making it a unique atomic policy.
- *Groups IV and V, Periods B.1, B.2, and B.3 (Temporary Policies)*: Temporary policies are generally not applicable to the existence of nodes (*AddNode*, *RemNode*) or to system-wide demand (κ). A node represents a permanent piece of infrastructure (an intersection, an area) that is not physically added or removed on a temporary basis. A temporary policy might use *TempAddLink* to model a temporary bypass, but not a temporary node. Likewise, the total demand, κ, is an external, long-term factor. A policy might reduce it, but the reduction itself is a molecular policy (e.g., a temporary fee), not a temporary atomic policy applied to κ itself. The concept is conceptually infeasible.

Having said the above, the "·" entries might also serve as a prompt for future research. Just as Mendeleev left gaps in his table for undiscovered elements, the ITA table's empty cells suggest areas where new policies or planning concepts could be discovered or rigorously defined. For example, can a temporary policy be applied to the *AddNode* atomic action? Perhaps a "pop-up city" or a temporary new destination, but this would likely be a molecular policy built on other atoms. The existence of "·" encourages researchers to either prove a policy's infeasibility or to conceive of a real-world scenario that would fill the empty cell.

The framework's rules for combining atomic policies help identify interventions that are logically and physically contradictory, making them infeasible. The negation operator is particularly useful here.

Example 14.6 (Simultaneous Capacity Addition and Reduction) A city proposes a policy to "simultaneously add a new high-speed bus lane to a corridor and reduce the number of general traffic lanes." This molecular policy can be written as (*AddCap* ∪ *RedCap*). According to the framework's conflict rules, applying these two policies to the same link at the same time is logically infeasible, as a link's capacity cannot both increase and decrease at once. The correct molecular representation is a single policy of *RedCap* on the original link (for general traffic) and *AddLink* for the new bus lane, showing that a new physical link is created with a different set of rules.

$$\mathcal{P}_{\text{bus\_lane}} = \Delta \mathsf{C}_{ij}^{-} \cup \Delta \mathsf{A}_{ij}^{+}$$

Example 14.7 (The Negation of Adding a Link) A proposal is made to build a new bridge. This is a straightforward atomic policy: $\mathcal{P}_{bridge} = AddLink$. A contradictory and logically infeasible policy would be to propose the "negation of adding a bridge." This policy, represented as $\neg(AddLink)$, is not a feasible action in the real world. In the framework, it is an invalid policy because it is a direct contradiction of a fundamental, creative action.

Similarly, a policy of $\mathcal{P}_{Contradictory} = (AddCap \cup \neg AddCap)$ is also infeasible. It states that a policy will both add capacity and not add capacity simultaneously, which is a logical contradiction.

The framework can be used to propose and explore novel policy interventions that do not yet exist in practice. The table's empty cells, marked "·," serve as a prompt for conceptual discovery.

Example 14.8 (The "Instant" Temporary Node) The table shows "·" for a temporary policy of adding a node (*TempAddNode*). While physically impossible to build a new intersection temporarily, this could represent a new type of intelligent policy. It is modeled as a temporary node and temporary links.

$$\mathcal{P}_{\text{hub}} = \delta \mathcal{V}_n^+ \cup \delta \mathsf{A}_{n,i}^+$$

For instance, a temporary "ride-sharing hub" could be created in a large parking lot for a major event. This location, which acts as a new origin/destination, could be modeled as a temporary node and connected to the network via temporary links. The framework's ability to model such a policy allows for proactive planning for large-scale events.

Example 14.9 (Dynamic, Congestion-Responsive Pricing) Traditional congestion pricing uses a static fee, but a dynamic system could use real-time data to adjust prices. This can be synthesized as a feedback loop.

$$\mathcal{P}_{Dynamic} = \big(Check(\max(\mathsf{G}) \geq \xi)\big) \rightarrow (TempRedDem \cup TempOptSig)$$

In shorthand notation, it becomes

$$\mathcal{P}_{\text{dynamic}} = \big(\text{Check}(\max(\mathsf{G}) \geq \xi)\big) \rightarrow \big(\delta\kappa^- \cup \delta\mathsf{S}\big)$$

This molecular policy is not a one-time action but as a continuous process. A trigger function, $Check(\cdot)$, continuously monitors the network's maximum congestion, $\max(\mathsf{G})$. When a critical threshold ξ is reached, the policy automatically synthesizes and applies a temporary reduction in demand (*TempRedDem*) by increasing the toll price and a temporary signal optimization (*TempOptSig*) to alleviate the bottleneck. Once congestion drops, the temporary policies are removed. This represents a proactive, responsive, and innovative approach to traffic management that is only possible through this kind of formal, computational framework.

To demonstrate the expressive power and optimization capabilities of the Ideal Traffic Assignment (ITA) framework, we introduce a novel class of theoretical molecular policies designed to perform lane reallocation across network links. These policies, denoted as $\mathcal{P}_{\text{Realloc}}^{(i \to j, -n)}$, represent the transfer of n lanes from a donor link i to a recipient link j, while preserving the total lane capacity of the network. This formulation enables the exploration of alternative network configurations under a fixed resource constraint, allowing for performance optimization without increasing infrastructure investment.

The motivation for this construct is twofold. First, it isolates the structural effects of capacity distribution from the confounding influence of total capacity changes. Second, it enables the use of computational optimization objective function to identify configurations that minimize system delay or travel time, maximize system speed, or improve equity metrics and balance (as measured by the VarPi index). In practice, such lane reallocation policies may be prohibitively expensive or infeasible due to physical, political, or economic constraints. However, within the ITA framework, they serve as theoretical instruments for scenario analysis, algorithmic synthesis, and policy benchmarking.

To encode these policies within the molecular grammar specification, we extend the formal syntax to include a reallocation operator. The following Extended Backus-Naur Form (EBNF) defines the structure of a lane reallocation policy:

This syntax is declarative, modular, and suitable for automated parsing. Each reallocation policy is translated into a transformation matrix $\Delta\mathsf{C}$, where:

- $\Delta C_{ij} = -n$ for the donor link $i \to j$
- $\Delta C_{kl} = +n$ for the recipient link $k \to l$

The following policy list encodes a sequence of lane reallocations applied to the premagic base matrix C_4 in Example 13.12:

- Realloc(cb $\to$ cd, -2)
- Realloc(cd $\to$ ad, -2)
- Realloc(ce $\to$ bc, -2)

These policies can be programmatically applied to generate new capacity matrices C_5 through C_8 in Example 13.12, each representing a distinct network configuration with identical total capacity but differing performance characteristics. This approach enables reproducible experimentation, algorithmic search for optimal configurations, and formal validation of policy effects within the ITA framework (Fig. 14.2).

```
policy ::= reduction | reallocation ;
reallocation ::= 'Realloc(' from_link '→' to_link ',' lanes ')' ;
from_link ::= node node ;
to_link ::= node node ;
lanes ::= INTEGER ;   // must satisfy capacity conservation
node ::= identifier ;
identifier ::= letter , { letter | digit | "_" } ;
letter ::= "A" | "B" | ... | "Z" | "a" | "b" | ... | "z" ;
digit ::= "0" | "1" | "2" | "3" | "4" | "5" | "6" | "7" | "8" | "9" ;
```

Fig. 14.2 A formal grammar for the lane reallocation policy, defined using an Extended Backus-Naur Form (EBNF) notation

14.6 Policy as a Combinatorial Discrete Optimization Problem

This section formalizes the generation and evaluation of policies into a computational process. It frames policy design not as an art, but as a discrete optimization problem on a finite set of atomic policies.

14.6.1 The Policy Space

The policy space, denoted by $\mathcal{P}_{space}$, is the set of all possible molecular policies that can be generated from the atomic policy table. A single molecular policy, $\mathcal{P}$, is a specific combination of atomic policies. The size of this space is finite, but it is combinatorially explosive.

Definition 14.23 (Policy Space) The policy space, $\mathcal{P}_{space}$, is the power set of all feasible molecular policies. A molecular policy $\mathcal{P}$ is a non-empty subset of the set of all atomic policies, $\mathcal{A}$, subject to the feasibility constraints:

$$\mathcal{P}_{space} = \{\mathcal{P} \mid \mathcal{P} \subseteq \mathcal{A} \wedge \text{IsFeasible}(\mathcal{P})\}$$

Here, $\mathcal{A}$ is the set of all atomic policies defined in the Periodic Table of Atomic Policies 14.5, and IsFeasible($\mathcal{P}$) is a boolean function that returns true if the molecular policy satisfies all feasibility conditions (non-contradictory, irreducible, etc.).

The policy space is simply the entire universe of possible policy combinations. Instead of a planner brainstorming three or four ideas, we're giving a computer the tools to explore every single possible combination, from adding one lane to a complex network-wide series of changes, and ensuring that each combination is logically sound.

14.6.2 The Automated Generation Algorithm

The task of automatically generating alternative policies can be implemented as a randomized algorithm that explores the policy space. This algorithm leverages the defined atomic policies and their combination rules.

Policy Generator

Input: An initial base network state $\mathcal{N}_0$, a set of atomic policies $\mathcal{A}$, and a target number of policies to generate, N. *Output:* A set $\{\mathcal{N}_i\}$ of randomly generated, feasible molecular policies. Begin by initializing an empty set of generated policies:

$\mathcal{P}_{\text{gen}} = \emptyset$. Then, for each iteration $i = 1$ to N, perform the following steps:

1. *Generate Policy Candidate:* Randomly select a non-empty subset of atomic policies from $\mathcal{A}$ to form a candidate molecular policy, denoted $\mathcal{P}_{\text{candidate}}$.
2. *Feasibility Check:* Evaluate whether $\mathcal{P}_{\text{candidate}}$ satisfies the feasibility constraints defined in Sect. 14.5.1. This includes checking for contradictory atomic policies.
3. *Apply Policy:* If feasible, apply the transformations encoded in $\mathcal{P}_{\text{candidate}}$ to the base network $\mathcal{N}_0$ to produce a new network state $\mathcal{N}_i$.
4. *Irreducibility Check:* Verify that the resulting network state $\mathcal{N}_i$ is irreducible. If not, discard the candidate policy.
5. *Add to Set:* If all checks pass, add $\mathcal{P}_{\text{candidate}}$ to the set $\mathcal{P}_{\text{gen}}$.

After completing all iterations, return the final set $\mathcal{P}_{\text{gen}}$.

This algorithm is like a computer trying out millions of different ideas. It randomly picks a bunch of atomic policies, like "add a lane here" and "change a speed limit there", and puts them together. Then, it quickly checks if the combination makes sense (e.g., it can't add and remove the same lane at the same time). If the combination is logical, it applies the changes and sees if the network still works (i.e., is not disconnected). If it passes all the tests, it's a valid policy, and the computer saves it for later evaluation.

14.6.3 The Policy Optimization Problem

The ITA framework culminates in a formal optimization problem. The final step is to find the optimal policy from the generated set. This is a classic discrete optimization problem, where the goal is to find the policy that minimizes the goal function $\omega(\mathcal{P})$.

Definition 14.24 (Policy Optimization Problem) The *policy optimization problem* is to find the optimal policy $\mathcal{P}^*$ from the set of all feasible policies, Π. The optimal policy is the one that minimizes the multi-objective goal function $\omega(\mathcal{P})$.

$$\mathcal{P}^* = \arg\min_{\mathcal{P}\in\Pi} \omega\left(\max(\mathsf{G}_{\mathcal{P}}), \mathrm{Cost}(\mathcal{P}), \mathrm{Time}(\mathcal{P})\right) \tag{14.4}$$

The set of feasible policies, Π, is constrained by the conditions of irreducibility, premagic property, and non-negativity as define in Sect. 14.5.1.

This concept is the central purpose of our framework. Instead of a planner coming up with a few ideas and testing them one by one, we can define all possible policy changes, adding lanes, removing roads, etc., and the computer can search through millions of combinations to find the one that best meets our goals. It might find a combination of a one-way street system here and a lane reduction there that costs less and works better than simply building a new, expensive highway. This transforms transportation planning into a systematic and data-driven process.

Proposition 14.7 (Existence of an Optimal Solution) *Given a finite set of feasible policies* Π*, there exists at least one optimal policy* $\mathcal{P}^* \in \Pi$ *that minimizes the goal function* $\omega(\mathcal{P})$.

Proof The goal function $\omega(\mathcal{P})$ is a scalar function that maps each policy in the finite set Π to a real number. By the properties of real numbers, every finite set of real numbers has a minimum value. Therefore, there must exist a policy $\mathcal{P}^*$ in Π that corresponds to this minimum value. □

Once the computer has a list of thousands of good policy ideas, it's time to choose the best one. The computer calculates the performance, cost, and time for each policy on its list. Because the list is finite, the computer is guaranteed to find the one that has the lowest overall cost and best performance, providing a clear, data-driven recommendation. This process replaces human intuition with computational certainty.

Proposition 14.8 (Two-Step Scenario Optimization) *The policy optimization problem can be solved in two distinct steps.*

1. *Feasibility Check: Generate a set of feasible alternative scenarios by applying molecular policies to the base network. A scenario is feasible if the resulting network is an irreducible IFN.*
2. *Performance Comparison: Compare the performance of all feasible alternative scenarios to the base scenario using the goal function* $\omega(\mathcal{P})$*. An alternative scenario is considered a viable improvement if its performance is better than the base scenario.*

Proof The policy optimization problem seeks to find the best policy $\mathcal{P}^*$ from the set of all feasible policies, Π. This requires that the policies are first vetted for feasibility (Step 1). A policy is only viable if it results in a network that can function as an IFN. Once the set of feasible policies is established, a comparative analysis can be performed. The performance of a policy is measured by its effect on the network's outputs, which can be quantified by the goal function $\omega(\mathcal{P})$. The best policy is the one that minimizes this goal function, thus completing Step 2. □

Note, however, the combinatorial explosion of the policy space means that for large, complex urban networks, the computational search for an optimal policy would be immense. Heuristics or constrained search algorithms would be necessary for practical implementation.

14.7 Application in Long-Term Planning and Space-Time Forecasting

The ITA framework provides a robust tool for long-term strategic planning by integrating policy interventions with traffic growth forecasts. This allows planners to anticipate future bottlenecks and design adaptive management strategies based on the principle of invariants.

Example 14.10 (Metropolis City (Reactive Policy)) Metropolis downtown road grid is shown in Fig. 14.3. The city's total traffic flow, κ_t, is a function of population and commercial vehicles. We use a predictive model with an annual growth rate to forecast future demand. The city's policy is to add a new lane to a link whenever its congestion level (flow-to-capacity ratio) reaches a critical threshold of $\xi = 0.9$. The model is built on three assumptions:

1. *assumption 1: Invariant Network Topology*. The fundamental network structure remains constant over time.
2. *assumption 2: Reactive Lane Expansion*. The city's strategy is to add one lane to a link whenever its congestion level reaches a critical threshold, $\xi = 0.9$. The construction time is considered negligible.
3. *assumption 3: Traffic Growth Model*. The total flow in the network is a function of population and car ownership, modeled as $\kappa_t = 3 \times \text{Pop}_t + 7 \times \text{Car}_t + 3530$. Population and car populations grow at 5% and 7% per year, respectively. Based on this model, we can project the human population and cars population to build the relationship between total flow in the network for each year as shown in Table 14.7.

Table 14.7 provides year-by-year projections of traffic flow, population, and car ownership, forming the basis for congestion, triggered lane expansion under the city's reactive policy model.

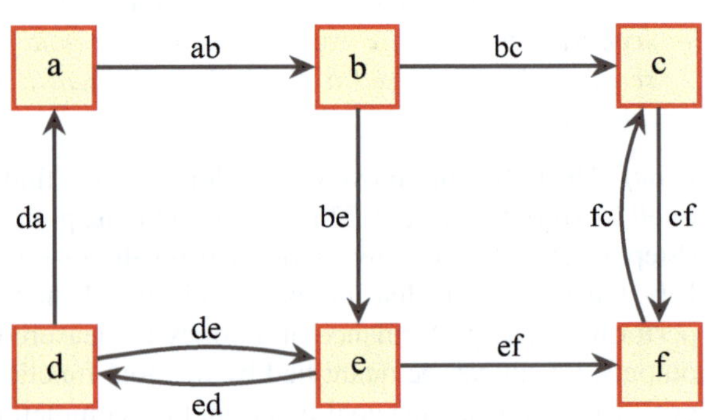

Fig. 14.3 Metroplis download grid with six nodes and nine links

Table 14.7 Rounded projections of kappa, population, and car ownership from 2030 to 2042

Year	Kappa	Population	*Car*
2030	12,650	1500	660
2031	13,198	1575	706
2032	13,781	1654	756
2033	14,399	1736	809
2034	15,056	1823	865
2035	15,753	1914	926
2036	16,494	2010	990
2037	17,281	2111	1060
2038	18,117	2216	1134
2039	19,005	2327	1213
2040	19,948	2443	1298
2041	20,951	2566	1389
2042	22,016	2694	1486

The city's policy is to reactively add a new lane to a link whenever its congestion reaches a critical threshold. This can be expressed as a continuous, conditional policy that triggers a specific atomic action.

$$\mathcal{P}_{\text{reactive}} = \text{if}(\max(g_{ij}) \geq \xi) \text{ then } \Delta \mathsf{C}_{ij}^{+}$$

or, we can write simple atomic policy:

$$\mathcal{P}_{\text{reactive}} = \Delta \mathsf{C}_{ij_{\xi \geq 0.9}}^{+}$$

We model a city's road network in the base year 2030 with relative link capacity

$$\mathsf{C}_{rel,base} = \left(\begin{array}{c|cccccc} \nearrow & a & b & c & d & e & f \\ \hline a & 0 & 1 & 0 & 0 & 0 & 0 \\ b & 0 & 0 & 2 & 0 & 2 & 0 \\ c & 0 & 2 & 0 & 0 & 0 & 2 \\ d & 1 & 0 & 0 & 0 & 1 & 0 \\ e & 0 & 0 & 0 & 1 & 0 & 1 \\ f & 0 & 0 & 2 & 0 & 0 & 0 \end{array} \right)$$

The capacity per lane is 1800 vehicles/hour. The city government policy is to add a new lane to a link whenever its congestion level reaches a critical threshold of $\xi = 0.9$. The policy involves adding a new lane to a congested link. This is a single permanent atomic policy applied at a specific time.

- Step 1: Base Scenario (2030):
- With an initial population of 1500 and 660 cars, the total flow is $\kappa_{2030} = 12{,}650$ veh/hr. The initial network flow $\mathsf{F}_{abs,base}$ has a total flow of $\kappa_{base} = 12{,}650$

veh/hr, with a most congested in link fc having a congestion of $2952/(2 \times 1800) = 81.99\%$.

$$
\mathsf{F}_{abs,base} = \left(\begin{array}{c|cccccc}
\nearrow & a & b & c & d & e & f \\
\hline
a & 0 & 422 & 0 & 0 & 0 & 0 \\
b & 0 & 0 & 1265 & 0 & 1265 & 0 \\
c & 0 & 2108 & 0 & 0 & 0 & 2108 \\
d & 422 & 0 & 0 & 0 & 422 & 0 \\
e & 0 & 0 & 0 & 843 & 0 & 843 \\
f & 0 & 0 & 2952 & 0 & 0 & 0
\end{array}\right)
$$

- Step 2: Forecasting Time for Policy Intervention:
  - We set our target maximum congestion $\xi = 0.9$ as the government policy stated.
  - We then calculate the scaling factor:

$$\varsigma = \frac{0.9}{0.8199} \approx 1.097695$$

  - The absolute total flow for our target year is then:

$$\kappa_{target} = \varsigma \times \kappa_{base} = 1.097695 \times 12{,}650 \approx 13{,}886 \text{ veh/hr}$$

  - The congestion on the most congested link now equals our target, 0.90. The model forecasts that the total flow will reach this target demand of $\kappa_{target} = 13{,}886$ by approximately the year 2032. By that year, the road network would exceed our benchmark allowable max congestion of 0.90.

- Step 3: Forecasting Space for Policy Intervention: Here is how we can create the Alternative Scenario, which is in year 2032.
  - We set $\kappa_{target} = 13{,}886$ veh/hr as our invariant total demand for the policy analysis.
  - We create a new network with the same configuration as the base network but only the total flow change to $\kappa_{target} = 13{,}886$ veh/hr. The most congested link with max congestion of 0.90 would happen in link *fc* with predicted flow $f_{fc} = 3240$ veh/hr. Notice that, using ITA framework, we can now predict not only the *time* on *when* the congestion would happen, we can also forecast the *space* on *where* the congestion might happen.

- Step 4: Policy Intervention and Analysis (2032): Now we simulate what happen if we add lane in the most congested link in the target year 2032. The city's policy is to add a lane to the most congested link to keep congestion below the threshold. We simulate this policy by increasing the capacity of the link. We model a city's road network in the base year 2030 with relative link capacity

$$\mathsf{C}_{rel,target} = \begin{pmatrix} \nearrow & a & b & c & d & e & f \\ \hline a & 0 & 1 & 0 & 0 & 0 & 0 \\ b & 0 & 0 & 2 & 0 & 2 & 0 \\ c & 0 & 2 & 0 & 0 & 0 & 2 \\ d & 1 & 0 & 0 & 0 & 1 & 0 \\ e & 0 & 0 & 0 & 1 & 0 & 1 \\ f & 0 & 0 & 3 & 0 & 0 & 0 \end{pmatrix}$$

$$\mathsf{F}_{abs,target} = \begin{pmatrix} \nearrow & a & b & c & d & e & f \\ \hline a & 0 & 463 & 0 & 0 & 0 & 0 \\ b & 0 & 0 & 1389 & 0 & 1389 & 0 \\ c & 0 & 2314 & 0 & 0 & 0 & 2314 \\ d & 463 & 0 & 0 & 0 & 463 & 0 \\ e & 0 & 0 & 0 & 926 & 0 & 926 \\ f & 0 & 0 & 3240 & 0 & 0 & 0 \end{pmatrix}$$

- We create a new network configuration by applying the atomic policy *AddCap* to the most congested links. The most congested link with max congestion of 0.90 was detected in the previous step in link *fc*.
- Evaluation: We now run the ITA framework on this new network, keeping the total demand constant at our benchmark, $\kappa = 13{,}886$ veh/hr.
- The analysis shows increasing number of lanes in link *fc* from 2 lanes to 3 lanes in target year 2032 would make the congestion in link *fc* improved, drops from 0.90 to 0.60. However, the network's internal flow rebalanced, causing induced traffic and the maximum congestion to shift to other links *cb* and *cf*, which now operate at a new maximum congestion of 0.6429.

To reach the critical congestion of 0.9, we calculate the scaling factor:

$$\varsigma = \frac{0.9}{0.6429} \approx 1.3999712$$

The absolute total flow for our target year is then:

$$\kappa_{target} = \varsigma \times \kappa_{base} = 1.3999712 \times 13{,}886 \approx 19{,}440 \text{ veh/hr}$$

Based on Table 14.7, the total flow of 19,440 vehicles/hour is projected to be reached in 2039. Continuing the same technique, we can forecast that links *cb* and *cf* would reach critical congestion of 0.9 at year 2039.

Using the same network, policy and the same model as previous Example 14.10, what will happen if the city government does proactively than reactive? In the following example, an analysis of a proactive policy is developed using the same network and growth model as the previous example. The proactive strategy involves implementing infrastructure changes in anticipation of future demand, rather than in response to existing critical congestion.

Example 14.11 (Metropolis City (Proactive Policy)) In this scenario, the city government adopts a proactive policy. Instead of waiting for congestion to reach a critical threshold, it uses the Ideal Traffic Assignment (ITA) framework to forecast future bottlenecks and intervenes in the base year (2030) to preemptively address the most vulnerable link. The goal is to maintain a lower level of congestion throughout the planning horizon.

The proactive policy can be expressed as:

$$\mathcal{P}_{\text{proactive}} = \Delta \mathsf{C}^{+}_{ij_{\text{forecast}}}$$

where the link ij_{forecast} is identified as the first link that will reach the critical threshold
xi in the future, based on the traffic growth model. The intervention is applied in the current year rather than the future year.

- Step 1: Base Scenario Analysis and Proactive Intervention (2030):
  - As established in Example 14.10, the base year is 2030 with a total flow of $\kappa_{2030} = 12{,}650$ veh/hr.
  - The initial analysis predicts that link fc will be the first to reach the critical congestion threshold of $\xi = 0.9$. This is forecast to occur in the year 2032.
  - The proactive policy dictates that this anticipated bottleneck should be addressed immediately. Therefore, the city adds one lane to link fc in the year 2030.
  - This intervention changes the relative capacity matrix in the base year itself.

$$\mathsf{C}_{rel,proactive} = \begin{pmatrix} \nearrow & a & b & c & d & e & f \\ \hline a & 0 & 1 & 0 & 0 & 0 & 0 \\ b & 0 & 0 & 2 & 0 & 2 & 0 \\ c & 0 & 2 & 0 & 0 & 0 & 2 \\ d & 1 & 0 & 0 & 0 & 1 & 0 \\ e & 0 & 0 & 0 & 1 & 0 & 1 \\ f & 0 & 0 & 3 & 0 & 0 & 0 \end{pmatrix}$$

- Step 2: Network Re-evaluation (2030):
  - With the new capacity on link fc, the ITA framework is used to re-calculate the flow distribution for the same total flow $\kappa_{2030} = 12{,}650$ veh/hr.

  - The traffic flow redistributes throughout the network due to the change in capacity. The resulting absolute flow matrix is:

$$
\mathsf{F}_{abs,proactive} = \begin{pmatrix} \nearrow & a & b & c & d & e & f \\ \hline a & 0 & 422 & 0 & 0 & 0 & 0 \\ b & 0 & 0 & 1265 & 0 & 1265 & 0 \\ c & 0 & 2108 & 0 & 0 & 0 & 2108 \\ d & 422 & 0 & 0 & 0 & 422 & 0 \\ e & 0 & 0 & 0 & 843 & 0 & 843 \\ f & 0 & 0 & 2952 & 0 & 0 & 0 \end{pmatrix}
$$

  - The congestion on link fc is now significantly reduced to $g_{fc} = 2952/(3 \times 1800) \approx 0.5466$.
  - However, the flow redistribution has increased traffic on other links. The new maximum congestion now occurs on links *cb* and cf, with a congestion level of $g_{max} = 2109/(2 \times 1800) \approx 0.58565$.
  - This proactive measure successfully prevents the imminent high congestion on link fc, resulting in a network where the maximum congestion in 2030 is 58.565%, a substantial improvement over the 81.99% in the reactive scenario's base case.

- Step 3: Forecasting the Next Policy Intervention:
  - The network is now stable until the new most congested links, *cb* and cf, reach the critical threshold of $\xi = 0.9$.
  - We calculate the scaling factor required for the current maximum congestion to reach the threshold:

$$
\varsigma = \frac{\xi}{g_{max}} = \frac{0.9}{0.58565} \approx 1.5367589
$$

  - The new target total flow at which the next intervention will be needed is:

$$
\kappa_{target} = \varsigma \times \kappa_{2030} = 1.5367589 \times 12{,}650 \approx 19{,}440 \text{ veh/hr}
$$

  - Referring to the projection data in Table 14.7, a total flow of 19,440 veh/hr is expected to be reached around the year 2039.

- Step 4: Policy Comparison and Conclusion:
  - *Reactive Policy Timeline*: Critical congestion (90%) is reached on link fc in *2032*. After intervention, the next critical point on links *cb* and cf is reached in *2039*. The network experiences high congestion between 2030 and 2032.
  - *Proactive Policy Timeline*: Intervention on link fc occurs in *2030*. The next critical point on links *cb* and cf is reached in *2039*. The network maintains a lower maximum congestion level (below 60%) for the entire period from 2030 to 2039.

- The proactive strategy requires an earlier capital investment (in 2030 instead of 2032) but provides a significantly better level of service for nine years. It effectively delays the onset of critical congestion anywhere in the network until 2039. In contrast, the reactive policy defers the initial cost but permits high levels of congestion to develop, necessitating intervention just two years into the forecast period.

The power of the Ideal Traffic Assignment framework to predict traffic phenomena in both space and time is thus demonstrated. This scenario analysis reveals trade-offs between infrastructure investment, congestion mitigation, and demand control. Reactive policies respond to symptoms but may induce cascading bottlenecks. Proactive strategies, like the one demonstrated, can forestall critical congestion for longer periods but require accurate forecasting and anticipatory capital expenditure. Proactive strategies reduce long-term interventions but require anticipatory planning. Demand management avoids expansion costs but may constrain mobility.

The iterative process of forecasting, intervention, and re-evaluation illustrates the principles of adaptive urban transportation planning. It highlights the importance of systemic thinking, where localized actions can produce network-wide effects, and long-term strategies must balance performance, cost, and equity.

14.8 Summary

The chapter establishes the Ideal Traffic Assignment (ITA) framework as a scientifically rigorous tool for transportation planning. It redefines policy formulation as a quantitative and optimizable procedure, moving it from a qualitative basis. The core of this methodology lies in defining policies as mathematical transformations on an Ideal Flow Network (IFN). The framework distinguishes between controllable variables (e.g., capacity, speed) and uncontrollable outputs (e.g., flow, congestion) , which separates the active role of a planner from the passive role of a forecaster.

The chapter introduces a formal language for policy interventions, categorizing them into primitive "atomic" policies. These policies are organized in a "periodic table" based on their function and permanence, providing a complete and exhaustive set of actions a planner can take. Complex, real-world interventions are then synthesized as "molecular policies" which are finite combinations of atomic policies. The framework's operators for union ($\cup$), negation ($\neg$), and sequencing ($\rightarrow$) enable the precise modeling of these interventions.

Policy feasibility is rigorously defined by constraints that ensure the network remains a valid IFN, specifically, that it remains strongly connected and premagic after an intervention. The framework also provides a formal "policy conflict theorem" to identify and discard logically contradictory policy combinations. This theoretical foundation transforms policy design into a combinatorial discrete optimization problem, where an automated algorithm can generate and evaluate all feasible policies against a user-defined goal function.

Finally, the chapter presents a practical method for scaling the relative IFN model to absolute, real-world flow values using a maximum allowable congestion level as a benchmark. This approach allows for meaningful scenario comparisons without requiring exhaustive real-world data, enabling planners to predict the systemic effects of their decisions and transition from reactive to proactive, data-driven planning.

Furthermore, the chapter illustrated the power of ITA for long-term forecasting. By integrating the model with growth projections and proactive/reactive policy assumptions, planners can anticipate future bottlenecks and understand the systemic, often non-local, effects of their decisions.

References

1. Mokhtar S. Bazaraa, Hanif D. Sherali, and C. M. Shetty. *Nonlinear Programming: Theory and Algorithms*. 3rd. Wiley, 2006. https://doi.org/10.1002/0471787773.
2. Noam Chomsky. "Three Models for the Description of Language". In: *IRE Transactions on Information Theory* 2.3 (1956), pp. 113–124. https://doi.org/10.1109/TIT.1956.1056813.
3. C. F. Daganzo and Y. Sheffi. "On Stochastic Models of Traffic Assignment". In: *Transportation Science* 11.3 (1977), pp. 253–274. https://doi.org/10.1287/trsc.11.3.253.
4. International Organization for Standardization and International Electrotechnical Commission. *ISO/IEC 14977:1996—Information technology—Syntactic metalanguage—Extended BNF*. https://www.iso.org/standard/26153.html. International Standard. Edition 1, confirmed in 2023. ISO/IEC JTC 1/SC 22, 1996.
5. John G. Kemeny and J. Laurie Snell. *Finite Markov Chains*. New York: Springer, 1976. https://link.springer.com/book/9780387901923.
6. H. W. Kuhn and A. W. Tucker. "Nonlinear Programming". In: *Proceedings of the Second Berkeley Symposium on Mathematical Statistics and Probability*. University of California Press, 1951, pp. 481–492.
7. Peter Naur, John Backus, et al. *Report on the Algorithmic Language ALGOL 60*. Tech. rep. Numerical Mathematics Group, Copenhagen, 1960. https://www.masswerk.at/algol60/report.htm.
8. Christos H. Papadimitriou and Mihalis Yannakakis. "Shortest paths without a map". In: *Theoretical Computer Science* 84.1 (1991), pp. 127–150. ISSN: 0304-3975. https://doi.org/10.1016/0304-3975(91)90263-2. https://www.sciencedirect.com/science/article/pii/0304397591902632.
9. M. Patriksson. *The traffic assignment problem: models and methods*. Courier Dover Publications, 2015.
10. Michael Patriksson. *The Traffic Assignment Problem: Models and Methods*. VSP, 1994. ISBN: 978-90-67641-63-0. https://www.math.chalmers.se/~mipat/LATEX/book_pub.pdf.
11. Eric R. Scerri. *The Periodic Table: Its Story and Its Significance*. 2nd ed. Oxford University Press, 2019. ISBN: 978-0190914363. https://doi.org/10.1093/oso/9780190914363.001.0001.
12. Yosef Sheffi. *Urban Transportation Networks: Equilibrium Analysis with Mathematical Programming Methods*. Englewood Cliffs, NJ: Prentice-Hall, 1985.
13. Kardi Teknomo. "Premagic and Ideal Flow Matrices". In: *arXiv preprint arXiv:1706.08856* (2017). arXiv: 1706.08856 [cs.DM]. https://arxiv.org/pdf/1706.08856.

Chapter 15
Conclusions, Practical Guidance, and Grand Challenges

15.1 Introduction

Urban traffic congestion is a pervasive problem that imposes significant social and economic costs. The conventional approach of widening roads to alleviate traffic is often a temporary solution, as it can lead to induced demand and a cycle of recurring congestion. We have reframed the problem of congestion, arguing that while it cannot be eliminated, it can be managed by optimizing a network's inherent structure.

In this book, we have discussed the Ideal Traffic Assignment (ITA) framework [11] and its core conclusion: that a network's physical design may determine its operational efficiency. We have learned how the principles of Premagic [10] Equilibrium, along with the axioms of Steady Link Characteristics and Identical Path Link Counts, provide a theoretical foundation for creating a perfectly balanced network. This framework will empower us to move beyond intuitive, ad-hoc solutions to a quantitative and objective approach, using metrics like the VarPi (ϖ) index to guide policy and investment decisions. Ultimately, this knowledge will equip us with a new paradigm for urban transportation planning that prioritizes network organization and balance over simple expansion.

15.2 Is Congestion Solvable?

The question of whether traffic congestion is solvable is a central theme in urban transportation planning.

Traffic congestion is a pervasive feature of urban life, imposing significant inefficiency costs. Many strategies exist to mitigate its effects, including congestion pricing, road network expansion, and various traffic management schemes. More

K. Teknomo, *Ideal Flow Network for Transportation Research*,
Lecture Notes in Intelligent Transportation and Infrastructure,
https://doi.org/10.1007/978-981-95-4727-2_15

permanent solutions may involve fundamental behavioral shifts, such as increased adoption of public transport or changes in land use patterns to reduce travel distances. However, these are ultimately delaying tactics. As population and economic activity continue to grow against a fixed network capacity, demand will inevitably exceed supply, leading to the recurrence of congestion.

Can we reduce traffic congestion by optimizing the network's structure and promoting its efficient utilization? That is, can we design a network that is inherently used in the most efficient manner possible? Through mathematical analysis of Ideal Flow Network (IFN), we have developed Ideal Traffic Assignment (ITA) framework. Ideal Traffic Assignment (ITA) framework is a specialized implementation of Ideal Flow Network theory for transportation systems that predicts traffic flow from network structure (road widths, lanes, connectivity). It applies Maximum Entropy Principle to handle data scarcity and uses steady state Markov Chain [5] with global network scaling to derive link-level flows. It can be calibrated and validated based on real world link flow, without requiring OD matrices. The strength of the Ideal Traffic Assignment (ITA) framework lies in its formal linkage between network topology and systemic performance metrics. This enables enabling the study of network designs that minimize congestion under equilibrium conditions.

Urban road networks are complex systems. Historically, transport planning has operated on the paradigm that traffic demand dictates the required supply. This represents past wisdom. When a road link becomes congested, the conventional response is to expand its capacity. This approach, however, faces physical and economic limitations in dense urban environments. Furthermore, the phenomenon of induced demand often means that new capacity is quickly consumed, leading to a cycle of recurring congestion.

Congestion is not solvable in the sense of permanent elimination when population and economic activity outpace feasible additions of right-of-way. The objective is to approach a lower bound on congestion, not to promise its disappearance. The network capacity needs to be expanded as demand grows but the question is what kind of expansion. If we simply think locally and short term by firefighting the congestion by keep expanding the link capacity (like road widening and adding more lanes), we may induce traffic and consequently exacerbate congestion in other areas.

The lower bound is a systems target: organize flows, distribute stress, and align individual choices with the collective good as closely as the physical city allows. We can approach that lower bound, and we can do so systematically, once the network is designed and operated to be balanced rather than merely bigger. The most effective way to expand the network is to increase its total capacity while ensuring that link capacities approach a premagic [10] configuration. Using premagic capacity and uniform link distance would make the network topology to be in the perfect harmony to keep the lowest price of anarchy [7].

15.3 The Principle of Premagic

15.3.1 Eulerian: Premagic Adjacency

Analytical findings indicate that the optimal network configuration, under ideal traffic assignment conditions, is one that exhibits Eulcrian structure, symmetry, and premagic capacity distribution, with uniform link lengths. Such properties are inherently characteristic of grid-based networks, which outperform concentric or radial designs in minimizing congestion and balancing flow. This suggests for urban planners that the ideal configuration aligns with a uniform grid network structure rather than a concentric, ring-based one. This theory posits that a symmetrical, grid-based network minimizes congestion and optimizes flow distribution, offering a clear theoretical foundation for urban design. In contrast, a similar mathematical theory supporting the concentric ring road pattern is not as evident, suggesting that the grid-based design has a stronger theoretical justification from a network flow perspective.

An Eulerian network is the adjacency-matrix face of the same idea that this book calls premagic structure.. An Eulerian circuit has equal in-degree and out-degree at each node, which is the adjacency version of "what flows in can flow out without bias." A network with a *premagic adjacency matrix* is one in which the total number of incoming links to a node equals the total number of outgoing links, reflecting a fundamental topological balance. A uniform grid system is the physical embodiment of this theoretical ideal. As proven in Corollary 12.1, an Eulerian circuit with uniform link capacities achieves the highest for both a microscopic average node entropy ratio and a macroscopic network entropy efficiency. This state represents the theoretical minimum congestion for any given number of nodes, as proven by the fact that it perfectly aligns Wardrop's First and Second Principles. Even when link capacities are not uniform, an Eulerian circuit in equilibrium still produces congestion levels for all outgoing links of a given node, which is a state superior to any other network structure.

15.3.2 Premagic Equilibrium: Unifying the Two Equilibria

We have discovered the principle of premagic equilibrium as a tool for *flow optimization for a fixed network*, not a tool for global network design optimization. This theory proves that for a given premagic network, a uniform congestion distribution is the most efficient way to operate it. It does not claim that this network is superior to all other possible designs in terms of total travel time. The proof of this principle uses Jensen's Inequality to demonstrate that for a given total flow, a uniform congestion ratio minimizes the total aggregate travel time. This establishes a clear and mathematically rigorous target for network management. The goal is to shape the physical capacity of a network so that it naturally distributes traffic in the most efficient and equitable manner possible.

The theory provides a new definition for what constitutes a "good" network design. A good network is not simply the one with the most capacity or the fastest average speed. A good network is one that is perfectly balanced, where the individual choices of drivers align with the collective good, and the Price of Anarchy is minimized. This design philosophy moves beyond localized problem-solving and toward a holistic optimization of the entire transport system.

The concept of *premagic capacity* extends this principle from topology to the link values themselves. Link capacities are dimensioned proportionally to their demand, ensuring the operating point is balanced rather than lopsided. Premagic Equilibrium is the operating state where the networks congestion is uniform across active links when links share steady characteristics. Premagic Equilibrium is not a slogan; it is the hidden organizing principle exists in nature that keeps user choices close to the systems best interest. Premagic capacity is therefore a design tool, and Premagic Equilibrium is the operating target.

The three axioms form a single hinge that closes the classic gap in transportation science. Steady Link Characteristics assert that used links are comparable in their physical and operational properties. Identical Path Link Counts assert that the competing paths between an origin and a destination are geometrically comparable in the number of links they traverse. Premagic Equilibrium asserts that the operating congestion is uniform across active links. This theory demonstrates that a balanced network, where congestion is uniform, aligns individual user choices with the system's overall best interest, thereby eliminating the Price of Anarchy In this aligned regime, the network is both individually stable and collectively efficient, without the usual price of anarchy penalty. This is a profound simplification of network behavior and a firm foundation for design and policy.

15.4 Practical Guidance and Policy

15.4.1 The Advantages of Grid-Based Networks

Grid-like, Eulerian-leaning networks with uniform link characteristics get us closest to the aligned regime under the books assumptions. At the node level, choice sets are balanced rather than skewed, which reduces the odds of chronic unfairness and instability. At the system level, capacity is used more evenly, which improves resilience and lowers the risk of brittle bottlenecks. Even when perfect uniformity is impossible, moving the topology and capacity toward the grids premagic signature improves real networks in predictable ways. This is why cities that invest in a grid structure age more gracefully under growth than cities that harden a few radial corridors.

15.4.2 How to Expand Without Making Things Worse

Expanding only the visibly congested link is firefighting that often pushes the fire down the road. Local widenings can invite induced demand and deepen imbalances that raise the systems worst congestion elsewhere. Balance first, widen second, and widen uniformly when we must widen at all. The guiding rule is simple: raise the uniformity of capacity relative to expected demand rather than deepening capacity peaks and valleys. Uniform expansion does not mean adding the same lanes everywhere; it means adding where the networks balance improves the most.

15.4.3 Policy Formulation and Quantitative Measurement: The Chemistry of Policy

The Ideal Traffic Assignment (ITA) framework transforms urban transportation planning from a qualitative process based on intuition to a quantitative, objective, and optimizable procedure. Policies are defined as irreducible [5] atomic transformations on the network's mathematical structure, which can be combined to form complex molecular policies.

The framework provides the tools to systematically quantify the effects of policy interventions. For example, a localized improvement, such as adding a lane, does not always guarantee a beneficial outcome and can have unintended, system-wide consequences. The ITA framework allows for the systematic evaluation of interventions by comparing a base network state to a feasible alternative scenario, highlighting the need for a holistic approach to network expansion. Atomic policies are irreducible edits to the networks structural matrices such as capacity, distances, controls, and routing constraints. Molecular policies are compositions of atomic policies, sequenced in time or combined in bundles. Order matters, conflicts exist, and feasible sequences must respect conservation, geometry, and operations. The chemistry of policy allows planners to enumerate, compare, and optimize scenarios with a computer rather than relying on intuition and isolated case studies. Policy making becomes programmable, auditable, and improvable.

15.4.4 Principle of Congestion Rebalancing

Average speed is not a health metric for a network because it is blind to the distribution of pain. Balance is the right health metric because it tells us whether the system is organized or brittle. This framework supports the *principle of congestion rebalancing* by actively managing supply. It also proposes that demand can be guided towards a more optimal equilibrium. The core objective is to reduce the maximum congestion level within a network. We have demonstrated that a counter-intuitive policy of reducing capacity can lead to a more balanced network where the most severe congestion is alleviated.

We introduced quantitative measures, such as the VarPi index, denoted as ϖ, to assess network uniformity. The ϖ index measures how close a network is to a perfectly balanced state and is the mathematically appropriate metric to guide policy. A policy that increases the ϖ index, even if it reduces total capacity, may lead to a more efficient and equitable network. This highlights the importance of measuring a network's performance not by a single metric like average speed, but by one that quantifies its balance. The VarPi index measures how close the network is to a balanced state and is portable across capacity and congestion views of the same reality. Raising VarPi, especially before adding gross capacity, reduces the worst congestion, improves reliability, and supports equity.

15.4.5 A Foundation for a New Paradigm

The discoveries presented herein establish a robust, structure-driven paradigm for traffic analysis and urban transportation planning. The IFN framework provides a rigorous mathematical foundation to investigate what kind of network produces the least congestion. The ideal network is an Eulerian circuit with symmetrical, premagic capacity and uniform link lengths, a topology that corresponds to a uniform grid network. This finding provides a powerful theoretical argument for grid-based urban transportation planning over concentric ring road patterns.

The ITA framework is a transformative instrument that empowers urban planners to make proactive, evidence-based decisions for sustainable and resilient urban traffic management. The theory redefines the objective of network design, moving it from a problem of maximizing capacity to one of optimizing balance. It provides a clear and mathematically rigorous target for design and management, shaping our understanding of a city's physical form and its functional performance.

A good network is not the one with the most lanes or the highest average speed. A good network is one that is organized. Organized means balanced across space, resilient to shocks, explainable to the public, and fair in how it carries demand. Premagic network structure which is an Eulerian-leaning grids provide the mathematics and the playbook to keep the network organized as the city grows.

15.5 Advanced Research Directions

The challenges identified below are not endpoints but rather invitations to new avenues of inquiry, designed to absorb focused research across multiple groups. They mark the frontier where the ITA framework can be extended, enriched, and integrated with other domains to create even more powerful models of urban mobility. These research directions represent exciting opportunities for graduate students, academics, and practitioners to build upon the foundation laid out in this book. They abstract from the strength of Premagic Equilibrium and the chemistry of policy within the Ideal Traffic Assignment framework.

Stochasticity and Uncertainty Integration

The current ITA framework is largely deterministic: given a set of inputs, it produces a single output. Reality, however, is inherently stochastic. Travel demand fluctuates unpredictably, link capacities can be temporarily reduced by minor incidents or adverse weather, and driver behavior is never perfectly uniform. A significant frontier lies in extending the IFN framework to explicitly account for this randomness. Instead of treating link capacities c_{ij} as fixed constants, a stochastic IFN model could represent them as random variables following a probability distribution. For example, a stable highway segment might have a capacity that is normally distributed, $c_{ij} \sim N(\mu, \sigma^2)$, where the variance σ^2 reflects its susceptibility to disruption.

Multi-Modal Integration

Urban transportation is a system of interacting modes, including private cars, public transit, cycling, and walking. The ITA framework, in its current form, is uni-modal, focused on the vehicular road network. A major area for future development is the extension of the framework to capture the interactions and choices between these different modes. A multi-modal ITA model could be conceptualized as a multi-layered graph. Each layer would represent a different transportation mode (e.g., a road layer, a transit layer, a pedestrian layer). The nodes in this graph would represent not only physical intersections but also crucial *transfer points*, such as bus stops, metro stations, and park-and-ride facilities. Using such a layered model, a researcher could investigate the system-wide impact of a *new public transit investment*, like a new metro line. The expanded stochastic matrix S would include mode-choice probabilities at the transfer nodes, perhaps modeled as a function of the relative travel times, costs, and convenience of each modal layer. The model could then predict how the new, faster metro service (a change in the transit layer) would induce a modal shift, pulling flow from the private vehicle layer. This would allow planners to quantify not just the ridership on the new line but also its secondary benefit: the corresponding reduction in road congestion and change in the network's equilibrium state.

A Universal Balance Metric for Multimodal Systems

Construct a single, interpretable balance measure for intrinsically coupled road, bus, rail, and pedestrian systems that respects each modes physics and user behavior. Require that the measure reduces to VarPi for unimodal roads, remains computable from partial data, and meaningfully compares cross-mode reallocations of priority and capacity.

Advanced Parameter Inference and Learning

The behavior of a transportation system is driven by the aggregate choices of its users. The ITA framework captures this with behavioral parameters, such as the sensitivity to travel time (γ) in the iterative Logit model. A key research frontier is to move beyond using single, static parameters for an entire population and instead leverage modern data and machine learning to infer more nuanced, dynamic, and heterogeneous behavioral models. The proliferation of GPS-enabled devices has created massive datasets of anonymized vehicle trajectories. This rich data provides an unprecedented window into real-world route choice behavior. A researcher could use a large trajectory dataset to *infer heterogeneous behavioral parameters*. Instead of calibrating one value of γ for the whole city, machine learning techniques (e.g., latent class analysis or inverse reinforcement learning) could be used to identify distinct clusters of driver behavior. The model might discover, for example, that commuters are highly sensitive to travel time (γ_{high}), while delivery drivers are more sensitive to travel time reliability ($\gamma_{variance}$). It might also find that behavior changes systematically by time of day or during special events. Integrating these heterogeneous, data-driven behavioral parameters would create a far more realistic IFN model, significantly improving its predictive accuracy and its value for policy analysis.

A Unified Theory of Premagic Networks Beyond Ideal Assumptions

Generalize the aligned regime by relaxing each axiom in turn while preserving the unification of User Equilibrium and System Optimum as a limit case. Characterize "near-premagic" networks and bound performance loss as Steady Link Characteristics or Identical Path Link Counts are violated in measured ways. Develop structural invariants and certificates that guarantee near-alignment in large networks without full path enumeration.

Dynamic Premagic: Control and Stability Under Time Variation

Extend the regime to time-varying demand, queue spillback, and incidents with explicit departure times and storage limits. Design feedback controllers that hold a city near premagic structure using sparse sensors and delayed information. Prove robustness and performance guarantees that survive worst-day stress rather than only average days.

The Non-commutative Algebra of Policy

Formalize an algebra where atomic policies compose into molecular policies with operators for sequencing, combination, inversion, and conflict resolution. Establish when policies commute and when order effects dominate, and provide normal forms so distinct bundles can be proven equivalent or not. Deliver decision procedures that certify feasibility and outcome classes before implementation.

Global Policy Optimization with Phasing and Irreversibility
Define long-horizon optimization of policy sequences under budget, construction staging, and legal irreversibility. Develop search methods that scale to metropolitan graphs, produce human-explainable plans, and carry approximation guarantees tied to balance and resilience.

Topology Discovery Under Premagic Guidance
Create generative design tools that search the space of buildable networks for forms that retain premagic virtues without ideal geometry [10]. Constrain the generation by engineering, urban design, and politics, and extract motifs that repeatedly deliver near-aligned performance.

Observability and Data Parsimony for Balance
Determine the minimal data that identifies a citys distance to balance with quantified uncertainty. Design sampling and estimation schemes that allow reliable decisions when privacy, cost, and platform silos limit observation.

Braess Immunity in the Wild
Define operational notions of immunity for non-ideal networks, and quantify how close a city must be to the aligned regime to suppress paradoxical outcomes [2, 3]. Design tests that cities can run before and after link changes to certify that new capacity will not backfire.

Incentives and Governance for Organized Networks
Design pricing and information policies that maintain near-balance without punitive measures and without brittle dependence on a single controller. Prove existence and fairness properties, and provide institutional patterns that persist through leadership changes.

Co-evolution of Network, Land Use, and Culture
Analyze and design the multi-decade feedback loop in which capacity edits change destinations, trips, and norms. Search for sequences of premagic-guided policies that keep balance while the city densifies, electrifies, and adapts to climate stress.

15.6 Summary

This book has traced a journey from the foundational axioms of network flow to the practical application of the Ideal Traffic Assignment (ITA) framework in urban transportation planning and policy. At its core lies a simple but profound insight: the physical structure of a road network encodes a hidden predictive logic for the traffic it carries. By modeling networks as closed systems governed by principles such as random walks [1, 6, 9], Markov chains [5], and maximum entropy [4, 8], we revealed the Ideal Flow as an intrinsic property of the network itself, independent of demand inputs.

This structure-driven paradigm marks a departure from traditional, demand-centric models. It offers a parsimonious, data-light alternative for understanding and predicting traffic, reducing reliance on costly Origin-Destination surveys. Through scenario analysis and design exercises, we demonstrated how planners and engineers can use this framework to evaluate interventions and guide the creation of inherently efficient networks, especially in new urban developments.

The central theoretical contribution is the concept of Premagic Equilibrium (PE), a state of uniform link congestion that aligns individual user choices (User Equilibrium) with the system-wide optimum (System Optimum). This alignment is not coincidental; it emerges from satisfying three structural axiomatic conditions: Premagic Capacity Matrix, Steady Link Characteristics (SLC), and Identical Path Link Counts (IPLC). When these conditions are met, the network achieves a Price of Anarchy of exactly one, meaning self-interested behavior yields socially optimal outcomes. This ideal state is not merely theoretical, it provides a rigorous benchmark for evaluating real-world networks.

The theory offers a clear architectural blueprint: networks with Eulerian structures, such as grid-based topologies, are more resilient and equitable in their congestion distribution. The VarPi index quantifies how close a network is to this ideal, enabling policy decisions to be auditable, systematic, and transparent. Importantly, the framework reframes congestion not as a problem to be eliminated, but as a phenomenon to be organized. A well-organized network ensures that no corridor fails for another to flow, that expansion preserves structural balance, and that changes are mathematically interpretable to the public. In this view, a good network is not one with the most capacity, but one that is balanced, transparent, and resilient.

We have been candid about the framework's boundaries and its limitations. ITA does not determine absolute demand magnitudes or capture time-varying dynamics of real-world traffic. These acknowledged boundaries are presented as essential first steps toward future innovation. However, these limitations open fertile ground for future research.

We also discussed key frontiers for future research, including

- Integrating stochastic uncertainty to model unpredictable events
- Expanding to multi-modal systems
- Using Ideal Flow Networks (IFN) to analyze resilience
- Applying machine learning for behavioral inference

These directions aim to bridge the gap between idealized theory and the complexity of real-world urban systems.

Ultimately, the ITA framework serves as a bridge, linking elegant mathematical theory with the urgent practical needs of urban mobility. It connects microscopic choices of individual drivers' behavior to macroscopic system emergent patterns, and a city's physical form to its functional performance. The "hidden secret" that inspired this work is the idea that the networks we build are not passive conduits; they are active systems that shape the flow within them.

By understanding and leveraging the deep structural logic of these systems, we are better equipped to design, manage, and adapt them for a future that demands more resilient, efficient, and equitable transportation for all. The Ideal Traffic Assignment framework is a crucial step on that path, offering not just a model, but a philosophy for how we build the next generation of cities.

References

1. David Aldous and James Allen Fill. *Reversible Markov Chains and Random Walks on Graphs*. Available at: https://www.stat.berkeley.edu/~aldous/RWG/book.html. 2002.
2. D. Braess, A. Nagurney, and T. Wakolbinger. "On a paradox of traffic planning". In: *Transportation Science* 39.4 (2005), pp. 446–450.
3. Dietrich Braess. "Über ein Paradoxon aus der Verkehrsplanung". In: *Unternehmensforschung* 12.1 (1968), pp. 258–268. https://doi.org/10.1007/BF01918335.
4. Edwin T. Jaynes. "Information Theory and Statistical Mechanics". In: *Physical Review* 106.4 (1957), pp. 620–630. https://doi.org/10.1103/PhysRev.106.620.
5. John G. Kemeny and J. Laurie Snell. *Finite Markov Chains*. New York: Springer, 1976. https://link.springer.com/book/9780387901923.
6. László Lovász. "Random walks on graphs: A survey". In: vol. 2. Budapest: János Bolyai Mathematical Society, 1993, pp. 1–46. https://www.cs.cmu.edu/~15859n/RelatedWork/random-walks-on-graphs.pdf.
7. T. Roughgarden and E. Tardos. "How bad is selfish routing?" In: *Proceedings 41st Annual Symposium on Foundations of Computer Science*. 2000, pp. 93–102. https://doi.org/10.1109/SFCS.2000.892069.
8. Roberta Sinatra et al. "Maximal-entropy random walks in complex networks with limited information". In: *Physical Review E* 83.3 (2011), p. 030103. https://doi.org/10.1103/PhysRevE.83.030103. https://journals.aps.org/pre/abstract/10.1103/PhysRevE.83.030103.
9. K. Teknomo. "Ideal Flow Based on Random Walk on Directed Graph". In: *Proceedings of the 9th International Collaboration Symposium on Information, Production and Systems*. Kitakyushu: Waseda University, 2015.
10. Kardi Teknomo. "Premagic and Ideal Flow Matrices". In: *arXiv preprint arXiv:1706.08856* (2017). arXiv: 1706.08856 [cs.DM]. https://arxiv.org/pdf/1706.08856.
11. Kardi Teknomo, R. Gardon, and C. Saloma. "Ideal Flow Traffic Analysis: A Case Study on a Campus Road Network". In: *Philippine Journal of Science* 148.1 (2019), pp. 51–62. https://archium.ateneo.edu/discsfaculty-pubs/271/.

Glossary

Active Link: A link (i, j) in the network for which its capacity $c_{ij} > 0$ and it carries a positive ideal flow $f_{ij} > 0$. The set of active links is denoted E'.

Atomic Policy: An irreducible and feasible transformation of a network attribute, such as capacity, control, or access, that can be composed into a molecular policy.

Behavioral Stochastic Matrix (BSM): The stochastic matrix S representing equilibrium microscopic turning movement probabilities at each node, reflecting drivers' aggregated local route choice behaviors.

Capacity Matrix (C): A matrix where c_{ij} denotes the maximum flow rate or capacity of the link from node i to node j.

Congestion Matrix (G): A matrix where g_{ij} represents the congestion ratio on link (i, j), defined as the Hadamard (element-wise) division of the Ideal Flow by Capacity (f_{ij}/c_{ij}).

Delay Matrix (Y): A matrix where y_{ij} represents the additional travel time incurred on link (i, j) due to congestion, calculated as the difference between actual travel time and free-flow travel time $(t_{ij} - t_{ij}^0)$.

Eulerian (Premagic Adjacency): A network is Eulerian if each node has equal in-degree and out-degree, allowing local inflow to equal outflow.

Free-Flow Travel Time (T^0): A matrix where t_{ij}^0 represents the minimum travel time on link (i, j) under uncongested conditions, typically calculated as link length divided by maximum speed (l_{ij}/u_{ij}).

Geodesic Gating at Splits: An operational restriction that prioritizes near-shortest branches at diverging nodes to preserve balanced routing. Used as a practical rule in policy design and control.

Hadamard Product ($\circ$): An element-wise multiplication of two matrices (e.g., $\mathsf{A} \circ \mathsf{B}$).

Hadamard Division ($\circ^{-1}$ or ./): An element-wise division of two matrices, where one matrix is multiplied by the element-wise inverse of the other (e.g., $\mathsf{A} \circ \mathsf{B}^{-1}$ or $\mathsf{A}./\mathsf{B}$).

K. Teknomo, *Ideal Flow Network for Transportation Research*,
Lecture Notes in Intelligent Transportation and Infrastructure,
https://doi.org/10.1007/978-981-95-4727-2

Ideal Flow Matrix (F)**:** A matrix where f_{ij} represents the predicted steady-state traffic flow on link (i, j), derived from the Stochastic Matrix S and the Total Flow parameter κ.

Ideal Flow Network (IFN): A strongly connected directed graph representing a transportation system, characterized by node-level flow conservation and the prediction of steady-state flows from network structure.

Ideal Traffic Assignment (ITA): A framework that applies Ideal Flow Network (IFN) principles to transportation research and traffic assignment. ITA links network structure and policy edits to balanced congestion outcomes, supporting scenario design through atomic and molecular policy compositions and VarPi-guided objectives.

Identical Path Link Counts (IPLC): An Idealized Network Axiom stating that for every Origin-Destination (O-D) pair, all active paths consist of exactly the same number of links.

Maximum Entropy Principle: A principle used to select the most unbiased probability distribution that is consistent with known constraints or partial information, applied in IFN to determine flow distributions.

Microscopic Turning Movement Probability (s_{ij})**:** The probability that a vehicle arriving at node i (or originating there) will choose to traverse outgoing link (i, j), reflecting local driver choice behavior.

Molecular Policy: A time-ordered bundle of atomic policies acting on the network, evaluated as a single scenario for feasibility, balance, and performance.

Nash Equilibrium (NE): A state where no individual player (driver) can improve their outcome (reduce travel time) by unilaterally changing their strategy (route choice), assuming all other players' strategies remain unchanged. Equivalent to User Equilibrium.

Network Flow Identity (NFI): A formula, $\kappa = \tilde{c}\varpi\xi$, that equates the total network flow (κ) to the product of the total network capacity ($\tilde{c}$), the efficiency-balance of congestion distribution (ϖ), and the maximum allowable congestion (ξ).

Origin-Destination (O-D) Matrix (D)**:** A matrix where D_{od} quantifies the demand (number of trips) from an origin o to a destination d.

Pareto Optimality: A state in which it is impossible to make any one individual better off without making at least one other individual worse off. System Optimum is a Pareto Optimal state.

Policy Algebra: A chemistry-inspired formalism used in ITA in which atomic policies are irreducible network edits and molecular policies are their compositions.

Premagic Capacity: A capacity matrix is premagic if its row sum vector equals the transpose of its column sum vector, indicating balanced aggregate capacity distribution.

Premagic Equilibrium (PE): An Idealized Network Axiom and a theoretical network state where the congestion ratio G_{ij} is a constant σ across all active links.

Price of Anarchy (PoA): The ratio of user equilibrium travel time or delay to system optimum travel time or delay, quantifying the efficiency loss due to decentralized decision-making.

Random Walk: A mathematical process describing a path consisting of a succession of random steps on some mathematical space, used as a foundation for flow generation in IFN.

Scaling Factor (ς): A scalar derived through calibration to convert Ideal Traffic Assignment's relative ideal flow values into absolute flow magnitudes, based on observed real link flows.

Steady Link Characteristics (SLC): An Idealized Network Axiom stating that all active links possess uniform physical properties (constant length l and constant maximum speed u).

Stochastic Matrix (S): A matrix where S_{ij} represents the transition probability from node i to node j via link (i, j), with rows summing to 1. In ITA, it often represents turning movement probabilities.

System Optimum (SO): A flow distribution that minimizes the total system aggregate travel time across the entire network.

Total Aggregate Travel Time (Z): The sum of flow on each link multiplied by its corresponding travel time, representing the total cost of travel experienced by all users in the network ($\sum f_{ij} \times t_{ij}$).

Total Capacity ($\tilde{c}$): The sum of capacities of all active links in the network ($\sum_{(i,j)\in E'} c_{ij}$).

Total Flow (κ): The scalar parameter representing the total flow circulating in the network, used to scale relative flows into absolute values.

Traffic Assignment: The process of allocating a given travel demand (O-D matrix or total flow) onto a transportation network.

Travel Time Matrix (T): A matrix where t_{ij} represents the actual (congested) travel time on link (i, j).

User Equilibrium (UE): See Nash Equilibrium (NE).

VarPi Index (ϖ): A unitless metric of network balance that increases as congestion ratios become more uniform across links. It serves as a primary policy objective for evaluating efficiency.

Wardrop's First Principle: See User Equilibrium (UE).

Wardrop's Second Principle: See System Optimum (SO).

Notations

Mathematical notations adhere to standard conventions: scalar values are represented by lowercase Latin or Greek letters; vectors are bold lowercase (e.g., $\mathbf{v}$ or $\boldsymbol{\pi}$) and matrices are typeset in uppercase (e.g., A). The following entities represent foundational notation standards.

Scalars

Symbol	Meaning
α, β	model parameters
γ, η	model parameters for BPR function
κ	Total flow in the network
$\tilde{c}$	Total capacity in the network
l	Constant link length under Steady Link Characteristics (SLC)
u	Constant link maximum speed under SLC
$\boldsymbol{\pi}_i$	the flow at node i
ϖ	VarPi index
ς	global scaling factor
ς^*	Optimal scaling factor for absolute flow calibration
ϵ	Small tolerance value for convergence criteria
P_i	Trip productions for zone i (Step 1).
A_j	Trip attractions for zone j (Step 1).

K. Teknomo, *Ideal Flow Network for Transportation Research*,
Lecture Notes in Intelligent Transportation and Infrastructure,
https://doi.org/10.1007/978-981-95-4727-2

Symbol	Meaning
T_{ij}	Trips from origin i to destination j (OD matrix entry).
p_{ijm}	Probability share for mode m between i and j .
$t_a(\cdot)$	Link generalized cost function for link a (time or generalized cost).
H^S	Shannon Entropy
H_i	Node Entropy
H	Network entropy

Vectors and Matrices

Symbol	Meaning
$\mathbf{j}$	a column vector of ones
$\boldsymbol{\pi}$	Node flow vector or stationary probability distribution vector
A	Adjacency matrix (used in general graph theory contexts)
C	Capacity matrix
D	Origin-Destination (O-D) matrix
Y	Delay matrix
F	Ideal Flow matrix
G	Congestion matrix
I	Identity matrix
J	All-ones matrix
L	Link length matrix
P	General probability matrix (e.g., link probability)
S	Stochastic matrix (microscopic turning movement / transition probabilities)
$\hat{\mathsf{S}}$	Auxiliary/target stochastic matrix in iterative updates
T	Travel time matrix (actual, congested)
T^0	Free-Flow Travel Time matrix
U	Link maximum speed matrix
V	Link speed matrix (actual)
$\mathsf{0}$	Zero matrix (or zero vector)

Sets

Symbol	Meaning
$\mathcal{G}$	A network
$\mathcal{V}$	Set of all nodes in the network
$\mathcal{E}$	Set of all directed links (edges) in the network
E'	Set of all active links (where $\mathsf{C}_{ij} > 0$ and $\mathsf{F}_{ij} > 0$)
K	Subset of monitored links used for calibration
$\text{in}(i)$	Set of links incoming to node i
$\text{out}(i)$	Set of links outgoing from node i
$\mathcal{N}_i$	A set of scenario i.
$\mathcal{P}_{ij}$	Set of admissible paths between i and j.

Operators

Symbol	Meaning
$\sum$	Summation operator
$\cdot$	Matrix or vector product
$\circ$	Hadamard product (element-wise multiplication)
$./$ or $\oslash$	Hadamard division (element-wise division)
$\exp(\cdot)$	Element-wise exponential function for matrices/vectors
$\ln(\cdot)$	Natural logarithm
$\lVert \cdot \rVert$	Matrix norm (e.g., Frobenius norm, L_1 norm)
$\frac{d(\cdot)}{d(\cdot)}$	Derivative operator
$f : X \to Y$	Function mapping elements from set X to set Y

Index

K. Teknomo, *Ideal Flow Network for Transportation Research*,
Lecture Notes in Intelligent Transportation and Infrastructure,
https://doi.org/10.1007/978-981-95-4727-2

W

MIX
Papier aus verantwortungsvollen Quellen
Paper from responsible sources
FSC® C105338

If you have any concerns about our products,
you can contact us on
ProductSafety@springernature.com

In case Publisher is established outside the EU,
the EU authorized representative is:
Springer Nature Customer Service Center GmbH
Europaplatz 3, 69115 Heidelberg, Germany

Printed by Libri Plureos GmbH
in Hamburg, Germany